THE
MORAL ARC

THE MORAL ARC

HOW SCIENCE AND
REASON LEAD
HUMANITY TOWARD
TRUTH, JUSTICE,
AND FREEDOM

MICHAEL SHERMER

HENRY HOLT AND COMPANY • NEW YORK

Henry Holt and Company, LLC
Publishers since 1866
175 Fifth Avenue
New York, New York 10010
www.henryholt.com

Henry Holt® and ® are registered trademarks of
Henry Holt and Company, LLC.

Distributed in Canada by H. B. Fenn and Company Ltd.

Library of Congress Cataloging-in-Publication Data

Shermer, Michael.
    The moral arc : how science and reason lead humanity toward truth, justice, and freedom /
Michael Shermer.—First edition.
        pages cm
    Includes bibliographical references and index.
    ISBN 978-0-8050-9691-0 (hardback)—ISBN 978-0-8050-9693-4 (electronic book)
  1. Science—Moral and ethical aspects—History.   2. Science—Social aspects—History.
I. Title.
    BJ57.S48 2015
    170.9—dc23   2014020084

Henry Holt books are available for special promotions and premiums.
For details contact: Director, Special Markets.

First Edition 2015

Designed by Kelly S. Too

Printed in the United States of America
10 9 8 7 6 5 4 3 2 1

To Jennifer

Unsterbliche Geliebte
Immortal Beloved

Forever thine
Forever mine
Forever us
∞

There must be no barriers to freedom of inquiry. There is no place for dogma in science. The scientist is free, and must be free to ask any question, to doubt any assertion, to seek for any evidence, to correct any errors. Our political life is also predicated on openness. We know that the only way to avoid error is to detect it and that the only way to detect it is to be free to inquire. And we know that as long as men are free to ask what they must, free to say what they think, free to think what they will, freedom can never be lost, and science can never regress.

—J. ROBERT OPPENHEIMER, 1949

CONTENTS

THE
MORAL ARC

Bending the Moral Arc

Sunday, March 21, 1965. Selma, Alabama.

About eight thousand people gather at Brown Chapel and begin to march from the town of Selma to the city of Montgomery, Alabama. The demonstrators are predominantly African American and they're marching on the capitol for one reason: *Justice.* They want simply to be given the right to vote. But they're not alone in their struggle. Demonstrators of "every race, religion, and class," representing almost every state, have come to march with their black brothers and sisters.[1] And at the front of the march is the Reverend Dr. Martin Luther King Jr., Nobel Prize winner, preacher, and civil rights activist, leading the march like Moses leading his people out of Egypt.

In the teeth of racial opposition backed by armed police and riot squads, they had tried to march twice before, but both times were met with violence by state troopers and a deputized posse. The first time—known as Bloody Sunday—the marchers were ordered to turn back but refused and, as onlookers cheered, they were met with tear gas, billy clubs, and rubber tubing wrapped in barbed wire. The second time they were again met by a line of state troopers and ordered to turn around, and after asking for permission to pray, King led them back.

But not this time. This time President Lyndon B. Johnson, finally having seen the writing on the wall, ordered that the marchers should be protected by two thousand National Guard troops and federal marshals. And so they marched. For five days, over a span of fifty-three miles, through biting cold and frequent rain, they marched. Word spread, the number of demonstrators grew, and by the time they reached the steps of the capitol in Montgomery on March 25, their numbers had swelled to at least twenty-five thousand.

But King wasn't allowed on the steps of the capitol—the marchers weren't allowed on state property. Sitting in the capitol dome like Pontius Pilate, Alabama governor George Wallace refused to come out and address the marchers, and Dr. King delivered his speech from a platform constructed on a flatbed truck parked on the street in front of the building.[2] And from that platform, King delivered his stirring anthem to freedom, first recalling how they had marched through "desolate valleys," rested on "rocky byways," were scorched by the sun, slept in mud, and were drenched by rains. The crowd, consisting of freedom-seeking people who had assembled from around the United States, listened intently as Dr. King implored them to remain committed to the nonviolent philosophy of civil disobedience, knowing that the patience of oppressed peoples wears thin and that our natural inclination is to hit back when struck. He asked, rhetorically, "How long will prejudice blind the visions of men, darken their understanding, and drive bright-eyed wisdom from her sacred throne?" And "How long will justice be crucified and truth bear it?" In response, Dr. King offered words of counsel, comfort, and assurance, saying that no matter the obstacles it wouldn't be long before freedom was realized because, he said, quoting religious and biblical tropes, "truth crushed to earth will rise again," "no lie can live forever," "you shall reap what you sow," and "the arc of the moral universe is long, but it bends toward justice."[3]

It was one of the greatest speeches of Dr. King's career, and arguably one of the greatest in the history of public oratory. And it worked. Less than five months later, on August 6, 1965, President Johnson signed the voting rights act into law. It was just as Dr. King had said—the arc of the moral universe is long but it bends toward justice.

. . .

Dr. King's reference—the title inspiration for this book—comes from the nineteenth-century abolitionist preacher Theodore Parker, who penned this piece of moral optimism in 1853, at a time when, if anything, pessimism would have been more appropriate as America was inexorably sliding toward civil war over the very institution Parker sought to abolish:

I do not pretend to understand the moral universe; the arc is a long one, my eye reaches but little ways; I cannot calculate the curve and complete the figure by the experience of sight; I can divine it by conscience. And from what I see I am sure it bends towards justice.[4]

In *The Moral Arc* my aim is to show that the Reverends Parker and King were right—that the arc of the moral universe does indeed bend toward justice. In addition to religious conscience and stirring rhetoric, however, we can trace the moral arc through science with data from many different lines of inquiry, all of which demonstrate that in general, as a species, we are becoming increasingly moral. As well, I argue that most of the moral development of the past several centuries has been the result of secular not religious forces, and that the most important of these that emerged from the Age of Reason and the Enlightenment are science and reason, terms that I use in the broadest sense to mean reasoning through a series of arguments and then confirming that the conclusions are true through empirical verification.

Further, I demonstrate that the arc of the moral universe bends not merely toward justice, but also toward truth and freedom, and that these positive outcomes have largely been the product of societies moving toward more secular forms of governance and politics, law and jurisprudence, moral reasoning and ethical analysis. Over time it has become less acceptable to argue that my beliefs, morals, and ways of life are better than yours simply because they are mine, or because they are traditional, or because my religion is better than your religion, or because my God is the One True God and yours is not, or because my nation can pound the crap out of your nation. It is no longer acceptable to simply *assert* your moral beliefs; you have to provide *reasons* for them, and those reasons had better be grounded in rational arguments and empirical evidence or else they will likely be ignored or rejected.

Historically, we can look back and see that we have been steadily—albeit at times haltingly—expanding the moral sphere to include more members of our species (and now even other species) as legitimate participants in the moral community. The burgeoning conscience of humanity has grown to the point where we no longer consider the well-being only of our family, extended family, and local community; rather, our consideration now extends to people quite unlike ourselves, with whom we gladly trade goods and ideas and exchange sentiments and genes rather than beating, enslaving, raping, or killing them (as our sorry species was wont to do with reckless abandon not so long ago). Nailing down the cause-and-effect relationship between human action and moral progress—that is, determining *why* it has happened—is the other primary theme of this book, with the implied application of what we can do to adjust the variables in the equation to continue expanding the moral sphere and push our civilization further along the moral arc.

Improvements in the domain of morality are evident in many areas of life: *governance* (the rise of liberal democracies and the decline of theocracies and autocracies); *economics* (broader property rights and the freedom to trade goods and services with others without oppressive restrictions); *rights* (to life, liberty, property, marriage, reproduction, voting, speech, worship, assembly, protest, autonomy, and the pursuit of happiness); *prosperity* (the explosion of wealth and increasing affluence for more people in more places; and the decline of poverty worldwide in which a smaller percentage of the world's people are impoverished than at any time in history); *health and longevity* (more people in more places more of the time live longer, healthier lives than at any time in the past); *war* (a smaller percentage of people die as a result of violent conflict today than at any time since our species began); *slavery* (outlawed everywhere in the world and practiced in only a few places in the form of sexual slavery and slave labor that are now being targeted for total abolition); *homicide* (rates have fallen precipitously from over 100 murders per 100,000 people in the Middle Ages to less than one per 100,000 today in the industrial West, and the chances of an individual dying violently is the lowest it has ever been in history); *rape and sexual assault* (trending downward, and while still too prevalent, it is outlawed by all Western states and increasingly prosecuted); *judicial restraint* (torture and the death penalty have been almost universally outlawed by states, and where it is still legal is less frequently practiced); *judicial equality* (citizens of nations are treated more equally under the law than at any time in the past); and *civility* (people are kinder, more civilized, and less violent to one another than ever before).

In short, we are living in the most moral period in our species' history.

I do not go so far as to argue that these favorable developments are inevitable or the result of an inexorable unfolding of a moral law of the universe—this is not an "end of history" argument—but there are identifiable causal relationships among social, political, and economic factors and moral outcomes. As Steven Pinker wrote in *The Better Angels of Our Nature*, a work of breathtaking erudition that was one of the inspirations for this book:

Man's inhumanity to man has long been a subject for moralization. With the knowledge that something has driven it down we can also treat it as a matter of cause and effect. Instead of asking "Why is there war?" we might ask "Why is there peace?" We can obsess not just over what we have been doing wrong but also what we have been doing right. Because we have been doing something right and it would be good to know what exactly it is.[5]

For tens of millennia moral *regress* best described our species, and hundreds of millions of people suffered as a result. But then something happened half a millennium ago. The Scientific Revolution led to the Age of Reason and the Enlightenment, and that changed everything. As a result, we ought to understand what happened, how and why these changes reversed our species' historical trend downward, and that we can do more to elevate humanity, extend the arc, and bend it ever upward.

. . .

During the years I spent researching and writing this book, when I told people that the subject was moral progress, to describe the responses I received as incredulous would be an understatement; most people thought I was hallucinatory. A quick rundown of the week's bad news would seem to confirm the diagnosis.

The reaction is understandable because our brains evolved to notice and remember immediate and emotionally salient events, short-term trends, and personal anecdotes. And our sense of time ranges from the psychological "now" of three seconds to the few decades of a human lifetime, which is far too short to track long-term incremental trends unfolding over centuries and millennia, such as evolution, climate change, and—to my thesis—moral progress. If you only ever watched the evening news you would soon have ample evidence that the antithesis of my thesis is true—that things are bad and getting worse. But news agencies are tasked with reporting only the bad news—the ten thousand acts of kindness that happen every day go unreported. But one act of violence—a mass public shooting, a violent murder, a terrorist suicide bombing—is covered in excruciating detail with reporters on the scene, exclusive interviews with eyewitnesses, long shots of ambulances and police squad cars, and the thwap-thwap-thwap of news choppers overhead providing an aerial perspective on the mayhem. Rarely do news anchors remind their viewers that school shootings are still incredibly rare, that crime rates are hovering around an all-time low, and that acts of terror almost always fail to achieve their objective and their victim death tolls are negligible compared to other forms of death.

News agencies also report what *happens*, not what *doesn't* happen. We will never see a headline that reads,

ANOTHER YEAR WITHOUT NUCLEAR WAR

This too is a sign of moral progress in that such negative news is still so uncommon that it is worth reporting. Were school shootings, murders, and terrorist attacks as commonplace as charity events, peacekeeping missions, and disease cures, our species would not be long for this world.

As well, not everyone shares my sanguine view of science and reason, which has found itself in recent decades under attack on many fronts: right-wing ideologues who do not understand science; religious-right conservatives who fear science; left-wing postmodernists who do not trust science when it doesn't support progressive tenets about human nature; extreme environmentalists who want to return to a prescientific and pre-industrial agrarian society; antivaxxers who wrongly imagine that vaccinations cause autism and other maladies; anti-GMO (genetically modified food) activists who worry about Frankenfoods; and educators of all stripes who cannot articulate why Science, Technology, Engineering, and Math (STEM) are so vital to a modern democratic nation.

Evidence-based reasoning is the hallmark of science today. It embodies the principles of objective data, theoretical explanation, experimental methodology, peer review, public transparency and open criticism, and trial and error as the most reliable means of determining who is right—not only about the natural world, but about the social and moral worlds as well. In this sense many apparently immoral beliefs are actually factual errors based on incorrect causal theories. Today we hold that it is immoral to burn women as witches, but the reason our European ancestors in the Middle Ages strapped women on a pyre and torched them was because they believed that witches caused crop failures, weather anomalies, diseases, and various other maladies and misfortunes. Now that we have a scientific understanding of agriculture, climate, disease, and other causal vectors—including the role of chance—the witch theory of causality has fallen into disuse; what was a seemingly moral matter was actually a factual mistake.

This conflation of facts and values explains a lot about our history, in which it was once (erroneously) believed that gods need animal and human sacrifices, that demons possess people and cause them to act crazy, that Jews cause plagues and poison wells, that African blacks are better off as slaves, that some races are inferior or superior to other races, that women want to be controlled or dominated by men, that animals are automata and feel no pain, that kings rule by divine right, and other beliefs no rational, scientifically literate person today would hold, much less proffer as a viable idea to be taken seriously. The Enlightenment philosopher Voltaire explicated the problem succinctly: "Those who can make you believe absurdities, can make you commit atrocities."[6]

One path (among many) to a more moral world is to get people to quit believing in absurdities. Science and reason are the best methods for doing that. As a methodology, science has no parallel; it is the ultimate means by which we can understand how the world works, including the moral world. Thus, employing science to determine the conditions that best expand the moral sphere is itself a moral act. The experimental methods and analytical reasoning of science—when applied to the social world toward an end of solving social problems and the betterment of humanity in a civilized state—created the modern world of liberal democracies, civil rights and civil liberties, equal justice under the law, open political and economic borders, free markets and free minds, and prosperity the likes of which no human society in history has ever enjoyed. More people in more places more of the time have more rights, freedoms, liberties, literacy, education, and prosperity than at any time in the past. We have many social and moral problems left to solve, to be sure, and the direction of the arc will hopefully continue upward long after our epoch so we are by no means at the apex, but there is much evidence of progress and many good reasons for optimism.

PART I

The Moral Arc Explained

Toward a Science
of Morality

Science has nothing to be ashamed of even in the ruins of Nagasaki. The shame is theirs who appeal to other values than the human imaginative values which science has evolved. The shame is ours if we do not make science part of our world. . . . For this is the lesson of science, that the concept is more profound than its laws.

—Jacob Bronowski, *Science and Human Values*, 1956[1]

The metaphor of the bending moral arc symbolizes what may be the most important and least appreciated trend in human history—moral progress—and its primary cause is one of the most understated sources: scientific rationalism.

By *progress* I accept the *Oxford English Dictionary*'s historical usage as "advancement to a further or higher stage; growth; development, usually to a better state or condition; improvement." By *moral* I mean "manner, character, proper behavior" (as from the Latin *moralitas*), in terms of intentions and actions that are right or wrong with regard to another moral agent.[2] Morality involves how we think and act toward other moral agents in terms of whether our thoughts and actions are right or wrong with regard to their *survival and flourishing*. By *survival* I mean the instinct to live, and by *flourishing* I mean having adequate sustenance, safety, shelter, bonding, and social relations for physical and mental health. Any organism subject to natural selection—which includes all organisms on this planet and most likely on any other planet as well—will by necessity have this drive to survive and flourish, for if they didn't they would not live long enough to reproduce and would therefore no longer be subject to natural selection.

Because I include animals (and, perhaps one day, extraterrestrial life-forms) in our sphere of moral consideration, by moral agent I mean *sentient beings*. By *sentient* I mean *emotive, perceptive, sensitive, responsive, conscious*, and therefore able to feel and to suffer. In addition to using criteria such as intelligence, language, tool use, reasoning power, and other cognitive skills, I am reaching deeper into our evolved brains toward more basic emotive capacities. Our moral consideration should be based not primarily on what sentient beings are *thinking*, but on what they are *feeling*. There is sound science behind this proposition. According to the Cambridge Declaration on Consciousness—a statement issued in 2012 by an international group of prominent cognitive neuroscientists, neuropharmacologists, neuroanatomists, and computational neuroscientists—there is a convergence of evidence to show the continuity between humans and nonhuman animals, and that *sentience* is the common characteristic across species.

The neural pathways of emotions, for example, are not confined to higher-level cortical structures in the brain, but are found in evolutionarily older subcortical regions. Artificially stimulating the same regions in human and nonhuman animals produces the same emotional reactions in both.[3] Further, attentiveness, sleep, and decision making are found across the branches of the evolutionary tree of life, including mammals, birds, and even some invertebrates, such as octopodes. In assessing all the evidence for sentience, these scientists declared, "Convergent evidence indicates that non-human animals have the neuroanatomical, neurochemical, and neurophysiological substrates of conscious states along with the capacity to exhibit intentional behaviors. Consequently, the weight of evidence indicates that humans are not unique in possessing the neurological substrates that generate consciousness."[4] Whether nonhuman animals are "conscious" depends on how one defines consciousness, but for my purposes the more narrowly restricted emotional capacity to *feel* and *suffer* is what brings many nonhuman animals into our moral sphere.

Given these reasons and this evidence, *the survival and flourishing of sentient beings* is my starting point, and the fundamental principle of this system of morality.[5] It is a system based on science and reason, and is grounded in principles that are themselves based on nature's laws and on human nature—principles that can be tested in both the laboratory and in the real world. Thus I take *moral progress* to mean *the improvement in the survival and flourishing of sentient beings*.

Here I am specifically referring to *individual* beings. It is the *individual* who is the primary moral agent—not the group, tribe, race, gender, state, nation, empire, society, or any other collective—because it is the *individual* who survives and flourishes, or who suffers and dies. It is individual sentient beings who perceive, emote, respond, love, feel, and suffer—not populations, races, genders, groups, or nations. Historically, immoral abuses have been most rampant, and body counts have run the highest, when the individual is sacrificed for the good of the group. It happens when people are judged by the color of their skin—or by their X/Y chromosomes, or by whom they prefer to sleep with, or by what accent they speak with, or by which political or religious group they belong to, or by any other distinguishing trait our species has identified to differentiate among members—instead of by the content of their *individual* character. The Rights Revolutions of the past three centuries have focused almost entirely on the freedom and autonomy of individuals, not collectives—on the rights of *persons*, not groups. Individuals vote, not races or genders. Individuals want to be treated equally, not races. Rights protect individuals, not groups; in fact, most rights (such as those enumerated in the Bill of Rights of the US Constitution) protect individuals from being discriminated against as members of a group, such as by race, creed, color, gender, and—soon—sexual orientation and gender preference.

The singular and separate organism is to biology and society what the atom is to physics—a fundamental unit of nature. (Here I am not including social insects such as worker drone bees, whose members are genetically nearly identical.) Thus the first principle of *the survival and flourishing of sentient beings* is grounded in the biological fact that the discrete organism is the principal target of natural selection and social evolution, not the group.[6] We are a social species—we need and enjoy the presence of others in groupings such as families, friends, and assorted social consortia—but we are first and foremost *individuals* within social groups and therefore ought not to be subservient to the collective.[7] Making sacrifices for one's social group is not the same as being sacrificed for the group.

This drive to survive is part of our essence, and therefore the freedom to pursue the fulfillment of that essence is a *natural right*, by which I mean it is universal and inalienable and thus not contingent only upon the laws and customs of a particular culture or government. Natural rights theory arose during the Enlightenment to counter the belief in the divine right of kings, and became the basis of the social contract that gave rise to democracy, a superior system for the protection of human rights. This is what the English philosopher John Locke had in mind in his 1690 *Second Treatise of Government* (written to rebut Sir Robert

Filmer's 1680 *Patriarcha*, which defended the divine right of kings[8])
when he wrote: "The state of nature has a law to govern it, which obliges
every one: and reason, which is that law, teaches all mankind, who will
but consult it, that being all equal and independent, no one ought to
harm another in his life, health, liberty, or possessions."[9] The social contract entered into freely, Locke argued, is the best way to ensure our
natural rights.[10]

In rights language, the individual is imbued with *personal autonomy*.
As a natural right, the personal autonomy of the individual gives us criteria by which we can judge actions as right or wrong: *do they increase or
decrease the survival and flourishing of individual sentient beings?* Morality is not arbitrary, relative, or completely culture-bound. Morality is universal. We are all born with a moral sense, with moral emotions that guide
us in our interactions with other people, and that are influenced by local
culture, customs, and upbringing. Nature endowed us with the capacity to
feel guilt for the violation of promises and social obligations, for example,
but nurture can tweak the guilt dial up or down. Thus morality is real,
discoverable, "out there" in nature, and "in here" as part of our human
nature. From these facts we can build a science of morality—a means of
determining the best conditions to expand the moral sphere and increase
moral progress through the tools of reason and science.

SCIENCE, REASON, AND THE MORAL ARC

Understanding the nature of things and the causes of effects is what science is designed to do, and ever since the Scientific Revolution there has
been a systematic effort by thinkers in all fields to apply the methods of
science—which include the philosophical tools of reason and critical
thinking—to understanding ourselves and the world in which we live,
including and especially the social, political, and economic worlds, with
an end toward the betterment of humanity. This effort has produced a
worldview known as Enlightenment Humanism (or secular humanism, or
simply humanism), which, unlike most other worldviews, is more a method
than an ideology; it is a means of solving problems more than it is a set of
doctrinaire beliefs. Humanism, as its name implies, is—*and ought to be*—
concerned with the survival and flourishing of humans, and its methods
of reason and science are directed at figuring out how best to do that. Thus
the goal of a science of morality is—*and ought to be*—to determine the

conditions under which humans and, by extension, other sentient beings best prosper. To that end I need to define what I mean by science and reason.

Science

Science is a set of methods that describes and interprets observed or inferred phenomena, past or present, and is aimed at testing hypotheses and building theories. By *set of methods* I mean to emphasize that science is more of a procedure than it is a set of facts, and to *describe and interpret* them means that the facts do not just speak for themselves. *Observed or inferred phenomena* means that there are some things in nature that we can see, such as elephants and stars, but other things that we must infer, such as the evolution of elephants and stars. *Past or present* means that the tools of science can be used to understand not only phenomena occurring in the present, but in the past as well. (The historical sciences include cosmology, paleontology, geology, archaeology, and history, including and especially human history.) *Testing hypotheses* means that for something to be truly scientifically sound it must be testable, such that we can confirm it as probably true or disconfirm it as probably false.[11] *Building theories* means that the aim of science is to explain the world by constructing comprehensive explanations from numerous tested hypotheses.

Defining the *scientific method* is not so easy. The process involves making observations and forming hypotheses from them, then making specific predictions based on those hypotheses, then making additional observations to test those predictions to confirm, disprove, or falsify the initial hypotheses. The process is a constant interaction of making observations, drawing conclusions, making predictions, and checking them against the evidence. But note that data-gathering observations are not made in a vacuum. The hypotheses shape what sort of observations a scientist will make, and these hypotheses are themselves shaped by education, culture, and the particular biases of the observer. Observation is key. The British astronomer Sir Arthur Stanley Eddington employed a legal metaphor to capture the sentiment: "For the truth of the conclusions of physical science, observation is the supreme court of appeal."[12] All facts in science are provisional and subject to challenge and change, therefore science is not a "thing" per se; rather it is a *method* of discovery that leads to *provisional* conclusions.

Reason

Reason is the cognitive capacity to establish and verify facts through the application of logic and rationality, and to make judgments and form beliefs based on those facts. *Rationality* is the application of reason to form beliefs based on facts and evidence, instead of guesswork, opinions, and feelings. That is to say, the rational thinker wants to know what is *really* true and not just what he or she would *like* to be true.[13]

However, as several decades of research in cognitive psychology has shown, we are not the rationally calculating beings we'd like to think we are, but are instead very much driven by our passions, blinded by our biases, and (for better or worse) moved by our moral emotions. The confirmation bias, the hindsight bias, the self-justification bias, the sunk-cost bias, the status-quo bias, anchoring effects, and the fundamental attribution error are just a few of the many ways that our brains work to convince us that what we *want* to be true *is* true—regardless of the evidence—in a general process called "motivated reasoning."[14] Nevertheless, the capacity for reason and rationality is within us as a feature of our brains that evolved to form patterns and make connections (it's called *learning*) in the service of survival and flourishing in the environment of our evolutionary ancestry. Reason is part of our cognitive makeup, and once it is in place it can be put to use in analyzing problems it did not originally evolve to consider. Pinker calls this an open-ended combinatorial reasoning system that "even if it evolved for mundane problems like preparing food and securing alliances, you can't keep it from entertaining propositions that are consequences of other propositions." This ability matters for morality because "if the members of species have the power to reason with one another, and enough opportunities to exercise that power, sooner or later they will stumble upon the mutual benefits of nonviolence and other forms of reciprocal consideration, and apply them more and more broadly."[15]

Drawing inferences about the movement of animals from their tracks—as hunter-gatherer trackers do—has obvious survival advantages, and we have been able to apply those inferential skills to everything from driving to the store to flying rockets to the moon. Historian of science and professional animal tracker Louis Liebenberg has, in fact, argued that our ability to reason scientifically is a by-product of fundamental skills for tracking game animals that our ancestors developed. Liebenberg's analogy between tracking and the *scientific method* is revealing: "As new factual information is gathered in the process of tracking, hypotheses may have to be revised or substituted by better ones. A hypothetical recon-struction of the animal's behaviors may enable trackers to anticipate and predict the animal's movements. These predictions provide ongoing test-ing of hypotheses."[16] Liebenberg distinguishes between *systematic track-ing* ("the systematic gathering of information from signs, until it provides a detailed indication of what the animal was doing and where it was going") and *speculative tracking* ("the creation of a working hypothesis on the basis of initial interpretation of signs, knowledge of the animal's behavior and knowledge of the terrain" that leads to hypotheses that are tested and, if not confirmed, to new hypothetical reconstructions of the animal's whereabouts). Speculative tracking also involves another cogni-tive process called "theory of mind," or "mind reading," in which trackers put themselves into the mind of the animal they are pursuing and imag-ine what it might be thinking in order to predict its actions.

Based on archaeological and anthropological evidence Liebenberg estimates that humans have been hunting and using systematic tracking for at least two million years (as far back as *Homo erectus*), and speculative tracking for at least one hundred thousand years.[17] Whenever these cogni-tive capacities arose, once the neural architecture is in place to deduce, say, that a lion slept here last night, a person can substitute lion with any other animal or object and can swap "here" with "there" and "last night" with "tomorrow night." The objects and time elements of the reasoning process are interchangeable. In a modern example, once you've mastered the multiplication tables and you know that $7 \times 5 = 35$, you can infer that 5×7 is also 35 because 5 and 7 are interchangeable in the equation. This interchangeability is a by-product of neural systems that evolved for basic reasoning abilities such as tracking animals for food.[18]

This is how a brain that evolved for one purpose can be put to other uses, and this cognitive capacity to substitute Xs and Ys in a representa-tional system that encompasses endless combinations and options—from prey to people—is what enables us to adopt the perspective of another moral agent, and is thus the cognitive architecture underlying moral reasoning.

THE EXPANDING MORAL SPHERE AND THE
PRINCIPLE OF INTERCHANGEABLE PERSPECTIVES

The *expanding moral sphere* is the metaphor I use to describe what has been pushing the moral arc upward, derived from the *expanding circle* metaphor first evoked in 1869 by the Irish historian William Edward Hartpole Lecky in his massive two-volume survey titled *History of European Morals*: "History tells us that, as civilisation advances, the charity of men becomes at once warmer and more expansive, their habitual conduct both more gentle and more temperate, and their love of truth more sincere." Such moral progress, however, is not built into our biology, Lecky says. "Men come into the world with their benevolent affections very inferior in power to their selfish ones, and the function of morals is to invert this order." After admitting that "The extinction of all selfish feeling is impossible for an individual, and if it were general, it would result in the dissolution of society," Lecky shows that moral progress is an incremental process: "The question of morals must always be a question of proportion or of degree. At one time the benevolent affections embrace merely the family, soon the circle expanding includes first a class, then a nation, then a coalition of nations, then all humanity, and finally, its influence is felt in the dealings of man with the animal world."[19] Expanding the moral circle to include animals, in nineteenth-century Europe? That was innovative for the time, and it shows what can happen once you start reasoning from basic moral principles.[20]

The philosopher Peter Singer too was ahead of the curve when he published *The Expanding Circle* in 1981, anticipating the developments in the sciences of evolutionary psychology and evolutionary ethics that unfolded in the 1990s and 2000s and out of which a science of morality could be developed. Singer makes the case for reason and science as providing rational arguments for *why* we should value the interests of X as much as we value our own interests, with X being racial minorities, gay people, women, children, and now animals. To explain the expanding circle Singer invokes what he calls "The principle of impartial consideration of interests": "In making ethical decisions I am trying to make decisions which can be defended to others. This requires me to take a perspective from which my own interests count no more, simply because they are my own, than the similar interests of others. Any preference for my own interests must be justified in terms of some broader impartial principle."[21]

Steven Pinker explains the logic like this: "If I appeal to you to do something that affects me then I can't do it in a way that privileges my interests over yours if I want you to take me seriously. I have to state my case in a way that would force me to treat you in kind. I can't act as if my interests are special just because I'm me and you're not, any more than I can persuade you that the spot I am standing on is a special place in the universe just because I happen to be standing on it."[22]

The reasoning process behind the expanding moral sphere (I prefer the three-dimensionality of a sphere instead of the two-dimensionality of a circle because I imagine it encompassing more range of variability within and across time and space and species) might more broadly be called the *principle of interchangeable perspectives*, and it applies not just to individuals within a group, tribe, or nation, but *between* groups, tribes, and nations as well. I cannot reasonably appeal to your nation to privilege my nation simply because it is my nation and not yours. (When I tell my European friends about a certain US conservative radio talk show host's routine refrain about America being "the greatest nation on God's green Earth"[23] they just roll their eyes.) Any preference for my group's interests over yours must be justified by some unbiased, disinterested ethic, which sounds simple, but given that we're dealing with humans, not Vulcans, it's sometimes difficult for two parties to agree on basic principles—especially parties who are unable or unwilling to switch points of view. This is the power of ethical reasoning, which, as Singer notes, "once begun, pushes against our initially limited ethical horizons, leading us always toward a more universal point of view."[24]

Reason and the *principle of interchangeable perspectives* put morals more on a par with scientific discoveries than cultural conventions. Scientists cannot just assert a claim without backing it up with reasoned arguments and empirical data (well, they can, but they'll be unceremoniously dismissed or publicly trounced by their colleagues). There really is a better way for people to live, and in principle we should be able to discover that way through the tools of science and reason. It is often said that you cannot reason someone out of a belief that they didn't reason themselves into in the first place, but when reasons are given for a belief we are entitled to counter those reasons with better reasons if they are available. And if no reasons are in the offing, we are entitled to dismiss them by invoking what I call *Hitchens's Dictum*, after my late friend and colleague Christopher Hitchens's observation, "What can be asserted without evidence can also be dismissed without evidence."[25]

The ability and, just as importantly, the willingness to change perspectives and points of view are major drivers of the expanding moral sphere, which is pictured in *figure 1-1*. The expansion outward from ourselves

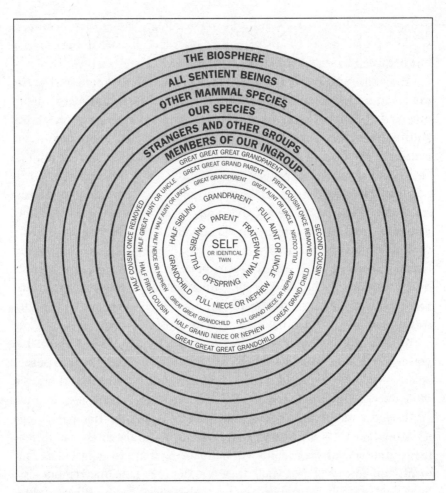

Figure 1-1. The Expanding Moral Sphere

shows that our moral concerns are most directly connected to those most closely related to us genetically, from an identical twin, to our siblings, parents, and offspring, to our grandparents, half siblings, aunts and uncles, nieces and nephews, to our great-grandparents, great-aunts and great-uncles, and great-grandchildren, all the way out to great-great-great-grandparents, to friends, acquaintances, and members of our in-group, to members of other groups, tribes, states, and nations, to all members of our species, to members of other mammal species, to all sentient beings, to the biosphere. Reflected in this ability to change perspectives and expand the moral sphere may be our expanding intelligence and abstract reasoning ability.

ABSTRACT REASONING AND MORAL INTELLIGENCE

Thinking scientifically requires the ability to reason abstractly, which itself is at the foundation of all morality. Consider the mental rotation required to implement the Golden Rule: Do unto others as you would have them do unto you. This necessitates one to change positions—to become the other—and then to extrapolate what action X would feel like as the receiver instead of the doer (or as the victim instead of the perpetrator). A case can be made that the type of conceptual ratiocination required for both scientific and moral reasoning not only is linked historically and psychologically, but also that it has been improving over time as we become better at nonconcrete, theoretical reflection.

In the 1980s the social scientist James Flynn discovered that IQ scores have been going up, on average, 3 points every decade for the past century. Now called the *Flynn Effect*, this is an astonishing increase of about 30 IQ points over 100 years. This translates into an improvement of two standard deviations of 15 points each from an "average" IQ of 100 to a "very superior" score of 130. (IQ test scores remain the same, however, as they are regularly "normed" upward to account for the Flynn Effect, which is how Flynn discovered it in the first place.) Were it just the case that we're all getting better at taking tests, then the scores should have been improving across the board. But this is not what has happened. The increases in IQ scores have been almost exclusively in the two subtests that most require abstract reasoning: Similarities and Matrices. The subtests of Information, Arithmetic, and Vocabulary have hardly budged at all.[26] *Figure 1-2* shows the trend lines since the late 1940s.

The subtest called *Similarities* asks questions such as "What do dogs and rabbits have in common?" If you answer, "Both are mammals," says Flynn, you are thinking like a scientist in classifying organisms by type, which is an abstraction. If you said, "You use dogs to hunt rabbits," you are thinking concretely, imagining a tangible use for a dog. According to Flynn, for the past century people have learned to think more abstractly than concretely.

Matrices are abstract figures that require determining a pattern and then deducing the missing piece in the pattern, as in *figure 1-3*.

The cause of the Flynn Effect is controversial. The hypothesis that the rising tide of standardized testing has lifted all boats is contradicted by the fact that the increase in IQ scores preceded the era of standardized test taking and they've continued their rise at a steady rate irrespective

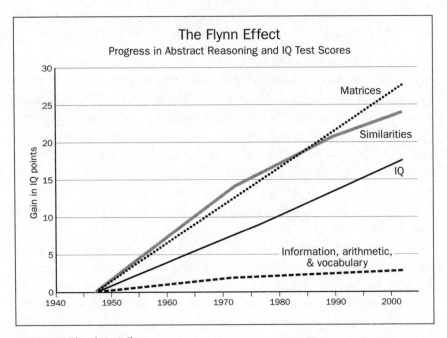

Figure 1-2. The Flynn Effect

The social scientist James Flynn discovered that IQ scores are increasing, on average, 3 points every decade, most noticeably in the two subtests that require abstract reasoning: Similarities and Matrices.[27]

of test-taking rates.[29] A more likely explanation is that the improvement is a function of more years in school, more technologies in society, more technical jobs, and a greater need for people to perform conceptual tasks as our economy has shifted from agrarian and industrial to information-based. Instead of manipulating plows, cows, and machinery, many of us are now manipulating words, numbers, and symbols. Even in science classes the trend has been shifting away from the rote memorization of facts about nature to reasoning about nature's laws and processes—content *and* process. And process thinking is a form of abstract reasoning.[30]

Flynn himself attributes the effect to an accelerating capacity for people to view the world through "scientific spectacles." He contrasts the "prescientific" world of his father with the "postscientific" world of today through a poignant anecdote about how he and his brother tried to mitigate their father's typical prejudice of his generation through a thought experiment: "What if you woke up one morning and discovered your skin had turned black? Would that make you any less of a human being?" The senior Flynn

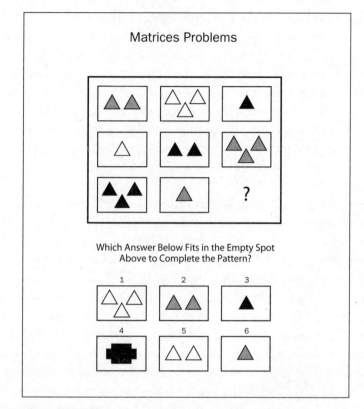

Figure 1-3. Matrices Problems
If you selected answer #5, then you are reasoning abstractly.[28]

shot back, "Now, that's the stupidest thing you've ever said. Who ever heard of a man's skin turning black overnight?" The Flynn patriarch was intelligent but uneducated, Flynn explained, in attributing the effect to nurture, not nature.[31] The anecdote is symbolic of larger social trends. Each generation is producing not only better *abstract* reasoners, but better *moral* reasoners as well. In an interview in *Skeptic* magazine, Flynn reflected on the early-twentieth-century research by the psychologist Alexander Luria on the reasoning abilities of Russian peasants:

The illiterate Russian peasants Luria studied were not willing to take the hypothetical seriously. He said, "Imagine that bears come from where there is always snow and imagine that if bears come from where there is always snow they are white. What color would the bears be at the North Pole?" and they would respond something like, "I've only seen brown bears. If an old man came from the North Pole and told me I might believe him." They were not interested in the hypothetical, or abstract categories. They were grounded in concrete reality. "There are no camels in Germany. B is in Germany. Are there camels there?" They said, "Well, it's big enough, there ought to be camels. Or maybe it's too small to have camels." We have wonderful data from the Raven's Progressive Matrices tests from 1950 and 2010 showing that the Raven's games are entirely correlated with freeing your mind of the concrete reference of the symbols in order to take the relationship between the symbols more seriously.[32]

Flynn and his colleague William Dickens suggest that the increases in cognitive reasoning may have started centuries ago with the Industrial Revolution, which saw an improvement in both the quantity and quality of education, better nutrition, disease control, and the manipulation of complex machinery. Then, after 1950, "IQ gains show a new and peculiar pattern. They are missing or small on the kind of IQ tests closest to school-taught material like reading and arithmetic. They are huge on tests that emphasize on-the-spot problem-solving, like seeing what verbal abstractions have in common, or finding the missing piece of a Matrices pattern, or making a pattern out of blocks, or arranging pictures to tell a story. Perhaps the Industrial Revolution stopped demanding progress in the basics and started demanding that people take abstract problem-solving more seriously."[33]

Whatever the cause of the Flynn Effect, it isn't genetic or biological, because there hasn't yet been enough time for natural selection to operate, and even though nutrition has improved over time, it more or less stabilized in the mid-twentieth century (and, if anything, may have gotten worse recently due to the prevalence of junk food), and yet IQ gains continue. Steven Johnson, in his intriguing book *Everything Bad Is Good for You*, makes the case for modern pop culture and media—even the "boob tube"—as drivers of abstract reasoning improvements, noting, for example, that the plot lines and character developments of today's television shows are far more complex than those of decades past.[34] Flynn theorizes a suite of cultural factors that came together over the past century:

Cognitively demanding jobs raise IQ. Look at how much more cognitively demanding a merchant banker's job is today than in 1900, when a banker just had to know who was a good risk for mortgages. Or cognitively demanding leisure: The video games may not be getting you to read good literature, but they do exercise your mind, rather than just playing sports. Leisure, occupation, and, of course, schooling too make a difference. The schooling of today has introduced into its curriculum much more intellectually challenging material. If you look at the exams given American school children in 1914 in Ohio, it was all about socially valuable material. "What are the capitals of the then 44 states?" Today it would be, "Why is the capital of a state rarely the largest city?"[35]

Being able to rattle off the capitals of the states does not require any abstract reasoning ability, but knowing that rural state legislatures controlled where the capital was put in a state, and that they disliked big cities and so put them in a county seat, leads to a deeper understanding of why Albany is the capital of New York rather than New York City, and why Harrisburg is the capital of Pennsylvania rather than Philadelphia. "So you see they are being asked to make hypotheses about fairly abstract concepts and propositions, and link those hypotheses through logic," Flynn explained. "So the demands conceptually placed on the schools have now changed."[36] And they have changed in the workplace. Flynn points out that in 1900 only 3 percent of Americans had cognitively demanding jobs, whereas that figure was 35 percent in 2000.

A case can be made that our improved ability to reason abstractly is the result of the spread of scientific thinking—that is, science in the broader sense of reason, rationality, empiricism, and skepticism. Thinking like a scientist means employing all our faculties to overcome our emotional, subjective, and instinctual brains to better understand the true nature of not only the physical and biological worlds, but the social world (politics and economics) and the moral world (abstracting how other people should be treated) as well. That is, the moral arc of the universe may be bending, in part, because of something like a *Moral Flynn Effect*, as Pinker calls it.[37] Pinker says "the idea is not crazy," but I would go farther. I claim that our improvement in abstract reasoning generally has translated into a specific improvement in abstract moral reasoning, particularly about other people who are not our immediate kith and kin. Evolution endowed us with a natural tendency to be kind to our genetic relations but to be xenophobic, suspicious, and even aggressive toward people in other tribes. As our brains become better equipped to reason abstractly in such tasks as lumping dogs and rabbits together into the category of "mammal," so too have we improved in our capacity to lump blacks and whites, men and women, straights and gays into the same category of "human." To employ a metaphor from evolutionary theory and the problem of defining a species, we are moving toward becoming "lumpers" instead of "splitters"—seeing similarities instead of differences.

As philosophers and scholars over the past two centuries have consciously adopted the methods of science to establish such abstract concepts as rights, liberty, and justice, successive generations have become schooled in thinking of these abstractions as applied to others in a Matrices-like mental rotation. Consider a number of studies and lines of evidence in support of the hypothesis that our moral intelligence is increasing[38]:

- Intelligence and education are negatively correlated with violent crime.[39] As intelligence and education increase, committing violent crimes and falling victim to them decrease, even when controlled for socioeconomic class, age, sex, and race.[40]
- Cognitive style predicts criminal justice attitudes. The psychologist Michael Sargent found a correlation between a high "need for cognition" (enjoying mental challenges such as those employed in intelligence tests) and a low demand for punitive justice, even when such attitudes are controlled for age, sex, race, education, income, and political orientation. In an aptly titled paper "Less Thought, More Punishment," Sargent's conclusions support the principle that the punishment should fit the crime, a principle that requires grasping the abstract concept of proportionality, a process fundamental to all scientific thought.[41]
- Abstract reasoning ability is positively correlated with cooperation in a game called Prisoner's Dilemma (a classic thought experiment in game theory that proves cooperation creates better outcomes even when a perfectly rational, self-interested actor would not want to do so). The economist Stephen Burks and his colleagues administered a thousand trainee truck drivers the Matrices IQ test and had them participate in a game of Prisoner's Dilemma in which they could either cooperate or defect with a game partner. Those wannabe truck drivers who scored high in ability to solve those Matrices figures were more likely to cooperate on the first move of a Prisoner's Dilemma game, even after the usual intervening variables were controlled for, such as age, race, gender, schooling, and income.[42] The economist Garrett Jones confirmed the connection in a meta-

analysis of thirty-six Prisoner's Dilemma experiments conducted between 1959 and 2003 in colleges and universities around the country, finding a positive correlation between a school's mean SAT score and the propensity of its students to respond cooperatively.[43]

- Intelligence predicts classical liberal attitudes toward helping others. An analysis of data from the National Longitudinal Study of Adolescent Health found that among twenty thousand young adults there was a positive correlation between IQ and liberalism, and data from the General Social Survey clarified the link in noting that the correlation was between intelligence and *classical* liberalism of the Enlightenment kind, in which smarter people were less likely to agree that the government should redistribute income from the rich to the poor but more likely to agree that the government should help African Americans to compensate for historical discrimination.[44] In other words, the effect was more in the moral dimension of how people are ethically treated instead of the more concrete dimension of economic adjustment, and thus its import for moral reasoning.

- The psychologist Ian Deary and his colleagues confirmed this link in a paper aptly titled "Bright Children Become Enlightened Adults." Deary found a positive correlation between the IQ of British children at age ten and their endorsement of antiracist, socially liberal, and pro-working-women attitudes at age thirty, holding the usual potentially intervening variables constant. The causal arrow from intelligence to moral abstraction is confirmed by the twenty-year gap between measures.[45] By "enlightened" Deary meant the values that came directly from the Enlightenment, the definition for which he adopted from the *Concise Oxford Dictionary*: "a philosophy emphasizing reason and individualism rather than tradition."

- Intelligence predicts economic attitudes, most notably abstract concepts such as how free trade is a form of positive-sum game that seems counterintuitive to our folk-economic intuitions that most economic exchanges are zero-sum in a fixed pie of wealth. The economists Bryan Caplan and Stephen Miller culled data from the General Social Survey and found a correlation between intelligence and openness to immigration, free markets, and free trade, and a reluctance to endorse government make-work projects, protectionist policies, and business interventionism.[46] Concrete thinking leads us to endorse economic tribalism along with populist and nationalist

zero-sum attitudes toward other tribes (nations in the modern
world). Abstract reasoning leads us to consider members of other
tribes (nations) as potential trading partners to be respected rather
than as potential enemies to be conquered or killed.

- Intelligence predicts democratic tendencies, most notably the rule of
law. The psychologist Heiner Rindermann ran correlational studies
on a number of datasets from many different countries, examining
their average scores on popular intelligence tests and measures of
academic achievement from 1960 to 1972, and found that these predicted the level of prosperity, democracy, and the rule of law found
in those countries in the subsequent period 1991 to 2003 (and this is
even when controlling for the country's prior level of prosperity).[47]
In other words, all other things being equal, a country that educates
its population in the ability to reason abstractly will be a more prosperous and moral country.

- Most encouraging for those of us who are citizens of the Republic of
Letters, evidence is now accumulating that shows a positive correlation between literacy and morality, and most particularly between
the reading of fiction and being able to take the perspective of others.[48]
Taking the perspective of characters in a novel requires a Matrices-
like rotation of relational positions, coupled to an emotional connection of what it would feel like if X happened to you, even though the
"you" in this case is a character in the novel. In a 2011 study, for
example, the Princeton University neuroscientist Uri Hasson and his
team scanned the brain of a woman while she told a story out loud
that the scientists recorded and subsequently played back for other
subjects while their brains were being scanned as they just listened.
Results: when the reader's emotional brain region called the insula lit
up during a certain portion of the story, so too did the listeners'
insulas; when the woman's frontal cortex became active during a different part of the story, the same region in listeners' brains were also
activated.[49] It's almost as if the fictional story synchronized the
brains of reader and listeners, enabling a form of mind reading and
moral perspective taking. (Such brain synchronicity was found in
another study by Hasson and his team when he scanned the brains
of subjects while they watched Sergio Leone's classic 1966 film *The
Good, the Bad, and the Ugly*, finding that "brain activity was similar
across viewers' brains," specifically that about 45 percent of the neo-

cortex across all five of the subjects' brains were lit up in the same areas during the same movie scenes.[50])

A 2013 study published in *Science* titled "Reading Literary Fiction Improves Theory of Mind" reported on the results of studies conducted by the psychologists David Comer Kidd and Emanuele Castano on the causal relationship between reading high-quality literary fiction and the ability to take the perspective of others, as measured by one of several well-tested tools, such as judging the emotions of others and eye-gaze directionality for interpreting what someone is thinking.[51] They found that participants who were assigned to read literary fiction performed significantly better on the Theory of Mind (ToM) tests than did participants assigned to the other experimental groups, who did not differ from one another.

This experiment is important because it nails down the direction of the causal arrow from reading literary fiction to perspective taking, instead of the other direction, in which people who are good at mind reading prefer fiction. That said, this research is in its infancy, and there are reasons to be skeptical about pushing the link between literacy and morality too far. Education in general and literature in particular may have morally salubrious effects for reasons we do not yet fully understand, but I am encouraged by these studies, and others that put theory into practice. For example, in a documentary film titled *The University of Sing Sing*, Tim Skousen documents the work of his parents—Jo Ann and Mark Skousen—and other teachers at the Sing Sing prison in New York, in which literature is used to broaden the critical thinking skills and expand the moral horizons of the prisoners there.[52]

Working through one of the few college degree-granting programs in the Department of Corrections of New York State, the psychologists interviewed cite statistics showing that the best predictor of success after prison is a college degree. As the psychologist Susan Weiner, who works with the program, noted, "These men and women will come back to the community. How do you want them to come back? This isn't just a gift for them. It's a gift for society. It's the way that we make society a better place." A prisoner named Denis Martinez, for example, explained what getting an education and learning to read deeply into subjects gave him in terms of perspective: "It's given me a new set of glasses. Before I wasn't able to see the things I see now. I was a nineteen-year old knucklehead going around and thinking I knew it all. The more I learned the more I could sense how wrong I was and how many things I didn't know." Inspired by his reading of René Descartes, Martinez reflected, "There are two ways to be in prison—physically and/or mentally. Being in prison mentally is to live in ignorance, closed-mindedness, and pessimism. You can confine me for as long as you want, but my mind will always be free." The title of a painting this prisoner made is revealing: *Cogito Ergo Sum Liber—I Think Therefore I Am Free.* (Now, *there's* a bumper sticker/T-shirt slogan for the modern Enlightenment thinker.)

THE VIRTUE OF CONTINUOUS THINKING

Thinking abstractly is not the only cognitive tool of the scientist that we can apply to moral reasoning. Thinking about concepts both on a *continuous scale* and as *categorical entities* illuminates—and sometimes eliminates—a number of moral problems. In my book *The Science of Good and Evil* I applied the idea of "fuzzy logic"[53] to show that "evil" and "good" are not simply black-and-white categories of reified "things" but are instead fuzzy shades of behavior along a continuous scale. *Good* and *evil* are descriptive terms for behaviors of moral actors that can be assessed along a continuum. Take altruism and selfishness: Like all behaviors, there is a broad range of expression of both types of behavior. Instead of categorizing someone as either altruistic or selfish in a 1 or 0 binary logic system, we might think of this person as 0.2 altruistic and 0.8 nonaltruistic (or selfish), or 0.6 cooperative and 0.4 noncooperative (or competitive).[54]

Most moral problems are better conceived as continuous rather than as categorical. The categorization of the world into cleanly cleaved boxes is a useful cognitive tool for some tasks, but it doesn't always serve us well in understanding social and moral problems. Does democracy decrease the probability of war? If you categorize nations into simple binary boxes of democracy or nondemocracy (1 or 0), you can find lots of exceptions to the democratic peace theory. But if you scale rank countries by degrees of democracy (on a 1–10 scale), and you also scale wars from minor conflicts to world wars, you find a significant negative correlation between the degree of democracy and the probability of fighting (more on this later).

Scientists tend to think about problems on a continuous scale. The evolutionary biologist Richard Dawkins, for example, points out that labeling a fossil as belonging to this or that species is falling prey to "the tyranny of the discontinuous mind," in which "Paleontologists will argue passionately about whether a particular fossil is, say, *Australopithecus* or *Homo*. But any evolutionist knows there must have existed individuals who were exactly intermediate." Darwin's theory of evolution, in fact, overturned the categorical "essentialism" of how organisms are conceived as fixed entities. "It's essentialist folly to insist on the necessity of shoehorning your fossil into one genus or the other," Dawkins notes. "There never was an *Australopithecus* mother who gave birth to a *Homo* child, for every child ever born belonged to the same species as its mother."[55]

Consider the category of "poverty." The World Bank defines poverty as making less than $1.25 a day, and the Bill and Melinda Gates Foundation has highlighted the fact that since 1990 the percentage of the world's population living in poverty has declined by 50 percent.[56] This is progress, and it should be duly noted what has been done to continue this positive trend, but the categorical thinking that puts people into two boxes of "poverty" and "nonpoverty" obscures the fact that making, say, $2.50 a day is still a serious detriment to one's survival and flourishing. A gradient scale indicating how much people make around the world more accurately depicts their economic well-being (or lack thereof).

Not all moral problems can be so conceived, but throughout this book I will show how thinking about issues on a continuous scale rather than as categorical entities is both instructive and enlightening, and if exceptions to generalizations come to mind it is helpful to consider whether they invalidate the generalization or fall on a continuous scale within the generalization.

FROM *IS* TO *OUGHT* IN MORAL PROGRESS

As we work our way through the evidence for moral progress and the many factors that help to bring it about, keep in mind that establishing the causes of moral progress tells us how to achieve it if that is our goal. But it does not explain *why* we would want to expand the moral sphere in the first place. One could just as well argue that science and reason show us how to *shrink* the moral sphere, and that too would be true. This difference between *how* and *why* gets at a vexing problem that has come to plague the study of morality ever since the philosopher David Hume identified what is known as the *Is-Ought Problem* (sometimes rendered as the *Naturalistic Fallacy,* or *Hume's Guillotine*), or the difference between *descriptive* statements (the way something *is*) and *prescriptive* statements (the way something *ought to be*). Here is how Hume described the problem:

In every system of morality, which I have hitherto met with, I have always remarked, that the author proceeds for some time in the ordinary ways of reasoning, and establishes the being of a God, or makes observations concerning human affairs; when all of a sudden I am surprised to find, that instead of the usual copulations of propositions, is, and is not, I meet with no proposition that is not connected with an ought, or an ought not. This change is imperceptible; but is however, of the last consequence. For as this ought, or ought not, expresses some new relation or affirmation, 'tis necessary that it should be observed and explained; and at the same time that *a reason should be given*; for what seems altogether inconceivable, how this new relation can be a deduction from others, which are entirely different from it. But as authors do not commonly use this precaution, I shall presume to recommend it to the readers; and am persuaded, that this small attention would subvert all the vulgar systems of morality, and let us see, that the distinction of vice and virtue is not founded merely on the relations of objects, nor is perceived by reason.[57]

Most people take Hume to mean that there is a wall separating is from ought, and that science can have nothing to say about determining human values and morals. But if morals and values should not be based on the way things are—reality—then on what should they be based? When I use the word "is" I do not just mean what is *natural*—as in biological nature alone—I mean the reality of the "is" under study.

When we undertake a study of war to understand its causes so that we may lessen its occurrence and attenuate its effects, this is an is-ought transition grounded in the true nature of war—and by nature I do not just mean the biological propensity (or not) of humans to fight. I mean all of the factors that go into the causes of war: biological, anthropological, psychological, sociological, political, economic, and the like.

Philosophers have grappled with the is-ought problem ever since Hume outlined it, and some have proposed solutions, such as John Searle's widely cited 1964 paper "How to Derive 'Ought' from 'Is,'" in which he proposes, for example, that the act of making a promise becomes the "is" and as such it becomes an obligation that one "ought" to fulfill.[58] In any case, note what Hume is actually saying: not that one *cannot* shift (however imperceptibly) from "is" to "ought," but that one *should not do so* without providing a *reason*. Fair enough. As with any claim in science, reasons and evidence must be provided in support, or else a claim is little more than an assertion. This appears to be what Hume is recommending to his readers when he suggests "that this small attention [to giving reasons] would subvert all the vulgar systems of morality." To make sure I am not misreading Hume, or reading into him what I think *ought* to be there instead of what *is* there (and thereby be guillotined), I queried one of the world's foremost Hume scholars—the Oxford philosopher Peter Millican—on the matter. He explained:

> It's true that Hume didn't explicitly say the Is/Ought gap was unbridgeable, but his analysis of morals as founded on sentiment did imply that in a sense—moral statements are not matters of fact (or mere relations of ideas, for that matter). I think he would go much the same way as you do—instead of making out that morals can be worked out by logical thinking, he would want to understand it as a natural human phenomenon—which requires scientific understanding, and then have that inform the decisions we make about how to foster it. Those decisions, of course, will be informed by our natural sentiment, so there's an element of circularity here (we're not simply inferring matters of fact from other matters of fact—ethical judgment is playing a role), but if there's enough agreement on basics (like war is bad, trust is good) that doesn't prevent progress.[59]

Not preventing progress is a worthy goal of any project, regardless of its degree of blending is and ought, but we can do better still by applying the knowledge of the causes of moral progress to help bring it about. It is in this sense that this book is *descriptive* in that it describes what has unfolded over time as we have become more moral, and it is *prescriptive* in that it prescribes what we ought to do if we want to continue the trend.

Take homosexuality and same-sex marriage, the latest of the Rights Revolutions that are unfolding in our time. *Descriptively,* science tells us that human beings have an evolved, innate drive to survive and to flourish, and that one of the most necessary and primal requirements among the many preconditions for life, health, and happiness for most people is a loving bond with another human being. *Prescriptively,* we can say that granting only a select group of privileged people the right to fulfill this evolved need—while simultaneously depriving others of the same basic right—is immoral because it robs them of the opportunity to fulfill their essence as evolved sentient beings. This is true even if the case could be made (as it has been by those who oppose same-sex marriage) that such discriminatory practices are better for the group (in a type of utilitarian calculus where the sacrifice of the few is justified if it leads to the greater happiness for the greater number). It is still wrong because the *individual* is the moral agent, not the group. It is the individual who feels the sharp pain of discrimination, the sting of being excluded, and the insult of being treated differently under the law. Science tells us why they feel this way and reason instructs us what to do about it if we want to continue the moral progress of the Rights Revolutions.

As well, social science shows that humans are naturally tribal and that we tend to exclude people simply because they're not one of us (however "us" is defined). So how can we thwart the mind's natural inclination to pigeonhole people into prejudicial categories that turn them into Others we can exclude, exploit, or kill? Scientific research gives us guidance. For example, studies show that straight people who know gay people as neighbors, friends, or work colleagues are less likely to hold bigoted and prejudicial views of homosexuality, and they are more likely to agree that gay people should be treated equally under the law and should be accorded equal rights, such as the right to marry. A 2009 Gallup study, for example, found that "when controlling for ideology, those who know someone who is gay or lesbian are significantly more supportive of gay marriage than are those of the same political persuasion who do not personally know someone who is gay or lesbian."[60] Thus the existence of plays, films, and television shows that feature LGBTQ actors; of literary and pop cultural references that portray homosexuality and non-normative gender expression in a positive light; of "coming out" campaigns; and of LGBTQ role models in politics, business, and sports—all of these are essential for awakening empathy and understanding, and thus for expanding the moral sphere ever outward.

A PUBLIC HEALTH MODEL OF MORAL SCIENCE

In proffering this rapprochement between the way things are and the way things ought to be—the interface of facts and values—I am doing nothing more than recognizing a trend that has been under way since the Enlightenment in taking the findings of science about the way the world is and applying them to the way we would like the world to be. There's a reason why social scientists—social psychologists, cognitive psychologists, evolutionary psychologists, anthropologists, sociologists, economists, political scientists, and criminologists—along with policymakers and politicians, have been amassing extensive databases and ethnographies, testing hypotheses, and crunching the numbers through models and theories related to violence, aggression, crime, war, terrorism, civil rights violations, and the like: we want to understand causes to effect changes.

This approach may be modeled on *public health*, defined as long ago as 1920 in a *Science* article as "the science and art of preventing disease, prolonging life, and promoting health through the organized efforts and informed choices of society, organizations, public and private, communities and individuals."[61] Public health science includes such fields as epidemiology, biostatistics, behavioral health, health economics, public policy, insurance medicine, occupational health, and others. If you want to know why the average person lives almost twice as long today as a century ago, look to public health. The *maximum life potential* (the age at death of the longest-lived member of the species) has not changed and remains at 120 years. *Life span* (the age at which the average person would die if there were no premature deaths from accidents or disease) has also not changed and remains at about 85 to 95 years. But *life expectancy* (the age at which the average individual would die when accidents and disease have been taken into consideration) has skyrocketed upward from 47 years in 1900 in the United States to 78.9 for all Americans born in 2010, and 85.8 for Asian American women.[62] The cause of this remarkable progress in both the quantity and quality of life is public health science and technology: flush toilets, sewers, and waste disposal technologies, clean water, hand washing, antiseptic surgery, vaccinations, pasteurization, road traffic safety, occupational safety, family planning, nutrition and diet, and other measures, coupled to the epidemiological study of infectious diseases such as smallpox and yellow fever, chronic diseases such as cancer and heart disease, and disease prevention through these many techniques. The survival and flourishing of humans have progressed in the past century more than in all previous centuries combined. If you agree that it is better that millions of people no longer die of yellow fever and smallpox, cholera and bronchitis, dysentery and diarrhea, consumption and tuberculosis, measles and mumps, gangrene and gastritis, and many other assaults on the human body, then you have offered your assent that the way something *is* (diseases such as yellow fever and smallpox kill people) means we *ought* to prevent them through vaccinations and other medical and public health technologies.

This analogy between social problems and public diseases is not at the level of cause—crime, violence, war, and terrorism are not diseases in the medical sense of being an abnormal condition caused by the equivalent of a virus or bacteria. Instead the analogy is at the level of methodology: how we approach solving the problem by using the best tools of science, technology, and social policy available. Most crimes and acts of violence, along with wars and acts of terrorism, are not abnormal responses in a diseased state; most are normal responses to specific situations and conditions. But the public health model of engaging numerous sciences to change the situations and conditions that cause them is a viable methodology toward the goal of making moral progress.

By way of example, in 2012 I undertook an extensive study of gun violence for a special issue of *Skeptic* magazine in response to the raft of mass public shootings in recent years, such as at the Sandy Hook elementary school.[63] Subsequently I engaged in a series of debates around the country on gun control with the economist John Lott,[64] whose book *More Guns, Less Crime* offers his own prescription in its title.[65] While conducting a literature search I was struck by how much data on gun violence are published in public health and medical journals. For example, Johns Hopkins University houses the Bloomberg School of Public Health, which is one of the leading centers in the country for research on gun violence, and in 2012 they published a scholarly study on the problem titled *Reducing Gun Violence in America.* The subtitle is illustrative of my point: *Informing Policy with Evidence and Analysis.* It showed, for example, that oversight of licensed gun dealers to make sure their sales were to customers without criminal records resulted in a 64 percent decrease in guns diverted to criminals.[66] A 1998 study in the *Journal of Trauma and Acute Care Surgery,* "Injuries and Deaths Due to Firearms in the Home," found that "every time a gun in the home was used in a self-defense or legally justifiable shooting, there were four unintentional shootings, seven criminal assaults or homicides, and 11 attempted or completed suicides." In other words, a gun is twenty-two times more likely to be used in a criminal assault, an accidental death or injury, a suicide attempt, or a homicide than it is for self-defense.[67] A 2009 study published in the *American Journal of Public Health* that was conducted by epidemiologists at the University of Pennsylvania School of Medicine found that on average not only did guns not protect those who possessed them from being shot in an assault, these scientists also determined that people with a gun were 4.5 times more likely to be shot in an assault than those not possessing a gun.[68]

The raw figures for deadly violence in general, and gun violence in particular, are staggering. If ever there were a public health problem that needs solving, this is it. According to the FBI's crime reports, between 2007 and 2011 the United States experienced an annual average of 13,700 homicides, with guns responsible for 67.8 percent of them.[69] That's an annual average of 9,289 people shot dead by a gun, or 774 a month, 178 a week, 25 a day, or a little more than 1 per hour. It's a disquieting thought that just in the United States, every hour of every day someone is shot to death. That fact alone should convince us of the value of understanding the causes that underlie it, but the problem is even worse: according to the National Center for Injury Prevention and Control, in 2010 a total of 19,392 US residents killed themselves with a firearm,[70] another 11,078 were shot to death, and 55,544 were injured by gunfire and treated in emergency rooms.[71]

The public health model can also be applied on a larger scale and time horizon by pulling back to see that, in fact, homicide rates have plummeted over the millennia, from almost 1,000 per 100,000 people per year in prehistoric times and in modern nonstate societies, to about 100 per 100,000 people per annum in Western societies through the Middle Ages, to about 10 per 100,000 each year by the time of the Enlightenment, to less than 1 per 100,000 today in Europe (and a little more than 5 per 100,000 in the United States), an improvement of four orders of magnitude. How do we know this? Science. Archaeologists can estimate the rates of violent deaths in prehistoric bands through skeletal remains (see chapter 2). The ethnographies of anthropologists record the rates of violence among modern prestate peoples from their oral accounts and histories. And historians have used old court and county records to calculate that homicide rates have, to give just one example, dropped from 110 homicides per 100,000 people per year in fourteenth-century Oxford to less than 1 homicide per 100,000 in mid-twentieth-century London. Similar patterns have been documented in Italy, Germany, Switzerland, the Netherlands, and Scandinavia, and by the same order of magnitude: from about 100 per 100,000 to less than 1 between the fourteenth century and the twenty-first century. *Figure 1-4* shows the unmistakable trend in the decline in homicides from the thirteenth century to the twentieth century in multiple countries.[72] The uptick in murders at the end of the long downward trends captures the crime wave of the 1970s and 1980s, but the rate returned to historical lows by the early years of the twenty-first century.

If that isn't moral progress as I have defined it—an increase in the survival and flourishing of sentient beings—I don't know what is. Regardless of your position on gun control, my point is that treating gun

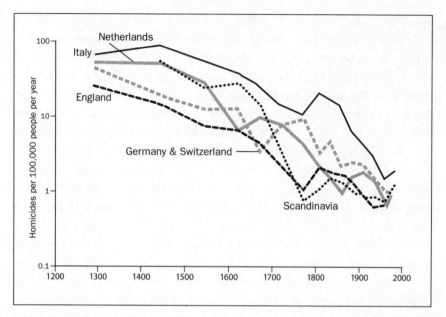

Figure 1-4. The Decline in Homicides

Homicide rates per 100,000 people per annum in five Western European regions from the thirteenth century to the twentieth century compiled by the criminologist Manuel Eisner.[73]

violence as a problem to be solved by better science and public policy is now common practice and part of the long-term trend to address moral issues pragmatically. If you agree that the hundreds of millions of lives that have been saved by public health sciences, technologies, and policies over the past two centuries is a moral good, then there is no reason why you might not also concur that applying social sciences to solving problems such as crime and violence is also something we ought to do. Why? Because saving lives is moral. Why is saving lives moral? Because the survival and flourishing of sentient beings is our moral starting point. But why should organisms *want* to survive and flourish in the first place? The answer may be found in the logic of the evolutionary process that created this drive.

THE EVOLUTIONARY LOGIC OF THE MORAL STARTING POINT

If you were a molecule, what would you do to survive? First you would need to build a substrate on which to generate a replication system inside a cell that contains machinery for energy consumption, maintenance, and repair, and other features that keep the molecule intact long enough to reproduce. Once such molecular machinery is up and running, the replicating molecules become immortal as long as there is energy to feed the system and an ecosystem in which these processes can take place. In time these replicating molecules will outsurvive nonreplicating molecules by virtue of the very process of replication—those that don't, die—thus the cells or bodies in which the replicators are housed are survival machines. In modern jargon, the replicators are called genes and the survival machines are called organisms, and this little thought experiment is what Richard Dawkins means by the "selfish gene" in his book of that title.[74] A cell, or body, or organism—a survival machine—is the gene's way of surviving and perpetuating itself. Genes that code for proteins that build survival machines that live long enough for them to reproduce will win out over genes that do not. Genes that code for proteins and enzymes that protect its survival machine from assaults such as disease help not just the organism to survive, but the genes as well. Survival, reproduction, flourishing: this is what survival machines do by their very nature. It is their—our—essence to strive to survive.

The problem is that survival machines scurrying around in, say, a liquid environment such as an ocean or pond will bump into other survival machines, all of whom are competing for the same limited resources. "To a survival machine, another survival machine (which is not its own child or another close relative) is part of its environment, like a rock or a river or a lump of food," says Dawkins. But there's a difference between a survival machine and a rock. A survival machine "is inclined to hit back" if exploited. "This is because it too is a machine that holds its immortal genes in trust for the future, and it too will stop at nothing to preserve them." Thus, Dawkins concludes, "Natural selection favors genes that control their survival machines in such a way that they make the best use of their environment. This includes making the best use of other survival machines, both of the same and of different species."[75] Survival machines could evolve to be completely selfish and self-centered, but there is something that keeps their pure selfishness in check, and that is the fact that other survival machines are inclined "to hit back" if attacked, to retaliate if exploited, or to attempt to use or abuse other survival machines first.

So in addition to selfish emotions that drive survival machines to want to hoard all resources for themselves, they evolved two additional pathways to survival in interacting with other survival machines: *kin altruism* ("blood is thicker than water") and *reciprocal altruism* ("I'll scratch your back if you'll scratch mine"). By helping its genetically related kin, and by extending a helping hand to those who will reciprocate its altruistic acts, a survival machine is helping itself. Thus there will be a selection for those who are inclined to be altruistic—to a point. With limited resources, a survival machine can't afford to help all other survival machines, so it must assess whom to help, whom to exploit, and whom to leave alone. It's a balancing act. If you're too selfish, other survival machines will punish you; if you're too selfless, other survival machines will exploit you. Thus, developing positive relationships—social bonds—with other survival machines is an adaptive strategy. If you are there to help your fellow group members when times are tough for them, they are more likely to be there when times are tough for you.

In this way survival machines develop networks and relationships that lead to interactions with one another that in addition to being neutral may be helpful or hurtful. From this we may derive the logic of moral emotions. In a social species such as ours, sometimes the most selfish thing you can do to help yourself is to help others, who will pay you back in kind, not necessarily out of some nebulous notion of "altruism" for its own sake, but because it pays to help others. Our legacy is a moral emotion system that includes the capacity for us to help and hurt other survival machines, depending on what they do. Sometimes it pays to be selfish, but other times it pays to be selfless as long as you're not a milquetoast who lies down and lets others run roughshod over your generosity.

THE EVOLUTIONARY LOGIC OF EMOTIONS

The language I'm using to describe these interactions makes it sound like it is a rational process, a moral calculation conducted by survival machines as they interact with one another. But that is not what is going on. Organisms are driven by their passions more than they are by rational calculations. Natural selection has done the calculating for organisms, who evolved emotions as proxies for those calculations. Let's drill down deeper into the brain to figure out why we evolved emotions in the first place, and then pull back to see how moral emotions work.

Emotions interact with our cognitive thought processes to guide our behaviors toward the goal of survival and reproduction. The neuroscientist Antonio Damasio has shown that at low levels of stimulation, emotions act in an advisory role, carrying additional information to the decision-making process along with input from higher-order cortical regions of the brain.

At medium levels of stimulation, conflicts can arise between high-road reason centers and low-road emotion centers. At high levels of stimulation, low-road emotions can so overrun high-road cognitive processes that people can no longer reason their way to a decision and report feeling "out of control" or "acting against their own self-interest."[76]

The emotion of *fear*, for example, directs an organism to steer clear of danger. The anthropologist Björn Grinde, a rock climber himself, uses the sport as an example of a situation where the positive emotion of the thrill of risk taking can quickly shift to the negative emotion of fear of death when a climber loses his or her grip: "The brain is designed to induce us to take some chances, otherwise we would never have laid down a large prey or ventured into uncharted land; but it is also designed to stop us from causing harm to ourselves, that is, to avoid hazards. The 'adrenaline kick' associated with climbing a mountain or riding a roller coaster may feel good, presumably because it improves the chance of survival if voluntarily encountered dangerous situations induce a positive mood and a high self-esteem. At the moment one loses the grip on the mountain, the unpleasant sensations devoted to harm avoidance kick in."[77] The survival value of fear has obvious adaptive value.

Consider *hunger* as a motivating drive that leads to such emotions as desire, longing, or lust for food. A little bit of hunger may be perceived as mildly pleasant in the anticipation of eating, and we've evolved to understand these small pangs to mean we should seek and find food. If too much time goes by, however, and our body becomes depleted to the point of feeling weak, hunger morphs into an unpleasant feeling. By this example, emotions act as a feedback mechanism to alert the brain when the body is out of balance. This is a *homeostatic* theory of emotions, in which the process operates like an emotional thermostat. When our bodies are low on energy we feel hungry, and that emotion is triggered by a number of internal and external feedback cues—such as shrinking or distension of the stomach, elevated or reduced blood glucose levels, or the sight or smell of food—which act as cues to drive us to raise the calorie thermostat, or bring the body back into homeostasis by eating.

In some cases it is literally like a thermostat, as when our core body temperature deviates above or below the 98.6 degree Fahrenheit set point and certain physiological systems kick in to correct the imbalance, such as sweating to cool the body or shivering to warm it. Departing from the set point of a homeostatic system *feels bad*, and this negative emotion motivates the organism to take action to correct the imbalance. Moving the out-of-balance system back toward homeostasis *feels good*, and behaviors that feel good tend to be repeated, which is the very definition of reinforcement (anything that causes an organism to repeat a behavior).

Thus our need to maintain homeostasis is caused by our emotions that direct us to avoid pain and to pursue pleasure, or to approach a stimulus that is reinforcing and to avoid a stimulus that is punishing. In this sense, in seeking something we call "pleasure," what we are really after is a homeostatic cue, an information signal to tell us what to do next. When those cues are unclear, or in conflict, it can set up a state that psychologists call approach-avoidance behavior. In the case of apparent moral dilemmas—such as what subjects experienced between obedience to authority and the discomfort of harming a fellow human in Stanley Milgram's famous shock experiments—the phenomenon becomes an *approach-avoidance conflict*. (More on this in chapter 9.)

THE EVOLUTIONARY LOGIC OF AGGRESSIVE EMOTIONS

Conflicts among survival machines are inevitable by-products of the evolutionary logic of the essence to survive and flourish and the many different ways there are to fulfill that need in an environment of limited resources. This approach helps us see that there is a certain evolutionary logic to violence and aggression, a taxonomy of which Steven Pinker classified into five types:[78]

1. *Predatory and instrumental*: violence as a means to an end, a way of getting something you want. Theft, for example, can grant the thief more resources necessary for survival and reproduction, and thus there evolved a capacity for cheating, stealing, and free riding (taking without giving in a social system) among some individuals in a group.
2. *Dominance and honor*: violence as a means of gaining status in a hierarchy; power over others; prestige in a group; or glory in sports, gangs, or war. Bullying, for example, can grant individuals higher status in the pecking order of social dominance.[79] A reputation for being aggressive can be a credible deterrent against other aggressors.
3. *Revenge and self-help justice*: violence as a means of punishment, retribution, or moralistic justice. Revenge murders, for example, are an evolved strategy for dealing with cheaters and free riders. *Jeal-*

ousy is another type of moralistic emotion that evolved to direct survival machines to mate guard against potential poachers of their sexual partner (and thus, for men, the bearer of their children who carry their genes), which when expressed violently can lead to spousal murders. Even *infanticide* has an evolutionary logic to it, as evidenced by the statistic that infants are fifty times more likely to be murdered by their stepfather than their biological father, an act far more common among species—including our own—than we care to admit.[80]

4. *Sadism*: violence as a means of gaining pleasure at someone else's suffering. Serial killers and rapists, for example, seem at least partially motived by the pain and suffering they cause, especially when it has no other apparent motive (such as instrumental, dominance, or revenge). It is not clear if sadism is adaptive or, more likely, is a by-product of something else in the brain that evolved for some other reason.

5. *Ideology*: violence as a means of attaining some political, social, or religious end that results in a utilitarian calculus whereby killing some for the sake of many is justified. (I deal with this cause of violence in detail in chapter 9.)

THE EVOLUTIONARY LOGIC OF MORAL EMOTIONS

On the platform of a subway station a woman and two men are standing a few feet away from the open track pit, when all of a sudden one of the men reaches forward and shoves the woman by her shoulders. She staggers backwards, loses her footing, and starts to fall toward the pit. The other man reaches out to catch her but he's too late—into the pit she goes. In an instant he reacts. But instead of reaching down to pull the woman to safety before a train arrives to crush her, the would-be rescuer turns on his heels and cold-cocks the perpetrator. It's a magnificent roundhouse blow right out of a Hollywood movie that snaps the perpetrator's head back with the sweet crash of fist against chin. Satisfied with this act of revenge, the rescuer steps back and pauses for a moment, then appears to remember what needs to be done next, dashes over to the pit, and pulls the woman to safety. He says something to her that appears reassuring, and then takes off after the perpetrator, who had beat a hasty retreat through an open door. The entire incident takes twenty seconds, and you can see it yourself in a viral video that includes a number of heroic rescues.[81] In that brief moment—too short a span for rational calculation—a conflict of pure emotional morality unfolds between revenge and rescue, hurting and helping. In a flash two neural networks in the rescuer's brain launched into action—help a fellow human in trouble or punish the person who caused the trouble. What's a morally motivated primate to do? In this case, the rescuer had time to do both as no train arrived to derail his problematic first choice to avenge the woman's maltreatment. Revenge is sweet and so is rescue. It doesn't always work out so well.

This vignette well illustrates our multifaceted moral nature that evolved to solve several problems at once in our ancestral environment—be nice to those who help us and our kin and kind, punish those who hurt. Evidence that these moral emotions are deeply entrenched in human nature may be found in a series of experiments with babies, succinctly synthesized in the psychologist Paul Bloom's book *Just Babies: The Origins of Good and Evil*.[82] Testing the theory that we have an innate moral sense as proposed by such Enlightenment thinkers as Adam Smith and Thomas Jefferson, Bloom provides experimental evidence that "our natural endowments" include "a moral sense—some capacity to distinguish between kind and cruel actions; empathy and compassion—suffering at the pain of those around us and the wish to make this pain go away; a rudimentary sense of fairness—a tendency to favor equal divisions of resources; a rudimentary sense of justice—a desire to see good actions rewarded and bad actions punished."[83] Consider an experiment conducted in Bloom's lab with a one-year-old baby who watched a puppet show in which one puppet rolls a ball to a second puppet, who passes the ball back to it. The first puppet then rolls the ball to a different puppet, who runs off with the ball. Next, the "nice" and the "naughty" puppets are placed before the baby, along with a treat in front of each; the baby is then given the choice of which puppet to take the treat away from. As Bloom predicted, the infant removed the treat from the naughty puppet—which is what most babies do in this experimental paradigm—but for this little moralist, removing a positive reinforcement (the treat) was not enough. In his inchoate moral mind, punishment was called for, as Bloom recounts: "The boy then leaned over and smacked this puppet on the head."[84]

Numerous permutations on this research paradigm (such as a puppet trying to roll a ball up a ramp, for which another puppet either helps or hinders it) show time and again that the moral sense of right (preferring helping puppets) and wrong (abjuring hurting puppets) emerges as early as three to ten months of age—far too early to attribute to learning and culture.[85] Young children who are exposed in a laboratory to an adult experiencing pain—the experimenter getting her finger caught in a clipboard, say, or the child's mother banging her knee—typically respond by soothing the injured party. Toddlers who see adults struggling to open a door because their arms are full, or to pick up an out-of-reach object, will spontaneously help without any prompting from the adults in question.[86] Another experiment involved three-year-old children who were asked, "Can you hand me the cup so that I can pour the water?" but the cup in question was broken. Remarkably, the youngsters spontaneously went in search of an intact cup to help the experimenter complete the task.[87]

However, children are not always so beneficent, particularly with other children, with whom they clearly show awareness of an unequal distribution of rewards after a shared task (in this case a candy treat), but are not always so eager to unselfishly right the wrong by redistributing the wealth.[88] But as children get older—from three-to-four-year-olds to seven-to-eight-year-olds—they are not only more aware of an unequal and unfair distribution of candy, they are also more likely to give away the extra unearned treat (50 percent of the three-to-four-year-olds did, compared to 80 percent of the seven-to-eight-year-olds), showing that while the moral sense is inborn and instinctive, it is a capacity that can be tuned by learning and culture and brought to bear (or not) in different environments that either encourage or discourage helping or hurting behavior.[89]

As well, research with infants shows how early in life xenophobia takes root. Babies become wary of strangers, or anyone who doesn't look like members of their family on whom they've imprinted, at a very early stage— days, in fact. In one experiment, three-day-old newborns were donned with headphones and special pacifiers that allowed them to pick audio tracks based on how rapidly they sucked on them. These infants not only figured out the connection between sucking and music selections, but also were able to transfer that learned skill to selecting a passage read to them from a Dr. Seuss book by their mother rather than a stranger. For newborns given the option to select among languages being spoken, results showed that "Russian babies prefer Russian, French babies prefer French, and American babies prefer English, and so on," and even more remarkably, says Bloom, "This effect shows up mere minutes after birth, suggesting that babies were becoming familiar with those muffled sounds that they heard in the womb."

This research confirms a classic experiment from the 1960s conducted by a third-grade schoolteacher named Jane Elliott on her class in the small, all-white rural town of Riceville, Iowa. Elliott began her experiment by dividing her class into two groups by eye color—blue and brown—then presented the kids with examples of blue-eyed good people and brown-eyed bad people. In addition, the blue-eyed kids in the class were told that they were superior and were given special privileges, while the brown-eyed kids were called inferior and treated as second-class citizens. Almost immediately a social division followed the physical classification. The blue-eyed kids quit playing with the brown-eyed kids, with some of the students even suggesting to Elliott that school officials should be alerted to the potential criminal behavior of brown-eyed youngsters. When a fight broke out between a blue-eyed and a brown-eyed boy, the latter justified his aggressive actions thusly: "He called me brown-eyes, like being a black person, like a Negro." By the second day of the experiment, the brown-eyed children already started to show signs of poorer performance in class and described themselves as feeling "sad," "bad," "stupid," and "mean."

As a control, the next day Mrs. Elliott reversed the conditions, explaining that she had been mistaken and that, in fact, it was the *brown-eyed* children who were superior, and the blue-eyed kids who were inferior. Just as quickly the self- and other perceptions were reversed, with the "happy," "good," "sweet," and "nice" labels previously used by the blue-eyed children to describe themselves now adopted by the brown-eyed children. "What had been marvelously cooperative, thoughtful children became nasty, vicious, discriminating little third-graders," Mrs. Elliott explained. "It was ghastly!"[90]

Bloom's conclusion about morality from this sizable body of research supports what I saw in the video vignette from the subway: "it entails certain feelings and motivations, such as a desire to help others in need, compassion for those in pain, anger towards the cruel, and guilt and pride about our own shameful and kind actions."[91] Of course, society's laws and customs can turn the moral dials up or down, but nature endowed us with the dials in the first place. It's as Voltaire said: "Man is born without principles, but with the faculty of receiving them. His natural disposition will incline him either to cruelty or kindness; his understanding will in time inform him that the square of twelve is a hundred and forty-four, and that he ought not to do to others what he would not that others should do to him."[92]

THE LOGIC OF MORAL DILEMMAS

The logic of our moral emotions has been worked out by game theorists in the aforementioned Prisoner's Dilemma paradigm. Here's the scenario: You and your partner are arrested for a crime, and you are held incommunicado in separate prison cells. Neither of you wants to confess or rat on the other, but the DA gives each of you the following options:

1. If you confess but the other prisoner does not, you go free and he gets three years in jail.
2. If the other prisoner confesses and you do not, you get three years and he goes free.
3. If you both confess, you each get two years.
4. If you both remain silent, you each get a year.

Figure 1-5, called a game matrix, summarizes the four outcomes.

With those outcomes, the logical choice is to defect and betray your partner. Why? Consider the choices from the first prisoner's point of view.

Prisoner's Dilemma

MY STRATEGY		MY OPPONENT'S STRATEGY: COOPERATE (remain silent)	MY OPPONENT'S STRATEGY: DEFECT (confess)
MY STRATEGY	COOPERATE (remain silent)	-1 (High payoff)	-3 (Sucker payoff)
MY STRATEGY	DEFECT (confess)	No jail time (Temptation payoff)	-2 (Low payoff)

Figure 1-5. Prisoner's Dilemma

The only thing the first prisoner cannot control about the outcome is the second prisoner's choice. Suppose the second prisoner remains silent. Then the first prisoner earns the "temptation" payoff (zero years in jail) by confessing but gets a year in jail (the "high" payoff) by remaining silent. The better outcome in this case for the first prisoner is to confess. But suppose, instead, that the second prisoner confesses. Then, once again, the first prisoner is better off confessing (the "low" payoff, or two years in jail) than remaining silent (the "sucker" payoff, or three years in jail). Because the circumstances from the second prisoner's point of view are entirely symmetrical to the ones described for the first, each prisoner is better off confessing no matter what the other prisoner decides to do.

Those preferences are not only theoretical. When test subjects play the game just once or for a fixed number of rounds without being allowed to communicate, defection by confessing is the common strategy. But when testers play the game for an unknown number of rounds, the most common strategy is tit for tat: each begins cooperating with the prior agreement by remaining silent, then mimics whatever the other player does. Even more mutual cooperation can emerge in a many-person prisoner's dilemma, provided the players are allowed to play enough repeated rounds to establish mutual trust. But the research shows that once defection by confessing builds momentum, it cascades throughout the game.

In an analysis for *Scientific American* I worked out the game matrix dynamics of why professional athletes use performance-enhancing drugs, and in particular why cyclists dope.[93] In cycling, as in any sport, the contestants compete according to a set of rules. The rules of cycling clearly prohibit the use of performance-enhancing drugs. But because the drugs are so effective and many of them are so difficult (if not impossible) to detect, and because the payoffs for success are so great, the incentive to use banned substances is powerful. Once a few elite riders defect from the rules by doping to gain an advantage, their rule-abiding competitors feel the need to defect as well—even if they don't want to—leading to a cascade of defection through the ranks. Because of the penalties for breaking the rules, however, a code of silence prevents any open communication about how to reverse the trend and return to abiding by the rules. *Figures 1-6* and *1-7* show the game matrices that favor cheating and that favor playing by the rules.

In game theory, if no player has anything to gain by unilaterally changing strategies, the game is said to be in a Nash equilibrium. The concept was developed by the mathematician John Forbes Nash Jr., who was por-

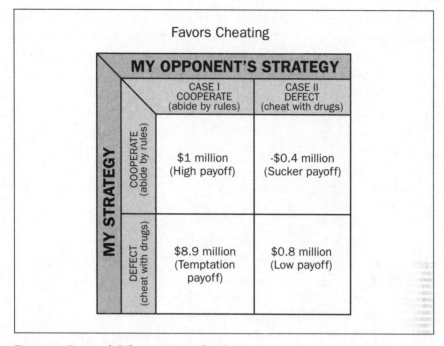

Figure 1-6. Prisoner's Dilemma Matrix for Cheating

The Cheating Matrix assumptions: Value of winning the Tour de France: $10 million. Likelihood that a doping rider will win the Tour de France against nondoping competitors: 100 percent. Value of cycling professionally for a year when the playing field is level: $1 million. Cost of getting caught cheating (penalties and lost income): $1 million. Likelihood of getting caught cheating: 10 percent. Cost of getting cut from a team (forgone earnings and loss of status): $1 million. Likelihood that a nondoping rider will get cut from a team for being noncompetitive: 50 percent. Under these conditions, in case 1 in which my opponent abides by the rules (he "cooperates"), if I also cooperate by not doping, the playing field is level and there is an expected payoff of $1 million. But if I cheat by doping and don't get caught, then I stand to make $8.9 million ($10 million × 90 percent − $0.1 million), which is more than $1 million, so I should cheat. In case 2, in which my opponent cheats by doping, if I play by the rules I'm a sucker and lose $0.4 million, but if I also cheat by doping then I too face the low payoff amount of $0.8 million, so my incentive is once again to cheat.

trayed in the film *A Beautiful Mind*. To end cheating in sports the doping game must be restructured so that competing clean is in a Nash equilibrium. That is, the governing bodies of each sport must change the payoff values of the expected outcomes identified in the abiding-by-the-rules matrix. First, when other players are playing by the rules, the payoff for doing likewise must be greater than the payoff for cheating. Second, and perhaps more importantly, even when other players are cheating, the payoff for playing fair must be greater than the payoff for cheating. Players

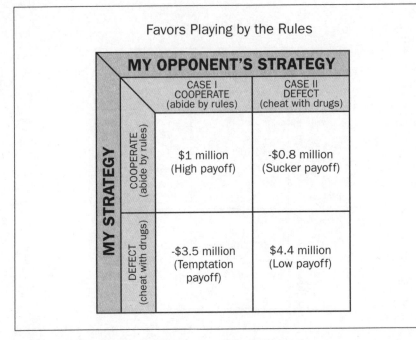

Figure 1-7. Prisoner's Dilemma Matrix for Abiding by the Rules

The Playing by the Rules Matrix assumptions: New, higher cost of getting caught cheating (penalties and lost income): $5 million. New, higher likelihood of getting caught cheating: 90 percent. Consequent new, lower likelihood that a nondoping rider will get cut from a team for being noncompetitive: 10 percent. Under these conditions, in case 1, in which my opponent abides by the rules (he "cooperates"), if I also cooperate by not doping, the playing field is level and there is an expected payoff of $1 million. But this time, if I cheat by doping there's a 90 percent chance I'll get popped in a drug test, so my expected payoff for cheating is now $1 million minus the expected penalty for cheating of $5 million × 90 percent = −$4.5 million, so I stand to lose $3.5 million, so the incentive is to play by the rules. Even in case 2, in which my opponent dopes and I'm a sucker for cooperating, I still come out on top with a net $0.8 million, compared to my also doping and getting caught and penalized, resulting in a net loss of $4.4 million. Either way, in this matrix, with these conditions, we should all play by the rules.

must not feel like suckers for following the rules. In the game of Prisoner's Dilemma, lowering the temptation to confess and raising the payoff for keeping silent if the other prisoner confesses increases cooperation. Giving players the chance to communicate before they play the game is the most effective way to increase their cooperation. In sports, that means breaking the code of silence. That will show each player that the payoff for playing fair is greater than the payoff for cheating, no matter what the other players do.

Whether this has happened in cycling is unclear, but I am encouraged by the startling events of 2012 and 2013 when Tyler Hamilton broke the code of silence in his book *The Secret Race*, and exposed the most sophisticated doping program in the history of sports, orchestrated by his teammate Lance Armstrong, the seven-time Tour de France winner, now stripped of his titles after a thorough investigation by the US Anti-Doping Association.[94] Hamilton revealed how such an elaborate system was maintained through a combination of a code of silence that led everyone to believe that everyone else believed that doping was the norm, which was then reinforced by the threat of punishment for speaking out or not complying. Since then it has been revealed that most athletes who have been caught doping say they didn't want to dope but that they did so out of the belief that *everyone else* was doping, and out of fear of retaliation if they didn't dope, and worse consequences still if they blew the whistle on the system.

As for real prisoners who find themselves in a Prisoner's Dilemma, the abiding-by-the-rules matrix also directs criminologists and policymakers to consider not just the size of the penalty (as most get-tough-on-crime politicians are wont to do) but the probability of getting caught as a factor as well. This harkens back to the eighteenth-century philosopher and reformer Cesare Beccaria, whose 1764 work *On Crimes and Punishments*— a high-water mark of the Italian Enlightenment—launched the movement to apply rational principles to criminal reform, such as adjusting the punishments to fit the crimes (proportionality) instead of, as was the custom of the day, the death penalty for such offenses as poaching, counterfeiting, theft, sodomy, bestiality, adultery, horse theft, being in the company of Gypsies, and two hundred other crimes and misdemeanors. Beccaria opposed the death penalty on two principles: (1) states do not possess the right over life and death, and (2) it doesn't work to deter crime because when would-be criminals are faced with a draconian but improbable penalty, they consider it a risk worth taking—another cost of doing business. To the principles of *proportional* and *probable* punishment, Beccaria added two more Ps: criminal proceedings should be *prompt* and *public*, the latter acting as a signal to other would-be criminals. Beccaria was an early game theorist applying rational principles born of Enlightenment values and observational data from real-world examples—with an end toward tilting the motivational matrices to incentivize the citizenry to commit fewer crimes.[95]

Another Enlightenment thinker we will meet later is Thomas Hobbes, who also proffered a game theoretic model for how people and nations

interact. All political theorists—both liberal and conservative—begin
with a Hobbesian premise that the state is a necessary evil to protect self-
motivated individuals from other self-motivated individuals (think of it as
two survival machines in Dawkins's thought experiment). This is sometimes known as the "Hobbesian trap." As Hobbes argued in his classic
work in political theory *Leviathan*, we are all motivated to seek pleasure
and avoid pain, so inevitably there will be conflict when interests between
people overlap. This leads to three forms of "quarrel": competition, diffi-
dence (fear), and glory (honor, status):

> The first maketh men invade for gain; the second, for safety; and the
> third, for reputation. The first use violence, to make themselves masters
> of other men's persons, wives, children, and cattle; the second, to defend
> them; the third, for trifles, as a word, a smile, a different opinion, and
> any other sign of undervalue, either direct in their persons or by reflec-
> tion in their kindred, their friends, their nation, their profession, or
> their name.[96]

As we saw in the Prisoner's Dilemma, for there to be cooperation
between competing agents in a game (or nations in the real world), there
need to be rules, and the rules must be enforced. With our complex moral
nature, people need to be encouraged to do the right thing and discouraged from doing the wrong thing—the proverbial carrots and sticks. The
psychology behind this interaction between inner psychological states
and external social conditions was explored by the economists Ernst Fehr
and Simon Gachter in a study on *moralistic punishment* in which subjects
were given the opportunity to punish others who refused to cooperate in
a group activity that calls for altruistic giving. They used a "common
goods" cooperation game in which the subjects were given money that
they then had the option to put into a shared commons that would then be
multiplied 1.5 times and divided equally among all the players. Let's say
the amount is $10 and there are four players. If everyone puts in the full
ten bucks, then $40 \times 1.5 = 60, which when split four ways equals $15
each. When this is done anonymously there is a temptation to game the
system by putting in less money. Let's say the other three players each put
in their full $10, but I put in only $5. The commons now has $35 \times 1.5 = 52.5,
which divided four ways equals $13.12 each. But I still have my original
$5, so I now have $18.12. Sweet! But it doesn't take long for the other players to catch on to the fact that someone is gaming the system, and under

these conditions cooperation quickly breaks down and the amount of money put into the commons collapses.

To remedy the free rider problem, in the seventh round Fehr and Gachter introduced a new condition in which contributions to the commons were no longer anonymous and the players were allowed to punish free riders by taking money from them, and this they did with impunity, which immediately triggered a rise in the levels of cooperation and giving by the former free riders.[97] The results, shown in *figure 1-8*, serve as a visual reminder of why people need rules, transparency, and the threat of punishment to be good. This role, the theory goes, is fulfilled by the Leviathan state.

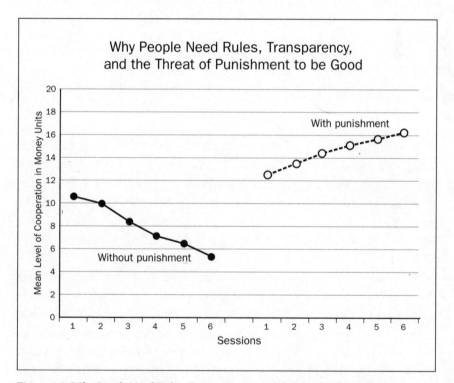

Figure 1-8. Why People Need Rules, Transparency, and the Threat of Punishment

The results of the Fehr and Gachter study on moralistic punishment. In a public goods game players are given a sum of money and have the choice of how much they would like to contribute into a common pool that will then be increased 1.5 times and returned to all the players evenly. A lack of transparency of how much everyone contributes leads to the temptation to reduce the amount given and thereby "free ride" on the others. Since all players face the same temptation, cooperation declines. When transparency is included, plus the opportunity to punish free riders who scrimp on their contributions, cooperation increases. The latter condition is an example of moralistic punishment. It works.

· · ·

In this chapter I have outlined the evolutionary origins of morality and
the logic of moral interactions, good and bad, and in the next chapter I
will show how these principles operate to lessen even the most dangerous
threats our species faces—violence, war, and terrorism—and will demonstrate that even here, there has been considerable moral progress.

2

The Morality of War, Terror, and Deterrence

To warn of an evil is justified only if, along with the warning, there is a way of escape.

—Cicero, *On Divination*, Book II, 44 BCE

In an episode of the original *Star Trek* television series, titled "Arena," an alien species called the Gorn attacks and destroys an Earth outpost at Cestus III, leading Captain Kirk and the *Enterprise* to give chase to avenge the unprovoked attack. Spock is not so sure about the aliens' motives, and wonders aloud about having "regard for another sentient being," but is interrupted by the more martial Kirk, who reminds him, "out here we're the only policemen around." Their moral quandary is interrupted when both ships are stopped by an advanced civilization called the Metrons, who explain, "We have analyzed you and have learned that your violent tendencies are inherent. So be it. We will control them. We will resolve your conflict in the way most suited to your limited mentalities." Kirk and the captain of the Gorn ship—a big-brained bipedal reptile—are transported to a neutral planet where they are instructed to fight to the death, at which point the loser's ship and crew will be destroyed. The Gorn is stronger than Kirk and is easily able to rebuff his assaults that escalate from tree branch strikes to a massive boulder impact. He tells Captain Kirk that if Kirk surrenders he will be "merciful and quick." "Like you were at Cestus III?" Kirk rejoins. "You were intruding! You established an outpost in our space," the Gorn counters. "You butchered helpless human beings," Kirk protests. "We destroyed invaders, as I shall destroy you!" the Gorn retorts. Back on the ship, where the crew is watching the battle unfold on the viewing screen, Dr. McCoy wonders aloud, "Can that be true? Was Cestus III an intrusion on their space?" "It may well be possible, Doctor," Spock reflects. "We know very little about that section of the galaxy." "Then we could be in the wrong," McCoy admits. "The Gorn might have been simply trying to protect themselves."

At the climax of the episode Kirk recalls the formula for making gunpowder after seeing the various elements readily available on the planet's surface—sulfur, charcoal, and potassium nitrate, with diamonds as the deadly projectile—elements that the Metrons had provided to see if judicious reason could triumph over brute strength. Kirk puts them all together into a lethal weapon that he fires at the Gorn as the latter closes in for the kill. Now incapacitated, the Gorn drops his stone dagger, which Kirk grabs and places at the throat of his opponent to deliver the coup de grâce. Then, at this moment of moral choice, Kirk opts for mercy. His reason drives him to take the moral perspective of his opponent. "No, I won't kill you. Maybe you thought you were protecting yourself when you attacked the outpost." Kirk tosses the dagger aside, at which point a Metron appears. "You surprise me, Captain. By sparing your helpless enemy who surely would have destroyed you, you demonstrated the advanced trait of mercy, something we hardly expected. We feel that there may be hope for your kind. Therefore, you will not be destroyed. Perhaps in several thousand years, your people and mine shall meet to reach an agreement. You are still half savage, but there is hope."[1]

The creator of *Star Trek*, Gene Roddenberry, invented a genre unto itself with the creation of the magnificent starship *Enterprise*, whose twenty-third-century mission was to expand humanity's horizons, both physically and morally, via extraterrestrial interactions with the starship's interracial, transnational, trans-species mixed crew. Each episode was both intrepid space adventure and thoughtful morality play, and many episodes explored the controversial issues of the age—war and conflict, imperialism and authoritarianism, duty and loyalty, racism and sexism, and how humanity might handle them centuries hence. Roddenberry made it clear that one of his goals with the series was to smuggle onto TV allegorical moral commentaries on current events. He said that by creating "a new world with new rules, I could make statements about sex, religion, Vietnam, politics, and intercontinental missiles. Indeed, we did make them on *Star Trek*: We were sending messages and fortunately they all got by the network."[2] This is one method, among many, of bringing about social change, and it is revealing to note that Roddenberry was personally well acquainted with war. In

1941, as a young man of only twenty, he'd enlisted in the US Army Air Corps and flew eighty-nine missions in the South Pacific, for which he was decorated with the Distinguished Flying Cross. So he knew of what he wrote: "The strength of a civilization is not measured by its ability to fight wars, but rather by its ability to prevent them."[3]

PIRATE MORALITY, OR CONFLICT AND COSTLY SIGNALING THEORY

Determining how civilizations have learned to prevent wars involves understanding the psychology of violence and deterrence—grounded in the logic of conflict and our moral emotions—that we examined in the previous chapter. Consider the following question: instead of using stealth and camouflage, why would a pirate choose to travel in a giant spotlight by flying the Jolly Roger flag with its ominous skull and crossbones, thereby signaling to potential prey that they're being pursued by a predator? This is an example of a phenomenon called *Costly Signaling Theory* (CST), which posits that organisms (including people) will sometimes do things that are costly to themselves to send a signal to others.[4] There are both positive and negative examples.

On the positive side, people sometimes act in certain ways not just to help those who are genetically related to them (explained by kin selection), and not just to help those who will return the favor (explained by reciprocal altruism), but to send a signal that says, in essence, "my altruistic and charitable acts demonstrate that I am so successful that I can afford to make such sacrifices for others." That is, some altruistic acts are a form of information that carries a signal to others of trust and status—*trust* that I can be counted on to help others when they need it so that I can expect others to do the same for me; and *status* that I have the health, intelligence, and resources to afford to be so kind and generous. This type of CST explains why some people make large donations to charity or drive expensive cars or wear expensive jewelry—as signals to others. And since the signal must be genuine, the signalers tend to believe in what they are doing, so to that extent the giving is done with honest motive.

The negative side of CST may be seen in high-risk-taking behavior by young males—behavior that occasionally ends very badly indeed, with the young male accidentally taking himself out of the gene pool entirely (for which Darwin Awards are given out annually[5]). Risky behavior may be a male's way of signaling to a female that his genes are *so* good and he is such a superior specimen that he can, for example, drink twelve beers and still drive home at more than one hundred miles per hour in total safety. He may be signaling to females that he is so genetically exceptional he can afford to risk life and limb, and would therefore be a good mate and an excellent genetic and resource provider for her and her future offspring with him. Dangerous and risky acts may also signal to other males that the danger seeker and risk taker is powerful and not to be messed with, as in Jim Croce's warning, "You don't tug on Superman's cape, you don't spit into the wind, you don't pull the mask off the old Lone Ranger, and you don't mess around with Jim."

With this background in mind we can see how Costly Signaling Theory explains why pirates flew the Jolly Roger flag on their ships. It was a signal that you—the innocent merchant ship—were about to be boarded by a lawless beastly bunch of wild and unruly maniacs hell-bent on murder and mayhem. This was especially clever because, according to the economist Peter Leeson in his pirate myth–busting book *The Invisible Hook*, pirates were not, in fact, the criminally insane, traitorous terrorists of popular lore, in which anarchy was the rule and the rule of law was nonexistent. This piratical mythology can't be true because ships packed full of riotous sociopaths, ruled by chaos and treachery, couldn't possibly be successful at anything for any length of time. The truth was much less exciting and mysterious; pirate communities were "orderly and honest," says Leeson, and had to be to meet buccaneers' economic goal of turning a profit. "To cooperate for mutual gain—indeed, to advance their criminal organization at all—pirates needed to prevent their outlaw society from degenerating into bedlam."[6] So there is, after all, honor among thieves. As Adam Smith noted in *The Wealth of Nations*, "Society cannot subsist among those who are at all times ready to hurt and injure one another. . . . If there is any society among robbers and murderers, they must at least . . . abstain from robbing and murdering one another."[7]

Pirate societies provide evidence for Smith's theory that economies are the result of bottom-up spontaneous self-organized order that naturally arises from social interactions. Leeson shows how pirate communities democratically elected their captains and quartermasters, and constructed constitutions that outlined rules about drinking, smoking, gambling, sex (no boys or women allowed on board), the use of fire and candles (a shipboard fire could prove disastrous for crew and cargo), fighting and disorderly conduct (the result of high-testosterone, risk-taking men confined in tight quarters for long stretches of time), desertion, and especially shirking one's duties during battle. Like any other society, pirates had to deal with the "free rider" problem because the equitable division of loot among inequitable efforts would inevitably lead to resentment, retaliation, and economic chaos.

Enforcement was key. Just as criminal courts required witnesses to swear on the Bible, pirate crews had to consent to the captain's code before sailing. In the words of one observer, "All swore to 'em, upon a Hatchet for want of a Bible. When ever any enter on board of these Ships voluntarily, they are obliged to sign all their Articles of Agreement to prevent Disputes and Ranglings afterwards." Leeson even tracked down the sharing of contractual arrangements between captains, made possible by the fact that "more than 70 percent of Anglo-American pirates active between 1716 and 1726, for example, can be connected back to one of three pirate captains." Thus, the pirate code "emerged from piratical interactions and information sharing, not from a pirate king who centrally designed and imposed a common code on all current and future sea bandits."[8]

Whence, then, did the myth of piratical lawlessness and anarchy arise? It arose from the pirates themselves, naturally, in whose best interests it was to perpetuate the myth to minimize losses and maximize profits. They flew the Jolly Roger to signal their reputation for mayhem but, in fact, the pirates didn't actually want a fight, because fighting is costly and dangerous and might result in economic loss. Pirates just want booty, and they prefer a low-risk surrender to a high-risk battle. From the merchant's perspective, the nonviolent surrender of their booty was also preferable to fighting back, because violence is costly to them too. Of course, to maintain a reputation that you are a badass, you actually have to occasionally be a badass, so pirates intermittently engaged in violence, reports of which they happily provided to newspaper editors, who duly published them in gory and exaggerated detail. As the eighteenth-century English pirate captain Sam Bellamy explained, "I scorn to do any one a Mischief, when it is not for my Advantage." Leeson concludes, "By signaling pirates' identity to potential targets, the Jolly Roger prevented bloody battle that would needlessly injure or kill not only pirates, but also innocent merchant seamen."[9]

The Jolly Roger effect also helps to explain why Somali pirates today typically receive ransom payoffs instead of violent resistance from shipping crews and their owners. It is in everyone's economic interest to negotiate the transactions as quickly and peacefully as possible. In Tom Hanks's film *Captain Phillips*—the true-life story of the 2009 Somali

hijacking of a tanker that resulted in the death of most of the Somali
pirates when the US Navy came to the rescue—viewers were baffled why
the ship's owners didn't just issue firearms to the captain and his crew to
fight the pirates off. The answer is obvious once you run through the cost-
benefit calculations. It's cheaper to just pay off the pirates than risk the
lives of a crew that's not combat-trained. Those who are, such as the US
Navy, may do so, but at an unrealistic financial cost given that they can't
patrol every shipping lane in such a vast expanse. (In fact, the number of
successful pirate hijackings dropped to zero in 2013, but at a cost considerably greater than the ransoms paid from 2005 to 2012, which totaled
$376 million. In 2013, shipping companies shelled out an additional $423
million in armed security guards over the $531 million from the year
before [totaling $954 million], plus incurred an additional $1.53 billion in
fuel costs to cruise at the higher rate of 18 knots.[10]) By killing every one of
the pirates who commandeered Captain Phillips's ship, the US Navy was
sending a powerful signal to Somali pirates to stop boarding US cargo
ships—or else. However, unless deterrence signals are consistent and long-
term, pirates will run through the same set of calculations and see that the
"or else" was a bluff and treat the risk as a cost-of-doing-business expense.

The long-term solution lies elsewhere—in Somalia itself. Markets
operating in a lawless society are more like black markets than free markets, and since the Somali government has lost control of its people, Somali
pirates are essentially free to take the law into their own hands. Until
Somalia establishes a rule of law and a lawful free market in which its citizens can find gainful employment, lawless black market piracy will remain
profitable. The pirates themselves may have their own set of rules by which
they organize themselves into mini Leviathans of order, but on the high
seas anarchy reigns.

THE EVOLUTIONARY LOGIC OF DETERRENCE

This brings us back to the prisoner's dilemma and the Hobbesian trap
discussed in the previous chapter. Pinker calls this the "other guy problem." The other guy may be nice, but you know that he also wants to "win"
(continuing the sports analogy), so he might be tempted to defect (cheat),
especially if he thinks that you too want to win and might be tempted to
cheat. And he knows that you know that he is running through the same
game matrix as you are, and you know that he knows that you know . . .

In international relations the "other guy" is another nation or state, and if they have nuclear weapons, and so do you, it can lead to an arms race resulting in something like a Nash equilibrium, such as the one that kept the United States and the Soviet Union in a Cold War nuclear freeze called a "balance of terror" or Mutual Assured Destruction (MAD) for more than half a century. Let's look at how this worked, and what more we can do to further reduce the risks of nuclear war based on what we know about human nature and the logic of deterrence. This exercise will serve as another example of both moral progress and how we can apply science and reason to solving a serious threat to our security.

When I was an undergraduate at Pepperdine University in 1974, the father of the hydrogen bomb—Edward Teller—spoke at our campus in conjunction with the awarding of an honorary doctorate. His message was that deterrence works, even though at the time I remember thinking— like so many politicos were saying—"yeah, but a single slipup is all it takes." Popular films such as *Fail-Safe* and *Dr. Strangelove* reinforced the point. But the slipup never came. MAD has worked because neither side has anything to gain by initiating a first strike against the other nation— the retaliatory capability of both is such that a first strike would most likely lead to the utter annihilation of both countries (along with much of the rest of the world). "It's not mad!" proclaimed Secretary of Defense Robert S. McNamara. "Mutual Assured Destruction is the foundation of deterrence. Nuclear weapons have no military utility whatsoever, except-ing only to deter one's opponent from their use. Which means you should never, never, never initiate their use against a nuclear-equipped opponent. If you do, it's suicide."[11]

The logic of deterrence was first articulated in 1946 by the American military strategist Bernard Brodie in his appropriately titled book *The Absolute Weapon*, in which he noted the break in history that atomic weapons brought with their development: "Thus far the chief purpose of our military establishment has been to win wars. From now on, its chief purpose must be to avert them. It can have almost no other purpose."[12] As Dr. Strangelove explained in Stanley Kubrick's classic Cold War film (in the famous war room scene where fighting is not allowed): "Deterrence is the art of producing in the mind of the enemy the *fear* to attack." Said enemy, of course, must know that you have at the ready such destructive devices, and that is why "The whole point of a doomsday machine is lost if you keep it a secret!"[13]

Dr. Strangelove was a black comedy that parodied MAD by showing what can happen when things go terribly wrong, in this case when General

Jack D. Ripper becomes unhinged at the thought of "Communist infiltra-
tion, Communist indoctrination, Communist subversion, and the international Communist conspiracy to sap and impurify all of our precious
bodily fluids"; thus he orders a nuclear first strike against the Soviet Union.
Given this unfortunate incident and knowing that the Russians know
about it and will therefore retaliate, General "Buck" Turgidson pleads with
the president to go all out and launch a full first strike. "Mr. President, I'm
not saying we wouldn't get our hair mussed, but I do say no more than ten
to twenty million killed, tops, uh, depending on the breaks."[14]

He wasn't far off from real projected casualties (Kubrick was a student
of Cold War strategy), as computed by Robert McNamara: "What kind of
amount of destruction must we be able to inflict upon the attacker in the
retaliation to ensure that he would indeed be deterred from initiating such
an attack? In the case of the Soviet Union, I would judge that a capability
on our part to destroy, say, one-fifth to one-fourth of their population and
one-half of her industrial capacity would serve as an effective deterrent."[15]
When he spoke these words in 1968, the population of the Soviet Union
was about 240 million, which translates to 48 million to 60 million dead. If
that doesn't make you shudder, Mao Zedong once said that he was willing
to sacrifice 50 percent of the Chinese population, which at the time was
about 600 million. "We have so many people. We can afford to lose a few.
What difference does it make?"[16]

The difference was fully appreciated by Harold Agnew, who was something of a real-life Dr. Strangelove. Director of the Los Alamos National
Laboratory for a decade, prior to that he worked on the Manhattan Project at Los Alamos, building the first atomic bombs—"Fat Man" and "Little
Boy"—and he flew in a B-29 plane parallel to the *Enola Gay* bomber to
observe and measure the yield of the explosion over Hiroshima; he even
sneaked onto the plane his own 16-millimeter movie camera and captured the only footage of the explosion that killed eighty thousand people.
Deterrence is what Agnew had in mind when he said that he would require
every world leader to witness an atomic blast every five years while standing in his underwear "so he feels the heat and understands just what he's
screwing around with because we're fast approaching an era where there
aren't any of us left that have ever seen a megaton bomb go off. And once
you've seen one, it's rather sobering."[17]

A 1979 report from the Office of Technology Assessment for the US
Congress, titled *The Effects of Nuclear War*, estimated that 155 million to
165 million Americans would die in an all-out Soviet first strike (unless

people used existing shelters near their homes, reducing fatalities to 110 to 120 million). The population of the United States at the time was 225 million, so the estimated percent that would be killed ranged from 49 percent to 73 percent. Staggering. The report then lays out a scenario for what would happen to one city the size of Detroit if it were hit by a 1-megaton (Mt) nuclear bomb. By comparison, Little Boy—the atomic bomb dropped on Hiroshima—had a yield of 16 kilotons. A 1-Mt bomb is a thousand kilotons, or the equivalent of 62.5 Little Boy bombs.

> A 1-Mt [megaton] explosion on the surface leaves a crater about 1,000 feet in diameter and 200 feet deep, surrounded by a rim of highly radioactive soil about twice this diameter thrown out of the crater. Out to a distance of 0.6 miles from the center there will be nothing recognizable remaining. . . . Of the 70,000 people in this area during nonworking hours, there will be virtually no survivors. . . . Individual residences in this region will be totally destroyed, with only foundations and basements remaining. . . . Whether fallout comes from the stem or the cap of the mushroom is a major concern in the general vicinity of the detonation because of the time element and its effect on general emergency operations. . . . The near half-million injured present a medical task of incredible magnitude. Hospitals and beds within 4 miles of the blast would be totally destroyed. Another 15 percent in the 4- to 8-mile distance range will be severely damaged, leaving 5,000 beds remaining outside the region of significant damage. Since this is only 1 percent of the number injured, these beds are incapable of providing significant medical assistance. . . . Burn victims will number in the tens of thousands; yet in 1977 there were only 85 specialized burn centers, with probably 1,000 to 2,000 beds, in the entire United States.

The report goes on like this for pages. Multiply those effects by 250 (the number of American cities believed to be targeted by the Soviet Union) and you get the report's stark conclusion: "The effects on U.S. society would be catastrophic."[18] The catastrophe would be no less so for the Soviet Union and its allies. A 1957 report by the Strategic Air Command (SAC) estimated that around 360 million casualties (killed and wounded) would be inflicted on both sides in the first week of a nuclear exchange with the Soviet bloc.[19] Such figures are so unimaginable that it is hard to wrap our minds around them.

Figure 2-1 shows a Civil Defense Air Raid card from the 1950s, instructing citizens to "drop and cover" in the event of a nuclear attack.[20]

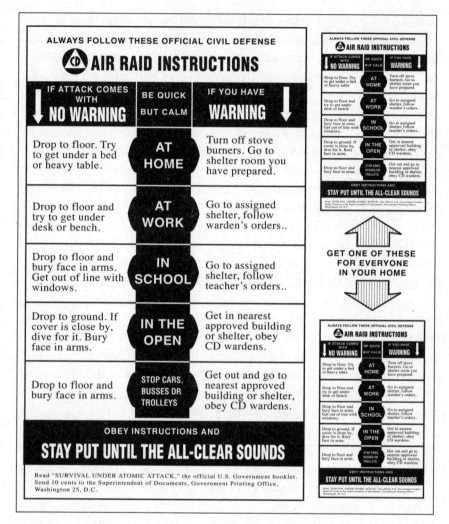

Figure 2-1. Civil Defense Air Raid card circa 1950s

As a child growing up in the early 1960s I recall our periodic Friday morning drills at Montrose Elementary School, trusting my teacher that our flimsy wooden desks would protect us from a thermonuclear blast over Los Angeles.

Deterrence has worked so far—no nuclear weapon has been detonated in a conflict of any kind since August 1945—but it would be foolish to think of deterrence as a permanent solution.[21] As long ago as 1795, in an essay titled *Perpetual Peace*, Immanuel Kant worked out what such deterrence ultimately leads to: "A war, therefore, which might cause the destruction of both parties at once . . . would permit the conclusion of a perpetual peace only upon the vast burial-ground of the human species."[22] (Kant's book title came from an innkeeper's sign featuring a cemetery—not the type of perpetual peace most of us strive for.) Deterrence acts as only a temporary solution to the Hobbesian temptation to strike first, allowing both Leviathans to go about their business in relative peace, settling for small proxy wars in swampy Third World countries.

In addition to the immediate deaths due to the explosion, heat, and radiation, there are possible long-term effects explored by the astronomer Carl Sagan and the atmospheric scientist Richard Turco in their book titled *A Path Where No Man Thought*[23] (based on a technical paper in *Science*[24])—in which the smoke, soot, and debris from the fires caused by an all-out thermonuclear war would render the planet nearly uninhabitable by blocking the sun's radiation and triggering another ice age. They called this scenario "nuclear winter," which has since been dismissed by most scientists as highly unlikely, or that at most it would lead to a "nuclear autumn" instead of winter.[25] As one critic noted, millions of people would die from hunger due to the disruption of the international food supply system rather than from climate change.[26] Well, that's a relief—only millions instead of billions. Regardless of the details of that particular debate, in terms of tracking moral progress toward a nuclear-free world, Sagan and Turco outlined a realistic proposal to reduce the world's nuclear stockpile to a level of Minimum Sufficiency Deterrence (MSD)—large enough to deter a nuclear first strike, but small enough so that if a mistake or a madman detonated a weapon it would not result in an all-out nuclear winter (or autumn).

It would appear that we are now well on our way to MSD, as evidenced in *figure 2-2*, showing a dramatic decline in nuclear stockpiles from a peak in 1986 of approximately 70,000, to about 16,400 to 17,200 total nuclear warheads in 2014.[27] That's still a long way from the figure of about 1,000 weapons that Sagan and Turco estimated for MSD,[28] but at the rate we're going we could get there by 2025. And since the end of the Cold War, it has become strategically less necessary and economically less desirable to retain so many nuclear weapons, resulting in a dramatic decline in stockpiles between the United States (7,315) and Russia (8,000), who together account for 93.4 percent of the total. Even more encouragingly, there are only about 4,200 operationally active nuclear warheads held by Russia (1,600), the United States (1,920), France (290), and the United Kingdom (160), making the world safer from being blown to smithereens by tens of thousands of nuclear warheads than it has been since 1945.[29]

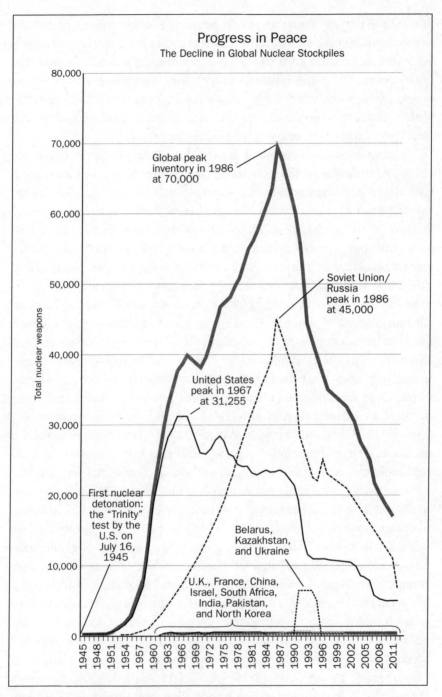

Figure 2-2. The Decline in Global Nuclear Stockpiles

Total nuclear warheads (16,400 to 17,200) and operationally active nuclear warheads (about 4,200) are the lowest since the 1950s.[30]

Can the nuclear stockpile get to zero? To find out I audited a class called "Perspectives on War and Peace" at Claremont Graduate University taught by the political scientist Jacek Kugler. His answer is no, for at least seven reasons: (1) Credible deterrence among nations that trust each other is stable and predictable. (2) Unstable or unpredictable states such as North Korea, which periodically rattle their nuclear swords inside their silo sheaths, require threat of retaliation. (3) There still exist rogue states such as Iran that need the threat of retaliation because they threaten to join the nuclear club but are not so keen on joining the club of nations. (4) States waging conventional wars that might escalate to using weapons of mass destruction require the threat of retaliation to keep them in check. (5) Nonstate entities such as terrorist groups that we either do not trust or do not know enough about also require the threat of retaliation. (6) There may be a taboo against *using* nuclear weapons, but as yet there is no taboo against *owning* them. (7) The nuclear genie of how to make an atomic bomb is out of the bottle, so there is always the chance that other nations or terrorists will obtain nuclear weapons and thereby destabilize deterrence as well as increase the probability of an accidental detonation.

Kugler thinks we can get to "regional zero"—nuclear-free zones such as South America and Australia—provided the major global nuclear powers (the United States, Russia, China, and perhaps the EU and India) provide a secure nonveto response to any preemptive use of nuclear weapons by potential rogue states or terrorist groups. But because of the trust problem, he says, global zero is unattainable. Kugler worries that a Middle East nuclear exchange or a nuclear terrorist attack on Israel is likely if current conditions persist. A major danger, he says, is that fissile material may well be available on the black market at a price that rogue states and terrorists can afford.

Analysts of various stripes make a compelling case that nuclear safety is an illusion and that we've come perilously close to a *Dr. Strangelove* ending of the world as we know it. Scientists with the Federation of Atomic Scientists and the *Bulletin of Atomic Scientists* sponsor the "Doomsday Clock," which seems to be perpetually set a few minutes to midnight and Armageddon. Popular authors such as Richard Rhodes in his nuclear tetralogy (*The Making of the Atomic Bomb, Dark Sun, Arsenals of Folly, and The Twilight of the Bombs*[31]) and Eric Schlosser in *Command and Control*[32] leave readers with vertigo knowing how many close calls there have been: the jettisoning of a Mark IV atomic bomb in British Columbia in 1950, the crash of a B-52 carrying two Mark 39 nuclear bombs in North

Carolina, the Cuban Missile Crisis, the Able Archer 83 Exercise in Western Europe that the Soviets misread as the buildup to a nuclear strike against them, and the Titan II missile explosion in Damascus, Arkansas, that narrowly avoided eradicating the entire city off the map. As Rhodes noted in reflecting on a career spent researching and writing about nuclear weapons:

> You know one of the sad things about nuclear weapons from the beginning is that they've been called weapons. They're vast destructive forces encompassed in this small, portable mechanism. They have no earthly use that I can see except to destroy whole cities full of human beings. This sort of mentality [has policymakers] thinking that nuclear weapons are like guns. There's a reason why no one has exploded one in anger since 1945. The risk is too great.[33]

According to a report published by the Global Zero US Nuclear Policy Commission chaired by the former Joint Chiefs of Staff general James E. Cartwright, the United States and Russia could maintain deterrence and still reduce their nuclear arsenals to nine hundred weapons each, ensuring that only half deploy at any one time and all with twenty-four to seventy-two-hour launch lag times to allow for fail-safe measures to prevent accidental strikes.[34] The global zero plan has been endorsed by such senior politicians as President Barack Obama, President Dmitry Medvedev of Russia, Prime Minister David Cameron of the United Kingdom, Prime Minister Manmohan Singh of India, Prime Minister Yoshihiko Noda of Japan, and Secretary-General Ban Ki-moon of the United Nations.[35] Of course, endorsement is one thing, action is another, but the global zero movement is gaining momentum.[36]

It is notable that of the 194 countries in the world, only 9 have nuclear weapons. That means 185 countries (95 percent) manage just fine without nukes. Some may want them but can't afford to produce the fissile (and other) materials, but it is noteworthy that since 1964 more nations have started and abandoned nuclear weapons programs than started and completed them, including Italy, West Germany, Switzerland, Sweden, Australia, South Korea, Taiwan, Brazil, Iraq, Algeria, Romania, South Africa, and Libya.[37] There are many good reasons not to own nuclear weapons, one of which is that they are very expensive. During the Cold War the United States and the Soviet Union spent an almost unfathomable $5.5 trillion to build 125,000 nuclear weapons, and the United States still spends $35 billion a year on its nuclear program.[38] Turning plowshares into swords is expensive. It also makes you a target. A 2012 study by the political scientist David Sobek and his colleagues tested the "conventional wisdom" hypothesis that possession of nuclear weapons confers many benefits to its owner by conducting a cross-national analysis of the relationship between a state initiating a nuclear weapons program and the onset of militarized conflict. They found that between 1945 and 2001 "the closer a state gets to acquiring nuclear weapons, the greater the risk it will be attacked." Why? "When a state initiates a nuclear weapons program, it signals its intent to fundamentally alter its bargaining environment. States that once had an advantage will now be disadvantaged." Once a state gets itself a nuke or two the risks of being the target of a first strike go down, but not below the level it was at before it set off to join the nuclear club.[39] In other words, either way it's better not to possess nuclear weapons.

That was the opinion of the preeminent Cold Warrior himself—the cowboy president Ronald Reagan—who called for the abolishment of "all nuclear weapons." According to Jack Matlock, the US ambassador to the Soviet Union in the late 1980s, President Reagan considered nuclear weapons to be "totally irrational, totally inhumane, good for nothing but killing, possibly destructive of life on earth and civilization." Kenneth Adelman, head of the US Arms Control and Disarmament Agency under Reagan, said that his boss would often "pop out with 'Let's abolish all nuclear weapons.'" As Adelman recalled, "I was surprised that for an anti-Communist hawk, how antinuclear he was. He would make comments that seemed to me to come from the far left rather than from the far right. He hated nuclear weapons." The whole point of the Strategic Defense Initiative (SDI, also known as "Star Wars"), in fact, was to eliminate the need for MAD. Matlock paraphrased what Reagan told him on the matter: "How can you tell me, the president of the United States, that the only way I can defend my people is by threatening other people and maybe civilization itself? That is unacceptable."[40]

Not everyone was on board with Reagan's vision of a nuclear-free world. Reagan's secretary of state George Shultz recalled an incident in which he got "handbagged" by British prime minister Margaret Thatcher when she found out that his boss suggested to Soviet premier Mikhail Gorbachev that they abolish nuclear weapons:

When we came back from [the 1986 U.S.-Soviet summit in] Reykjavik with an agreement that it was desirable to get rid of all nuclear weapons, she came over to Washington, and I was summoned to the British Embassy. It was then that I discovered the meaning of the British expression to be "handbagged." She said: "George, how could you sit there and allow the president to agree to abolish nuclear weapons?" To which I said: "Margaret, he's the president." She replied, "But you're supposed to be the one with the feet on the ground." My answer was, "But, Margaret, I agreed with him."[41]

Today, not only is George Shultz a nuclear abolitionist, but so too are his Cold Warrior colleagues former secretary of state Henry Kissinger, former senator Sam Nunn, and former secretary of defense William Perry. The four of them went on record calling for "a world free of nuclear weapons" in, of all places, the *Wall Street Journal*.[42] There (and elsewhere) they outlined the realpolitik difficulties of achieving nuclear zero, which they equated to climbing a mountain: "From the vantage point of our troubled world today, we can't even see the top of the mountain, and it is tempting and easy to say we can't get there from here. But the risks from continuing to go down the mountain or standing pat are too real to ignore. We must chart a course to higher ground where the mountaintop becomes more visible."[43]

Some theorists think that the path to peace is more deterrence. The late political scientist Kenneth Waltz, for example, thought that a nuclear Iran would bring stability to the Middle East because "in no other region of the world does a lone, unchecked nuclear state exist. It is Israel's nuclear arsenal, not Iran's desire for one, that has contributed most to the current crisis. Power, after all, begs to be balanced."[44] Except for when it doesn't, as in the post-1991 period after the collapse of the Soviet Union and the unipolar dominance of the United States—no other medium-size power rose to fill the vacuum, no rising power started wars of conquest to consolidate more power, and the only other candidate, China, has remained war-free for almost four decades. Plus, as Jacek Kugler points out, the Islamic Republic of Iran doesn't play by the rules of the international system; it has no formal diplomatic relations with the United States or Israel, thereby making communication in the event of an emergency problematic; and it is so close to Israel that it lowers the warning time of a missile launch to minutes, thereby limiting the effectiveness of countermeasures such as anti-ballistic missiles, and makes sneaking a dirty bomb into the country more likely.[45]

To that I would add: Iran has a history of training terrorist groups such as Hamas and Hezbollah, both of whom are opposed to the United

States and Israel, and its leaders have repeatedly and clearly expressed their anti-Semitic views, as happened in 2005 when the new president, Mahmoud Ahmadinejad, said that Israel must be "wiped off the map," a view he expressed before an audience of about four thousand students at a program starkly titled "The World Without Zionism."[46] Given what unfolded after another state leader proclaimed on numerous occasions in the 1930s that he wanted to rid the world of Jews—and almost did—we can hardly fault Israel for not fully embracing the image of an *Allahu Akbar*-bellowing imam with his finger on a nuclear trigger.

As the political scientist Christopher Fettweis notes in his book *Dangerous Times?*, despite the popularity of such intuitive notions as the "balance of power"—based on a small number of nongeneralizable cases from the past that are in any case no longer applicable to the present—"clashes of civilization" like the world wars of the twentieth century are extremely unlikely to happen in the highly interdependent world of the twenty-first century. In fact, he shows, never in history has such a high percentage of the world's population lived in peace, conflicts of all forms have been steadily dropping since the early 1990s, and even terrorism can bring states together in international cooperation to combat a common enemy.[47]

Fat Man Morality and Little Boy Brawls

In addition to all this moral deliberation over the possible future use of nuclear weapons, there is an ongoing debate about the past use of the only atomic bombs dropped on cities—Little Boy, which obliterated Hiroshima, and Fat Man, which annihilated Nagasaki. Over the past couple of decades a cadre of critics have put forth the claim that neither bomb was necessary to bring about the end of the Second World War and thus their use was immoral, illegal, or even a crime against humanity. In 1946 the Federal Council of Churches issued a statement declaring, "As American Christians, we are deeply penitent for the irresponsible use already made of the atomic bomb. We are agreed that, whatever be one's judgment of the war in principle, the surprise bombings of Hiroshima and Nagasaki are morally indefensible."[48] In 1967 the linguist and contrarian politico Noam Chomsky called the two bombings "the most unspeakable crimes in history."[49]

More recently, in an otherwise deeply insightful history of genocide titled *Worse Than War*, the historian Daniel Goldhagen opens his analysis by calling US president Harry Truman "a mass murderer" because in ordering the use of atomic weapons he "chose to snuff out the lives of approximately 300,000 men, women and children." Goldhagen proceeds to opine that "it is hard to understand how any right-thinking person could fail to call slaughtering unthreatening Japanese *mass murder*."[50] In morally equating Harry Truman with Adolf Hitler, Joseph Stalin, Mao Zedong, and Pol Pot, Goldhagen allows himself to be constrained by the categorical thinking that prevents one from discerning the different kinds, levels, and motives of genocide (although he does this for other mass killings). If one defines "genocide" broadly enough, as when Goldhagen equates it with "mass murder" (without ever defining what, exactly, that means) then nearly every act of killing large numbers of people could be considered genocidal because there are only two categories—mass murder and non-mass murder. The virtue of continuous thinking allows us to distinguish the differences between types of mass killings (some scholars define genocide as one-sided killing by armed people of unarmed people), their context (during a state war, civil war, "ethnic cleansing"), motivations (termination of hostilities or extermination of a people), and quantities (hundreds to millions) along a sliding scale. In 1946 the Polish jurist Raphael Lemkin created the term *genocide* and defined it as "a conspiracy to exterminate national, religious or racial groups."[51] In that same year the UN General Assembly defined genocide as "a denial of the right of existence of entire human groups."[52] More recently, in 1994 the highly respected philosopher Steven Katz defined genocide as "the actualization of the intent, however successfully carried out, to murder in its totality any national, ethnic, racial, religious, political, social, gender or economic group."[53]

By these definitions, the dropping of Fat Man and Little Boy were not acts of genocide, and the difference between Truman and the others is in the context and motivation of the act, apparent in the subtitle of Goldhagen's book, *Genocide, Eliminationism, and the Ongoing Assault on Humanity*. In their genocidal actions against targeted people, Hitler, Stalin, Mao, and Pol Pot had as their objective the total elimination of a group. The killing would only stop when every last pursued person was exterminated (or if the perpetrators were stopped or defeated). Truman's goal in dropping the bombs was to end the war, not to eliminate the Japanese people. If eliminationism was the goal, then why did the United States lead the Marshall Plan after the war to rebuild both Japan and (West) Germany such that within twenty years both were world economic powers?[54] This would seem to be the very opposite of an eliminationist program.

As to the morality of dropping the bombs, by the definition of morality in this book—the survival and flourishing of sentient beings—then not only did Fat Man and Little Boy end the war and stop the killings, they also saved lives, possibly millions of lives, both Japanese and American. My father was possibly one such survivor. During the Second World War he was aboard the USS *Wren*, a navy destroyer assigned to escort aircraft carriers and other large capital ships to protect them from Japanese submarines and especially for shooting down kamikaze planes. He was part of the larger fleet that was working its way toward Japan for a planned invasion of the home islands. He told me that everyone on board dreaded that day because they had heard of the horrific carnage resulting from the invasion of just two tiny islands held by the Japanese—Iwo Jima and Okinawa. During the invasion of Iwo Jima there were approximately 26,000 American casualties, which included 6,821 dead in the 36-day battle. How fiercely did the Japanese defend that little volcanic rock 700 miles from Japan? Of the 22,060 Japanese soldiers assigned to fight to the bitter end, only 216 were captured during the battle.[55] The subsequent battle for Okinawa, only 340 miles from the Japanese mainland, was fought even more ferociously, resulting in a staggering body count of 240,931 dead, including 77,166 Japanese soldiers, 14,009 American soldiers, plus an additional 149,193 Japanese civilians living on the island who either died fighting or committed suicide rather than let themselves be captured.[56] No wonder, as my father told me, when the atomic bombs were dropped there was an emotional relief among the crew.[57] With an estimated 2.3 million Japanese soldiers and 28 million Japanese civilian militia prepared to defend their island nation to the end,[58] it was clear to all what an invasion of the Japanese mainland would mean.

It is from these cold, hard facts that Truman's advisers estimated that between 250,000 and 1 million American lives would be lost in an invasion of Japan.[59] General Douglas MacArthur estimated that there could be a 22:1 ratio of Japanese to American deaths, which translates to a minimum death toll of 5.5 million Japanese.[60] By comparison (cold though it may sound), the body count from both atomic bombs—about 200,000 to 300,000 total (Hiroshima: 90,000 to 166,000 deaths, Nagasaki: 60,000 to 80,000 deaths[61])—was a bargain. In any case, if Truman hadn't ordered the bombs dropped, General Curtis LeMay and his fleet of B-29 bombers would have continued pummeling Tokyo and other Japanese cities into rubble before the invasion, and the death toll from conventional bombing would have been just as high as that produced by the two atomic bombs, if not higher, given the fact that previous mass bombing raids had produced

Hiroshima-level death rates, plus the likelihood that more than just two cities would have been destroyed before the Japanese surrendered. For example, Little Boy was the energy equivalent of 16,000 tons of TNT. By comparison, the US Strategic Bombing Survey estimated that this was the equivalent of 220 B-29s carrying 1,200 tons of incendiary bombs, 400 tons of high-explosive bombs, and 500 tons of antipersonnel fragmentation bombs, with an equivalent number of casualties.[62] In fact, on the night of March 9–10, 1945, 279 B-29s dropped 1,665 tons of bombs on Tokyo, leveling 15.8 square miles of the city, killing 88,000 people, injuring another 41,000, and leaving another 1 million homeless.[63]

On balance, then, dropping the atomic bombs was the least destructive of the options on the table. Although we wouldn't want to call it a moral act, it was in the context of the time the least immoral act by the criteria of lives saved. That said, we should also recognize that the several hundred thousand killed is still a colossal loss of life, and the fact that the invisible killer of radiation continued its effects long after the bombings should dissuade us from ever using such weapons again. Along that sliding scale of evil, in the context of one of the worst wars in human history that included the singularly destructive Holocaust of 6 million murdered, it was not, pace Chomsky, the most unspeakable crime in history—not even close—but it was an event in the annals of humanity never to be forgotten and, hopefully, never to be repeated.

A Path to Nuclear Zero

The abolition of nuclear weapons is an exceptionally complex and difficult puzzle that has been studied extensively by scholars and scientists for more than half a century. The problems and permutations of getting from here to there are legion, and there is no single surefire pathway to zero, but the following steps that have been proposed by various experts and organizations seem reasonable and realistic long-term goals.[64]

1. *Continue nuclear stockpile reduction.* Following the trend lines in *figure 2-2*, work to reduce the global stockpile of nuclear weapons from more than 10,000 now to 1,000 by 2020, and to less than 100 by 2030. This is enough nuclear firepower to maintain Minimum Sufficiency Deterrence to keep the peace among nuclear states, and yet in the event of a mistake or a madman a nuclear war will not result in the annihilation of civilization.[65] The global zero campaign

calls for "the phased, verified, proportional reduction of all nuclear arsenals to zero total warheads by 2030," noting that if this seems unrealistic "the United States and Russia retired and destroyed twice as many nuclear warheads (40,000+) as this action plan proposes (20,000+) over the next twenty years (2009–2030)."[66] That is encouraging, but it is far easier to go from 70,000 to 10,000 than it is from 10,000 to 1,000, and even harder to go from 1,000 to 0 because of the security dilemma that will always exist until and unless the rest of the steps below are met.

2. *No first use.* Make all "first strike" strategies illegal by international law. Nuclear weapons should only be used defensively, in a retaliatory function. Any nation that violates the law and initiates a first strike will be subject to global condemnation, economic sanctions, nuclear retaliation, and possible invasion, toppling their government and putting their leaders on trial for crimes against humanity. China and India have both signed on to No First Use (NFU), but NATO, Russia, and the United States have not. Russian military doctrine calls for the right to use nukes "in response to a large-scale conventional aggression."[67] France, Pakistan, the United Kingdom, and the United States have stated that they will only use nukes defensively, although Pakistan said it would strike India even if the latter did not use nuclear weapons first,[68] and the United Kingdom said it would use nuclear weapons against "rogue states" such as Iraq were the latter to use weapons of mass destruction against British troops in the field.[69] For its part the United States reiterated its long-term policy that "The fundamental role of US nuclear weapons, which will continue as long as nuclear weapons exist, is to deter nuclear attack on the United States, our allies, and partners," adding that it "will not use or threaten to use nuclear weapons against non-nuclear weapons states that are party to the NPT [Non-Proliferation Treaty] and in compliance with their nuclear non-proliferation obligations."[70]

3. *Form a nuclear great power pact.* Such an alliance would hold strong against small powers and terrorists who either have nuclear weapons or are trying to obtain them with an intent to use them. Jacek Kugler has outlined a model for reducing nuclear stockpiles worldwide while maintaining deterrence that would include, in addition to the no first use policy, a proviso that "nuclear great powers must guarantee that any first nuclear strike by small nuclear powers will

face automatic nuclear devastating retaliation from a member of the nuclear club." Kugler and his colleague Kyungkook Kang propose a "Nuclear Security Council" composed of the four great powers—the United States, China, Russia, and the European Union (France and the United Kingdom)—that have enough nuclear weapons for a second strike should any of the smaller nuclear powers, or a terrorist group, launch a first strike against them or anyone else.[71] For example, if North Korea attacked either South Korea or Japan, the United States would retaliate. Or if Iran acquired nuclear weapons and used them against Israel, the United States (and maybe the European Union, or maybe not) would counterstrike.

4. *Shift the taboo from* using *nuclear weapons to* owning *nuclear weapons.* This is what the Nobel Foundation had in mind in 2009 when it awarded President Barack Obama the Peace Prize: "The Committee has attached special importance to Obama's vision of and work for a world without nuclear weapons."[72] Taboos are effective psychological mechanisms for deterring all sorts of human behaviors, and they worked well in keeping poison gas from being used in the Second World War, although other states (England and Germany) have used them at other times (in the First World War), and sometimes even on their own people (Saddam Hussein on the Kurds in Iraq). But across the board and over time, the taboo against using chemical and biological weapons has grown stronger, and their use is considered by most nations and international law to be a crime against humanity (it was one of the crimes for which Saddam Hussein was hanged[73]), although the taboo was not instantaneous.

Nukes began as sexy weapons, as evidenced by the fact that the bikini bathing suit was so named by its French designer, Louis Réard, because he hoped its revealing design would create an explosive reaction not unlike that of the two atomic bombs detonated earlier that summer of 1946 on the atoll of Bikini in the South Pacific.[74] As the political scientist Nina Tannenwald writes in her history of the origin of the nuclear taboo, throughout the 1950s everyone accepted nuclear weapons as conventional and articulated "a view with a long tradition in the history of weapons and warfare: a weapon once introduced inevitably comes to be widely accepted as legitimate." But that didn't happen. Instead, "nuclear weapons have come to be defined as abhorrent and unacceptable weapons of mass destruction, with a taboo on their use. This taboo is associated with a widespread revulsion toward nuclear weapons and broadly held inhibitions on their use. The opprobrium has come to apply to all nuclear weapons, not just to large bombs or to certain types or uses of nuclear weapons." The taboo developed as a result of three forces, Tannenwald argues: "a global grassroots antinuclear weapons movement, the role of Cold War power politics, and the ongoing efforts of nonnuclear states to delegitimize nuclear weapons."[75]

The psychology behind the taboo against chemical and biological weapons transfers readily to that of nuclear weapons. Deadly heat and radiation—like poison gas and lethal diseases—are invisible killers that are indiscriminate in the carnage they wreak. This is a psychological break from traditional warfare between two armies bearing spears, swords, firearms, grenades, or even cannons and rocket launchers. The moral emotion of moralistic punishment that evolved to deter free riders and bullies brings no satisfaction if one's enemy simply disappears in a white flash on some other continent.[76] Also, the revulsion people feel toward nuclear weapons may be linked in the brain to the emotion of disgust that psychologists have identified as being associated with invisible disease contagions, toxic poisons, and revolting materials (such as vomit and feces) that carry them—reactions that evolved to direct organisms away from these substances for survival reasons.[77]

5. *Nuclear weapons should no longer be seen as a deterrence solution.* Former foreign minister of Australia Gareth Evans, who is also chair of the International Commission on Nuclear Non-Proliferation and Disarmament, makes a convincing case that nuclear weapons no longer make sense for deterrence. It is not at all clear, Evans argues, that it was nuclear weapons that kept the great powers at a standoff throughout the Cold War, given the fact that prior to their invention in the mid-1940s, the great powers fought in spite of the existence of weapons of mass destruction. "Concern about being on the receiving end of the extreme destructive power of nuclear weapons may simply not be, in itself, as decisive for decision-makers as usually presumed," says Evans. Instead, the long peace since 1945 may be the result of "a realisation, after the experience of World War II and in the light of all the rapid technological advances that followed it, that the damage that would be inflicted by *any* war would be unbelievably horrific, and far outweighing, in today's economically interdependent world, any conceivable benefit to be derived."[78]

6. *Evolution instead of revolution.* All of these changes should be implemented gradually and incrementally with "trust but verify" strategy and with as much transparency as possible. Gareth Evans proposes a two-stage process: minimization, then elimination, "with some inevitable discontinuity between them." He uses the target date of 2025 for the achievement of a minimization objective set by the International Commission on Nuclear Non-Proliferation and Disarmament that "would involve reducing the global stockpile of all existing warheads to no more than 2,000 (a maximum of 500 each for the United States and Russia and 1,000 for the other nuclear-armed states combined), with all states being committed by then to 'No First Use'—and with these doctrinal declarations being given real credibility by dramatically reduced weapons deployments and launch readiness."[79]

7. *Reduce spending on nuclear weapons and research.* Nuclear weapons are indefensibly costly, with estimates coming in at more than $100 billion a year for maintenance by the nine nuclear states, and $1 trillion projected over the next decade.[80] Building budgets now that wind down monies allocated for all nuclear weapons–related agencies over the next twenty years will drive nations to consider other solutions to the problems that nuclear weapons have historically been created to solve.

8. *Revise twentieth-century nuclear plans and policies for the twenty-first century.* In their manifesto "A World Free of Nuclear Weapons," the aforementioned Shultz, Perry, Kissinger, and Nunn proposed that we "discard any existing operational plans for massive attacks that still remain from the Cold War days"; "increase the warning and decision times for the launch of all nuclear-armed ballistic missiles, thereby reducing the risks of accidental or unauthorized attacks"; "undertake negotiations toward developing cooperative multilateral ballistic-missile defense and early warning systems"; and "dramatically accelerate work to provide the highest possible standards of security for nuclear weapons . . . to prevent terrorists from acquiring a nuclear bomb."[81]

9. *Economic interdependency.* The more two countries trade with one another the less likely they are to fight. It's not a perfect correlation—there are exceptions—but countries that are economically interdependent are less likely to allow political tensions to escalate to the point of

conflict. Wars are expensive—economic duties, sanctions, embargoes, and blockades are costly; and business often suffers on both sides of a conflict (except for weapons manufacturers, of course). In democracies, for better or for worse, politicians are more beholden to monied interests, who generally prefer to keep their transaction costs as low as possible, and those go way up in wartime. Thus, the sooner nations such as North Korea and Iran can be brought into economic trading blocs that make them codependent with the nuclear great powers, the less likely they are to feel the need to develop nuclear weapons in the first place, much less use them.

10. *Democratic governance.* The more democratic two nations are, the less likely they are to fight. As with economic interdependency, democratic peace is a general trend, not a law of nature, but the effect is found in the transparency of a political system that includes checks and balances on power and the ability to change leaders so that a liar, lunatic, or lord wannabe obsessed with power, intent on revenge, or fixated on a nation's racial purity or precious bodily fluids will not be allowed to escalate tensions to the point of launching nuclear-tipped missiles.

With nuclear weapons there is no easy way out of the security dilemma. Even though Reagan said he wanted to go to zero, in Iceland he refused offers from Gorbachev to reduce nuclear weapons drastically precisely because he did not trust the Russians and was only willing to trust and verify—which is a form of distrust. I am hopeful that we can get to zero before we annihilate ourselves, but it's going to be a long row to hoe. One place to begin is to apply the *principle of interchangeable perspectives* when negotiating—for example, a spin-down of arms, which is what Gareth Evans recommends as the first step: "In each case the key to progress, as in all diplomacy, is to try to understand the interests and perspectives of the other side, and to find ways of accommodating them by all means short of putting at real risk genuinely vital interests of one's own."

For example, Evans cites the United States' failure to give Russia "an acceptable response to its concerns...about BMD [Ballistic Missile Defense] and new long-range conventional weapons systems" in Europe, which, he says, seriously diminished "its second-strike retaliatory capability."[82] Encouragingly, in 2013, the United States took an important step by canceling part of its BMD program in Europe, even though it was done for budgetary reasons and to bolster BMD in Asia over concerns about North Korea.[83] Or, to keep China in its current "minimal deterrence" posture instead of ramping up into MAD deterrence, Evans writes, the United States should acknowledge "that its nuclear relationship with China is one of 'mutual vulnerability,' meaning in practice that the United States 'should plan and posture its force and base its own policy on the assumption that an attempted US disarming first strike, combined with US missile defences, could not reliably deny a Chinese nuclear retaliatory strike on the United States.' "[84] For example, the United States already has a strong enough BMD system in place in the Pacific region to counter any North Korean missile launches, so any further development could be perceived as threatening by the Chinese.

There are dozens of such scenarios that are played out in search of what—in the spirit of creative acronyms so common in this field—we might call a Minimally Dangerous Pathway to Zero (MDPZ). I do not believe that the deterrence trap is one from which we can never extricate ourselves, and the remaining threats should direct us to work toward nuclear zero sooner rather than later. In the meantime, minimum is the best we can hope for given the complexities of international relations, but given enough time, as Shakespeare poetically observed,

> Time's glory is to calm contending kings,
> To unmask falsehood and bring truth to light,
> To stamp the seal of time in aged things,
> To wake the morn and sentinel the night . . .
> To slay the tiger that doth live by slaughter . . .
> To cheer the ploughman with increased crops,
> And waste huge stones with little water-drops.[85]

WHAT ABOUT TERRORISM?

All this game-theoretic computation assumes that humans are rational actors. As the international relations scholar Hedley Bull noted, "mutual nuclear deterrence . . . does not make nuclear war impossible, but simply renders it irrational," but then added that a rational strategist is one "who on further acquaintance reveals himself as a university professor of unusual intellectual subtlety."[86]

Are terrorists rational actors? How rational is it for a Muslim terrorist to look forward to martyrdom and a reward of seventy-two virgins in heaven? (That's if you're a man, of course; a female terrorist has no equivalent consolation.) At least the godless Communists didn't harbor such delusions. MAD deters you from launching a first strike if you think your target has retaliatory capability and you don't want to die—the drive to stay alive and all that. But if your religion has convinced you that you're not really going to die, and that the next life is spectacularly better than this life, and that you'll be a hero among those whom you've left behind . . . it changes the calculation. As the nuclear zero advocate Sam Nunn said, "I'm much more concerned about a terrorist without a return address that cannot be deterred than I am about deliberate war between nuclear powers. You can't deter a group who is willing to commit suicide."[87] Nevertheless, I lean toward optimism because of the dismal history of terrorism to achieve its stated goals through violence. Despite the seemingly constant barrage of media stories of suicide bombers blowing themselves up, the long-term trends in social change over the past half century are in the direction of less violence and more moral action, even with terrorism.

Terrorism is a form of asymmetrical warfare by nonstate actors against innocent, noncombatant civilians. As its name suggests, it does so by evoking terror. This exercises our alarmist emotions, which in turn confound our reasoning, making clear thinking about terrorism well nigh impossible. As such, I suggest that there are at least seven myths that have arisen that need to be debunked to properly understand the causes of terrorism in order to continue to reduce its frequency and effectiveness.

1. *Terrorists are pure evil.* This first myth took root in September 2001 when President George W. Bush announced, "We will rid the world of the evil-doers" because they hate us for "our freedom of religion, our freedom of speech, our freedom to vote and assemble and disagree with each other."[88] This sentiment embodies what the social psychologist Roy Baumeister calls "the myth of pure evil" (more on this in chapter 9, on moral regress), which holds that perpetrators of violence act only to commit senseless injury and pointless death for no rational reason. The "terrorists-as-evil-doers" myth is busted through the scientific study of violence, of which at least four types motivate terrorists: *instrumental, dominance/honor, revenge,* and *ideology.*

    In a study of fifty-two cases of Islamic extremists who have targeted the United States, for instance, the political scientist John Mueller concluded that terrorist motives include *instrumental violence* and *revenge*: "a simmering, and more commonly boiling, outrage at U.S. foreign policy—the wars in Iraq and Afghanistan, in particular, and the country's support for Israel in the Palestinian conflict." *Ideology* in the form of religion "was a part of the consideration for most," Mueller suggests, "but not because they wished to spread Sharia law or to establish caliphates (few of the culprits would be able to spell either word). Rather they wanted to protect their co-religionists against what was commonly seen to be a concentrated war upon them in the Middle East by the U.S. government."[89] As for *dominance and honor* as drivers of violence, through his extensive ethnography of terrorist cells the anthropologist Scott Atran has demonstrated that suicide bombers (and their families) are showered with status and honor in this life (and, secondarily, the promise of virgins in the next life), and that most "belong to loose, homegrown networks of family and friends who die not just for a cause, but for each other." Most terrorists are in their late teens or early twenties, especially students and immigrants "who are especially prone to movements that promise a meaningful cause, camaraderie, adventure, and glory."[90] All of these motives are on display in the 2013 documentary film by Jeremy Scahill called *Dirty Wars*, a sobering look at the effects of US drone attacks and assassinations in foreign countries such as Somalia and Yemen—countries with whom the United States is not at war—in which we see citizens swearing revenge against Americans for these violations of their honor and ideology.[91]

2. *Terrorists are organized.* This myth depicts terrorists as part of a vast global network of top-down, centrally controlled conspiracies against the West. But as Atran shows, terrorism is "a decentralized, self-organizing, and constantly evolving complex of social networks," often organized through social groups and sports organizations such as soccer clubs.[92]

3. *Terrorists are diabolical geniuses.* This myth began with the 9/11 Commission report that described the terrorists as "sophisticated, patient, disciplined, and lethal."[93] But according to the political scientist Max Abrahms, after the decapitation of the leadership of the top terrorist organizations, "terrorists targeting the American homeland have been neither sophisticated nor masterminds, but incompetent fools."[94] Examples abound: The 2001 airplane shoe bomber Richard Reid was unable to ignite the fuse because it was wet from the rain and his own foot perspiration; the 2009 underwear bomber

Umar Farouk Abdulmutallab succeeded only in setting his pants ablaze, burning his hands, inner thighs, and genitals, and getting himself arrested; the 2010 Times Square bomber Faisal Shahzad managed merely to torch the inside of his 1993 Nissan Pathfinder; the 2012 model airplane bomber Rezwan Ferdaus purchased C-4 explosives for his rig from FBI agents who promptly arrested him; and the 2013 Boston Marathon bombers were equipped with only one gun for defense and had no money and no exit strategy beyond hijacking a car with no gas in it that Dzhokhar Tsarnaev used to run over his brother Tamerlan, followed by a failed suicide attempt inside a land-based boat. Evidently terrorism is a race to the bottom.

4. *Terrorists are poor and uneducated.* This myth appeals to many in the West who like to think that if we throw enough money at a problem it will go away, or if only everyone went to college they'd be like us. The economist Alan Krueger, in his book *What Makes a Terrorist,* writes: "Instead of being drawn from the ranks of the poor, numerous academic and government studies find that terrorists tend to be drawn from well-educated, middle-class or high-income families. Among those who have seriously and impartially studied the issue, there is not much question that poverty has little to do with terrorism."[95]

5. *Terrorism is a deadly problem.* In comparison to homicides in America, deaths from terrorism are in the statistical noise, barely a blip on a graph compared to the 13,700 homicides a year. By comparison, after the 3,000 deaths on 9/11, the total number of people killed by terrorists in the 38 years before totals 340, and the number killed after 9/11 and including the Boston bombing is 33, and that includes the 13 soldiers killed in the Fort Hood massacre by Nidal Hasan in 2009.[96] That's a total of 373 killed, or 7.8 per year. Even if we include the 3,000 people who perished on 9/11, that brings the average annual total to 70.3, compared to that of the annual homicide rate of 13,700. No comparison.

6. *Terrorists will obtain and use a nuclear weapon or a dirty bomb.* Osama bin Laden said he wanted to use such weapons if he could get them, and Secretary of Homeland Security Tom Ridge pressed the point in calling for more support for his agency: "Weapons of mass destruction, including those containing chemical, biological or radiological agents or materials, cannot be discounted."[97] But as

Michael Levi of the Council on Foreign Relations reminds us, "Politicians love to scare the wits out of people, and nothing suits that purpose better than talking about nuclear terrorism. From President Bush warning in 2002 that the 'smoking gun' might be a mushroom cloud, to John Kerry in 2004 conjuring 'shadowy figures' with a 'finger on the nuclear button' and Mitt Romney invoking the specter of 'radical nuclear jihad' last spring, the pattern is impossible to miss."[98] But most experts agree that acquiring the necessary materials and knowledge for building either weapon is far beyond the reach of most (if not all) terrorists. George Harper's delightful 1979 article in *Analog* titled "Build Your Own A-Bomb and Wake Up the Neighborhood" is revealing in showing just how difficult it is to actually make a bomb:

> As a terrorist one of the best methods for your purposes is the gaseous diffusion approach. This was the one used for the earliest A-bombs, and in many respects it is the most reliable and requires the least sophisticated technology. It is, however, a bit expensive and does require certain chemicals apt to raise a few eyebrows. You have to start with something on the order of a dozen miles of special glass-lined steel tubing and about sixty tons of hydrofluoric acid which can be employed to create the compound uranium-hexafluoride. Once your uranium has been converted into hexafluoride it can be blown up against a number of special low-porosity membranes. The molecules of uranium hexafluoride which contain an atom of U-238 are somewhat heavier than those containing an atom of U-235. As the gas is blown across the membranes more of the heavier molecules are trapped than the light ones. The area on the other side of the membrane is thus further enriched with the U-235 containing material; possibly by as much as $1/2$% per pass. Repeat this enough times and you wind up with uranium hexafluoride containing virtually 100% core atoms of U-235. You then separate the fluorine from the uranium and arrive at a nice little pile of domesticated U-235. From there it's all downhill.[99]

In his book *On Nuclear Terrorism*, Levi invokes what he calls "Murphy's Law of Nuclear Terrorism: What Can Go Wrong Might Go Wrong," and recounts numerous failed terrorist attacks due to sheer incompetence by the terrorists to build and detonate even the simplest of chemical weapons.[100] In this context it is important to note that no dirty bomb has ever been successfully deployed resulting in casualties by anyone anywhere, and that according to the US Nuclear Regulatory Commission—which tracks fissile materials—"most reports of lost or stolen material involve small or short-lived radioactive sources that are not useful for a RDD [radiological disbursal device, or dirty bomb]. Past experience suggests there has not been a pattern of collecting such sources for the purpose of assembling a RDD. It is important to note that the radioactivity of the combined total of all unrecovered sources over the past 5 years would not reach the threshold for one high-risk radioactive source."[101] In short, the chances of terrorists successfully building and launching a nuclear device of any sort are so low that we would be far better off investing our limited resources in diffusing the problem of terrorism in other areas.

7. *Terrorism works.* In a study of forty-two foreign terrorist organizations active for several decades, Max Abrahms concluded that only two achieved their stated goals—Hezbollah achieved control over southern Lebanon in 1984 and 2000, and the Tamil Tigers took over parts of Sri Lanka in 1990, which they then lost in 2009. That results in a success rate of less than 5 percent.[102] In a subsequent study, Abrahms and his colleague Matthew Gottfried found that when terrorists kill civilians or take captives it significantly lowers the likelihood of bargaining success with states, because violence begets violence and public sentiments turn against the perpetrators of violence. Further, they found that when terrorists did get what they want it is more likely to be money or the release of political prisoners, not political objectives. They also found that liberal democracies are more resilient to terrorism, despite the perception that because of their commitment to civil liberties democracies tend to shy away from harsh countermeasures against terrorists.[103] Finally, in terms of the overall effectiveness of terrorism as a means to an end, in an analysis of 457 terrorist campaigns since 1968 the political scientist Audrey Cronin found that not one terrorism group had conquered a state and that a full 94 percent had failed to gain even *one* of their strategic political goals. And the number of terrorist groups who accomplished all of their objectives? *Zero.*

Cronin's book is titled *How Terrorism Ends*. It ends swiftly (groups survive only five to nine years on average) and badly (the deaths of its leaders).[104]

A rejoinder I often hear when recounting these studies is that terrorism has worked in terms of terrorizing the government into expending enormous resources into combating its threat, and along the way sacrificed our freedom and privacy. It's a valid point. The United States alone has spent upwards of $6 trillion since 9/11 on two wars and a bloated bureaucracy in response to the loss of 3,000 lives,[105] less than a tenth of the number of people who die annually on American highways. The explosive revelations by Edward Snowden about the National Security Agency's surveillance programs launched a national conversation about the balance between privacy and transparency, freedom and security. As Snowden told the 2014 TED audience in Vancouver via video link from an undisclosed location in Moscow:

> Terrorism provokes an emotional response that allows people to rationalize and authorize programs they wouldn't have otherwise. The U.S. asked for this authority in the 1990s; it asked the FBI to make the case in Congress, and they said no, it's not worth the risk to the economy, it would do too much damage to society to justify gains. But in the post 9/11 era, they used secrecy and justification of terrorism to start programs in secret without asking Congress or the American people. Government behind closed doors is what we must guard against. We don't have to give up privacy to have good government, we don't have to give up liberty to have security.[106]

That balance between liberty and security is one all governments contend with in many areas of society.[107] We must be vigilant always, of course, but these seven myths point to the unavoidable conclusion that in the course of history terrorism fails utterly to achieve its goals or divert civilization from its path toward greater justice and freedom unless we fall victim to fear itself.

VIOLENT VS. NONVIOLENT CHANGE

Violence as a means of attaining political change is a problematic strategy. What about nonviolent social change? In a classic work in political philosophy published in 1970 titled *Exit, Voice, and Loyalty*, the Harvard economist Albert Hirschman observed that when organizations such as firms and nations begin to stagnate and decline, members and interested parties can employ one of two nonviolent strategies to turn things around: *voice* their opinions by making suggestions, proposing changes, filing grievances, or protesting; or *exit* and start a new organization that incorporates their ideas for change.[108] In response to political repression, for example, citizens of a nation can either protest (*voice*) or emigrate (*exit*); employees or customers of a company can either file a complaint or take their business elsewhere. In both cases, people can vote with their voices and their feet (and dollars). *Loyalty* keeps *exit* in check so that nations and companies are not constantly failing or going bankrupt. A certain amount of stability is required for progress and profits, so in attenuating the *exit* strategy *loyalty* enables *voice* to be a more effective (and nonviolent) means of bringing about change. When people feel that their voices are heard—and can see real change made—they are less likely to exit. Conversely, when voices are not heard—as when nations silence political dissenters by locking them up or executing them—exit becomes the only viable strategy for change, and that can lead to violence.

In terms of marking moral progress, which strategy is better, voice or exit? It depends on how the change is brought about—through nonviolent resistance or violent response. Historically, political regime change has often come about by butchery and bloodshed. Regicide, for example, was a common method of regime change throughout most of European history. In a study of 1,513 monarchs in 45 monarchies across Europe between AD 600 and 1800, for example, the criminologist Manuel Eisner found that about 15 percent (227) were assassinated, corresponding to a homicide rate of about 1,000 per 100,000 ruler-years—10 times the background rate of homicide during those centuries.[109] Mao Zedong was being a realist when he proclaimed in 1938, "Political power grows out of the barrel of a gun."[110] But that is changing.

Like the many other forms of moral progress tracked in this book, nonviolent resistance has now overtaken violent response. The political scientists Erica Chenoweth and Maria Stephan entered all forms of both nonviolent and violent revolutions and reforms since 1900 into a database and then crunched the numbers.[111] Results: "From 1900 to 2006, nonviolent campaigns worldwide were twice as likely to succeed outright as violent insurgencies." Chenoweth added that "this trend has been increasing over time—in the last 50 years civil resistance has become increasingly frequent and effective, whereas violent insurgencies have become increasingly rare and unsuccessful. This is true even in extremely repressive, authoritarian conditions where we might expect nonviolent resistance to fail." Why does nonviolence trump violence in the long run as a means to an end? "People power," Chenoweth says. How many people? According to her data, "no campaigns failed once they'd achieved the active and sustained participation of just 3.5 percent of the population—and lots of them succeeded with far less than that." Further, she notes, "Every single campaign that did surpass that 3.5 percent threshold was a nonviolent one. In fact, campaigns that relied solely on nonviolent methods were on average four times larger than the average violent campaign. And they were often much more representative in terms of gender, age, race, political party, class, and urban-rural distinctions."[112]

How does this nonviolent strategy translate into political change? If your movement is based on violence, you are necessarily going to be limiting yourself to mostly young, strong, violence-prone males who have a propensity for boozing and brawling, whereas, Chenoweth explains, "Civil resistance allows people of all different levels of physical ability to participate—including the elderly, people with disabilities, women, children, and virtually anyone else who wants to." It's a faster track to the magic 3.5 percent number when you're more inclusive and participation barriers are low. Plus, you don't need expensive guns and weapons systems. Civil disobedience often takes the form of strikes, boycotts, stay-at-home demonstrations, banging on pots and pans and other noise generators, and—like a scene out of the 1951 film *The Day the Earth Stood Still*—shutting off the electricity at a designated time of the day. A diffuse group of isolated individuals scattered about a city employing such measures is difficult for oppressive regimes to stop. Plus, by including the mainstream instead of the marginalized in your movement, your shock troops are more likely to know people on the other side. In the case of Serbia and its dictator Slobodan Milosevic, Chenoweth notes that "once it became clear that hundreds of thousands of Serbs were descending on Belgrade to demand that Milosevic leave office, policemen ignored the order to shoot on demonstrators. When asked why he did so, one of them said: 'I knew my kids were in the crowd.'"[113]

There is one more benefit to nonviolent resistance: what you're left with afterward. Nonviolent campaigns of change are far more likely to result in democratic institutions than are violent insurgencies, and they are 15 percent less likely to relapse into civil war. "The data are clear," Chenoweth

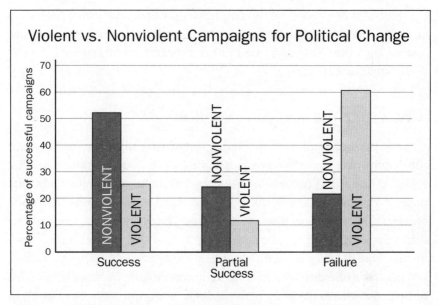

Figure 2-3. Nonviolent Campaigns for Political Change

Success rate of campaigns for political change since the 1940s comparing violent and nonviolent methods reveals that violence is a failed strategy and nonviolence is the method of choice.[115]

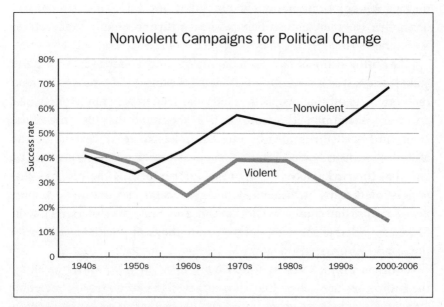

Figure 2-4. Progress in Nonviolent Campaigns for Political Change

The percentage of successful campaigns for political change comparing violent and nonviolent methods.[116]

concludes: "When people rely on civil resistance, their size grows. And when large numbers of people withdraw their cooperation from an oppressive system, the odds are ever in their favor."[114] *Figure 2-3* and *figure 2-4* show these remarkable trends.

WAR, VIOLENCE, AND MORAL PROGRESS

As with the many myths surrounding terrorism, there are myths about the origins and causes of war that have clouded our thinking, starting with the myth that humans are by nature relatively nonviolent and that prestate peoples were peaceful and lived in relative harmony with each other and their environment. A convergence of evidence from multiple lines of scientific inquiry, however, tells us that this view of human pre-history is at the very least misleading and very probably wrong. The reason has less to do with the character of human nature being either pacifist or bellicose and more to do with the logic of how organisms respond to free riders, bullies, challenges, and threats to our survival and flourishing. That is, the data I review below are meant less to settle the debate that has long raged about what humans were like in a state of nature as either noble savages or the war of all against all; instead, I am building on the logic of our moral emotions and how they direct us to respond one way or another to other sentient beings who, in turn, respond to our actions accordingly.

The ecologist Bobbi Low used data from the Standard Cross-Cultural Sample to analyze 186 Hunting-Fishing-Gathering (HFG) societies around the world to show that people in traditional societies are not living in balanced eco-harmony with nature. In fact, she found that their use of the environment is constrained by limited ecological resources and not by attitudes (such as sacred prohibitions against harming Mother Earth), and that their relatively low environmental impact is the result of low population density, inefficient technology, and the lack of profitable markets, not from any conscious effort at conservation. Low also found that in 32 percent of HFG societies, not only were they not practicing conservation, environmental degradation was severe.[117]

In his book *Sick Societies: Challenging the Myth of Primitive Harmony*, the anthropologist Robert Edgerton surveyed the ethnographic record of traditional societies that have not been exposed to Western civilization and found clear evidence of drug addiction, abuse of women and children, bodily mutilation, economic exploitation of the group by political leaders, suicide, and mental illness.[118]

In *War Before Civilization: The Myth of the Peaceful Savage*, the archaeologist Lawrence Keeley tested the hypothesis that prehistoric warfare was rare, harmless, and little more than ritualized sport. Surveying primitive and civilized societies, he found that prehistoric war was—relative to population densities and fighting technologies—at least as frequent (as measured in years at war versus peace), as deadly (as measured by percentage of conflict deaths), and as ruthless (as measured by killing and maiming of noncombatant women and children) as modern warfare. One prehistoric mass grave in South Dakota, for example, yielded the remains of five hundred scalped and mutilated men, women, and children, and this happened half a century before Europeans arrived on the continent. Overall, says Keeley, "Fossil evidence of human warfare dates back at least 200,000 years, and it is estimated that as many as 20–30% of ancestral men died from intergroup violence."[119]

Archaeologist Steven LeBlanc's *Constant Battles: The Myth of the Peaceful, Noble Savage* documented his title's description of the state of nature with examples such as a ten-thousand-year-old grave site along the Nile River containing "the remains of fifty-nine people, at least twenty-four of whom showed direct evidence of violent death, including stone points from arrows or spears within the body cavity, and many contained several points. There were six multiple burials, and almost all those individuals had points in them, indicating that the people in each mass grave were killed in a single event and then buried together." At another site, in Utah, ninety-seven bodies were unearthed, revealing that "six had stone spear heads in them . . . several breast bones shot through with arrows and many broken heads and arms. . . . Individuals of all ages and both sexes were killed, and individuals were shot with atlatl darts, stabbed, and bludgeoned, suggesting that fighting was at close quarters." Another half dozen archaeological sites in Mexico, Fiji, Spain, and other parts of Europe showed human bones broken open lengthwise and cooked, and a pre-Columbian Native American coprolite was laced with the human muscle protein myoglobin, all of which adds up to the fact that humans once ate one another.[120] LeBlanc identified ten societies that did not show "constant battles" between groups, but he noted that "some of these same 'peaceful' societies have extremely high homicide rates. Among the Copper Eskimo and the New Guinea Gebusi, for example, a third of all adult deaths were from homicide." So, he asked rhetorically, "Which killing is considered a homicide and which killing is an act of warfare? Such questions and answers become somewhat fuzzy. So some of this so-called peacefulness is more dependent on the definition of homicide and warfare than on reality."[121]

A visual reminder of the reality of what life was often like for many of our ancestors may be seen in *figure 2-5*, featuring two skulls of people who died violent deaths some 8,500 to 10,700 years ago in northern Europe, putting a face on the numbers of our violent past.

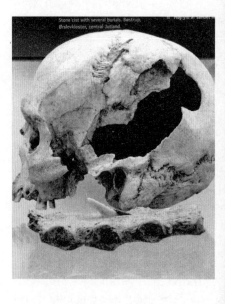

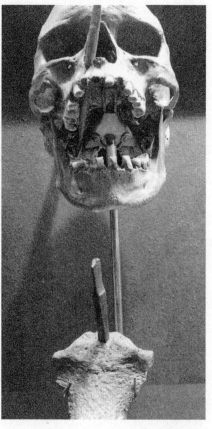

Figure 2-5. The Face of Violent Death

Two people who died violently are part of an exhibition of what life was like in northern Europe from 6,500 to 8,700 BCE, from the National Museum of Denmark collection in Copenhagen. A blow to the head that shattered the skull and left a gaping hole and an arrow point in the sternum ensured that the man on the left succumbed, and a dagger in the chest and an arrow through the face terminated the life of the man on the right.[122] Although undoubtedly traditional societies varied greatly in their rates of violence—as do modern societies—in general, if you were a man living at that time, there was about a one in four chance that you would die violently.

Like historic societies, prehistoric societies varied considerably in their rates of violence, but statistically speaking, the chances of dying violently in a prestate society versus a state society is as unmistakable as it is terrifying. Compared to modern humans, prehistoric people were far more murderous in terms of the percentage of the population slaughtered in combat and offed by one another, as Steven Pinker explained to me in an interview in which I asked him to summarize the massive datasets he compiled for his book *The Better Angels of Our Nature*. "Violent deaths of all kinds have declined, from around 500 per 100,000 people per year in pre-state societies to around 50 in the Middle Ages, to around 6 to 8 today worldwide, and fewer than 1 in most of Europe." What about gun-toting Americans and our inordinate rate of homicides (currently about 5 per 100,000 per year)? In 2005, Pinker computes, a grand total of 0.008, or eight tenths of 1 percent of all Americans died of domestic homicides and in two foreign wars combined. In the world as a whole that year, in fact, the rate of violence from war, terrorism, genocide, and killings by warlords and militias was 0.0003 of the total population of 6.5 billion, or three hundredths of 1 percent.[123]

What about wars? Surely more people have died due to state-sponsored conflicts than in prestate battles? This isn't the case if you compute it as a percentage of the population killed, says Pinker: "On average, non-state societies kill around 15 percent of their people in wars, whereas today's states kill a few hundredths of a percent." Pinker calculates that even in the murderous twentieth century, about 40 million people died in direct battle deaths (including civilians caught in the crossfire) out of the approximately six billion people who lived, or 0.7 percent. Even if we include war-related deaths of citizens from disease, famine, and genocide, that brings the death toll up to 180 million deaths, or about 3 percent. But what about the two world wars and the Holocaust, Stalin's gulags, and Mao's purges? "A very pessimistic estimate of the human damage from all wars, genocides, and war-induced and man-made famines in the 20th century would be 60 per 100,000 per year—still an order of magnitude less than tribal warfare. And of course those numbers are dominated by 1914–1950 in Europe and 1920–1980 in East Asia, both of which have since calmed down."[124] In a moment of comic relief in reviewing such grave matters, when Pinker appeared on *The Colbert Report*, the comedian Stephen Colbert wondered how he could say that violence is declining when the twentieth century was the most violent in human history. Pinker responded with a wry smile, "a century lasts for a hundred years, and the last 55 years of the twentieth century had unusually low rates of death in warfare so after that spike of wars between 1914 and 1918 and '39 to '45, the rate of killing in war went down."[125] The second half of the twentieth century, continuing into the twenty-first century—the long peace, as it is called—is the real mystery to be explained.

Figure 2-6 presents the aggregated data compiled by Pinker from multiple sources for the percentage of deaths in warfare for prehistoric people vs. modern hunter-gatherers vs. modern hunter-horticulturalists and other tribal groups vs. modern states. The difference is visually striking and unambiguous in its conclusion because so many datasets all point in the same direction. While one might be skeptical of how the numbers were computed for any one dataset, it is highly unlikely that all of these studies could be wrong so consistently.

The reason for computing the rates of death as a percentage of the population or as a number per hundred thousand—instead of the raw numbers of total killed—is threefold: (1) this is customary among scholars of war and violence, (2) raw numbers will increase over time as a result of larger populations, larger armed forces, and improved technologies for killing, thereby distorting what we really want to know, which is . . . (3) determining the chances that a given *individual* (you or I) will die violently. This takes us back to the first principle of moral consideration of this book: *the survival and flourishing of individual sentient beings.* The violent death of an individual is what I'm focusing on here because it is the individual who suffers the ultimate loss—not a group, race, nation, or statistical collective. Although big Leviathans can put up big numbers in both armies and war deaths, if you had to choose a time in history in which it was safest for you personally to survive and thrive, by these criteria alone there is no time like the present.

Given the implications that these data have for one's view of human nature and the causes and future of war and violence, the scientific debate has turned ideological, even tribal. On one side are the "peace and harmony mafia"[127]—those who hold that war is a recent learned cultural phenomenon and that humans are by nature peaceful, a view they defend with vigor and even ferocity. On the other side are what the peace and harmony mafia pejoratively call the "Harvard Hawks" (an invidious smear meant to imply that they favor war over peace)[128]—Richard Wrangham, Steven LeBlanc, Edward O. Wilson, and Steven Pinker—who contend that war is the outcome of the logic of evolutionary dynamics. The "evolution wars" (as the "anthropologists of peace" call it) have been going on since the 1970s, which I have documented in a previous book.[129]

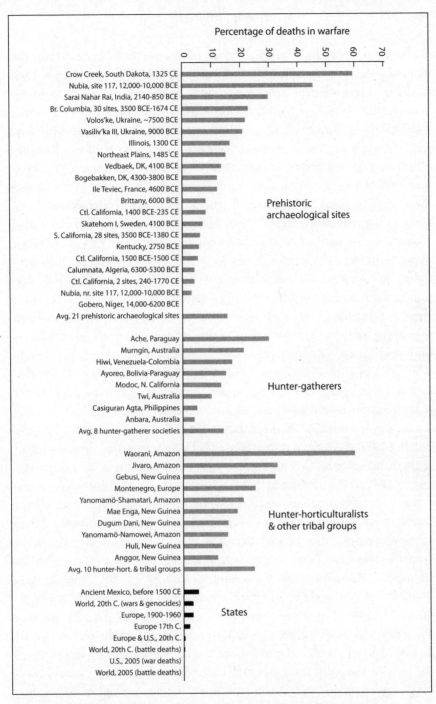

Figure 2-6. The Decline of War Deaths from Prehistoric Bands to Modern States

The percentage of deaths in warfare for prehistoric people vs. modern hunter-gatherers vs. hunter-horticulturalists and other tribal groups vs. modern states. This long-term precipitous fall in violence occurred even while the quantity and efficiency of deadly weapons evolved into the killing tools they are today.[126]

In the latest round of debates over the nature of human nature, the extensive datasets compiled by others and coalesced by Pinker have been challenged in both scholarly and popular publications. In a 2013 edited volume titled *War, Peace, and Human Nature*, Brian Ferguson claims that the datasets in "Pinker's List" greatly exaggerate prehistoric war mortality; unfortunately, he then confuses *frequency* (rates) with *tendency* (inevitability) when he challenges "the idea that deadly intergroup violence has been common enough in our species' evolutionary history to act as a selection force shaping human psychological tendencies toward either external violence or internal cooperation."[130] But Pinker is not arguing that the rate of violence in the past acts as a selective force in human evolution; quite the opposite, in fact—the logic of game theoretic interactions (like the Prisoner's Dilemma matrices I outlined in chapter 1) between players means that a certain amount of defection (in games) or violence (in life) is inevitable when there is no outside governing body (an association in sports, a government in society) to tilt the matrix toward more cooperative and peaceful choices via both rewards and punishments. Over the millennia we have learned how to adjust the conditions of the matrix of life to make people interact in less violent and more peaceable and cooperative ways, and it is those adjustments that have led to a decline of violence and to moral progress.[131]

In a 2013 paper in the prestigious journal *Science*, Douglas Fry and Patrik Söderberg disputed the theory that war is prevalent in mobile foraging band societies (MFBS) by claiming that in a sample of 148 episodes from 21 MFBS "more than half of the lethal aggression events were perpetrated by lone individuals, and almost two-thirds resulted from accidents, interfamilial disputes, within-group executions, or interpersonal motives such as competition over a particular woman." From this they conclude, "most incidents of lethal aggression among MFBS may be classified as homicides, a few others as feuds, and a minority as war."[132] Well, that's comforting! So the skulls in *figure 2-5* were of guys who either got bludgeoned to death by friends within their tribes rather than by enemies from other tribes, or they accidentally, savagely shot themselves in the face or brutally stabbed themselves in the chest with arrows and daggers. Addressing the point about what the data actually indicate, Fry and Söderberg target Samuel Bowles (whose data are included in *figure 2-6*), accusing him of claiming that "war is prevalent in MFBS" and that "war has been pervasive during human evolution." Bowles responded, saying, "I did not make the two claims that the authors attribute to me for the simple reason that I was answering a different question." The question Bowles was trying to answer is, "Did warfare among ancestral hunter-gatherer groups affect the evolution of human social behaviors?" To that end, Bowles explains, "I needed data on the fraction of all deaths that were due to intergroup conflict, not the evidence that Fry and Söderberg present, namely data on whether 'war' is 'prevalent' or 'pervasive' or the major source of violent deaths."[133]

Here again we see how *categorical* and *binary thinking* cloud the issue. Forcing a continuum of violence into a category of "prevalent" or "pervasive" misses the point of what we're interested in knowing here: whatever the rate of violence in the past—by whatever the means and whatever the cause—was it enough to affect human evolution? If you insist that the rate must be high enough to be called "prevalent" or "pervasive," then you have to operationally define these terms with a quantity, including the term "war" that by today's definition has no meaning for the type of intergroup conflicts that happened during the Late Pleistocene epoch in which our species came of age. As Bowles explains, "In my models of the evolution of human behaviour, the appropriate usage of the term [war] is 'events in which coalitions of members of a group seek to inflict bodily harm on one or more members of another group'; and I have included 'ambushes, revenge murders and other kinds of hostilities' analogizing human intergroup conflict during the Late Pleistocene to 'boundary conflicts among chimpanzees' rather than 'pitched battles of modern warfare.'"[134]

Modern urban gangs, for example, engage in violent intergroup conflict that cumulatively can rack up significant body counts. Think of the ongoing Mexican drug wars among rival cartels in which more than one hundred thousand people have been killed and more than a million displaced since 2006.[135] Yet scholars would not classify these events as "wars" because the motives are more commonly related to honor, revenge, feuding, or turf conflicts. But as Bowles points out, urban gangs meet the criteria for what constitutes an MFBS: small group size, fluctuating group membership, multilocal residence, and a type of egalitarianism with no authority to order others to fight. Bowles carefully examined the data that Fry and Söderberg published in their paper and noted that "motives such as 'revenge' or killing 'over a particular man' or the fact that a killing was 'interpersonal' mean that the event did not fall under the heading of 'war.'" And yet, Bowles concludes, "From the standpoint of evolutionary biology these aspects of the killing are irrelevant: what matters for the dynamics of population composition is that members of a group (more than one) cooperated in killing a member of another group, for any reason whatever."[136] To that end, the data in *figure 2-6* showing the dramatic decline in violence over time remain indisputable, and that is real moral progress however one defines any particular kind of violence.

It's a point well made in *The Arc of War* by the political scientists Jack Levy and William Thompson, who begin by adopting a continuum rather than a categorical style of reasoning: "War is a persistent feature of world politics, but it is not a constant. It varies over time and space in frequency, duration, severity, causes, consequences, and other dimensions. War is a social practice adopted to achieve specific purposes, but those practices vary with changing political, economic, and social environments and with the goals and constraints induced by those environments."[137] When nuanced in this continuous rather than categorical manner, we can see both how and when rates of warfare change. By defining war as "sustained, coordinated violence between political organizations,"[138] however, Levy and Thompson have defined away prehistoric group conflicts that don't at all resemble political organizations of today. As such, "war" cannot even begin until there are political organizations of a substantive size, which necessarily means that what we think of as war, by definition, was impossible before civilization began.

Nevertheless, Levy and Thompson acknowledge that the rudimentary foundations for war as they define it were already there in our earliest ancestors—even suggesting that "border skirmishes" with Neanderthals in northern Europe may account for the latter's extinction some thirty-five thousand years ago—including "the observation that hunting and homicide skills made suitable weaponry, tactics, and rudimentary military organization available" and that "group segmentation helped define group identities and enemies, thereby also facilitating the potential for organizing politically and militarily."[139] Thus they endorse "an early if infrequent start for warfare among hunter-gatherers," which then increased over time in lethality with improved weapons and increased population sizes, and this continued throughout the history of civilization as states increased in size until nations fought nations, leading to an increase in the total number of deaths but a decrease in the total number of conflicts.

Working in prehistoric South America, archaeologists Elizabeth Arkush and Charles Stanish use a convergence-of-evidence approach from many different sources to compile extensive evidence showing indisputably that "Late Andean prehistory was profoundly shaped by warfare." And by "warfare" they don't just mean "ritual battle" in which "real" violence was rare. Archaeological remains of defensive walls and fortifications match the records of Spanish conquistadors who reported, for example, that they "encountered huge Inca armies supported by a superb logistical framework of roads, supply depots, secondary centers, and forts." Spanish chroniclers and Incan oral histories also make clear that "military might was a cornerstone of imperial power. The empire had emerged from military victories over some groups, the peaceful submission of others persuaded by the threat of military reprisals, and the violent suppression of several rebellions. Inca histories also describe a period of frequent warfare before the empire arose in which local war leaders battled each other for plunder or political dominance."[140]

Archaeologist George Milner includes a photograph of an antler point embedded in a lumbar vertebra from prehistoric Kentucky as a visual example of the numerous datapoints indicating that, if anything, "surely underestimate warfare casualties." But even if the estimates are not exaggerated, "even low frequencies of deaths from intergroup conflicts added just one more element of uncertainty to lives already full of uncertainties. Warfare is likely to have had a broader impact than the immediate loss of life alone. Sudden and unexpected deaths of people who played critical roles in the survival of small groups increased the risk of death for remaining household and community members."[141] Time travel fantasists and postmodernists who bewail modern society and long for simpler times would do themselves a favor by examining this evidence a little more closely.

In his 1996 book *Demonic Males*, Richard Wrangham traced the origins of patriarchy and violence all the way back to our hominid origins millions of years before the Neolithic Revolution.[142] In 2012 and 2013 Wrangham published two papers with his graduate student Luke Glowacki that painted a much more nuanced portrait of hunter-gatherers (HG) as highly risk-averse when it comes to violence and war. Most rational agents don't *want* to get maimed or killed, so they only risk going to war when, Wrangham and Glowacki write, "cultural systems of reward, punishment, and coercion rather than evolved adaptations to greater risk-taking" are in place. These cultural systems include the "teaching of specific war skills, apprenticeship, games and contests, pain endurance tests, other endurance tests, and the use of legends and stories."[143] Cultural systems also shower would-be warriors with promises of honor and glory—for themselves and their families—and since death may preclude fallen warriors from cashing in the promises, seeing your late comrade's family so rewarded acts as a social signal to help individuals overcome their natural risk aversion to dangerous and deadly conflict. Wrangham and Glowacki call this the "cultural rewards war-risk hypothesis" and predict that the greater the risk—the higher the probability of being maimed or killed in battle—the greater the number of benefits that accrue to participating individuals. An assessment of the ethnographic literature on simple warfare among small-scale societies found just that.[144]

Far from portraying humans as innately violent and warlike, Wrangham admits that "whether humans have evolved specific psychological adaptations for war is uncertain."[145] Instead, a review of all the available literature on group conflicts in both humans and chimpanzees shows that, like their chimp cousins, hunter-gatherers follow a game theoretic strategy of an imbalance of power: *if we outnumber them, invade; if they outnumber us, evade.* As Lawrence Keeley concluded his extensive studies on war and conflict in HG bands: "The most elementary form of warfare is a raid (or type of raid) in which a small group of men endeavour to enter enemy territory undetected in order to ambush and kill an unsuspecting isolated individual, and to then withdraw rapidly without suffering any casualties."[146]

Apocalypto, Mel Gibson's film about the collapse of the Mayan civilization, portrays a visually striking example of a typical prehistoric raid in which one tribe of Mesoamericans launches an early morning attack on the hero's village while they are asleep, setting fire to their huts and delivering devastating blows before they can organize a concerted counter-strike. By the time the defending warriors come out of their slumberous state it is all over but the shouting. The writers of the script did their homework and offered a more realistic visage of what life was like before civilization than, say, Kevin Costner's *Dances with Wolves*. According to archaeologist Michael Coe, "Maya civilization in the Central Area reached its full glory in the early eighth century, but it must have contained the seeds of its own destruction, for in the century and a half that followed, all its magnificent cities had fallen into decline and ultimately suffered abandonment. This was surely one of the most profound social and demographic catastrophes of all human history."[147] And it unfolded long before the arrival of European guns, germs, and steel at the end of the fifteenth century, which is why Gibson opened the film with a quote from the historian Will Durant: "A great civilization is not conquered from without until it has destroyed itself from within."[148]

· · ·

The military historian John Keegan once reflected, "War, it seems to me, after a lifetime of reading about the subject, mingling with men of war, visiting the sites of war and observing its effects, may well be ceasing to commend itself to human beings as a desirable, or productive, let alone rational, means of reconciling their discontents."[149] Joshua Goldstein's 2011 book *Winning the War on War* compiled massive datasets to support that conclusion when he wrote, "We have avoided nuclear wars, left behind world war, nearly extinguished interstate war, and reduced civil wars to fewer countries with fewer casualties."[150] The political scientist Richard Ned Lebow drew similar conclusions in his 2010 book *Why Nations Fight*, delineating four motives behind wars of the past 350 years, all of which are in decline: fear, interest, status/standing, and revenge.[151] According to Lebow, none of these motives is any longer effectively served by going to war, and more and more nations' leaders are finding ways to avoid conflict when these motives arise, especially the one that he identifies as the most common motive, status and standing. "I contend that standing has been the most common cause of war historically and that war has declined in large part because it no longer confers standing." Lebow makes the compelling point that the mechanistic and unheroic destructiveness of the two world wars effectively ended the notion of bravery and heroism that wars allegedly bestowed on their survivors, which gave individuals and states higher status and standing among their peers. Referencing the First World War, for example, Lebow writes, "It is worth considering the counterfactual that opposition to war would not have been nearly so pronounced if the war had been more like its Napoleonic predecessor, a war of maneuver that encouraged individual, recognizable acts of bravery that might have more than minor tactical consequences. War deprived of its heroic and romantic associations, and considered instead an irrational source of slaughter, destruction and suffering, was no longer able to win honor for its combatants or standing for the states that sent them to their deaths."[152]

A comprehensive 2014 report by a team of social scientists at Simon Fraser University tested the "declinist" hypothesis by reviewing all of the available data, concluding, "There are now compelling reasons for believing that the historical decline in violence is both real and remarkably large—and also that the future may well be less violent than the past." About that future, they note that "there are ample grounds for cautious optimism but absolutely none for complacency."[153]

The underlying goal in the study of the nature and causes of violence and war—whatever the blend of biology, culture, and circumstance turns out to be—is to attenuate them. (See, for example, the group Vision of

Humanity, which tracks trends toward or away from peace and ranks coun-tries in a peace index.[154]) Because the stakes are so high, emotions in those who conduct such studies run deep. Samuel Bowles said it best in a casual remark to me: "It seems to be a highly ideologically charged debate, which is unfortunate, because finding that war was frequent in the past, or that out-group hostility might have a genetic basis says something about our legacy, not our destiny."[155]

Science is—and ought to be—concerned with understanding both our legacy and our destiny, for as Cicero noted in the epigram to this chapter, warnings of evil are only justified if there is a way of escape. In the next chapter we will consider how science and reason are not only two of the main drivers of moral progress, but they also show us the way of escape from the traps we have set for ourselves.

3

Why Science and Reason Are the Drivers of Moral Progress

Science, the partisan of no country, but the beneficent patroness of all, has liberally opened a temple where all may meet. Her influence on the mind, like the sun on the chilled earth, has long been preparing it for higher cultivation and further improvement. The philosopher of one country sees not an enemy in the philosophy of another: he takes his seat in the temple of science, and asks not who sits beside him.

—Thomas Paine, 1778[1]

In the 1970s, the NBC comedy series *Saturday Night Live* featured skits by the comedian and author Steve Martin, whose recurring roles as Theodoric of York, Medieval Barber and Medieval Judge, tapped into the tacit knowledge of moral progress since the Middle Ages—assumed to be held by even late-night viewers. Martin played a barber surgeon who employed bloodletting and other barbaric practices to cure any and all illnesses; as he explained to the mother of one of his patients, "You know, medicine is not an exact science, but we are learning all the time. Why, just fifty years ago they thought a disease like your daughter's was caused by demonic possession or witchcraft. But nowadays we know that Isabelle is suffering from an imbalance of bodily humors, perhaps caused by a toad or a small dwarf living in her stomach." The mother didn't buy it and blasted him for his still-barbaric bloodletting ways until Theodoric had a moment of scientific enlightenment . . . almost:

Wait a minute. Perhaps she's right. Perhaps I've been wrong to blindly follow the medical traditions and superstitions of past centuries. Maybe

we barbers should test these assumptions analytically, through experimentation and a "scientific method." Maybe this scientific method could
be extended to other fields of learning: the natural sciences, art, architecture, navigation. Perhaps I could lead the way to a new age, an age of
rebirth, a Renaissance. . . . Nah!

As the Medieval Judge, Theodoric of York had a similar near-awakening
after passing judgment on an accused witch based on the classic trial by
ordeal—specifically, in this instance, the ordeal by water. This particular
test involved tying up the accused and then dunking her into a body of
water. If the accused sank (and drowned) that meant she was innocent;
but if she managed to float, she was obviously guilty—either because the
pure element of water naturally expels evil or because, in the words of an
observer at the time, "the witch, having made a compact with the devil,
hath renounced her baptism, hence the antipathy between her and water,"[2]
or because only by employing her demonic powers could she overcome
the weight of the stones with which some of the hapless accused were
unfairly burdened. In the case of the accused in Theodoric's court, the
woman proves to be innocent, and thus sinks to her death. The mother,
naturally, is furious, proclaiming, "You call this justice!? An innocent girl
dead?" Theodoric ponders the mother's protestations, thinking:

> Wait a minute—perhaps she's right. Maybe the King doesn't have a
> monopoly on the truth. Maybe he should be judged by his peers. Oh! A
> jury! A jury of his peers . . . everyone should be tried by a jury of their
> peers and be equal before the law. And perhaps, every person should be
> free from cruel and unusual punishment. . . . Nah![3]

THE WITCH THEORY OF CAUSALITY

Compressed into these comedic vignettes are centuries of intellectual
advancement, from the medieval worldview of magic and superstition to
the modern age of reason and science. It is evident that most of what we
think of as our medieval ancestors' barbaric practices were based on mistaken beliefs about how the laws of nature actually operate. If you—and
everyone around you, including ecclesiastical and political authorities—
truly believe that witches cause disease, crop failures, sickness, catastrophes, and accidents, then it is not only a rational act to burn witches, it is
also a moral duty. This is what Voltaire meant when he wrote that people

who believe absurdities are more likely to commit atrocities. An even more pertinent translation of his famous quote is relevant here: "Truly, whoever is able to make you absurd is able to make you unjust."[4]

Consider a popular thought experiment and how you would respond in the following scenario: You are standing next to a fork in a railroad line and a switch to divert a trolley car that is about to kill five workers on the track unless you throw the switch and divert the trolley down a side track, where it will kill one worker. Would you throw the switch to kill one but save five? Most people say that they would.[5] We should not be surprised, then, that our medieval ancestors performed the same kind of moral calculation in the case of witches. Medieval witch burners torched women primarily out of a utilitarian calculus—better to kill the few to save the many. Other motives were present as well, of course, including scapegoating, the settling of personal scores, revenge against enemies, property confiscation, the elimination of marginalized and powerless people, and misogyny and gender politics.[6] But these were secondary incentives grafted on to a system already in place that was based on a faulty understanding of causality.

The primary difference between these premodern people and us is, in a word, science. Frankly, they often had not even the *slightest* clue what they were doing, operating as they were in an information vacuum, and they had no systematic method to determine the correct course of action either. The witch theory of causality, and how it was debunked through science, encapsulates the larger trend in the improvement of humanity through the centuries by the gradual replacement of religious supernaturalism with scientific naturalism. In his sweeping survey of traditional societies, *The World Until Yesterday*, the evolutionary biologist and geographer Jared Diamond explains how our prescientific ancestors dealt with the problem of understanding causality:

> An original function of religion was explanation. Pre-scientific traditional peoples offer explanations for everything they encounter, of course without the prophetic ability to distinguish between those explanations that scientists today consider natural and scientific, and those others that scientists now consider supernatural and religious. To traditional peoples, they are all explanations, and those explanations that subsequently became viewed as religious aren't something separate. For instance, the New Guinea societies in which I have lived offer many explanations for bird behavior that modern ornithologists consider perceptive and still

accurate (e.g., the multiple functions of bird calls), along with other explanations that ornithologists no longer accept and now dismiss as supernatural (e.g., that songs of certain bird species are voices of former people who became transformed into birds).[7]

In my book *Why People Believe Weird Things* I review the extensive scientific literature concerning the role of superstition in pre- or nonscientific societies. The Trobrianders, for example, who live on islands near Papua New Guinea, employed weather magic, healing magic, garden magic, dance magic, love magic, sailing and canoe magic, and especially fishing magic. In the calm waters of the inner lagoon where a catch is much more likely and safer, there are few superstitious rituals. By contrast, in preparation for the uncertain waters of deep sea fishing the Trobrianders perform many magical rituals, including whispering and murmuring magical formulas. The anthropologist Gunter Senft has cataloged many such verbal utterances, such as one called Yoya's fish magic, which he recorded in 1989 and that involved the repetition of certain phrases:[8]

Totwaieee	Tokwai
kubusi kuma kulova	come down, come, come inside
o bwalita bavaga	to the sea I will return
kubusi kuma kulova	come down, come, come inside
o bwalita a'ulova	at sea I put a spell in [it]

A seven-second pause is then followed by another series of related phrases, all toward an end of "ordering and commanding their addressees to do or change something, or by foretelling changes, processes, and developments that are necessary for reaching these aims," Senft writes. The anthropologist who published the first and definitive ethnography on the Trobriand Islanders, Bronislaw Malinowski, concluded that his charges were misinformed, not ignorant. Magic, he said, is "to be expected and generally to be found whenever man comes to an unbridgeable gap, a hiatus in his knowledge or in his powers of practical control, and yet has to continue in his pursuit."[9] The solution to magical thinking, then, is to close those unbridgeable gaps with scientific thinking.

Other anthropologists have made similar discoveries about their subjects, such as E. E. Evans-Pritchard in his classic study *Witchcraft, Oracles and Magic Among the Azande*, a traditional society in the southern Sudan in Africa. After a survey of the many bizarre beliefs about witches held by the Azande, Evans-Pritchard explains the psychology behind witchcraft beliefs, starting with the fact that "Witches, as the Azande conceive them, clearly cannot exist. Nonetheless, the concept of witchcraft provides them with a natural philosophy by which the relations between men and unfortunate events are explained and a ready and stereotype means of reacting to such events." Here we see in a premodern society what happens when magical thinking goes unchecked by critical thinking:

> Witchcraft is ubiquitous. It plays its part in every activity of Zande life; in agricultural, fishing, and hunting pursuits; in domestic life of homesteads as well as in communal life of district and court; it is an important theme of mental life in which it forms the background of a vast panorama of oracles and magic; . . . there is no niche or corner of Zande culture into which it does not twist itself. If blight seizes the ground-nut crop it is witchcraft; if the bush is vainly scoured for game it is witchcraft; if women laboriously bale water out of a pool and are rewarded by but a few small fish it is witchcraft; if a wife is sulky and unresponsive to her husband it is witchcraft; if a prince is cold and distant with his subject it is witchcraft; if, in fact, any failure or misfortune falls upon anyone at any time and in relation to any of the manifold activities of his life it may be due to witchcraft.[10]

Nowadays, science has all of these problems covered. We know that crops can fail due to disease, which we study through the science of agronomy and the etiology of disease; or they fail due to insects that we can investigate through the science of entomology and further control through chemistry; or they fail due to inclement weather that we can understand through the science of meteorology. Ecologists and biologists can tell us why populations of fish rise and fall and what we can do to prevent a region being fished out or decimated by disease or climate change. Psychologists specializing in marital counseling can explain why a wife might not be as responsive as her husband may wish (and vice versa); and though there may not be a big call for this sort of thing these days, psychologists who study personality and temperament could explain why some princes are cold and distant while others are warm and connected to their subjects. Statisticians and risk analysts can assess the rates of failure and misfortune that might befall anyone at any time in relation to any number of activities of life, well captured in the ne plus ultra of pop-culture, bumper-sticker philosophy—"Shit Happens."

Tellingly, Evans-Pritchard notes that the Zande do not attribute *everything* that happens to witchcraft—only those things for which they do not have a plausible-sounding causal explanation. "In Zandeland sometimes an old granary collapses. There is nothing remarkable in this. Every Zande knows that termites eat the supports in course of time and that even the hardest woods decay after years of service." But when a group of people are sitting inside the granary when it collapses and they are injured, the Zande wonder, in Evans-Pritchard's description, "why should these particular people have been sitting under this particular granary at the particular moment when it collapsed? That it should collapse is easily intelligible, but why should it have collapsed at the particular moment when these particular people were sitting beneath it?" Evans-Pritchard then draws the distinction between prescientific and scientific worldviews:

> To our minds the only relationship between these two independently caused facts is their coincidence in time and space. We have no explanation of why the two chains of causation intersected at a certain time and in a certain place, for there is no interdependence between them. Zande philosophy can supply the missing link. The Zande knows that the supports were undermined by termites and that people were sitting beneath the granary in order to escape the heat and glare of the sun. But he knows besides why these two events occurred at a precisely similar moment in time and space. It was due to the action of witchcraft. If there had been no witchcraft people would have been sitting under the granary and it would not have fallen on them, or it would have collapsed but the people would not have been sheltering under it at the time. Witchcraft explains the coincidence of these two happenings.[11]

A witch is a causal theory of explanation. And it's fair to say that if your causal theory to explain why bad things happen is that your neighbor flies around on a broom and cavorts with the devil at night, inflicting people, crops, and cattle with disease, preventing cows from giving milk, beer from fermenting, and butter from churning—and that the proper way to cure the problem is to burn her at the stake—then either you are insane or you lived in Europe six centuries ago, and you even had biblical support, specifically Exodus 22:18: "Thou shalt not suffer a witch to live." Witches were believed to be able to inflict harm on others just by staring at them—giving them "the evil eye"—thereby releasing an invisible but potent emanation, especially if she was menstruating.

Figure 3-1.
Four women being interrogated for witchcraft in the attempted murder of King James I.

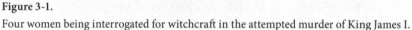

Figure 3-1 shows four women being interrogated for witchcraft in the attempted murder of King James I. One of them, a woman named Agnes Sampson, confessed under torture that she had danced counterclockwise, which was believed to lead to disaster. Witch-hunters developed techniques to determine the guilt or innocence of accused witches, including searching their bodies for telltale marks of their cavorting with the devil.

Again, the point is not that our medieval ancestors were irrational in their magical thinking. To the contrary, they fully believed that what they were doing in employing various magical incantations, spells, and superstitions of various sorts would have the desired effect. As the medieval historian Richard Kieckhefer notes, the people of medieval Europe thought of magic as rational for two reasons: "first of all, that it could actually work (that its efficacy was shown by evidence recognized within the culture as authentic) and, secondly, that its workings were governed by principles (of theology or of physics) that could be coherently articulated."[12] It was the Roman Catholic Church that first articulated the witch theory of causality in medieval Europe with the Papal Bull of Innocent VIII in 1484, *Summis Desiderantes Affectibus* (*Desiring with Supreme Ardor*), followed two years later with the Catholic clergyman Heinrich Kramer's *Malleus*

Maleficarum (*Hammer of the Witch*). The latter was a how-to manual on finding and prosecuting witches, who, it alleged, were able to copulate with the devil, steal men's penises, wreck ships, ruin crops, eat babies, turn men into frogs, shed no tears, cast no shadow in the sun, have hair that could not be cut, and pretty much anything considered to be "devilish" and "wicked."

The manual instructed investigators on how to look for the witch's mark—a spot or excrescence on her body that supposedly did not bleed when pricked (giving rise, as one can imagine, to inappropriate touching by the mostly male investigators of the mostly female suspects). Finding witches not only explained evil, it also was tangible evidence of God's existence. As the sixteenth-century Cambridge theologian Roger Hutchinson argued, in a polished bit of circular reasoning, "If there be a God, as we most steadfastly must believe, verily there is a Devil also; and if there be a Devil, there is no surer argument, no stronger proof, no plainer evidence, that there is a God."[13] And, conversely, as noted in a seventeenth-century witch trial, "Atheists abound in these days and witchcraft is called into question. If neither possession nor witchcraft [exists], why should we think that there are devils? If no devils, no God."[14]

Figure 3-2 is a woodcut illustration from the title page of a pamphlet titled *Witches Apprehended, Examined and Executed*, published in 1613. It shows the classic "trial by ordeal" or "ordeal by water." Pictured is a woman named Mary Sutton being put to the test in 1612.

Today, the witch theory of causality has fallen into disuse, with the exception of a few isolated pockets in Papua New Guinea, India, Nepal, Saudi Arabia, Nigeria, Ghana, Gambia, Tanzania, Kenya, or Sierra Leone, where "witches" are still burned to death. A 2002 World Health Organization study, for example, reported that every year more than 500 elderly women in Tanzania alone are killed for being "witches." In Nigeria, children by the thousands are being rounded up and torched as "witches," and in response the Nigerian government arrested a self-styled bishop named Okon Williams, who it accused of killing 110 such children.[15] Another study found that as many as 55 percent of sub-Sahara Africans believe in witches.[16] And such wrong beliefs can kill. On February 6, 2013, a twenty-year-old woman and mother of two named Kepari Leniata was burned alive in the Western Highlands of Papua New Guinea because she was accused of sorcery by the relatives of a six-year-old boy who died on February 5.[17] As in witch hunts of old, the conflagration on a pile of rubbish was preceded by torture with a hot iron rod, after which Kepari was bound

Figure 3-2.
From the title page of *Witches Apprehended, Examined and Executed*, published in 1613. It shows the classic "ordeal by water."

and doused in gasoline and ignited while surrounded by gawking crowds that prevented police and authorities from rescuing her. A 2010 Oxfam study explains why sorcery and witchcraft are not uncommon in this part of the world in which many people still "do not accept natural causes as an explanation for misfortune, illness, accidents or death," and instead place the blame for their problems on supernatural sorcery and black magic.[18]

Are these people evil or misinformed? By modern Western moral standards, of course, they're morally reprehensible and, if living in places where witchcraft and the burning of alleged witches is illegal, they are also criminal. But given the fact that Europeans and Americans abandoned their belief in witches when science supplanted superstition as a better explanation for evil (and it was outlawed), the generous assessment is that these witch-hunters are merely misinformed. In short, they hold a wrong theory of causality. That said, it is not just a matter of improving science education, although an education of any sort would be a good start for all sorts of reasons, both moral and practical. For starters, governments must outlaw the burning of witches. Jared Diamond tells me that in Papua New Guinea the extraordinarily high rate of violence—including witch burning—was greatly attenuated by government agents going from village to village and outlawing such diabolical practices, confiscating weapons, and laying down the law.

A poignant example of what it often takes to bring about an end to a superstitious barbaric act may be seen in the Indian practice of suttee, or the burning of widows. The British government abolished suttee by outlawing it, and followed up by severely punishing transgressors. As the nineteenth-century British commander in chief in India, General Charles Napier, told his charges who complained that suttee was their cultural custom that the British should respect: "Be it so. This burning of widows is your custom; prepare the funeral pile. But my nation has also a custom. When men burn women alive we hang them, and confiscate all their property. My carpenters shall therefore erect gibbets on which to hang all concerned when the widow is consumed. Let us all act according to national customs."[19]

In the long run, however, external restrictions in the form of laws must be supplemented with internal controls in the form of ideas. An example of how the witch theory of causality was tested and debunked in Germany is recounted by Charles MacKay in his classic work *Extraordinary Popular Delusions and the Madness of Crowds*. At the height of the witch craze the Duke of Brunswick invited two learned and famous Jesuits—both of whom believed in witchcraft and in torture as a means of eliciting a confession—to join him in the Brunswick dungeon to witness the torture of a woman accused of witchcraft. Suspecting that people will say anything to stop the pain, the duke told the woman on the rack that he had reasons to believe that the two men accompanying him were warlocks and that he wanted to know what she thought, instructing her torturers to jack up the pain a little more. The woman promptly "confessed" that she had seen both men turn themselves into goats, wolves, and other animals, that they had sexual relations with other witches, and that they had fathered many children with heads like toads and legs like spiders. "The Duke of Brunswick led his astounded friends away," MacKay narrates. "This was convincing proof to both of them that thousands of persons had suffered unjustly; they knew their own innocence, and shuddered to think what their fate might have been if an enemy instead of a friend had put such a confession into the mouth of a criminal."[20]

One of these Jesuits was Friedrich Spee, who in response to this shocking display of induced false confessions published a book in 1631 called *Cautio Criminalis*, which exposed the horrors of the torturous witch trials.

This led the archbishop and elector of Menz, Schonbrunn, to abolish torture entirely, which in turn led to the abolition of torture for witchcraft elsewhere—a catalyst in the cascading effect that caused the witch mania to collapse. "This was the beginning of the dawn after the long-protracted darkness," Mackay writes. "The tribunals no longer condemned witches to execution by hundreds in a year. A truer philosophy had gradually disabused the public mind. Learned men freed themselves from the trammels of a debasing superstition, and governments, both civil and ecclesiastical, repressed the popular delusion they had so long encouraged."[21]

Before the dawn broke, however, thousands of people were senselessly murdered. Precise figures are hard to come by—given the spottiness of the record—but the historian Brian Levack puts the figure at sixty thousand, based on the number of trials and the rate of convictions (often close to 50 percent),[22] while the medieval historian Anne Llewellyn Barstow adjusts that figure upward to one hundred thousand, based on lost records.[23] *Figure 3-3* is a depiction by the Dutch artist Johannes Jan Luyken

Figure 3-3.
Anneken Hendriks, a woman accused of witchcraft, about to be burned to death in 1571.

of a woman named Anneken Hendriks who is about to be burned to death as a witch in 1571.

Whatever the tally, it was a tragically high number, and after the immediate solution of banning it, the ultimate solution to ending witch-craft everywhere proved to be a better understanding of causality: science. As the historian Keith Thomas concludes in his sweeping history *Religion and the Decline of Magic*, the first and most important factor in the decline "was the series of intellectual changes which constituted the scientific and philosophical revolution of the seventeenth century. These changes had a decisive influence upon the thinking of the intellectual élite and in due course percolated down to influence the thought and behavior of the people at large. The essence of the revolution was the triumph of the mechanical philosophy."[24]

By "mechanical philosophy" Thomas means the Newtonian clockwork universe, the worldview that holds that all effects have natural causes and the universe is governed by natural laws that can be examined and under-stood. In this worldview there is no place for the supernatural, and that is what ultimately doomed the witch theory of causality, along with other supernatural explanations. "The notion that the universe was subject to immutable natural laws killed the concept of miracles, weakened the belief in the physical efficacy of prayer, and diminished faith in the possibility of direct divine inspiration," Thomas concludes. "The triumph of the mechan-ical philosophy meant the end of the animistic conception of the universe which had constituted the basic rationale for magical thinking."[25]

There were other factors at work besides reason and science that I dis-cuss below, but my point here is that beliefs such as witchcraft are not immoral so much as they are mistaken. In the West, science debunked the witch theory of causality, as it has and continues to discredit other super-stitious and religious ideas. We refrain from burning women as witches not because our government prohibits it, but because we do not believe in witches and therefore the thought of incinerating someone for such prac-tices never even enters our minds. What was once a moral issue is now a nonissue, pushed out of our consciousness—and our conscience—by a naturalistic, science- and reason-based worldview.

LIFE BEFORE SCIENCE

The witch theory of causality—a catchall explanation for the miseries of life—was hardly up to the formidable task of elucidation given that there was just so much misery to explain. To fully feel the change let's go back to a time when civilization was lit only by fire, five centuries ago when populations were sparse and 80 percent of people lived in the countryside and were engaged in food production, largely for themselves. Cottage industries were the only industries in this preindustrial and highly stratified society, in which a third to half of the population lived at subsistence level and were chronically underemployed, underpaid, and undernourished. Food supplies were unpredictable, and plagues decimated weakened populations. In the century spanning 1563 to 1665, for example, there were no fewer than six major epidemics that swept through London, each of which annihilated between a tenth and a sixth of the population. The death tolls are almost unimaginable by today's standards: 20,000 in 1563, 15,000 in 1593, 36,000 in 1603, 41,000 in 1625, 10,000 in 1636, and 68,000 in 1665, all in one of the world's major metropolitan cities whose population was around 120,000 in 1550, 200,000 in 1600, and 400,000 in 1650, so the percentage of deaths during each plague was substantial. Childhood diseases were unforgiving, felling 60 percent of children before age 17. As one observer noted in 1635, "We shall find more who have died within thirty or thirty-five years of age than passed it."[26] The historian Charles de La Roncière provides examples from fifteenth century Tuscany in which lives were routinely cut short:

> Many died at home: children like Alberto (aged ten) and Orsino Lanfredini (six or seven); adolescents like Michele Verini (nineteen) and Lucrezia Lanfredini, Orsino's sister (twelve); young women like beautiful Mea with the ivory hands (aged twenty-three, eight days after giving birth to her fourth child, who lived no longer than the other three, all of whom died before they reached the age of two); and of course adults and elderly people.[27]

And this does not include, La Roncière adds parenthetically, the deaths of newborns, which historians estimate could have been as high as 30 to 50 percent.[28]

Since magical thinking is positively correlated with uncertainty and unpredictability,[29] we should not be surprised at the level of superstition given the grim vagaries of premodern life. There were no banks for people to set up personal savings accounts during times of plenty to provide a cushion of comfort during times of scarcity. There were no insurance policies for risk management, and few people had much personal property to insure anyway. With homes constructed of thatched roofs and wooden chimneys in a night lit only by candles, fires would routinely devastate entire neighborhoods. As one chronicler noted, "He which at one o'clock was worth five thousand pounds and, as the prophet saith, drank his wine in bowls of fine silver plate, had not by two o'clock so much as a wooden dish left to eat his meat in, nor a house to cover his sorrowful head."[30] Alcohol and tobacco became essential anesthetics for the easing of pain and discomfort that people employed as a form of self-medication, along with the belief in magic and superstition to mitigate misfortune.

Under such conditions it's no wonder that almost everyone believed in sorcery; werewolves; hobgoblins; astrology; black magic; demons; prayer; providence; and, of course, witches and witchcraft. As Bishop Hugh Latimer of Worcester explained in 1552, "A great many of us, when we be in trouble, or sickness, or lose anything, we run hither and thither to witches, or sorcerers, whom we call wise men . . . seeking aid and comfort at their hands."[31] Saints were invoked and liturgical books provided rituals for blessing cattle, crops, houses, tools, ships, wells, and kilns, along with special prayers for sterile animals, the sick and infirm, and even infertile couples. In his 1621 book *Anatomy of Melancholy*, Robert Burton noted, "Sorcerers are too common; cunning men, wizards, and white witches, as they call them, in every village, which, if they be sought unto, will help almost all infirmities of body and mind."[32]

Was everyone in the prescientific world so superstitious? They were. As the historian Keith Thomas notes, "No one denied the influence of the heavens upon the weather or disputed the relevance of astrology to medicine or agriculture. Before the seventeenth century, total skepticism about astrological doctrine was highly exceptional, whether in England or elsewhere." And it wasn't just astrology. "Religion, astrology and magic all purported to help men with their daily problems by teaching them how to avoid misfortune and how to account for it when it struck." With such sweeping power over people, Thomas concludes, "If magic is to be defined as the employment of ineffective techniques to allay anxiety when effective ones are not available, then we must recognize that no society will ever be free from it."[33]

That may well be, but the rise of science diminished this near universality of magical thinking by proffering natural explanations where before there were predominately supernatural ones. The decline of magic and the rise of science was a linear ascent out of the darkness and into the light. As empiricism gained status there arose a drive to find empirical evidence for superstitious beliefs that previously needed no propping up with facts.

This attempt to naturalize the supernatural carried on for some time and spilled over into other areas. The analysis of portents was often done meticulously and quantitatively, albeit for purposes both natural and supernatural. As one diarist privately opined on the nature and meaning of comets, "I am not ignorant that such meteors proceed from natural causes, yet are frequently also the presages of imminent calamities."[34] Yet the propensity to portend the future through magic led to more formalized methods of ascertaining causality by connecting events in nature—the very basis of science. In time, natural theology became wedded to natural philosophy and science arose out of magical beliefs, which it ultimately displaced. By the eighteenth and nineteenth centuries, astronomy replaced astrology, chemistry succeeded alchemy, probability theory displaced luck and fortune, insurance attenuated anxiety, banks replaced mattresses as the repository of people's savings, city planning reduced the risks from fires, social hygiene and the germ theory dislodged disease, and the vagaries of life became considerably less vague.

FROM THE PHYSICAL SCIENCES TO THE MORAL SCIENCES

To the debunking of the witch theory of causality and the general improvement in living conditions, we can add as promoters of moral progress the general application of reason and science to all fields, including governance and the economy. This shift was the result of two intellectual revolutions: (1) the Scientific Revolution, dated roughly from the publication of Copernicus's *On the Revolutions of the Heavenly Spheres* in 1543 to the publication of Isaac Newton's *Principia* in 1687; and (2) the Age of Reason and the Enlightenment, dated from approximately 1687 to 1795 (Newton to the French Revolution). The Scientific Revolution led directly to the Enlightenment, as intellectuals in the eighteenth century sought to emulate the great scientists of the previous centuries in applying the rigorous methods of the natural sciences and philosophy to explaining phenomena and solving problems. This marriage of philosophies resulted in Enlightenment ideals that placed supreme value on reason, scientific inquiry, human natural rights, liberty, equality, freedom of thought and expression, and on a diverse, cosmopolitan worldview that most people today embrace—a "science of man," as the great Scottish Enlightenment philosopher David Hume called it.

From an intellectual history perspective, I have described this shift as the "battle of the books"—the book of authority vs. the book of nature.[35] The *book of authority*—whether it was the Bible or Aristotle in the Western world—is grounded in the cognitive process called *deduction*, or making specific claims from generalized principles, or arguing from the general to the specific. By contrast, the *book of nature* is grounded in *induction*, or the cognitive process of drawing generalized principles from specific facts, or arguing from the specific to the general. None of us—nor any tradition—practices pure induction or pure deduction, but the Scientific Revolution was revolting against the overemphasis on the book of authority, and instead promoted an insistence on checking one's assumptions with the book of nature.

For example, one of the giants of the Scientific Revolution—Galileo Galilei—got himself in hot water with the Church partly for insisting on observation as the gold standard, rather than blind acceptance, in order to determine if the ancient authorities were correct in their conjectures. "The authority of Archimedes was of no more importance than that of Aristotle," he said. "Archimedes was right because his conclusions agreed with experiment."[36]

It's a matter of balance between deduction and induction—between reason and empiricism—and in 1620 the English philosopher Francis Bacon published his *Novum Organum*, or "new instrument," which described science as a blend of sensory data and reasoned theory. Ideally, Bacon argued, one should begin with observations, then formulate a general theory from which logical predictions can be made, then check the predictions against experiment.[37] If you don't give yourself a reality check you end up with half-baked (and often fully baked) ideas, such as the ancient Roman philosopher Pliny the Elder's "Remedies for Ulcerous Sores and Wounds" from his first-century CE book *The Natural History*, which reads like a skit from Monty Python:

> With sheep's dung, warmed beneath an earthen pan and kneaded, the swellings attendant upon wounds are reduced, and fistulous sores and epinyctis are cleansed and made to heal. But it is in the ashes of a burnt dog's head that the greatest efficacy is found; as it quite equals spodium in its property of cauterizing all kinds of fleshy excrescences, and causing sores to heal. Mouse-dung, too, is used as a cautery, and weasels' dung, burnt to ashes.[38]

In France, the philosopher and mathematician René Descartes—considered to be the founder of modern philosophy—set himself the momentous (and one should have thought impossible) task of unifying all knowledge so as to "give the public . . . a completely new science which would resolve generally all questions of quantity." In his 1637 skeptical work *Discourse on Method*, Descartes instructed readers to take as false what was probable, to take as probable what was certain, and to discard everything else that relied on old books and authorities.

Doubting everything, Descartes famously concluded that there was one thing he could not doubt, and that was his own thinking mind: "*Cogito, ergo sum*—I think, therefore I am." Building from this first principle he turned to mathematical reasoning and from there built not only a new branch of mathematics (the Cartesian coordinate system in common use today) but also a new and powerful science that could be applied to any subject. Descartes's work generated an *esprit géométrique* and an *esprit du mechanism* (a spirit of geometry and a spirit of mechanical causation) to find mathematical and mechanical explanations for everything. This mechanical philosophy gained international credibility through Newton's clockwork universe. This embrace of mathematical precision is still visible today in the gardens of France with their geometric regularity. There Descartes became fascinated with the mechanical automata he saw operating under hydraulic pressure, and this esprit led to his (and others) turning to mechanical explanations for animals and humans.[39]

The watershed event that changed everything, however, was the publication in 1687 of Isaac Newton's *Principia Mathematica*, which synthesized the physical sciences and which his contemporaries declared to be "the premier production of the human mind" (Joseph-Louis Lagrange) and a work that "has a pre-eminence above all other productions of the human intellect" (Pierre-Simon Laplace). Upon Newton's death, Alexander Pope eulogized Newton thusly: "Nature and Nature's laws lay hid in night. God said let Newton be! And all was light." The Enlightenment luminary David Hume described Newton as "the greatest and rarest genius that ever rose for the adornment and instruction of the species."[40]

Newton showed that the rigorous methods of mathematics and science could be applied to all fields. And he practiced what he preached, making significant contributions to pure and applied mathematics (he invented the calculus), optics, the law of universal gravity, tides, heat, the chemistry and theory of matter, alchemy, chronology, interpretation of Scripture, the design of scientific instruments, and even the minting of money. After the precisely predicted return of Halley's comet confirmed Newton's theory of universal gravitation, the race was on to apply Newtonian methods of science to all fields. "Men and women everywhere saw a promise that all of human knowledge and the regulation of human affairs would yield to a similar rational system of deduction and mathematical inference coupled with experiment and critical observation," notes the great historian of science Bernard Cohen. "Newton was the symbol of successful science, the ideal for all thought—in philosophy, psychology, government, and the science of society."[41]

The Scientific Revolution that culminated in Newtonian science led scientists in diverse fields to strive to be the Newton of their own particular science. In his 1748 work *Esprit des Lois* (*The Spirit of the Laws*), for example, the French *philosophe* Charles-Louis de Secondat, Baron de Montesquieu—known simply as Montesquieu—consciously invoked Newton when he compared a well-functioning monarchy to "the system of the universe" that includes "a power of gravitation" that "attracts" all bodies to "the center" (the monarch). And his method was the deductive method of Descartes: "I have laid down first principles and have found that the particular cases follow naturally from them." By "spirit" Montesquieu meant "causes" from which one could derive "laws" that govern society. "Laws in their most general signification, are the necessary relations derived from the nature of things," he wrote. "In this sense all beings have their laws, the Deity has his laws, the material World its laws, the intelligences superior to man have their laws, the beast their laws, man his laws."

As a young man Montesquieu published a number of scientific papers on a wide range of topics—tides, fossil oysters, the function of kidneys, causes of echoes—and in *Esprit des Lois* he applied his naturalistic talents to produce a theory of the natural conditions that led to the development of the different governments and legal systems throughout the world, such as the climate, the quality of the soil, the religion and occupation of the inhabitants, their numbers, commerce, manners, customs, and the like. His typology included four types of societies: hunting, herding, agriculture, and trade or commerce, with legal systems becoming ever more complex. "The laws have a very great relation to the manner in which the several nations procure their subsistence," he wrote. "There should be a code of laws of a much larger extent for a nation attached to trade and navigation than for people who are content with cultivating the earth. There should be a much greater for the latter than for those who subsist by their flocks and herds. There must be a still greater for these than for such as live by hunting." This led Montesquieu to become one of the earliest proponents of the trade theory of peace when he observed that hunting and herding nations often found themselves in conflict and wars, whereas trading nations "became reciprocally dependent," making peace "the natural effect of trade." The psychology behind the effect, Montesquieu speculated, was exposure of different societies to customs and manners different from their own, which leads to "a cure for the most destructive prejudices." Thus, he concluded, "we see that in countries where the people move only by the spirit of commerce, they make a traffic of all the humane, all the moral virtues."[42]

Following in the natural law tradition of Montesquieu, a group of French scientists and scholars known as the physiocrats declared that all "social facts are linked together in necessary bonds eternal, by immutable, ineluctable, and inevitable laws" that should be obeyed by people and governments "if they were once made known to them" and that human societies are "regulated by *natural laws* . . . the same laws that govern the physical world, animal societies, and even the internal life of every organism." One of these physiocrats, François Quesnay—a physician to the king of France who later served as an emissary to Napoleon for Thomas Jefferson—modeled the economy after the human body, in which money flowed through a nation like blood flows through a body, and ruinous government policies were like diseases that impeded economic health.[43] He argued that even though people have unequal abilities, they have equal natural rights, and so it was the government's duty to protect the rights of individuals from being usurped by other individuals, while at the same time enabling people to pursue their own best interests. This led them to advocate for private property and a free market. It was, in fact, the physiocrats who gave us the term "laissez-faire"—translated as "leave alone"—the economic practice of minimum government interference in the economic interests of citizens and society.

The physiocrats asserted that people operating in a society were subject to knowable laws of both human and economic nature not unlike those discovered by Galileo and Newton, and this movement grew into the school of classical economics championed by David Hume, Adam Smith, and others and that forms the basis of all economic policy today. The very title of Adam Smith's monumental 1776 work reveals its scientific emphasis: *An Inquiry into the Nature and Causes of the Wealth of Nations*. Smith employed the terms "nature" and "causes" in the scientific sense of identifying and understanding the cause-and-effect relationships in the natural system of an economy, with the underlying premise that natural laws govern economies, that humans are rationally calculating economic actors whose behaviors can be understood, and that markets are self-regulating by an "invisible hand." The origin of Smith's famous metaphor was astronomical in nature. As Smith wrote in his little-known work on the history of astronomy,

> For it may be observed, that in all Polytheistic religions, among savages, as well as in the early ages of heathen antiquity, it is the irregular events of nature only that are ascribed to the agency and power of the gods. Fire burns, and water refreshes; heavy bodies descend, and lighter substances fly upwards, by the necessity of their own nature; nor was the *invisible hand* of Jupiter ever apprehended to be employed in those matters.[44]

Here Smith was describing the invisible hand of gravity, but his later application of the metaphor in the *Wealth of Nations* implied that an invisible hand appears to guide markets and economies. Smith, it should be noted, was a professor of moral philosophy, and his first great work, published in 1759, was titled *A Theory of Moral Sentiments*, in which he laid the foundation for the theory that we have an innate sense of morality: "How selfish soever man may be supposed, there are evidently some principles in his nature, which interest him in the fortune of others, and render their happiness necessary to him, though he derives nothing from it except the pleasure of seeing it. Of this kind is pity or compassion, the emotion which we feel for the misery of others." The emotion of empathy—what Smith called sympathy—allows us to feel someone else's joy or agony by imagining ourselves as that person and sensing how we would feel: "As we have no immediate experience of what other men feel, we can form no idea of the manner in which they are affected, but by conceiving what we ourselves should feel in the like situation."[45] This is the principle of interchangeable perspectives at work.

In the arena of governance, another Enlightenment luminary who consciously applied the principles and methods of the physical sciences to the moral sciences was the English philosopher Thomas Hobbes, whose 1651 book *Leviathan* is considered to be one of the most influential works in the history of political thought. In it Hobbes deliberately modeled his analysis of the social world after the work of Galileo and the English physician William Harvey, whose 1628 *De Motu Cordis* (*On the Motion of the Heart and the Blood*) outlined a mechanical model of the workings of the human body. As Hobbes later immodestly reflected, "Galileus . . . was the first that opened to us the gate of natural philosophy universal, which is the knowledge of the nature of motion. . . . The science of man's body, the most profitable part of natural science, was first discovered with admirable sagacity by our countryman, Doctor Harvey. Natural philosophy is therefore but young; but civil philosophy is yet much younger, as being no older . . . than my own *de Cive*."

Hobbes even patterned his *Elements of Law* after Euclid's *Elements of Geometry*, classifying all previous philosophers into two camps: the *dogmatici*, who for two millennia had failed to create a viable moral or political philosophy; and the *mathematici*, who proceeded "from most low and humble principles . . . going on slowly, and with most scrupulous ratiocinations" to create a system of useful knowledge about the social world. And this new system of thought is not "that which makes philosophers' stones, nor that which is found in the metaphysic codes," he proclaimed in an epistle to his readers, "but that it is the natural reason of man, busily flying up and down among the creatures, and bringing back a true report of their order, causes and effects."[46]

Hobbes self-consciously applied both the esprit géométrique and the esprit du mechanism to the study of nature, man, and "civil governments and the duties of subjects."[47] Here we see both the connection from the physical and biological sciences to the social sciences, and also the point of my focusing on this period in the history of science—our modern concepts of governance arose out of this drive to apply reason and science to any and all problems, including human social problems.

According to the historian of science Richard Olson (my doctoral adviser who first introduced me to these links between science and society), "Hobbes's theories of nature, man, and society clearly derived their form from a Hobbesian version of the Cartesian *esprit géométrique*." Not only that, Olson continues, "Hobbes believed that the sciences of man and society could, like the science of inanimate natural bodies, be constructed on the geometrical or hypothetico-deductive model."[48] The latter is the science philosopher's term for the modern scientific method, which can be summarized in three steps: (1) formulating a hypothesis based on initial observations, (2) making a prediction based on the hypothesis, and (3) checking or testing whether the prediction is accurate.

Hobbes's theory of how to build a civil society is a purely naturalistic argument that employed the best science of his day, and Hobbes, along with his Enlightenment colleagues, considered themselves to be practicing what today we call science (but what they called natural philosophy).[49]

He begins with the assumption that the universe is composed only of material objects that are in motion (such as atoms and planets). The brain detects the movement of these objects through the senses—either directly through, say, touch, or indirectly via the transmission of some energy, as in vision—and all ideas come from these basic sense movements. Humans can sense matter in motion, and humans themselves are in motion (like "the motion of the blood, perpetually circulating" he notes, citing William Harvey), constantly driven by the passions: appetites (pleasures) and aversions (pain). When motion ceases (e.g., blood circulation), life ceases, so all human action is directed toward maintaining the vital motions of life. Pleasure (or delight or contentment), he says, "is nothing really but motion about the heart, as conception is nothing but motion about the head, and the objects which cause it are called pleasant or delightful."

What we think of as good and bad, then, are directly related to a person's desires or fears in response to a given stimulus. To gain pleasure and avoid pain one needs power: "The power of a man is his present means to obtain some future apparent good," Hobbes continues. In a state of nature everyone is free to exert their power over others to gain greater pleasures. This Hobbes calls the *Right of Nature*. Even though humans have equal ability, they have unequal passions that, he says, "during the time men live without a common power to keep them all in awe, they are in that condition which is called war; and such a war as is of every man against every man." By war, Hobbes does not just mean actual fighting, but a constant state of *fear* of fighting, which makes it impossible to plan for the future. As he concluded in one of the most famous (and oft-quoted) passages in all of political theory:

> In such condition there is no place for industry, because the fruit thereof is uncertain: and consequently no culture of the earth; no navigation, nor use of the commodities that may be imported by sea; no commodious building; no instruments of moving and removing such things as require much force; no knowledge of the face of the earth; no account of time; no arts; no letters; no society; and which is worst of all, continual fear, and danger of violent death; and the life of man, solitary, poor, nasty, brutish, and short.[50]

But we do not live in a state of nature, says Hobbes, because humans have one more mental property that enables us to rise above the Right of Nature, and that is *reason*. It is reason that led people to realize that to be free they must surrender all rights to a sovereign in a *social contract*. This sovereign Hobbes calls the *Leviathan*, after the powerful Old Testament sea monster.[51]

Half a century after Hobbes, no scholar of political or economic thought was taken seriously unless they overtly employed a scientific approach to their study, that is, some combination of reason and empiricism to derive conclusions about how humans behave (*descriptive observations*) and how humans *should* behave (*prescriptive morals*) in society. As the Scottish Enlightenment philosopher David Hume colorfully declared toward the end of his classic 1749 work *An Inquiry Concerning Human Understanding*: "If we take in our hand any volume; of divinity or school metaphysics, for instance; let us ask, Does it contain any abstract reasoning concerning quantity or number? No. Does it contain any experimental reasoning concerning matter of fact and existence? No. Commit it then to the flames: for it can contain nothing but sophistry and illusion."[52] As well, Hobbes's mechanical model imagined people as atoms—interchangeable particles in a social universe governed by natural laws that can be studied in the same manner as physicists measure atoms or astronomers track planets, and from which general theories can be derived to explain their motions. The eminent modern political philosopher Michael Walzer clarifies what this new way of studying the social world meant: "For two hundred years there is hardly an English writer, hardly a coffee house conversationalist, who is not a successor to Hobbes."[53]

FROM IS TO OUGHT: SOCIAL SCIENCE AND MORAL PROGRESS

Whether or not Hobbes was right about the Leviathan origins of the social contract (it's a mixed history because humans are social creatures and have never lived in isolation), the fact is that the Leviathan state is what emerged over the past half millennium as thousands of tiny municipalities, duchies, baronies, and the like coalesced into ever larger political organizations during the early modern period of state building.

Political scientists estimate that in Europe there were about five thousand political units in the fifteenth century, five hundred in the seventeenth century, two hundred by the eighteenth century, and less than fifty in the twentieth century.[54] This coalescence resulted in two major trends: (1) the decline of individual violence in which the percentage of people who die violently in states is significantly lower than that of traditional, prestate societies; and (2) an increase and a decrease in total death counts from 1500 until 1950, in that there was a decrease in the number and duration of great-power wars, but an increase in their intensity (the number of people killed per country per year)—with these two trends pushing in opposite directions, the total death count rose and fell. After the Second World War, however, both the frequency and the intensity of war decreased until, essentially, the world's great powers quit fighting.

Let's look at the logic of how the Leviathan works to reduce violence, and transition here from *is* (how science and reason developed historically) to *ought* (how this knowledge was—and ought to be—used to bend the moral arc). The Leviathan reduces violence by asserting a monopoly on the legitimate use of force, thereby replacing what criminologists call "self-help justice"—in which individuals settle their own scores and disputes, often violently (such as the Mafia)—with criminal justice, leading overall to a decrease in violence. But there are other factors at work as well.

Trade, Commerce, and Conflict

Hobbes was only partially right in advocating top-down state controls to keep the inner demons of our nature in check. Trade and commerce were also major factors, given the moral and practical benefits of trading for what you need instead of killing to get it. I call this Bastiat's Principle (after the nineteenth-century French economist Frédéric Bastiat, who first articulated the concept): *where goods do not cross frontiers, armies will, but where goods do cross frontiers, armies will not.*[55] I call it a principle instead of a law because there are exceptions both historically and today. Trade does not prevent war and interstate violence, but it attenuates its likelihood.

As I documented in *The Mind of the Market*, trade breaks down the natural animosity between strangers while simultaneously elevating trust between them, and as the economist Paul Zak has demonstrated, trust is among the most powerful factors affecting economic growth. In his neuroeconomics lab at Claremont Graduate University, for example, Zak has shown that the trust hormone oxytocin is released during exchange games between strangers, thereby enhancing trust and setting off a positive feedback loop. In addition, the neurotransmitter dopamine is released, a chemical that controls the brain's motivation, reward, and pleasure centers, thereby encouraging an organism to repeat a particular behavior. Thus the learned behavior of exchange is reinforced through a chemical pleasure hit. When playing Prisoner's Dilemma games, the brain scans of the subjects revealed that when they were cooperating, the same areas of the brain were activated as in response to stimuli such as sweets, money, cocaine, and attractive faces. The neurons that were most responsive were those rich in dopamine located in the *anteroventral striatum* in the "pleasure center" of the brain.[56]

The effects of trade have been documented in the real world as well as in the lab. In a 2010 study published in *Science* titled "Markets, Religion, Community Size, and the Evolution of Fairness and Punishment," the psychologist Joseph Henrich and his colleagues engaged more than two thousand people in fifteen small communities around the world in two-player exchange games in which one subject is given a sum of money equivalent to a day's pay and is allowed to keep or share some of it, or all of it, with another person. You might think that most people would just keep all the money, but in fact the scientists discovered that people in hunter-gatherer communities shared about 25 percent, while people in societies who regularly engage in trade gave away about 45 percent. Although religion was a modest factor in making people more generous, the strongest predictor was "market integration," defined as "the percentage of a household's total calories that were purchased from the market, as opposed to homegrown, hunted, or fished." Why? Because, the authors conclude, trust and cooperation with strangers lowers transaction costs and generates greater prosperity for all involved, and thus market fairness norms "evolved as part of an overall process of societal evolution to sustain mutually beneficial exchanges in contexts where established social relationships (for example, kin, reciprocity, and status) were insufficient."[57]

Trade, Democracy, and Conflict

Instead of configuring the issue of the complex interaction between trade and politics at the global level in terms of categorical binary logic—either you trade or you do not trade, either you are a democracy or you are not a democracy—using a continuous scale reveals more subtle but very real effects. A particular country may trade with other countries very little, some, or a lot, and it may be less or more democratic. This continuous rather than categorical approach enables researchers to treat each case as a datapoint on a continuum rather than as an example or an exception in an artificial choice in which one is tempted to cherry-pick the data to force fit them into preconceived models.

Employing a continuous style of analysis to address this question are the political scientists Bruce Russett and John Oneal in their book *Triangulating*

Peace, in which they use a multiple logistic regression model on data from the Correlates of War Project, which recorded twenty-three hundred militarized interstate disputes between 1816 and 2001.[58] Assigning each country a democracy score between 1 and 10 (based on the Polity Project, which measures how competitive its political process is, how openly its leaders are chosen, how many constraints on a leader's power are in place, the transparency of the democratic process, the fairness of its elections, etc.), Russett and Oneal found that when two countries are fully democratic (that is, they score high on the Polity scale), disputes between them decrease by 50 percent; but when one member of a county pair was either a low-scoring democracy or a full autocracy, it doubled the chance of a quarrel between them.[59]

When you add a market economy and international trade into the equation it decreases the likelihood of conflict between nations. Russett and Oneal found that for every pair of at-risk nations, when they entered the amount of trade (as a proportion of GDP) they found that countries that depended more on trade in a given year were less likely to have a militarized dispute in the subsequent year, controlling for democracy, power ratio, great power status, and economic growth. In general, the data show that liberal democracies with market economies are more prosperous, more peaceful, and fairer than any other form of governance and economic system. In particular, they found that democratic peace happens only when both members of a pair are democratic, but that trade works when *either* member of the pair has a market economy.[60] In other words, trade was even more important than democracy (although the latter is important for other reasons as well).

Finally, the third vertex of Russett and Oneal's triangle of peace is membership in the international community, a proxy for transparency. Evil is more likely to thrive in secret. Openness and transparency make it harder for dictators and demagogues to commit violence and genocide. To test this hypothesis, Russett and Oneal counted the number of Intergovernmental Organizations (IGOs) that every pair of nations jointly belonged to and ran a regression analysis with democracy and trade scores, finding that, overall, democracy, trade, and membership in IGOs all favor peace, and that a pair of countries that are in the top tenth of the scale on all three variables are 83 percent less likely than an average pair of countries to have a militarized dispute in a given year.[61]

Figure 3-4 presents data showing that as democracies increase and autocracies decrease, war declines.[62] *Figure 3-5* shows the number of nations scoring 8 or more on the Polity IV scale from 1800 to 2003, revealing a hockey-stick-like improvement in the number of nations after the Second World War who made the transition from autocracies or corrupt democracies to fair and transparent liberal democracies.[63] *Figure 3-6* shows membership in Intergovernmental Organizations shared by a pair of countries from 1885 to 2000.[64] *Figure 3-7* brings all these datasets together into a "Trifecta of Peace": Democracy + Economic Interdependence + Membership in Intergovernmental Organizations = More Peace.

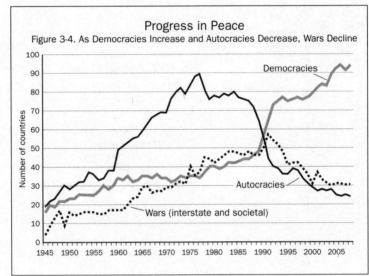

Figure 3-4

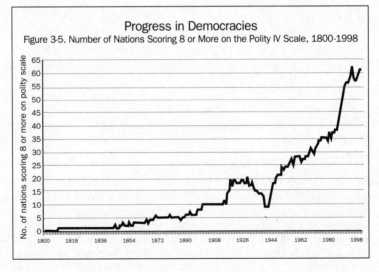

Figure 3-5

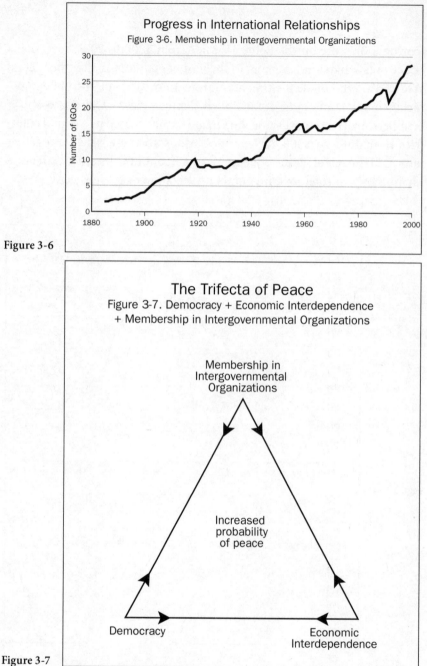

Progress in International Relationships

Figure 3-6. Membership in Intergovernmental Organizations

Figure 3-6

The Trifecta of Peace

Figure 3-7. Democracy + Economic Interdependence + Membership in Intergovernmental Organizations

Membership in Intergovernmental Organizations

Increased probability of peace

Democracy

Economic Interdependence

Figure 3-7

Figures 3-4 to 3-7. The Trifecta of Peace: Liberal Democracy, Trade, Transparency.

Figure 3-4 presents data showing that as democracies increase and autocracies decrease, wars decline. *Figure 3-5* shows the number of nations scoring 8 or more on the Polity IV scale from 1800 to 2003, showing a hockey-stick-shaped improvement in the number of nations after the Second World War who made the transition from autocracies or corrupt democracies to fair and transparent liberal democracies. *Figure 3-6* shows membership in Intergovernmental Organizations shared by a pair of countries from 1885 to 2000. *Figure 3-7* brings all these datasets together into a "Trifecta of Peace": Democracy + Economic Interdependence + Membership in International Organizations = More Peace.

In a 1989 essay on "The Causes of War" Jack Levy noted that the "absence of war between democratic states comes as close as anything we have to an empirical law in international relations."[65] In 2010 Russett and Oneal updated their research and concluded that "the period since World War II has seen progressive realization of the classical-liberal ideal of a security community of trading states." Since 2010 there has been much conflict around the world, so how has the democratic peace theory held up? In a 2014 special issue of the *Journal of Peace Research*, the Uppsala University political scientist Håvard Hegre reassessed all of the evidence on "Democracy and Armed Conflict," concluding "the empirical findings that pairs of democratic states have a lower risk of interstate conflict than other pairs holds up, as does the conclusion that consolidated democracies have less conflict than semi-democracies."[66]

. . .

Thinking about the problem of how to reduce conflict between Leviathans in a continuous rather than categorical way also enables us to deal with the apparent exceptions in a more scientific manner because science traffics in continuums and probabilities more than it does in black-and-white categories. For example, when the argument is made that no two democracies ever go to war (the democratic peace theory) or that no two countries who trade with one another ever fight (the McDonald's peace theory), skeptics reach into the bin of history for the exceptions, such as the United States vs. Great Britain in the War of 1812, the American Civil War, or the India-Pakistan wars—all democracies of a sort—or the great powers on the eve of the First World War, all of whom traded with one another just before the guns of August in 1914. This leads proponents to counter that, say, the United States was not really a democracy because at the time of the War of 1812, and also during the Civil War, slavery was practiced and women couldn't vote, so it doesn't count as a true democracy. But treating all historical examples as datapoints on a continuum allows us to perceive nuances in the cause-and-effect relationships at work in the messiness of the real world.

The misreading of Nobel Peace Prize laureate Norman Angell is a case in point. His 1910 book *The Great Illusion*—in which he argued for the futility of war as a means to greater economic prosperity compared to trade—was pilloried as a fool's errand in prediction. In 1915, with the Great War ramping up to a full head of steam, the *New York Times* opined that Angell had "written books in the endeavor to prove that war has been made impossible by modern economic conditions . . . [but] events have shown their fallacy. Ten nations, more or less closely bound a short time ago by economic ties, are now involved in war." Nearly a century later, in a 2013 article in the *National Interest,* Jacob Heilbrunn wrote, "Angell had wrongly deprecated the centrality of power in international relations. In 1914, for example, he announced that 'There will never be another war between European powers.'"[67] And Heilbrunn was defending Angell!

In fact, notes Ali Wyne in a rebuttal in *War on the Rocks*, what Angell actually said in his pre–World War I edition of the book was that "War is not impossible, and no responsible [p]acifist ever said it was; it is not the likelihood of war which is the illusion, but its benefits." Angell further clarified his position in a 1913 letter to the *Sunday Review* (which was included in his 1921 sequel titled *The Fruits of Victory*) by noting, "not only do I regard war [between Britain and Germany] as possible, but extremely likely." As Wyne notes, this misreading of Angell blinded subsequent analysts to his additional observations that have relevance to the topic of moral progress: "At least two of them merit reexamination today: 'national honor' should not be invoked to justify war, and 'human nature' does not make it inevitable."[68] The second observation was especially prescient given what science has learned about the malleability of human behavior, which Angell expressed in his 1935 Peace Prize acceptance speech as clearly as any scientist working today:

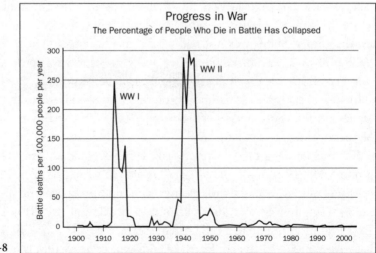

Figure 3-8

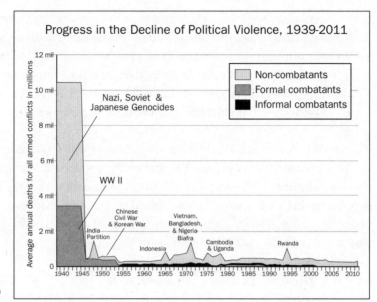

Figure 3-9

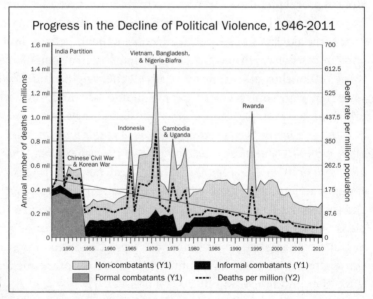

Figure 3-10

Figures 3-8 to 3-10. Progress in the Decline of War

Figure 3-8 presents data showing the percentage of people who die in battle decreased dramatically in the second half of the twentieth century.[70] *Figure 3-9* puts that second half century into perspective by tracking the average annual deaths for all armed conflicts in millions, showing how small the spikes are in comparison to the Korean and Vietnam Wars, and the genocides in Cambodia, Uganda, and Rwanda.[71] *Figure 3-10* expands the death spikes to show the decline even in these wars and genocides between 1946 and 2010.[72]

Perhaps you cannot "change human nature"—I don't indeed know what
the phrase means. But you can certainly change human behavior, which
is what matters, as the whole panorama of history shows. . . . The more it
is true to say that certain impulses, like those of certain forms of nation-
alism, are destructive, the greater is the obligation to subject them to the
direction of conscious intelligence and of social organization.[69]

Exactly. Whatever your view of human nature—blank slate, genetic determinism, or a realistic nature-nurture interactive model—it is human *action*
that should most concern us when it comes to morality. The way in which
people interact with others is what ultimately matters and, when it comes
to conflict, at long last we are beginning to understand the conditions under
which we can drive war into extinction. *Figures 3-8* through *3-10* show
just how much progress we have made toward this end.

Leviathans Better and Worse

Hobbes's Leviathan (or sovereign) required a degree of control over its
"subjects" that when put into practice in the twentieth century by various
totalitarian regimes failed utterly. Hobbes's theory fell victim to the tyranny of categorical thinking: either people live in a state of anarchy in a
war of all against all, or they give up all their rights and freedoms to a
sovereign who controls what people should or should not do. Hobbes
proposed, for example, that "every man should say to every man *I authorize and give up my right of governing myself to this man, or to this assembly
of men, on this condition, that thou give up the right to him, and authorize
all his actions in like manner.*" This "great Leviathan . . . may use the
strength and means of them all, as he shall think expedient, for their peace
and common defence." The rights and power of this sovereign over the people is almost absolute: subjects cannot change their form of government
or transfer their rights to another sovereign. Dissenting minorities must
consent to the sovereign and submit to the decrees of the majority "or be
left in the condition of war he was before."

The sovereign is "unpunishable" by his subjects and he is the sole judge
in determining the "Peace and Defence" of his subjects, which includes
"what opinions and doctrines are averse" to the Commonwealth and thus
what may or may not be published. "The sovereignty has the whole power
of prescribing the rules whereby every man may know what goods he may
enjoy, and what actions he may do, without being molested by any of his

fellow subjects; These rules of propriety . . . and of good, evil, lawful, and unlawful in the actions of subjects are the civil laws." The sovereign alone decides when and where to wage war, whom it will be waged against, the size of the armies to be deployed, which weapons will be used, and of course the sovereign has the right to tax the subjects to finance the whole operation.

If that's not extreme enough for our modern Western taste for freedom, Hobbes proposed that the sovereign ought even to control the subjects' "liberty to buy, and sell, and otherwise contract with one another; to choose their own abode, their own diet, their own trade of life, and institute their children as they themselves think fit, and the like."[73]

The problem of giving up so much control and autonomy to a state is that the people running it have the same flaws, biases, prejudices, aspirations, and temptations to cheat as everyone else. The "Hobbesian trap" of the Prisoner's Dilemma exists in government no less than in business and sports. Granting someone—anyone—too much power leads to the temptation to take advantage and to treat other people like suckers, a temptation that seems to be just too much for most people to resist. An excess of power led to the profound abuses by European monarchs against which the American and French revolutionaries revolted. This is what James Madison had in mind in *Federalist Paper Number 51* when he explained why checks and balances between the different branches of government were needed: "If men were angels, no government would be necessary. If angels were to govern men, neither external nor internal controls on government would be necessary."[74] It is what Edmund Burke was thinking when he reflected on the French Revolution: "The restraints on men, as well as their liberties, are to be reckoned among their rights."[75]

Democracies developed in response to the monarchic autocracies of the eighteenth and nineteenth centuries and to the dictatorship regimes of the twentieth century because democracies empower individuals with a methodology instead of an ideology, and it is to this extent that we can see that the scientific values of reason, empiricism, and antiauthoritarianism are not the *products* of liberal democracy but the *producers* of it. Democratic elections are analogous to scientific experiments: every couple of years you carefully alter the variables with an election and observe the results. If you want different results, change the variables.[76] The political system in the United States is often called the "American experiment," and the founding patriarchs referred to it as such, and thought of this experiment in democracy as a means to an end, not an end in itself.

Many of the Founding Fathers were, in fact, scientists who deliberately adapted the method of data gathering, hypothesis testing, and theory formation to their nation building. Their understanding of the provisional nature of findings led them to develop a social system in which doubt and dispute were the centerpieces of a functional polity. Jefferson, Franklin, Paine, and the others thought of social governance as a *problem to be solved* rather than as power to be grabbed. They thought of democracy in the same way that they thought of science—as a method, not an ideology. They argued, in essence, that no one knows how to govern a nation, so we have to set up a system that allows for experimentation. Try this. Try that. Check the results. Repeat. That is the very heart of science. As Thomas Jefferson wrote in 1804, "No experiment can be more interesting than that we are now trying, and which we trust will end in establishing the fact, that man may be governed by reason and truth." However, he noted, as in science, where open peer criticism and the freedom to debate increase the probability of discovering provisional truths, Jefferson added that this daring new political experiment being tried in the New World depended unconditionally on open access to knowledge and the freedom of its citizens to see and to think for themselves: "Our first object should therefore be, to leave open to him all the avenues to truth. The most effectual hitherto found, is the freedom of the press. It is therefore, the first shut up by those who fear the investigation of their actions."[77]

Even the fundamental principles underlying the Declaration of Independence, which is usually thought of as a statement of political philosophy, were in fact grounded in the type of scientific reasoning that Jefferson and Franklin employed in all the other sciences in which they worked. Consider the foundational line, "We hold these truths to be self-evident, that all men are created equal. . . ." In his biography of Benjamin Franklin, Walter Isaacson recounts the story of how the term "self-evident" came to be added to Jefferson's original draft by Franklin, on Friday, June 21, 1776.

The most important of his edits was small but resounding. He crossed out, using heavy backslashes that he often employed, the last three words of Jefferson's phrase, "We hold these truths to be sacred and undeniable" and changed them to the words now enshrined in history: "We hold these truths to be self-evident."

The idea of "self-evident" truths was one that drew less on John Locke, who was Jefferson's favored philosopher, than on the scientific determinism espoused by Isaac Newton and on the analytic empiricism of Frank-

lin's close friend David Hume. In what became known as "Hume's fork,"
the great Scottish philosopher, along with Leibniz and others, had developed a theory that distinguished between synthetic truths that describe
matters of fact (such as "London is bigger than Philadelphia") and analytic truths that are self-evident by virtue of reason and definition ("The
angles of a triangle equal 180 degrees"; "All bachelors are unmarried.") By
using the word "sacred," Jefferson had asserted, intentionally or not, that
the principle in question—the equality of men and their endowment by
their creator with inalienable rights—was an assertion of religion. Franklin's edit turned it instead into an assertion of rationality.[78]

The hypothesis that reason-based Enlightenment thinking leads to
moral progress is one that can be tested through historical comparison
and by examining what happens to countries that hold anti-Enlightenment
values. Countries that quash free inquiry, distrust reason, and practice
pseudoscience, such as Revolutionary France, Nazi Germany, Stalinist
Russia, Maoist China, and, more recently, fundamentalist Islamist states,
stagnate, regress, and often collapse. Theists and postmodernist critics of
science and reason often label the disastrous Soviet and Nazi utopias as
"scientific," but their science was a thin patina covering a deep layer of
counter-Enlightenment, pastoral, paradisiacal fantasies of racial ideology
grounded in ethnicity and geography, as documented in Claudia Koonz's
book *The Nazi Conscience*[79] and in Ben Kiernan's book *Blood and Soil*.[80]

These utopian, ideologically driven states occasionally rack up incredibly high body counts with a utilitarian calculus in which everyone is presumed to be happy forever, and thus dissenting individuals are labeled as
enemies of the state to be destroyed in the name of the collective. If even
rational people in democratic nations today agree that it is acceptable to
kill one person to save five by diverting a runaway trolley, imagine how
easy it is to convince people living in totalitarian and collectivist states,
steeped in utopian thinking, that they could save five million people by
killing one million; the ratio is the same but the raw numbers are genocidal. Add to that the anti-Enlightenment belief in the inequality of peoples and races and the unequal treatment under the law of those with a
different perspective or a different face (as in the Nazi political theorist
Carl Schmitt's claim that "Not every being with a human face is human"[81])
and you have a formula for genocide.

By contrast, in a reason-based worldview like that of Enlightenment
humanism in which the *principle of interchangeable perspectives* means

that no one can reasonably argue for special privilege over others, morality shifts from the vantage point of the group to that of the individual, and instead of working toward some unfounded and unattainable utopian ideology in the distant future, the political system is designed to solve specific problems that are obtainable in the here and now.

LEFT, RIGHT, AND CENTER

So much of politics comes down to finding the right balance between individual liberty and social order. And so governance reduces to this question: do we want to preserve or to change the social order? Ideologies lead us to choose sides in answer to this question: preserve (conservative) or change (liberal)? Which side you fall on is not random, nor is it a quirk of your environment or upbringing. Research on identical twins separated at birth and raised in different environments shows that about 40 to 50 percent of the variance among people in their political attitudes is due to inheritance, and these data come from multiple studies all deriving similar findings: a 1990 Australian sample involving 6,894 individuals from 3,516 families, a 2008 Australian sample of 1,160 related individuals from 635 families, and a 2010 Swedish sample involving 3,334 individuals from 2,607 families.[82] The effect was presciently captured in an 1894 poem by W. S. Gilbert (he of Gilbert and Sullivan fame) in which he declares himself "an intellectual chap" who can "think of things that would astonish you" such as:

> I often think it's comical
> How Nature always does contrive
> That every boy and every gal
> That's born into the world alive
> Is either a little Liberal,
> Or else a little Conservative![83]

Of course, there is no gene (or gene complex) for being a liberal or a conservative. Instead, genes code for temperament and people tend to sort themselves into the left and right clusters of moral values based on their personality preferences, moral emotions, hormones, and even brain structures. In *Predisposed: Liberals, Conservatives, and the Biology of Political Differences*, the political scientist John Hibbing and his colleagues report on a study they conducted in which they found that higher levels of sensitivity to pictures that trigger a feeling of disgust (such as eating worms) is predictive of political conservatism and disapproval of same-sex marriage.[84] Such findings help to explain why people are so predictable in their beliefs on such a wide range of seemingly unconnected issues—why someone who believes that the government should stay out of the private bedroom nevertheless believes that the government should be deeply involved in private business; or why someone who believes that the government should be small, taxes kept low, and spending maintained at a minimum nonetheless votes to raise taxes and increase spending on the military and the police.

The evidence that much of what divides us is rooted in our biology was compiled by the evolutionary anthropologist (and Peruvian political adviser) Avi Tuschman, in his transdisciplinary work *Our Political Nature*, in which he identifies three primary and relatively permanent personality traits running throughout political beliefs: *tribalism, tolerance for inequality,* and one's view of *human nature.* Xenophobia as a form of tribalism, for example, may be the result of breeding preferences in our evolutionary ancestors, where in cases in which infectious diseases are common—such as in warmer climes—people tend to be more sexually conservative and thereby tend not to prefer sexual partners from ethnic groups that differ from their own. Tuschman argues that conservatism encourages ethnocentrism, which in turn reinforces tribalism in a feedback loop between biology and culture; in like manner, liberalism encourages xenophilia and the desire to interact (and interbreed) with people from other groups. And, telling for the left-right division on religious matters, Tuschman shows a correlation between religiosity and ethnocentrism and conservatism, and that where rates of religion are high, so too is fecundity—religious people have more babies and thereby propagate their conservative-religious genes and culture. Tuschman sums up the evolutionary effects on our political personalities—left, right, and the moderate middle—this way:

> We're here with the political orientations we have because our ancestors' personalities helped them survive and reproduce successfully over thousands of generations. Their political personalities were instrumental in the regulation of inbreeding and outbreeding. These dispositions helped them mediate biological conflicts between parents, offspring, and siblings. And their moral emotions also balanced various types of altruism against self-interest in countless social interactions. In some types of social or ecological environments, more extreme personality traits were adaptive. In most cases, moderate personality solutions proved fit. That's one

reason why there are many moderates among us. Another reason for moderates and flexibility is that environments change, so it wouldn't make sense for our genes to rigidly determine our personalities. They just influence them based on the "memory" of our ancestors' success.[85]

It's not all biology, of course, and Tuschman shows that the home environment influences and interacts with biology. Here the effects are also multiplicative, however, as in something called "assortative mating," in which like-minded (and like-bodied) people tend to mate with one another due to biological preferences, and so the home environment effect is multiplied by the fact that the parents are not randomly sorted together.[86]

The left-right divide also depends heavily on the vision of human nature that you hold—as either *constrained* (right-wing) or *unconstrained* (left-wing) in Thomas Sowell's classification system from his work *A Conflict of Visions*,[87] or as *utopian* (left-wing) or *tragic* (right-wing) in Steven Pinker's taxonomy from *The Blank Slate*.[88] Left-wingers lean toward believing that human nature is largely unconstrained by biology, and thus utopian-like social engineering schemes to overcome poverty, unemployment, and other social ills are appealing in their logic and feasibility. Right-wingers lean toward believing that human nature is largely constrained by biology and thus social, political, and economic policies must necessarily be limited in their scope and ambition. In *The Believing Brain* I showed that the *constrained-tragic* vision better matches the data we have from the sciences that illuminate human nature, and I pressed for a *Realistic Vision* of human nature that can be embraced by both the left and the right as one that is relatively constrained by our biology and evolutionary history but can be modified by social and political systems that adjust the dials up and down.[89]

A *Realistic Vision* rejects the blank slate model that suggests people are so malleable and responsive to social programs that governments can actually design and engineer their lives into the ultimate society (the mistake made by all utopias). The *Realistic Vision* acknowledges that people vary widely both physically and intellectually—in large part because of natural inherited differences—and therefore will rise (or fall) to their natural levels. Family, custom, law, and traditional institutions are all sources for social harmony because they direct us toward internal forms of self-control over our passions. Short-term solutions include top-down rules and regulations; long-term solutions include internalizing values that reinforce the logic of honor and integrity and playing by the rules. We need external nudges on our behavior, but the long-term goal for moral improvement must come from within so that it becomes second nature.

Finally, the left-right divide and its biological roots can be linked to ideology through psychology and history, starting with the definition of ideology as "a set of beliefs about the proper order of society and how it can be achieved."[90] The origin of the modern labels of left and right can be traced to the French Assembly of 1789 on the eve of the French Revolution, in which the delegates who favored preservation of the *Ancien Régime* (the Old Order) of France sat on the right side of the chamber, while those who favored change sat on the left.[91] Since then, the terms *left* and *right* have stood for liberalism and conservatism, respectively.

In political and economic beliefs, our evolved tribal instincts to be seen as a reliable group member whom others can count on as being consistent from day to day lead us to punish "flip-floppers" who change their minds willy-nilly and thus may betray our tribe to an enemy or defect on the internal social contract that binds us into a cohesive clan. Consistency is not the hobgoblin of small minds, but a signal to others that we can be trusted. The evolved overarching moral emotion is to be consistent in our beliefs in the interest of group unity. Thus it is that specific beliefs may themselves be contradictory even while we convince ourselves that we are being consistent. This explains why conservatives can righteously claim to be in favor of freedom even while legislating what people do in the privacy of their bedrooms, and liberals can righteously proclaim government has no business telling us how to run our private lives even while legislating our guns and money. The inconsistency arises from the fact that we have competing motives within our minds that evolved for different purposes. Such predictability can be seen in the political narratives the left and the right spin to their moral righteousness. For example, which of these narratives best fits your own political beliefs?

> *Once upon a time people lived in societies that were unequal and oppressive, where the rich got richer and the poor got exploited. Chattel slavery, child labor, economic inequality, racism, sexism, and discriminations of all types abounded until the liberal tradition of fairness, justice, care, and equality brought about a free and fair society. And now conservatives want to turn back the clock in the name of greed and God.*

> *Once upon a time people lived in societies that embraced values and tradition, where people took personal responsibility, worked hard, enjoyed the*

fruits of their labor, and through charity helped those in need. Marriage, family, faith, honor, loyalty, sanctity, and respect for authority and the rule of law brought about a free and fair society. And now liberals want to over-turn these institutions in the name of utopian social engineering.

Although we may quibble over the details, political science research shows that most people fall onto a left-right spectrum, with these two grand narratives as bookends. In his book *Moral, Believing Animals* the sociolo-gist Christian Smith has constructed similar composite narratives that capture the moral foundations that most concern each side and that reflect the ancient tradition of "once upon a time things were bad and now they're good thanks to our party" or "once upon a time things were good but now they're bad thanks to the other party."[92] So consistent are we in our beliefs that if you identify with the first narrative I predict that if you live in the United States, you read the *New York Times*, listen to progressive talk radio, watch CNN, are pro-choice, anti-gun, adhere to the separation of church and state, are in favor of universal health care, and vote for mea-sures to redistribute wealth and tax the rich. If you lean toward the second narrative I predict that you read the *Wall Street Journal*, listen to conser-vative talk radio, watch Fox News, are pro-life, antigun control, believe that America is a Christian nation that should not ban religious expres-sions in the public sphere, are against universal health care, and vote against measures to redistribute wealth and tax the rich.

This political duopoly means that we need two parties in competition to reach a livable middle ground. Bertrand Russell reached deep into human history to identify the divide: "From 600 B.C. to the present day, philosophers have been divided into those who wished to tighten social bonds and those who wished to relax them. . . . It is clear that each party to this dispute—as to all that persist through long periods of time—is partly right and partly wrong. Social cohesion is a necessity, and mankind has never yet succeeded in enforcing cohesion by merely rational argu-ments. Every community is exposed to two opposite dangers: ossification through too much discipline and reverence for tradition, on the one hand; on the other hand, dissolution, or subjection to foreign conquest, through the growth of an individualism and personal independence that makes cooperation impossible."[93]

In fact, the American and French revolutions that launched the Rights Revolutions mark the point at which the duopoly was played out most forcefully both on the battlefield and in the battle of the books. In *The Great Debate: Edmund Burke, Thomas Paine, and the Birth of Left and Right*, the political analyst Yuval Levin allows these two intellectual giants to speak for each side through their works, and shows how most of today's political debates between liberals and conservatives have their roots in these foundational positions.[94] Burke has long been associated with and quoted by conservatives, who themselves foment against fomenting revolution because it too often descends into chaos, anarchy, and violence. If political change is called for it should happen gradually and only after due deliberation, because human history, he wrote, "consists for the most part of the miseries brought upon the world by pride, ambition, avarice, revenge, lust, sedition, hypocrisy, ungoverned zeal, and all the train of disorderly appetites."[95] Thus, when his colonial colleagues asserted in their Declaration of Independence that "Prudence, indeed, will dictate that Governments long established should not be changed for light and transient causes," Burke cast his support their way. By contrast, he did not support the revolution in France, "where the Elements which compose Human Society seem all to be dissolved and a world of Monsters to be produced in the place of it."[96] Burke wrote to his son in 1789, describing the country later that month as "a country undone." As France descended into chaos and bloodshed, in 1790 Burke told the British Parliament that "The French had shewn themselves the ablest architects of ruin that had hitherto existed in the world. In that very short space of time they had completely pulled down to the ground, their monarchy; their church; their nobility; their law; their revenue; their army; their navy; their commerce; their arts; and their manufactures."[97] There are good and bad ways to reform a government or a society, and by Burke's reckoning the Americans did it the right way and the French most emphatically did not.

On the other side of the political fence was Thomas Paine, whose 1776 political pamphlet *Common Sense* was a best-selling call to revolution that garnered him the title "Father of the American Revolution." There, in the section "On the Origin and Design of Government in General," Paine explained that "Society is produced by our wants, and government by our wickedness; the former promotes our happiness positively by uniting our affections, the latter negatively by restraining our vices."[98] But unlike Burke and his respect for religion as a force for teaching people to control their passions, Paine was a Deist—perhaps even an atheist—who had nothing but disdain for organized religion, as is evident when he wrote in

The Age of Reason: "Of all the systems of religion that ever were invented, there is none more derogatory to the Almighty, more unedifying to the man, more repugnant to reason, and more contradictory in itself, than this thing called Christianity." His worldview was that of an Enlightenment humanist, and he once said, "I believe in the equality of man; and I believe that religious duties consist of doing justice, loving mercy, and endeavoring to make our fellow-creatures happy. . . . I do not believe in the creed professed by the Jewish church, by the Roman church, by the Greek church, by the Turkish church, by the Protestant church, nor by any church that I know of. My own mind is my own church. All national institutions of churches whether Jewish, Christian or Turkish, appear to me no other than human inventions, set up to terrify and enslave mankind, and monopolize power and profit."[99]

Whence, then, come morality and civil society? They come from reason, Paine argued, and in his own way he applied the *principle of interchangeable perspectives* when he wrote in his 1795 *Dissertation on First Principles of Government*: "He that would make his own liberty secure, must guard even his enemy from oppression; for if he violates this duty, he establishes a precedent that will reach to himself."[100]

Who was right, Burke or Paine? Your answer may depend on your temperament and its concomitant political preference, but we would do well to heed the wisdom of one of the most brilliant political thinkers of the nineteenth century, John Stuart Mill, who condensed all these debates into a single observation: "A party of order or stability, and a party of progress or reform, are both necessary elements of a healthy state of political life."[101]

TESTING LIBERTY VS. LEVIATHAN

Most people most of the time in most circumstances are honest and fair and cooperative and they want to do the right thing for their community and society. But most people are also competitive, aggressive, and self-interested and they want to do the right thing for themselves and their family. This evolved disposition sets up two potential areas of conflict: (1) within ourselves, our selfish desire for self-improvement conflicts with our altruistic desire for social enhancement; and (2) our competitive desire to better our lot in life sometimes comes into conflict with the same desire that others have in themselves. The deeply insightful and always provocative H. L. Mencken captured the essence of the Janus-faced nature of the Leviathan in a 1927 essay titled simply "Why Liberty?":

I believe liberty is the only genuinely valuable thing that men have invented, at least in the field of government. I believe it is better to be free than not to be free, even when the former is dangerous and the latter appears to be safe. I believe that the finest qualities of men can flourish only in free air—that progress made under the shadow of the policeman's club is false and of no permanent value. I believe that any man who takes the liberty of another into his keeping is bound to become a tyrant, and any man who yields up his liberty, in however the slightest measure, is bound to become a slave.[102]

Liberty is not just an ideal. It has practical real-world results. Look at the striking difference between a democracy with open borders and free trade (South Korea) and a dictatorship with closed borders and next to no trade (North Korea) in *figure 3-11*. Measured by per capita GDP in 1990 dollars over the past half century, as a historical experiment using the *comparative method* the results could not be more dramatic. The experiment began in August 1945, when the two countries were partitioned at the thirty-eighth parallel, dividing one of the most homogeneous societies in the world. Both countries began the experiment with an annual average per capita GDP of $854 and were in lockstep parallel until the 1970s, when South Korea implemented economic measures to grow its economy and North Korea turned into a dictatorship, with North Koreans goose-stepping to the tyrannical Great Leader and Eternal President (Kim Il-sung).

By 2013 South Korea ranked 12th out of 186 countries on the United Nations' Human Development Index, with a life expectancy at birth of 80.6. The Bertelsmann Foundation Transformation Index (BTI) ranked South Korea 11th in the world on political and economic development.[103] By contrast, North Korea was ranked 125th out of 128 countries on the BTI, with a life expectancy of 68.8. And it was shrinking not only economically but physically as well, and by millimeters every year. According to Daniel Schwekendiek from Sungkyunkwan University in Seoul, who conducted a study of the height of North Korean refugees that he measured when they crossed the border into South Korea, North Korean men averaged 3 to 8 centimeters (1.2 to 3.1 inches) shorter than their South Korean counterparts. In children, he found, the height gap between North and South Koreans averages 4 centimeters (1.6 inches) among preschool boys and 3 centimeters (1.2 inches) among preschool girls.[104] Economically, with a per capita GDP difference of $19,614 vs. $1,122, the two countries have diverged by 1,748 percent.

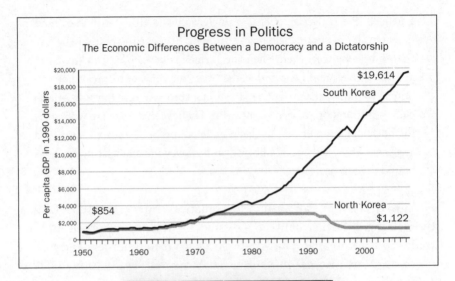

Progress in Politics

The Economic Differences Between a Democracy and a Dictatorship

Figure 3-11. The Economic Differences Between a Democracy and a Dictatorship

The average annual per capita GDP difference in 1990 dollars between North and South Korea by 2010 was a staggering 1,748 percent, or a net difference of $18,492.[106] What does that buy you? For starters, an average range in height differences among men of 1.2 to 3.1 inches, the direct result of nutrition. The differences are also visible from space in this satellite photograph from NASA's Earth Observatory in which South Korea's prosperity ends at the border and North Korea fades into darkness.[107]

Is this moral progress? Ask the South Korean citizens who have an extra $18,492 every year to spend on food, clothing, housing, and luxury goods. What does that buy you? For starters—height. And electricity. The per capita power consumption of South Korea is 10,162 kilowatt-hours compared to North Korea's 739 kilowatt-hours.[105] The difference is visible from space in satellite photographs.

SCIENCE, REASON, AND VALUES

The point of this exercise in testing historical hypotheses is that in addition to making philosophical arguments, we can make a *scientific case* for liberal democracy, market economies, and international transparency as means of increasing prosperity, health, and happiness. Although there may be many peaks on the moral landscape, as Sam Harris argues,[108] science can help us sort through them quantitatively. Perhaps liberals, conservatives, libertarians, Tea Partiers, Green Partiers, and others can coexist on different peaks within a range of possibilities on that landscape that are themselves vastly superior to a different cluster of peaks that emphasize autocratic rule. And we can measure those differences quantitatively through science.

The fact that there may be many types of democracies (e.g., direct and representative) and economies (with various trade agreements or membership in trading blocs) only suggests that human well-being is multifaceted and multicausal, and not that the existence of more than one way to survive implies that all political, economic, and social systems are equal. They are not equal, and we have the scientific data and historical examples to prove it. When we have to squint to see the differences between or among political parties, then that is what elections are for, where the winners get their way until the next election and the losers lump it; either way, the differences are minuscule.

In other words, we can ground human values and morals not just in philosophical principles such as Aristotle's virtue ethics, Kant's categorical imperative, Mill's utilitarianism, or Rawls's fairness ethics, but in science as well. From the Scientific Revolution through the Enlightenment, reason and science slowly but systematically replaced superstition, dogmatism, and religious authority. As the German philosopher Immanuel Kant proclaimed, "*Sapere Aude!*—dare to know! Have the courage to use your own understanding." As he explained, "Enlightenment is man's emergence from his self-imposed immaturity. Immaturity is the inability to use one's understanding without guidance from another." The Age of Reason, then, was the age when humanity was born again, not from original sin, but from original ignorance and dependence on authority and superstition. Never again should we allow ourselves to be the intellectual slaves of those who would bind our minds with the chains of dogma and authority. In its stead we use reason and science as the arbiters of truth and knowledge.

Instead of divining truth through the authority of an ancient holy book or philosophical treatise, people began to explore the book of nature for themselves.

Instead of looking at illustrations in illuminated botanical books, scholars went out into nature to see what was actually growing out of the ground.

Instead of relying on the woodcuts of dissected bodies in old medical texts, physicians opened bodies themselves to see with their own eyes what was there.

Instead of human sacrifices to assuage the angry weather gods, naturalists made measurements of temperature, barometric pressure, and winds to create the meteorological sciences.

Instead of enslaving people because they were a lesser species, we expanded our knowledge to include all humans as members of the species through evolutionary sciences.

Instead of treating women as inferiors because a certain book says it is men's right to do so, we discovered natural rights that dictate that all people should be treated equally through the moral sciences.

Instead of the supernatural belief in the divine right of kings, people employed a natural belief in the legal right of democracy, and this gave us the political sciences.

Instead of a tiny handful of elites holding most of the political power by keeping their citizens illiterate, uneducated, and unenlightened, through science, literacy, and education people could see for themselves the power and corruption that held them down and began to throw off their chains of bondage and demand rights.

Instead of labeling homosexuality as an abomination, or atheists and nonbelievers as immoral noncitizens, or animals as automatons to be exploited in any way we choose, we are today engaged in a great legal struggle to make these final legal hurdles in the long rights revolution.

The constitutions of nations should be grounded in the constitution of humanity, which science and reason are best equipped to understand.

4

Why Religion Is Not the Source of Moral Progress

All the world's major religions, with their emphasis on love, compassion, patience, tolerance, and forgiveness can and do promote inner values. But the reality of the world today is that grounding ethics in religion is no longer adequate. This is why I am increasingly convinced that the time has come to find a way of thinking about spirituality and ethics beyond religion altogether.

—The Dalai Lama's Facebook statement for September 10, 2012

The claim that religion cannot be the driver of moral progress will surprise—and in many cases offend—some readers, who may assume that advancement in the realm of morality has been primarily due to the guiding light of religious teachings.[1] The reason for this misunderstanding is twofold. One, religion has had a monopoly on morality for millennia, and so we have grown accustomed to associating any moral progress with the one institution most closely associated with it. Two, religious institutions take credit for moral progress while ignoring or glossing over moral regress. Before I turn to what the data show, a brief history of religious morality is illuminating for my thesis.

On the good side of the moral scale, it was Jesus who said to help the poor, to turn the other cheek, to love thine enemies, to judge not lest ye be judged, to forgive sinners, and to give people a second chance. In the name of their religion, people have helped the poor and needy in developed nations around the world, and in America they are the leading supporters of food banks for the hungry and postdisaster relief. Many Christian theologians, along with Christian churches and preachers, advocated the abolition of the slave trade, and continued to press for justice in modern times. Some civil rights leaders were motivated by their religion, most notably the Reverend Martin Luther King Jr., whose speeches were filled with passionate religious tropes and quotes. I have deeply religious friends who are highly driven to do good and, though they may have a complex variety of motives, they often act in the name of their particular religion.

So religion can and does motivate people to do good works, and we should always acknowledge any person who or institution that pushes humanity farther along the path of progress, expands the moral sphere, or even just makes the life of one other person a little better. To that end we would do well to emulate the ecumenicalism of the late astronomer Carl Sagan, who appealed to all religious faiths to join scientists in working to preserve the environment and to end the nuclear arms race. He did so because, he said, we are all in this together; our problems are "transnational, transgenerational and transideological. So are all conceivable solutions. To escape these traps requires a perspective that embraces the peoples of the planet and all the generations yet to come."[2] That stirring rhetoric urges all of us—secularists and believers—to work together toward the common goal of making the world a better place.

But for too long the scales of morality have been weighed down by the religious thumb pressing on the side of the scale marked "Good." Religion has also promoted, or justified, such catastrophic moral blunders as the Crusades (the People's Crusade, the Northern Crusade, the Albigensian Crusade, and Crusades One through Nine); the Inquisitions (Spanish, Portuguese, and Roman); witch hunts (products, in part, of the Inquisitions that ran from the Middle Ages through the Early Modern Period and executed tens of thousands of people, mostly women); Christian conquistadors who exterminated native peoples by the millions through their guns, germs, and steel; the endless European wars of religion (the Nine Years' War, the Thirty Years' War, the Eighty Years' War, the French Wars of Religion, the Wars of the Three Kingdoms, the English Civil War, to name just a few); the American Civil War, in which Northern Christians and Southern Christians slaughtered one another over the issue of slavery and states' rights; and the First World War, in which German Christians fought French, British, and American Christians, all of whom believed that God was on *their* side. (German soldiers had *Gott mit uns—God with us*—embossed in the metal of their belt buckles.) And that's just in the Western world. There are the seemingly endless religious conflicts in India, Indonesia, Afghanistan, Pakistan, Iraq, Sudan, and numerous countries in

Africa, the Coptic Christian persecution in Egypt, and of course Islamist terrorism has been a scourge on societal peace and security in recent decades, and a day doesn't go by without some act of violence committed in the name of Islam.

All of these events have political, economic, and social causes, but the underlying justification they share is religion. Once moral progress in a particular area is under way, most religions eventually get on board—as in the abolition of slavery in the nineteenth century, women's rights in the twentieth century, and gay rights in the twenty-first century—but this often happens after a shamefully protracted lag time. In this chapter I will focus primarily on the effects of religion in the Western world, especially Christianity, since it has been so influential in the history of the West and because it lays claim more than any other to being the driver of moral progress.

WHY RELIGION *CANNOT* BE THE DRIVER OF MORAL PROGRESS

The rules that were dreamed up and enshrined by the various religions over the millennia did not have as their goal the expansion of the moral sphere. Including other sentient beings within the circle of moral concern was not on their radar. Moses did not come down from the mountain with a detailed list of the ways in which the Israelites could make life better for the Moabites, the Edomites, the Midianites, or for any other tribe of people that happened not to be *them*. One justification for this constricted sphere can be found in the Old Testament injunction to "Love thy neighbor." Who, precisely, is thy neighbor? According to the anthropologist John Hartung, thy neighbor was understood to be one's immediate kin and kind, which was admittedly an evolutionary stratagem appropriate for the time:

> The phrase *Love thy neighbor as thyself* comes from the Torah. The word *Torah* means law and *the* Torah is *the* Law. If Moses had been transmitting the word of his god to modern biologists he might have said "Love your neighbor as if $r = 1$—as if all of your genes are identical." According to the ancient Israelites' autobiographical ethnography, this was the general principle from which prohibitions against murder, theft, and lying were derived. But who qualifies for this apex of morality? Who is thy *neighbor*?[3]

Hartung notes that "Most contemporary Jews and Christians, who both have the highest regard for the god of the Torah, answer that the law applies to everybody," but as I have outlined above, this is because Jews and Christians have inculcated into their moral thinking the modern Enlightenment goal of broadening and redefining the parameters of moral consideration. But that is not what the authors of the Old Testament had in mind, as Hartung explains:

> [W]hen the Israelites received the love law, they were isolated in a desert. According to the account, they lived in tents clustered by extended families, they had no non-Israelite neighbors, and dissention was rife. Internecine fighting became rather vicious, with about 3,000 killed in a single episode (Exodus 32:26–28). Most of the troops wanted to "choose a [new] captain and go back to Egypt" (Numbers 14:4). But their old captain, Moses, preferred group cohesion. If we want to know who Moses thought his god meant by neighbor, the law must be put into context, and the minimum context that makes sense is the biblical verse from which the love law is so frequently extracted.

The biblical passage in question is Leviticus 19:18, which reads, "Thou shalt not avenge, nor bear any grudge against the children of thy people, but thou shalt love thy neighbor as thyself." As Hartung notes, "In context, neighbor meant 'the children of thy people,' . . . in other words, fellow in-group members."

Again, from an evolutionary perspective this makes perfect sense. Indeed, it would be suicidal to love thy neighbor as thyself when thy neighbor would like nothing better than to exterminate you, which was often the case for the Bronze Age peoples of the Old Testament. What good would have come of the Israelites loving, for example, the Midianites as themselves? The results would have been catastrophic given that the Midianites were allied with the Moabites in their desire to see the Israelites wiped off the face of the earth. That's why Moses assembled an army of twelve thousand troops as recorded in Numbers 31:7–12:

> They warred against Mid'ian, as the LORD commanded Moses, and slew every male. They slew the kings of Mid'ian. . . . And the people of Israel took captive the women of Mid'ian and their little ones; and they took as booty all their cattle, their flocks, and all their goods. All their cities in the places where they dwelt, and all their encampments, they burned

with fire, and took all the spoil and all the booty, both of man and of beast. Then they brought the captives and the booty and the spoil to Moses.

That sounds like a good day's pillaging, but when the troops got back, Moses was furious. "What do you mean you didn't kill the *women*?" he asked, exasperated, since it was apparently the women who had enticed the Israelites to be unfaithful with another god. Moses then ordered them to kill all the women who had slept with a man, and the boys. "But save for yourselves every girl who has never slept with a man," he commanded, predictably, at which point one can imagine the thirty-two *thousand* virgins who'd been taken captive rolling their eyes and saying, "Oh, *God* told you to do that, did he? Right." Was the instruction to "keep the virgins for yourselves" what God had in mind by the word "love" in the "love thy neighbor" command? I think not. Of course, the Israelites knew *exactly* what God meant (this is the advantage of writing scripture yourself—you get to say what God meant) and they acted accordingly, fighting for the survival of their people. *With a vengeance.*

The world's religions are tribal and xenophobic by nature, serving to regulate moral rules within the community but not seeking to embrace humanity outside their circle. Religion, by definition, forms an identity of *those like us*, in sharp distinction from *those not us, those heathens, those unbelievers.* Most religions were pulled into the modern Enlightenment with their fingernails dug into the past. Change in religious beliefs and practices, when it happens at all, is slow and cumbersome, and it is almost always in response to the church or its leaders facing outside political or cultural forces.

The history of Mormonism is a case in point. In the 1830s the church's founder, Joseph Smith, received a revelation from God to enact what he euphemistically called "celestial marriage," more accurately described as "plural marriage"—the rest of the world calls it polygamy—just about the time he found a new love interest while married to another woman. Once Smith caught the Solomonic fever for multiple wives (King Solomon had at least seven hundred), he couldn't stop himself or his brethren from spreading their seed, along with the practice, which in 1852 was codified into Mormon law through its sacred *Doctrines and Covenants.* Until 1890, that is, when the people of Utah—desirous for their territory to become a state in the Union—were told by the US federal government that polygamy would not be tolerated; it was outlawed in all other states.

Conveniently, God issued a new revelation to the Mormon leaders, instructing them that a plurality of wives was no longer a celestial blessing, and that instead monogamy was now the One True Way. As well, Mormon policy forbade African Americans to be priests in the church. The reason, Joseph Smith had decreed, was that they are not actually from Africa but instead are descendants of the evil Lamanites, whom God cursed by making their skin black after they lost the war against the good Nephites, both clans of which were descendants of two of the lost tribes of Israel. Naturally, since the evil Lamanites were prohibited from having sexual relations with the good Nephites, interracial marriage was also banned. (Smith claimed to know this story because it had been dictated in an ancient language onto gold plates by the angel Moroni, who buried them in Smith's backyard near Palmyra, New York. Smith translated the plates into English by burying his face in a hat containing magic stones.) This racist nonsense lasted a century and a half until it came into contact with the civil rights movement of the 1960s and 1970s. Even then, it took the church a while to notice that the times they were a-changing. In 1978, the church head, Spencer W. Kimball, announced that he had received a new revelation from God instructing him to drop the racial restrictions and adopt a more inclusive attitude.[4]

There are three reasons for the sclerotic nature of religion: (1) The foundation of the belief in an absolute morality is the belief in an absolute religion grounded in the One True God. This inexorably leads to the conclusion that anyone who believes differently has departed from this truth and thus is unprotected by our moral obligations. (2) Unlike science, religion has no systematic process and no empirical method to employ to determine the verisimilitude of its claims and beliefs, much less right and wrong. (3) The morality of holy books—most notably the Bible—is not the morality any of us would wish to live by, and thus it is not possible for the religious doctrines derived from holy books to be the catalyst for moral evolution. Let's look at this last point in more detail to understand why.

THE MORALITY OF THE BIBLE

The Bible is one of the most immoral works in all literature. Woven throughout begats and chronicles, laws and customs, is a narrative of accounts written by, and about, a bunch of Middle Eastern tribal warlords who constantly fight over land and women, with the victors taking dominion over both. It features a jealous and vengeful God named Yahweh who decides to punish women for all eternity with the often intolerable pain of childbirth, and further condemns them to be little more than beasts of burden and sex slaves for the victorious warlords. Why were women to be chastened this way? Why did they deserve an eternity of misery and submission? It was all for that one terrible sin, the first crime ever recorded in the history of humanity—a thought crime, no less—when that audacious autodidact Eve dared to educate herself by partaking of the fruit of the tree of the knowledge of good and evil. Worse, she inveigled the first man—the unsuspecting Adam—to join her in choosing knowledge over ignorance. For the appalling crime of hearkening unto the voice of his wife, Yahweh condemned Adam to toil in thorn- and thistle-infested fields, and further condemned him to death, to return to the dust whence he came.

Yahweh then cast his first two delinquent children out of paradise, setting an angel and a flaming sword at the entrance to be certain that they could never return. Then, in one of the many foul moods he was wont to fall into, Yahweh committed an epic hemoclysm of genocidal proportions by killing every sentient being on Earth—including unsuspecting adults, innocent children, and all the land animals—in a massive flood. To repopulate the planet after he decimated it of all life save those spared in the Ark, Yahweh commanded the survivors—numerous times—to "be fruitful and multiply," and rewarded his favorite warlords with as many wives as they desired. Thus was born the practice of polygamy and the keeping of harems, fully embraced and endorsed—along with slavery—in the "good book."

As an exercise in moral casuistry, this perspective-taking question comes to mind: did anyone ask the *women* how *they* felt about this arrangement? What about the millions of people living in other parts of the world who had never heard of Yahweh? What about the animals and the innocent children who drowned in the Flood? What did they do to deserve such a final solution to Yahweh's anger problem?

Many Christians say that they get their morality from the Bible, but this cannot be true because as holy books go the Bible is possibly the most unhelpful guide ever written for determining right from wrong. It's chock-full of bizarre stories about dysfunctional families, advice about how to beat your slaves, how to kill your headstrong kids, how to sell your virgin daughters, and other clearly outdated practices that most cultures gave up centuries ago.

Consider the morality of the biblical warlords who had no qualms about taking multiple wives, adultery, keeping concubines, and fathering countless children from their many polygamous arrangements. The anthropologist Laura Betzig has put these stories into an evolutionary context in noting that Darwin predicted that successful competition leads to successful reproduction. In his 1871 book *The Descent of Man and Selection in Relation to Sex*, Darwin showed how natural selection works when members of a species compete for reproductive resources: "It is certain that amongst almost all animals there is a struggle between the males for the possession of the females," adding this anthropological observation: "With savages, the women are the constant cause of war."[5]

With this framework in mind Betzig analyzed the Old Testament and found no less than forty-one named polygamists, not one of which was a powerless man. "In the Old Testament, powerful men—patriarchs, judges, and kings—have sex with more wives; they have more sex with other men's women; they have sex with more concubines, servants, and slaves; and they father many children."[6] And not just the big names. According to Betzig's analysis, "men with bigger herds of sheep and goats tend to have sex with more women, then to father more children."[7] Most of the polygynous patriarchs, judges, and kings had two, three, or four wives with a corresponding number of children, although King David had more than eight wives and twenty children, King Abijah had fourteen wives and thirty-eight children, and King Rehoboam had eighteen wives (and sixty other women) who bore him no fewer than eighty-eight offspring. But they were all lightweights compared to King Solomon, who married at least seven hundred women. There were Moabite, Ammonite, Edomite, Sidonian, and Hittite women he married, then for good measure added three hundred concubines, which he called "man's delight."[8] (What Solomon's concubines called *him* was never recorded.)

Although many of these stories are fiction (there is no evidence, for example, that Moses ever existed, much less led his people for forty years in the desert but leaving behind not a single archaeological artifact), what these biblical patriarchs purportedly did to women was, in fact, how most men treated women at that time, *and that's the point*. Put into context, the Bible's moral prescriptions were for another time for another people and have little relevance for us today.

To make the Bible relevant, believers must pick and choose biblical passages that suit their needs; thus the game of cherry picking from the Bible generally works to the advantage of the pickers. In the Old Testament, the believer might find guidance in Deuteronomy 5:17, which says, explicitly, "Thou shalt not kill"; or in Exodus 22:21, a verse that delivers a straightforward and indisputable prohibition: "You shall not wrong a stranger or oppress him, for you were strangers in the land of Egypt."

These verses seem to set a high moral bar, but the handful of positive moral commands in the Old Testament are desultory and scattered among a sea of violent stories of murder, rape, torture, slavery, and all manner of violence, such as occurs in Deuteronomy 20:10–18, in which Yahweh instructs the Israelites on the precise etiquette of conquering another tribe:

> When you draw near to a city to fight against it, offer terms of peace to it. And if its answer to you is peace and it opens to you, then all the people who are found in it shall do forced labor for you and shall serve you. But if it makes no peace with you, but makes war against you, then you shall besiege it; and when the LORD your God gives it into your hand you shall put all its males to the sword, but the women and the little ones, the cattle, and everything else in the city, all its spoil, you shall take as booty for yourselves. . . . But in the cities of these peoples that the LORD your God gives you for an inheritance you shall save alive nothing that breathes, but you shall utterly destroy them, the Hittites and the Amorites, the Canaanites and the Perizzites, the Hivites and the Jebusites, as the LORD your God has commanded.

Today, as the death penalty fades into history, in the Old Testament Yahweh offers this list of actions punishable by death:

- *Blaspheming or cursing the Lord*: "And he that blasphemeth the name of the LORD, he shall surely be put to death, and all the congregation shall certainly stone him: as well the stranger, as he that is born in the land, when he blasphemeth the name of the Lord, shall be put to death." (Leviticus 24:13–16)
- *Worshiping another god*: "He that sacrificeth unto any god, save unto the LORD only, he shall be utterly destroyed." (Exodus 22:20)
- *Witchcraft and wizardry*: "Thou shalt not suffer a witch to live." (Exodus 22:18) "A man also or woman that hath a familiar spirit, or that is a wizard, shall surely be put to death: they shall stone them with stones: their blood shall be upon them." (Leviticus 20:27)
- *Female loss of virginity before marriage*: "If any man take a wife [and find] her not a maid . . . Then they shall bring out the damsel to the door of her father's house, and the men of her city shall stone her with stones that she die." (Deuteronomy 22:13–21)

- *Homosexuality*: "If a man also lie with mankind, as he lieth with a woman, both of them have committed an abomination: they shall surely be put to death; their blood shall be upon them." (Leviticus 20:13)
- *Working on the Sabbath*: "Six days shall work be done, but on the seventh day there shall be to you an holy day, a sabbath of rest to the LORD: whosoever doeth work therein shall be put to death." (Exodus 35:2)

The book considered by more than two billion people to be the greatest moral guide ever produced—inspired as it was by an all-knowing, totally benevolent deity—recommends the death penalty for saying the Lord's name at the wrong moment or in the wrong context, for imaginary crimes such as witchcraft, for commonplace sexual relations (adultery, fornication, homosexuality), and for the especially heinous crime of not resting on the Sabbath. How many of today's two billion Christians agree with their own holy book on the application of capital punishment?

And how many would agree with this gem of moral turpitude from Deuteronomy 22:28–29: "If a man meets a virgin who is not engaged, and seizes her and lies with her, and they are caught in the act, the man who lay with her shall give fifty shekels of silver to the young woman's father, and she shall become his wife. Because he violated her he shall not be permitted to divorce her as long as he lives." I dare say no Christian today would follow this moral directive. No one today—Jew, Christian, atheist, or otherwise—would even *think* of such draconian punishment for such acts. That is how far the moral arc has bent in four millennia.

The comedian Julia Sweeney, in her luminous monologue "Letting Go of God," makes the point when she recalls rereading a familiar story she learned in her Catholic childhood upbringing:

> This Old Testament God makes the grizzliest tests of people's loyalty. Like when he asks Abraham to murder his son, Isaac. As a kid, we were taught to admire it. I caught my breath reading it. We were taught to admire it? What kind of sadistic test of loyalty is that, to ask someone to kill his or her own child? And isn't the proper answer, "No! I will not kill my child, or any child, even if it means eternal punishment in hell!"?[9]

Like so many other comedians who've struck the Bible's rich vein of unintended comedic stories, Sweeney allows the material to write itself. Here she continues her tour through the Old Testament with its preposterous commandments:

> Like if a man has sex with an animal, both the man and the animal should be killed. Which I could almost understand for the man, but the animal? Because the animal was a willing participant? Because now the animal's had the taste of human sex and won't be satisfied without it? Or my personal favorite law in the Bible: in Deuteronomy, it says if you're a woman, married to a man, who gets into a fight with another man, and you try to help him out by grabbing onto the genitals of his opponent, the Bible says you immediately have to have your hand chopped off.[10]

Richard Dawkins memorably characterized this God of the Old Testament as "arguably the most unpleasant character in all fiction: jealous and proud of it; a petty, unjust, unforgiving control-freak; a vindictive, bloodthirsty ethnic cleanser; a misogynistic, homophobic, racist, infanticidal, genocidal, filicidal, pestilential, megalomaniacal, sadomasochistic, capriciously malevolent bully."[11] Most modern Christians, however, respond to arguments like mine and Dawkins's by saying that the Old Testament's cruel and fortunately outdated laws have nothing to do with how they live their lives or the moral precepts that guide them today. The angry, vengeful God Yahweh of the Old Testament, Christians claim, was displaced by the kinder, gentler *New* Testament God in the form of Jesus, who two millennia ago introduced a new and improved moral code. Turning the other cheek, loving one's enemies, forgiving sinners, and giving to the poor is a great leap forward from the capricious commands and copious capital punishment found in the Old Testament.

That may be, but nowhere in the New Testament does Jesus revoke God's death sentences or ludicrous laws. In fact, quite the opposite (Matthew 5:17–30 passim): "Think not that I am come to destroy the law, or the prophets: I am not come to destroy, but to fulfill." He doesn't even try to edit the commandments or soften them up: "Whosoever therefore shall break one of these least commandments, and shall teach men so, he shall be called the least in the kingdom of heaven." In fact, if anything, Jesus's morality is even more draconian than that of the Old Testament: "Ye have heard that it was said by them of old time, Thou shalt not kill; and whosoever shall kill shall be in danger of the judgment: But I say unto you, That whosoever is angry with his brother without a cause shall be in danger of the judgment."

In other words, even *thinking* about killing someone is a capital offense. In fact, Jesus elevated *thought crimes* to an Orwellian new level (Matthew 9:27–29): "Ye have heard it was said by them of old time, Thou shalt not commit adultery: But I say unto you, That whosoever looketh on a woman to lust after her hath committed adultery with her already in his heart." And if you don't think you can control your sexual impulses Jesus has a practical solution: "If thy right eye offend thee, pluck it out, and cast it from thee: for it is profitable for thee that one of thy members should perish, and not that thy whole body should be cast into hell." President Bill Clinton may have physically sinned in the White House with an intern, but by Jesus's moral code even the evangelical Christian Jimmy Carter sinned when he famously admitted in a 1976 *Playboy* magazine interview while running for president, "I've looked on a lot of women with lust. I've committed adultery in my heart many times."[12]

As for Jesus's own family values, he never married, never had children, and he turned away his own mother time and again. For example, at a wedding feast Jesus says to her (John 2:4): "Woman, what have I to do with you?" One biblical anecdote recounts the time that Mary waited patiently off to the side for Jesus to finish speaking so that she could have a moment with him, but Jesus told his disciples, "Send her away, you are my family now," adding (Luke 14:26): "Whoever comes to me and does not hate father and mother, wife and children, brothers and sisters, yes, and even life itself, cannot be my disciple."

Charming. This is what cultists do when they separate followers from their families to control both their thoughts and their actions, as when Jesus calls to his flock to follow him *or else* (John 15:4–7): "Abide in me as I abide in you. Just as the branch cannot bear fruit by itself unless it abides in the vine, neither can you unless you abide in me. I am the vine, you are the branches. Those who abide in me and I in them bear much fruit, because apart from me you can do nothing. Whoever does not abide in me is thrown away like a branch and withers; such branches are gathered, thrown into the fire, and burned." But if a believer abandons his family and gives away his belongings (Mark 10:30), "he shall receive an hundredfold now in this time, houses, and brethren, and sisters, and mothers, and children, and lands." In other passages Jesus also sounds like the tribal warlords of the Old Testament:

> Do not think that I have come to bring peace to the earth; I have not come
> to bring peace, but a sword. For I have come to set a man against his

father, and a daughter against her mother, and a daughter-in-law against
her mother-in-law; and one's foes will be members of one's own household.
Whoever loves father or mother more than me is not worthy of me; and
whoever loves son or daughter more than me is not worthy of me; and whoever does not take up the cross and follow me is not worthy of me. (Matthew 10:34–39)

Even sincere Christians cannot agree on Jesus's morality and the moral
codes in the New Testament, holding legitimate differences of opinion on
a number of moral issues that remain unresolved based on biblical scripture alone. These include dietary restrictions and the use of alcohol, tobacco,
and caffeine; masturbation, premarital sex, contraception, and abortion;
marriage, divorce, and sexuality; the role of women; capital punishment
and voluntary euthanasia; gambling and other vices; international and civil
wars; and many other matters of contention that were nowhere in sight
when the Bible was written, such as stem-cell research, gay marriage, and
the like. Indeed, the fact that Christians, as a community, keep arguing
over their own contemporary question "WWJD?" (What Would Jesus Do?)
is evidence that the New Testament is silent on the answer.

IS RELIGION RESPONSIBLE FOR WESTERN CIVILIZATION?

Even if our morality does not originate in the Bible, religious believers will
often argue that Christianity gave Western civilization its most precious
assets: art, architecture, literature, music, science, technology, capitalism,
democracy, equal rights, and the rule of law. In the United States, you hear
it in the boosterism of everything from conservative talk radio to presidential speeches. According to the conservative president Ronald Reagan, for
example, America is "a shining city on a hill,"[13] a metaphor he got from the
liberal president John F. Kennedy, who quoted the seventeenth-century
cofounder of the Massachusetts Bay Colony, John Winthrop, who proclaimed, "We must always consider that we shall be as a city upon a hill—
the eyes of all people are upon us."[14] The reference originates from Jesus's
statement in the Sermon on the Mount to his followers: "You are the light
of the world. A city that is set on a hill cannot be hidden." (Matthew 5:14)

The question of religion's role in the shaping of the West is an empirical one, which the popular conservative Christian apologist Dinesh
D'Souza attempts to answer in the affirmative in his 2008 book *What's So*

Great About Christianity, sans question mark to affirm the case.[15] "Western civilization was built by Christianity," D'Souza proclaims. "The West was built on two pillars: Athens and Jerusalem. By Athens I mean classical civilization, the civilization of Greece and pre-Christian Rome. By Jerusalem I mean Judaism and Christianity. Of these two, Jerusalem is more important."

After the Dark Ages, in which marauding hordes of Huns, Goths, Vandals, and Visigoths overturned the progress that arose from Athens and Jerusalem and transformed Europe into a cultural backwater, Christianity illuminated the dark continent "with learning and order, stability and dignity. The monks copied and studied the manuscripts that preserved the learning of late antiquity."[16] To support his case, D'Souza quotes the historian J. M. Roberts, who opined in his book *The Triumph of the West*, "We could none of us today be what we are if a handful of Jews nearly two thousand years ago had not believed that they had known a great teacher, seen him crucified, dead, and buried, and then rise again."[17] Dante, Milton, Shakespeare, Mozart, Handel, Bach, Leonardo, Michelangelo, Rembrandt, and all the other geniuses throughout the past half millennium, D'Souza asserts, were inspired by the great "Christian themes of suffering, transformation, and redemption." D'Souza's point is not just that all these great artists were Christian. "Rather, it is that their great works would not have been produced without Christianity. Would they have produced other great works? We don't know. What we do know is that their Christianity gives their genius its distinctive expression. Nowhere has human aspiration reached so high or more deeply touched the heart and spirit than in the works of Christian art, architecture, literature, and music."[18]

This last claim is both preposterous and parochial. Were Homer and Sappho Christians? Were the Seven Wonders of the Ancient World inspired by Christ's great gift of salvation? Clearly we *do* know what geniuses in the past have done without Christianity, in the great ancient pre-Christian civilizations of Sumeria, Babylonia, Akkadia, Assyria, Egypt, and Greece in the West; the ancient civilizations that arose in the Indus Valley in modern-day Pakistan and India; in the Yellow River and Yangtze River valleys in modern-day China; and in many others besides. Every one of these peoples produced magnificent works of art and architecture, music and literature, science and technology—though it should be noted that both Christians and Muslims often did their best to annihilate all evidence of these achievements with innumerable acts of cultural vandalism, pillaging, and censorship.

No one disputes the magnificence of countless works of art, architecture, literature, and music that were inspired by Christianity: the great cathedrals that make the spirit soar, the requiems that capture the heartache of loss, the psalms of joy that unite listeners, the paintings that dazzle with light and human emotion. But artists who live in a Christian world, who are surrounded by other Christians, who understand next to nothing beyond Christianity, and who are likely being supported by Christian patrons are going to produce Christian work. Christianity was the dominant religion at a moment in history when Europe was going through a Renaissance and an explosion of discoveries of new lands and new sources of energy; no wonder it ended up being the great patron. Thus the fact that artists living in Christendom were inspired by the life and death of Jesus by crucifixion, and not, say, by the life and death of the Buddha by mushrooms is not a surprise. Christianity was the only game in town.

RELIGION AND CAPITALISM

In his 2005 book *The Victory of Reason: How Christianity Led to Freedom, Capitalism, and Western Success*, Rodney Stark writes of Christianity, "It is this commitment to manual labor that so distinguishes Christian asceticism from that found in other great religious cultures, where piety is associated with rejection of the world and its activities. In contrast with Eastern holy men, for example, who specialize in meditation and live by charity, medieval Christian monastics lived by their own labor, sustaining highly productive estates. This ... sustained a healthy concern with economic affairs. Although the Protestant ethic thesis is wrong, it is entirely legitimate to link capitalism to a Christian ethic."[19]

Once again we can run the historical experiment and make a prediction: if this hypothesis is true, then societies in which Christianity is or was the dominant religion should show Western-like forms of democracy and capitalism. They don't. The Byzantine Empire, for example, was predominantly Eastern Orthodox Christian from the early AD 300s, and for seven centuries produced nothing remotely like democracy and capitalism as practiced in modern America. Even early America wasn't like it is today when, a mere two centuries ago, women couldn't vote, slavery was legal and widely practiced, and capitalism's wealth was vouchsafed to only a tiny minority of landholders or factory owners. Throughout the late Middle Ages and well into the Early Modern Period, all the nation-states, city-states, and various political conglomerates of western and central Europe were not only Christian but also *Western* Christian, and yet as late as the nineteenth century the only quasi-democratic republics in Europe were England, Holland, and Switzerland.[20] In Christian Europe, both England and Spain profited handsomely from their overseas colonial empires, made more profitable yet by the decimation of the native populations and the looting of their troves of precious metals, gems, and other natural resources—actions that by today's moral standards would be condemned.[21]

In any case, how modern conservatives turned Jesus into a free-market capitalist is baffling given what he had to say on the matter in the Bible: In Matthew 19:21 Jesus told his flock: "And again I say unto you, It is easier for a camel to go through the eye of a needle, than for a rich man to enter into the kingdom of God." In Matthew 19:24 Jesus told a disciple: "If thou wilt be perfect, go and sell that thou hast, and give to the poor, and thou shalt have treasure in heaven: and come and follow me." In Luke 6:24–25 the Messiah admonished the wealthy (along with the fulfilled and happy): "Woe unto you that are rich! for ye have received your consolation. Woe unto you that are full! for ye shall hunger. Woe unto you that laugh now! For ye shall mourn and weep." And in Luke 16 Jesus recounts a moral homily about a rich man "clothed in purple and fine linen, and fared sumptuously every day" and "a certain beggar named Lazarus, which was laid at his gate, full of sores, and desiring to be fed with the crumbs which fell from the rich man's table." When the beggar died he "was carried by the angels into Abraham's bosom" but when the rich man died he was buried, "and in hell he lift up his eyes, being in torments, and seeth Abraham afar off, and Lazarus in his bosom."

Religion has a mixed effect on the psychology of wealth—how the poor feel about being poor, and how the rich feel about being rich. According to a 2013 study by Humboldt University of Berlin psychologist Jochen Gebauer and his colleagues, religion makes being poor less onerous, but it makes being rich less salubrious: "In comforting the poor, religious teachings de-emphasize the importance of money, which would buffer low-income's psychological harms." The study involved a massive dataset of 187,957 respondents from eleven religiously diverse countries (Austria, France, Germany, Italy, Poland, Russia, Spain, Sweden, Switzerland, the Netherlands, and Turkey), who took the Trait Psychological Adjustment Scale, which measures adaptability, calmness, cheerfulness, contentedness, energy, health, optimism, positivity, resilience, and stability.

They found that believers who are rich are better adjusted than believers who are poor, but only in less religious cultures. In more religious cultures, rich believers were more poorly adjusted than rich nonbelievers. And richer nonbelievers were better adjusted than poorer nonbelievers in both highly religious and less religious cultures.[22] The study, however, just examined religiosity in general, and thus it does not distinguish, say, the Protestant prosperity preachers of the United States—such as Oral Roberts, the Reverend Ike, and Joel Osteen—who tell their aspirants that God wants them to be rich (and give plenty back to the church),[23] from the Catholic Mother Teresa–like nuns of India, who tell their supplicants that prosperity is to be found in the next life as a salve for their misery in this one.[24] So here again religion is a mixed bag: offering consolation to the poor for what they don't have and justification to the rich for what they do have.

RELIGION AND EQUAL RIGHTS

If God really believes in equal rights for all of his people, one would think that he would have said something about them in his holy book. But such sentiments are nowhere to be found in the Bible.

In an entire book on the subject, for example, Dinesh D'Souza manages to find only one biblical passage that supports anything like a modern moral value—in Galatians 3:28—when the Apostle Paul says, "There is neither Jew nor Greek, there is neither bond nor free, there is neither male nor female: for ye are all one in Christ Jesus." D'Souza imagines that this Bible verse is the foundation of the famous line in the preamble of the American Declaration of Independence—"All men are created equal." D'Souza opines, "Here Christian individualism is combined with Christian universalism, and the two together are responsible for one of the great political miracles of our day, a global agreement on rights held to be inviolable."[25]

I'm afraid not. D'Souza has yanked the passage out of context, and the surrounding verses demonstrate clearly what Paul is up to (Galatians 3:1): "O foolish Galatians, who hath bewitched you, that ye should not obey the truth, before whose eyes Jesus Christ hath been evidently set forth, crucified among you?" And what is this truth, according to Paul? The truth is that "[T]he Jew in becoming a Christian did not need to become a Greek, nor the Greek a Jew. The slave might continue to serve his master, and 'male' and 'female' retained each its function in the ongoing stream of life."[26] In other words, Paul is saying that you can carry on as you are. If you're Greek, there's no need to become a Jew—a significant dispensation, given that a man converting to Judaism often had to submit to adult circumcision, and this is just the kind of thing that puts a guy off the whole idea. (Rule number one for getting lots of converts, Paul realized, is let the guys keep their foreskins. Let circumcision be spiritual—"of the heart"—instead of physical, as Paul writes in Romans 2:29, and a lot more men are going to sign up and become card-carrying members of the faith.) Paul was not a revolutionary advocating violence,[27] and he most assuredly wasn't ghostwriting the US Constitution. He was saying that if you're a slave, you must keep on being a slave; if you're a wife, you must continue being regarded as property; no matter who you are, you can still worship Jesus Christ and be abused by your culture in whatever manner is customary for someone of your breeding and station.

D'Souza's claim that the Bible points toward equality is especially non-sensical in light of the fact that slaves remained slaves for eighteen more centuries, and women remained little more than property for nineteen more centuries in Christian countries around the world. Clearly, even if Paul's message were interpreted to mean that we're all equal, absolutely no one took it seriously. But what Paul's passage really meant was that anyone can go to heaven by accepting Jesus as the Christ (as instructed in John 3:16), and that's the message of universalism—not equal treatment in *this* world, but in the *next* world.[28]

Finally, as for Thomas Jefferson's declaration that "All men are created equal," far from the Bible being the source of this greatest of all moral precepts, Jefferson explained its inspiration a half century after he wrote it. In a letter to Henry Lee in 1825, he wrote: "Neither aiming at originality of principle or sentiment, nor yet copied from any particular and previous writing, it was intended to be an expression of the American mind, and to give to that expression the proper tone and spirit called for by the occasion. All its authority rests then on the harmonizing sentiments of the day, whether expressed in conversation, in letters, printed essays, or in the elementary books of public right as Aristotle, Cicero, Locke, Sidney, &c."[29]

IS RELIGION GOOD FOR SOCIETAL HEALTH AND HAPPINESS?

If religion is not the source of morality nor the foundation of Western civilization, is it nonetheless good for societal well-being? This is a hotly contested question for which the data are complex and often conflicting—in part because different scholars have different definitions of societal health—so it is easy to data-snoop studies to support one conclusion or the other. I will expound on the many possible criteria of societal health below, but let's start with something relatively simple: charity.

According to the social scientist Arthur C. Brooks, in his book *Who Really Cares: The Surprising Truth About Compassionate Conservatism* (another book title that boldly carries no question mark), when it comes to charitable giving and volunteering, numerous quantitative measures debunk the myths of "bleeding heart liberals" and "heartless conservatives."[30] Conservatives donate 30 percent more money than liberals (even when controlled for income), give more blood, and log more volunteer hours. Religious people are four times more generous than secularists in giving to all charities, 10 percent more munificent to nonreligious charities, and 57 percent more likely to help a homeless person.[31] Those raised in intact, religious families are more charitable than those who are not. Charitable givers are 43 percent more likely to say they are "very happy" than nongivers, and 25 percent more likely than nongivers to say their health is "excellent" or "very good."[32] The working poor give a substantially higher percentage of their incomes to charity than any other income group, and three times more than those on public assistance of comparable income; in other words, poverty is not a barrier to charity, but welfare is.[33] "For many people," Brooks explains, "the desire to donate other people's money displaces the act of giving one's own."[34] This, he concludes, has led to a "bright cultural line inside our nation":

> On one side are the majority of citizens who are charitable in all sorts of formal and informal ways—so charitable that they make America exceptional by international standards. On the other side of the line, however, is a sizeable minority who are conspicuously uncharitable. We have identified the reasons these two groups are so different, and they are controversial reasons: One group is religious, the other secular; one supports government income redistribution, the other does not; one works, the other accepts income from the government; one has strong, intact families, the other does not.[35]

A major explanation for these findings, however, is that they don't reflect *religious* belief but *political* belief. People who think it is the government's job to take care of the poor through public programs (liberals and many secularists) feel less of a need to give privately, since they already give through their taxes, whereas people who think taking care of the poor should be done privately (many religious conservatives) feel the need to step up and deliver.

Northwestern University law professor James Lindgren observes that Brooks focuses too little on what he calls "the forgotten middle: moderates." Liberals donate significantly more money than moderates, even while conservatives give significantly more than both moderates and liberals. Thus, Lindgren concludes, "moderates would seem to be the ungenerous ones, not liberals." Yet, to the point of what beliefs lead to an expansion of the moral sphere to include more people, Lindgren adds, "Those who oppose income redistribution tend to be less racist, more tolerant of unpopular groups, happier, less vengeful, and more likely to report generous charitable donations."[36] But conservatives hold that government income redistribution of other people's money is not the same as charitable giving of one's own money. When the government takes money from one person and gives it to another person, the moral motivation is removed and shifted into the realm of politics. You may think this is where it belongs (as many liberals do), but note that this difference between rates of giving by liberals and conservatives may reflect a personal versus political perspective.

On the negative side of the moral ledger, social scientist Gregory S. Paul conducted an in-depth statistical analysis involving seventeen First World prosperous democracies (those with a population of four million or more and a per capita GDP of $23,000 or more in 2000 dollars) in the Successful Societies Scale database (Australia, Austria, Canada, Denmark, England, France, Germany, Holland, Ireland, Italy, Japan, New Zealand, Norway, Spain, Sweden, Switzerland, United States). He wanted to know how they score on a wide range of twenty-five different indicators of social health and well-being, including homicides, incarceration, suicides, life expectancy, gonorrhea and syphilis infections, abortions, teen births (fifteen- to seventeen-year-olds), fertility, marriage, divorce, alcohol consumption, life satisfaction, corruption indices, adjusted per capita income, income inequality, poverty, employment levels, and others, on a 1–9 scale from dysfunction to healthy. Paul also quantified the religiosity of each of the seventeen countries by measuring to what extent the citizens in each believe in God, are biblical literalists, attend religious services at least several times a month, pray at least several times a week, believe in an afterlife, and believe in heaven and hell, ranking them on a 1–10 scale.[37]

The results were striking . . . and disturbing. Far and away—without having a close second—the United States is not only the most religious of the seventeen nations but also the most dysfunctional. You can see it in *figure 4, 1–7* below: *figure 4-1*: religiosity and overall societal health; *figure 4-2*: religiosity and annual homicides per 100,000; *figure 4-3*: religiosity and incarcerations per 100,000; *figure 4-4*: religiosity and suicides per 100,000; *figure 4-5*: religiosity and teen pregnancies per 1,000; *figure 4-6*: religiosity and abortions per 1,000 women aged 15–19; and *figure 4-7*: religiosity and divorces per 100.

Since this 2009 study Paul has accumulated additional supporting data that he shared with me. "I have since doubled the SSS [Successful Societies Scale] to a massive four dozen indicators of socioeconomic success and dysfunction, everything including the kitchen sink is in this thing," he wrote in an email, concluding even more definitively:

> In both the original and upgraded SSS the U.S. scores the worst, sometimes severely so, in the most factors, and despite some strengths it scores the lowest on the 0–10 SSS at around 3. It is not just social ills—income growth is mediocre (1995–2010), social mobility is so low that we have become a rigid class society, private and public debt loads are unusually high. The laissez-faire World Economic Forum used to rank the U.S. as the most competitive, now we are behind five of those progressive Euro countries the right endlessly denounces as "Entitlement Societies" and falling. This cannot be explained away by inherent extraneous factors such as high ethnic-racial diversity or immigration because the correlations are too weak. . . . The most successful democracies are the most progressive, scoring up to 7 on the SSS.[39]

Correlation is not causation, of course. But if religion is such a powerful force for societal health, then why is America—the most religious nation in the Western world—also the unhealthiest on all of these social measures? If religion makes people more moral, then why is America seemingly so immoral in its lack of concern for its poorest, most troubled citizens, notably its children?

It may well be that the left and the right are not so cleanly cleaved, and the moderate middle is muddying the statistical waters. The right and the left do seem to cluster along religious lines. Political scientists Pippa Norris and Ronald Inglehart examined data from the Comparative Study of Electoral Systems, analyzing thirty-seven presidential and parliamentary elections in thirty-two nations over the previous decade. They found that 70 percent of the devout (those who attend religious services at least once

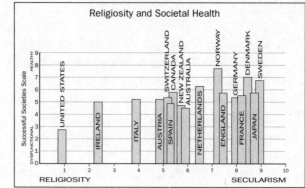

Figure 4-1

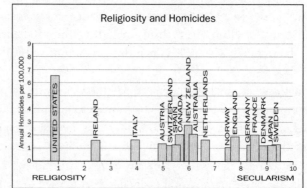

Figure 4-2

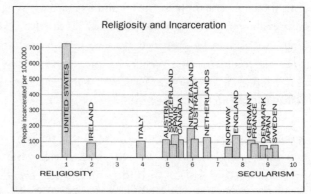

Figure 4-3

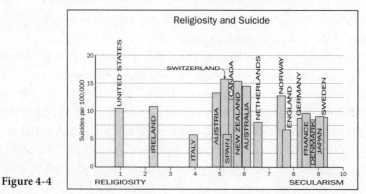

Figure 4-4

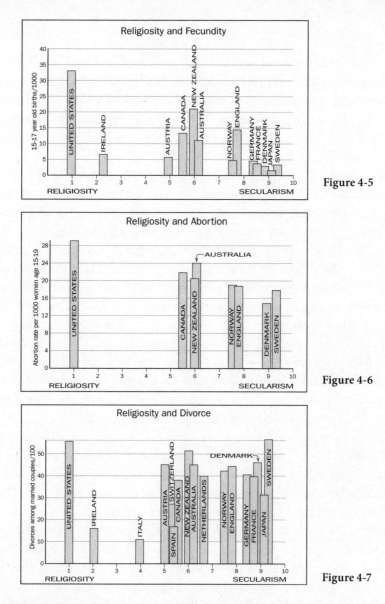

Figure 4-5

Figure 4-6

Figure 4-7

Figure 4, 1–7. The Relationship Between Religion and Societal Health

Figure 4-1: religiosity and overall societal health; *figure 4-2*: religiosity and annual homicides per 100,000; *figure 4-3*: religiosity and incarcerations per 100,000; *figure 4-4*: religiosity and suicides per 100,000; *figure 4-5*: religiosity and teen pregnancies per 1,000; *figure 4-6*: religiosity and abortions per 1,000 women aged 15–19; and *figure 4-7*: religiosity and divorces per 100.[38]

per week) voted for parties of the right, compared with only 45 percent of
the secular (those who never attend religious services). The effect is especially striking in America. For example, in the 2000 US presidential election, they write, "religion was by far the strongest predictor of who voted
for Bush and who voted for Gore—dwarfing the explanatory power of
social class, occupation, or region."[40]

The theory of "social capital" may help explain the benefits of religious
community. As defined by Robert Putnam in his 2000 book *Bowling
Alone*, social capital means "connections among individuals—social networks and the norms of reciprocity and trustworthiness that arise from
them." This is something different from individual "civic virtue," Putnam
qualifies. "The difference is that 'social capital' calls attention to the fact
that civic virtue is most powerful when embedded in a dense network of
reciprocal social relations. A society of many virtuous but isolated individuals is not necessarily rich in social capital." The reason is that social
capital has what economists call "positive externalities," or the unintended
consequences accruing to others or to society at large. "If the crime rate in
my neighborhood is lowered by neighbors keeping an eye on one another's homes, I benefit even if I personally spend most of my time on the
road and never even nod to another resident on the street."

Thus, Putnam continues, social capital can be both a private and a
public good. "Some of the benefit from an investment in social capital
goes to bystanders, while some of the benefit redounds to the immediate
interest of the person making the investment. For example, service clubs,
like Rotary or Lions, mobilize local energies to raise scholarships or fight
disease at the same time that they provide members with friendships and
business connections that pay off personally."[41] For our purposes here,
however, social capital has implications for moral progress by defining
relationships of reciprocity and enforcing rules of conduct in communities that may or may not be religious in nature. This is an example of
reciprocal altruism discussed in chapter 1, and social capital generalizes
the principle to communities at large, as Putnam expounds:

> I'll do this for you without expecting anything specific back from you, in
> the confident expectation that someone else will do something for me down
> the road. The Golden Rule is one formulation of generalized reciprocity.
> Equally instructive is the T-shirt slogan used by the Gold Beach, Oregon,
> Volunteer Fire Department to publicize their annual fund-raising effort:

"Come to our Breakfast, we'll come to your fire." "We act on a norm of specific reciprocity," the firefighters seem to be saying, but onlookers smile because they recognize the underlying norm of generalized reciprocity—the firefighters will come even if *you* don't.[42]

Consider Putnam's "social capital" in the context of Norris and Inglehart's analysis of data from the World Values Survey, in which they found a positive correlation between "religious participation" and membership in "non-religious community associations," including women's, youth, peace, social welfare, human rights, and environmental conservation groups and, apparently, bowling leagues (evidently, religious people are less likely to bowl alone). As Norris and Inglehart note, "By providing community meeting places, linking neighbors together, and fostering altruism, in many (but not all) faiths, religious institutions seem to bolster the ties of belonging to civic life." The data, Norris and Inglehart conclude, support one theory that explains religion's endurance in the face of secularism: "This pattern confirms social capital theory's claim that the social networks and personal communications derived from regular churchgoing play an important role, not just in promoting activism within religious-related organizations, but also in strengthening community associations more generally."[43]

Integrating all of these studies and their mixed findings, I hypothesize that in less religious democracies, it is secular institutions that produce the social capital that leads to societal health. In America, sacred social capital leads to charitable generosity but does comparatively less well when it comes to such social ills as homicides, STDs, abortions, and teen pregnancies. Two reasons suggest themselves: (1) these problems have other causes entirely; (2) secular social capital works better for such problems. The social scientist Frank Sulloway proposed another explanation when I queried him about these studies. "The first and most important influence is as follows: people who live in awful situations, with rampant disease, war, crime, and general insecurity, turn to religion. With reference to *this* relationship, bad health and a threatening existence are causing religiosity, not vice versa." The causal relationship, says Sulloway, is not "religion leads to bad health," but rather "bad health leads to religion" and "good health promotes liberalism."[44]

On balance, then, my conclusion is that religion does not contribute significantly to a nation's overall well-being. Religion doesn't "poison everything," as Christopher Hitchens famously concluded in his book *God Is Not Great*,[45] but it is noxious enough to conclude that it is not needed to create a healthy society.

IS RELIGION GOOD FOR PERSONAL HEALTH AND HAPPINESS?

If religion does not give us the moral values we hold today, and it did not produce Western civilization, and it does not make societies healthier and happier, does it at least make individual believers healthier and happier? Here again the results are mixed. On the one hand, certain religious beliefs are blatantly bad for health: Jehovah's Witnesses, who refuse life-saving blood transfusions; Christian Scientists, who believe God will heal their disease or sickness, thus there is no need for medical treatment and as a consequence some die from treatable diseases; Pentecostal Followers of Christ, who reject all medical treatment and as a consequence have a childhood mortality rate twenty-six times greater than that of the general population; Hindus, who believe in reincarnation and who believe illness is the result of karmic revenge from bad action in a prior life and thus do not seek medical care.[46]

On the other hand, in a meta-analysis of more than three dozen studies published in 2000, the psychologist Michael McCullough and his colleagues found a strong correlation between religiosity and health, well-being, and longevity: highly religious people were 29 percent more likely to be alive at any given follow-up point than were less religious people.[47] When this study was widely publicized, skeptics were challenged by believers to explain the results, as if to say, *see, there is a God and this is the payoff for believing.*

In science, however, "God did it" is not a testable hypothesis. Inquiring minds want to know *how* God did it and what forces or mechanisms were at work. "God works in mysterious ways" will not pass peer review. Even such explanations as "belief in God" or "religiosity" must be broken down into their component parts to find possible causal mechanisms for the links between belief and behavior that lead to health, well-being, and longevity. This is what McCullough and his University of Miami colleague Brian Willoughby did in a meta-analysis of hundreds of studies. They found that religious people are more likely to engage in healthy behaviors such as exercise, visiting dentists, and wearing seat belts, and they are less likely to smoke, drink, take recreational drugs, and engage in risky sex.[48] Why? Religion provides a tight social network that reinforces positive behaviors and discourages negative habits, and it leads to greater self-regulation for goal achievement and self-control over negative temptations—another form of social capital that can be constructive whether it is religious or secular.

Even the term "self-control" needs to be operationally defined and broken down into its component parts to see how it works. This is precisely what the Florida State University psychologist Roy Baumeister did in his 2011 book coauthored with the science writer John Tierney and appropriately titled *Willpower*.[49] Self-control is the employment of one's power to will a behavioral outcome, and research shows that young children who delay gratification (for example, forgoing one marshmallow now for two later[50]) score higher on measures of academic achievement and social adjustment later in life. Religions offer the ultimate delay of gratification strategy (eternal life), and Baumeister and Tierney cite research showing that "students who spent more time in Sunday school scored higher on laboratory tests of self-discipline," and that "religiously devout children were rated relatively low in impulsiveness by both parents and teachers."[51] Of course, many religions require a certain level of self-discipline to even become a member (through required membership rituals, sacraments, tithing, and the like), so when measured later by social scientists they may be a self-selected group who are already high in self-control and willpower.

The underlying mechanisms of setting goals and monitoring one's progress, however, can be tapped by anyone, religious or not. Meditation, in which you count your breaths up to ten and then do it over and over, Baumeister and Tierney note, "builds mental discipline. So does saying the rosary, chanting Hebrew psalms, repeating Hindu mantras." Brain scans of people conducting such rituals show strong activity in areas associated with self-regulation and control of attention. McCullough, in fact, describes prayers and meditation rituals as "a kind of anaerobic workout for self-control."

In his lab, Baumeister has demonstrated that self-control can be increased with the practice of resisting temptation, but you have to pace yourself because, like a muscle, self-control can become depleted after excessive effort, leaving you more likely to succumb to a subsequent temptation. Finally, Baumeister and Tierney add that religion acts as a monitor of behavior, a feedback system, and it gives people a sense that someone is watching over them. For believers, that someone may be God or other members of their religion.[52] For nonbelievers, family, friends, and colleagues serve as the watchers—those who will look upon misbehaviors with disapproval.

The world is full of temptations, and as Oscar Wilde said, "I can resist everything except temptation." Religion is one path to resisting temptation, but there are others. We could follow the secular path of the nineteenth-century African explorer Henry Morton Stanley, who proclaimed, "self-control is more indispensable than gunpowder," especially if we have a "sacred task," as Stanley called it; his was the abolition of slavery. We would do well, in our darker moments, to reflect on Stanley's admission that "This poor body of mine has suffered terribly . . . it has been degraded, pained, wearied & sickened, and has well nigh sunk under the task imposed on it; but this was but a small portion of myself. For my real self lay darkly encased, & was ever too haughty & soaring for such miserable environments as the body that encumbered it daily."[53]

Select your sacred task, monitor and pace your progress toward that goal, eat and sleep regularly to increase your willpower, sit and stand up straight and be organized and well groomed (Stanley shaved every day in the jungle), and surround yourself with a supportive social network that reinforces your efforts. Such sacred salubriousness is the province of everyone—believers and nonbelievers—who will themselves to loftier purposes.

DECONSTRUCTING THE DECALOGUE

There is arguably no better known set of moral precepts than the Ten Commandments, but they were written by and for people whose culture and customs were so different from ours as to make them either irrelevant to modern peoples or immoral were they to be obeyed. As an exercise in moral casuistry, let's consider them again in the context of how far the moral arc has bent since they were decreed more than three millennia ago, and then reconstruct them from the perspective of a science- and reason-based moral system.[54]

I. *Thou shalt have no other gods before me.* First, this commandment reveals that polytheism was commonplace at the time and that Yahweh was, among other things, a jealous god (see God's own clarification in Commandment II). Second, it violates the First Amendment of the US Constitution in that it restricts freedom of religious expression ("Congress shall make no law respecting an establishment of religion, or prohibiting the free exercise thereof"), making the posting of the Ten Commandments in public places such as schools and courthouses unconstitutional.

II. *Thou shalt not make unto thee any graven image, or any likeness of any thing that is in heaven above, or that is in the earth beneath, or that is in the water under the earth. Thou shalt not bow down thyself to them, nor serve them: for I the LORD thy God am a jealous God, visiting the iniquity of the fathers upon the children unto the third and fourth generation of them that hate me.* This commandment is also in violation of the First Amendment's guarantee of the freedom of speech, of which artistic expression is included by precedence of many Supreme Court cases ("Congress shall make no law . . . abridging the freedom of speech"). It also brings to mind what the Taliban did in Afghanistan when they destroyed ancient religious relics not approved by their Islamist masters. Elsewhere in the Bible, the word "idol" is synonymously used, with the Hebrew word *pesel* translated as an object carved or hewn out of stone, wood, or metal. What, then, are we to make of the crucifix, worn by millions of Christians as an image, an idol, a symbol of what Jesus suffered for their sins? The crucifix is a graven image of torture as it was commonly practiced by the Romans. If Jews today were suddenly to start sporting little gas chambers on gold necklaces the shocked public reaction would be as unsurprising as it would be unmistakable.

I the LORD thy God am a jealous God. That might explain the genocides, wars, conquests, and mass exterminations commanded by the deity of the Old Testament. These humanlike emotions reveal Yahweh to be more like a Greek god, and much like an adolescent who lacks the wisdom to control his passions. The last part of this commandment—*visiting the iniquity of the fathers upon the children unto the third and fourth generation of them that hate me*—violates the most fundamental principle of Western jurisprudence developed over centuries of legal precedence that one can only be guilty of one's own sins and not the sins of one's parents, grandparents, great-grandparents, or anyone else, for that matter.

III. *Thou shalt not take the name of the Lord thy God in vain, for the LORD will not hold him guiltless who takes his name in vain.* This commandment is once again an infringement on our Constitutionally guaranteed right to free speech and religious expression, and another indication of Yahweh's petty jealousies and un-Godlike ways.

IV. *Remember the Sabbath day, to keep it holy.* Again, freedom of speech and of religious expression mean we may or may not choose to treat the Sabbath as holy, and the rest of this commandment— *For in six days the LORD made heaven and earth, the sea, and all that is in them, and rested on the seventh day. Therefore the LORD blessed the Sabbath day and made it holy*—makes it clear that its purpose is to once again pay homage to Yahweh.

Thus far, the first four commandments have nothing whatsoever to do with morality as we understand it today in terms of how we are to interact with others, resolve conflicts, or improve the survival and flourishing of other sentient beings. At this point the Decalogue is entirely concerned with the relationship of humans and God, not humans and humans.

V. *Honor thy father and thy mother.* As a father myself, this commandment feels right and reasonable, since most of us parents appreciate being honored by our children, especially because we've invested considerable love, attention, and resources into them. But "commanding" honor—much less love—doesn't ring true to me as a parent, since such sentiments usually come naturally anyway. Plus, commanding honor is an oxymoron, made all the worse by the hint of a reward for so doing, as in the rest of that commandment: *that thy days may be long upon the land which the LORD thy God giveth thee.* Honor either happens naturally as a result of a loving and fulfilling relationship between parents and offspring, or it doesn't. For a precept to be moral, it must involve an element of choice between doing something entirely self-serving and doing something that helps another, even at the cost of oneself.

VI. *Thou shalt not kill.* Finally, we get a genuine moral principle worth our attention and respect. Yet even here, much ink has been spilled by biblical scholars and theologians about the difference between murder and killing (such as in self-defense), not to mention all the different types of killing, from first-degree murder to manslaughter, along with mitigating circumstances and exclusions, such as self-defense, provocation, accidental killings, capital punishment, euthanasia, and, of course, war. Many Hebrew scholars believe that the prohibition is against murder only. But what are we to make of the story in Exodus (32:27–28) in which Moses brought down from the mountaintop the first set of tablets, which he smashed in anger, and then commanded the Levites: "Thus saith the LORD God of Israel, put every man his sword by his side, and go in and out from gate to gate throughout the camp, and slay every man his brother, and every man his companion, and every man his neighbor. And the children of Levi did according to the word of Moses: and there fell of the people that day about three thousand men"? How can we reconcile God's commandment not to kill anyone with his commandment to kill everyone? In light of this account, and many others like it, the Sixth Commandment should perhaps read thus: *Thou shalt not kill—not unless the Lord thy God says so. Then shalt thou slaughter thine enemies with abandon.*

VII. *Thou shalt not commit adultery.* Coming from a deity who impregnated somebody else's fiancée, that's a bit rich. However, the bigger issue is that this commandment, like all the others, is a blunt instrument that doesn't take into account the wide variety of circumstances in which people find themselves. Surely grown-ups in intimate relationships can and should negotiate the details of their relationship for themselves, and one hopes that they'll act honorably toward their partner out of a sense of integrity, and not because a deity told them to.

VIII. *Thou shalt not steal.* Again, do we really need a deity to command this? All cultures had and have moral rules and legal codes about theft.

IX. *Thou shalt not bear false witness against thy neighbor.* Anyone who has been lied to or gossiped about can explain why this moral commandment makes sense and is needed, so chalk one up for the Bible's authors, whose insights here were spot-on.

X. *Thou shalt not covet thy neighbor's house, thou shalt not covet thy neighbor's wife, nor his manservant, nor his maidservant, nor his cattle, nor anything that is thy neighbor's.* Consider what it means to covet something—to crave or want or desire it—so this commandment is the world's first thought crime, which goes against centuries of Western legal codes. More to the point, the very foundation of capitalism is the coveting or desire for things and, ironically, it is Bible-quoting Christian conservatives who most defend the very coveting forbidden in this final mandate. The late Christopher Hitchens best summed up the implications of taking this commandment seriously: "Leaving aside the many jokes about whether or not it's okay or kosher to covet thy neighbor's wife's ass, you are bound to notice once again that, like the Sabbath order, it's addressed to the servant-owning and property-owning class. Moreover, it lumps the wife in with the rest of the chattel (and in that epoch could have been rendered as 'thy neighbor's wives,' to boot)."[55]

A PROVISIONAL RATIONAL DECALOGUE

The problem with any religious moral code that is set in stone is just that—*it is set in stone.* Anything that can never be changed has within its DNA the seeds of its own extinction. A science-based morality has the virtue of having built into it a self-correcting mechanism that does not just allow redaction, correction, and improvement; it insists upon it. Science and reason can be employed to inform—and in some cases even determine—moral values.

Science thrives on change, on improvement, on updating and upgrading its methods and conclusions. So it should be for a science of morality. No one knows for sure what is right and wrong in all circumstances for all people everywhere, so the goal of a science-based morality should be to construct a set of provisional moral precepts that are true for most people in most circumstances most of the time—as assessed by empirical inquiry and rational analysis—but admit exceptions and revisions where appropriate. Indeed, as we have seen, as humanity's concept of "who and what is human, and entitled to protection" has expanded over the centuries, so we have extended moral protection to categories once thought beneath our notice. Here are ten provisional moral principles to consider:

1. *The Golden Rule Principle: Behave toward others as you would desire that they behave toward you.*
The Golden Rule is a derivative of the basic principle of exchange reciprocity and reciprocal altruism, and thus evolved in our Paleolithic ancestors as one of the primary moral sentiments. In this principle there are two moral agents: the *moral doer* and the *moral receiver.* A moral question arises when the moral doer is uncertain how the moral receiver will accept and respond to the action in question. In its essence

this is what the Golden Rule is telling us to do. By asking yourself, "How would I feel if this were done unto me?" you are asking "How would others feel if I did it unto them?"

2. *The Ask-First Principle: To find out whether an action is right or wrong, ask first.*
The Golden Rule Principle has a limitation to it: what if the moral receiver thinks differently from the moral doer? What if you would not mind having action X done unto you, but someone else would mind it? Smokers cannot ask themselves how they would feel if other people smoked in a restaurant where they were dining because they probably wouldn't mind. It's the *non*smokers who must be asked how they feel. That is, the moral doer should ask the moral receiver whether the behavior in question is moral or immoral. In other words, the Golden Rule is still about *you*. But morality is more than just about you, and the Ask-First Principle makes morality about *others*.

3. *The Happiness Principle: It is a higher moral principle to always seek happiness with someone else's happiness in mind, and never seek happiness when it leads to someone else's unhappiness through force or fraud.*
Humans have a host of moral and immoral passions, including being selfless and selfish, cooperative and competitive, nice and nasty. It is natural and normal to try to increase our own happiness by whatever means available, even if that means being selfish, competitive, and nasty. Fortunately, evolution created both sets of passions, such that by nature we also seek to increase our own happiness by being selfless, cooperative, and nice. Since we have within us both moral and immoral sentiments, and we have the capacity to think rationally and intuitively to override our baser instincts, and we have the freedom to choose to do so, at the core of morality is choosing to do the right thing by acting morally and applying the happiness principle. (The modifier "force or fraud" was added to clarify that there are many activities that do not involve morality, such as a sporting contest, in which the goal is not to seek happiness with your opponent's happiness in mind, but simply to win.)

4. *The Liberty Principle: It is a higher moral principle to always seek liberty with someone else's liberty in mind, and never seek liberty when it leads to someone else's loss of liberty through force or fraud.*

The Liberty Principle is an extrapolation from the fundamental principle of all liberty as practiced in Western society: *The freedom to believe and act as we choose so long as our beliefs and actions do not infringe on the equal freedom of others.* What makes the Liberty Principle a moral principle is that in addition to asking the moral receiver how he or she might respond to a moral action, and considering how that action might lead to your own and the moral receiver's happiness or unhappiness, there is an even higher moral level toward which we can strive, and that is the freedom and autonomy of yourself and the moral receiver, or what we shall simply refer to here as liberty. Liberty is the freedom to pursue happiness and the autonomy to make decisions and act on them to achieve that happiness.

Only in the past couple of centuries have we witnessed the worldwide spread of liberty as a concept that applies to all peoples everywhere, regardless of their race, religion, rank, or social and political status in the power hierarchy. Liberty has yet to achieve worldwide status, particularly among those states dominated by theocracies that encourage intolerance, and dictate that only some people deserve liberty, but the overall trend since the Enlightenment has been to grant greater liberty, for more people, everywhere (see *figure 4-8*). Although there are setbacks still, and periodically violations of liberties disrupt the overall historical flow from less to more liberty for all, the general trajectory of increasing liberty for all continues, so every time you apply the Liberty Principle you have advanced humanity one small step forward.

5. *The Fairness Principle: When contemplating a moral action imagine that you do not know if you will be the moral doer or receiver, and when in doubt err on the side of the other person.*
This is based on the philosopher John Rawls's concepts of the "veil of ignorance" and the "original position" in which moral actors are ignorant of their position in society when determining rules and laws that affect everyone, because of the self-serving bias in human decision making. Given a choice, most people who enact moral rules and legislative laws would do so based on their position in society (their gender, race, sexual orientation, religion, political party, etc.) in a way that would most benefit themselves and their kin and kind. Not knowing ahead of time how the moral precept or legal law will affect you pushes you to strive for greater fairness for all. A simpler version

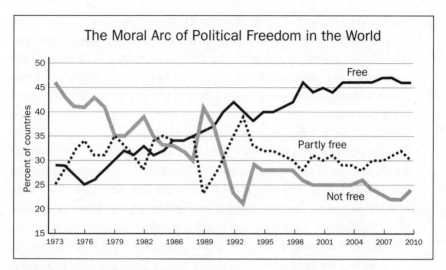

Figure 4-8. The Moral Arc of Political Freedom
The percentage of countries throughout the world that are free has been on the rise since
the 1970s, while the percentage of those that are not free has been on the decline.[56]

is in the example of cutting a cake fairly: if I cut the cake, you choose
which piece you want, and if you cut the cake, I choose which piece
I want.

*6. The Reason Principle: Try to find rational reasons for your moral
actions that are not self-justifications or rationalizations by consulting
others first.*
Ever since the Enlightenment the study of morality has shifted from
considering moral principles as based on God-given, divinely inspired,
holy book–derived, authority-dictated precepts from the top down,
to bottom-up, individually considered, reason-based, rationally constructed, science-grounded propositions in which one is expected to
have reasons for one's moral actions, especially reasons that consider
the other person affected by the moral act. This is an especially difficult moral principle to carry out because of the all-too-natural propensity to slip from rationality to rationalization, from justification to
self-justification, from reason to emotion. As in the second principle,
to "ask first," whenever possible one should consult others about one's
reasons for a moral action in order to get constructive feedback and to
pull oneself out of a moral bubble in which whatever you want to do
happens to be the most moral thing to do.

7. The Responsibility and Forgiveness Principle: Take full responsibility for your own moral actions and be prepared to be genuinely sorry and make restitution for your own wrongdoing to others; hold others fully accountable for their moral actions and be open to forgiving moral transgressors who are genuinely sorry and prepared to make restitution for their wrongdoing.

This is another difficult principle to uphold in both directions. First, there is the "moralization gap" between victims and perpetrators, in which victims almost always perceive themselves as innocent and thus any injustice committed against them must be the result of nothing more than evil by the perpetrator; and in which perpetrators may perceive themselves to have been acting morally in righting a wrong, redressing an immoral act, or defending the honor of oneself or family and friends. The self-serving bias, the hindsight bias, and the confirmation bias practically ensure that we all feel we didn't do anything wrong, and whatever we did was justified, and thus there is no need to apologize and ask for forgiveness.

As well, the sense of justice and revenge is a deeply evolved moral emotion that serves three primary purposes: (1) to right wrongs committed by transgressors, (2) as a deterrent to possible future bad behavior, and (3) to serve as a social signal to others that should they commit a similar moral transgression the same fate of your moral indignation and revenge awaits them.

8. The Defend Others Principle: Stand up to evil people and moral transgressors, and defend the defenseless when they are victimized.

There are people in the world who will commit moral transgressions against us and our fellow group members, either through the logic of violence and aggression in which perpetrators of evil always feel justified in their acts or through such conditions as psychopathy, in which a non-negligible portion of a population will commit selfish or cruel acts. We must stand up against them.

9. The Expanding Moral Category Principle: Try to consider other people not of your family, tribe, race, religion, nation, gender, or sexual orientation as an honorary group member equal to you in moral standing.

We have a moral obligation not only to ourselves, our kin and kind, our family and friends, and our fellow in-group members; we also owe it to those people who are different from us in a variety of ways, who in the past have been discriminated against for no other reason than that they were different in some measurable way. Even though our first moral obligation is to take care of ourselves and our immediate family and friends, it is a higher moral value to consider the moral values of others, and in the long run it is better for yourself, your kin and kind, and your in-group to consider members of other groups to be honorary members of your own group, as long as they so honor you and your group.

10. *The Biophilia Principle: Try to contribute to the survival and flourishing of other sentient beings, their ecosystems, and the biosphere as a whole.*
Biophilia is the love of nature, of which we are a part. Expanding the moral sphere to include the environments that sustain sentient beings is a lofty moral principle.

If by fiat I had to reduce these ten principles to just one it would be this: *Try to expand the moral sphere and to push the arc of the moral universe just a bit farther toward truth, justice, and freedom for more sentient beings in more places more of the time.*

PART II

The
Moral Arc
Applied

5

Slavery and a
Moral Science of Freedom

In *giving* freedom to the *slave*, we *assure* freedom to the *free*—honorable alike in what we give, and what we preserve. We shall nobly save, or meanly lose, the last best hope of earth. Other means may succeed; this could not fail. The way is plain, peaceful, generous, just—a way which, if followed, the world will forever applaud.

—Abraham Lincoln, Annual Message to Congress, December 1, 1862

As a young man in the late 1970s and early 1980s I spent a lot of time on the road exploring California and the other western states, initially in my 1966 Ford Mustang, and then on a bicycle after I seriously took up cycling, riding long distances all over and across America. When I wasn't listening to lectures and books on cassette tapes on my Sony Walkman, I would while away the time by counting the number of Denny's vs. Sambo's restaurants along any given stretch of highway. For a while it was close, but then suddenly Sambo's went bankrupt and Denny's Grand Slam Breakfast became my meal of choice at any time of the day or night.

What happened to Sambo's is emblematic of what happened across all of Western culture in the second half of the twentieth century: we became aware of the power of language and logos, symbols and gestures, to affect how we see and treat others—especially those of a different race. Sambo's became embroiled in controversy over the restaurant chain's name. The company was founded in 1957 by *Sam* Battistone Sr. and Newell *Bohnett*, who said that they derived the company's name from the combined letters of their first and last names; but the duo also took advantage of the popularity of *The Story of Little Black Sambo*,[1] and featured scenes from the book in menus and on the walls of their restaurants. The original story is about a dark-skinned Indian boy named Little Black Sambo who ventures out for a walk in the jungle and, in an unlikely turn of events, has his clothing expropriated by tigers. The tigers wind up quarreling over which of them is most attractive in Sambo's clothes and eventually they chase each other around a tree so fast that they turn into butter, which Sambo's mother, Black Mumbo, makes into pancakes. The restaurant marketers stayed true to the story in that they made the Sambo mascot Indian, but they portrayed him as a light-skinned Indian boy whose skin got lighter still from the fifties to the sixties until he was essentially an apple-cheeked white boy in a turban. The story was also modified—gone were Black Mumbo and Sambo's father, Black Jumbo, and the unfortunate tigers no longer turned into butter but instead were treated to a stack of "the finest, lightest pancakes they ever ate"[2] in exchange for returning Sambo's clothes.

The trouble was that the Little Black Sambo featured on the cover of the US edition of the book was unmistakably African, and in time the two images—Indian and African black—morphed into a single likeness, reminiscent of "darky" or blackface iconography that was popular during the end of the nineteenth century and the beginning of the twentieth. Increasingly, the word "Sambo" came to be understood as a pejorative label and a racist stereotype. There was a media feeding frenzy when the National Association for the Advancement of Colored People (NAACP) launched a formal protest and filed lawsuits against Sambo's to block the opening of new restaurants in several northeastern states. The corporate suits at Sambo's headquarters scrambled to explain, rationalize, and then respond by changing the names of some of the restaurants in those parts of the country where it was found offensive, to such innocuous titles as Jolly Tiger, No Place Like Sam's, Season's Friendly Eating, and even just Sam's. As Nan Ellison, president of the Brockton, Massachusetts, NAACP, said, "I think I can go to Sam's to eat without feeling degraded. I could not feel comfortable at Sambo's when I reflected on the sacrifices made by demonstrators at sit-ins during the 1960s."[3] But it was all to no avail. By 1982 all but 1 of the 1,117 Sambo's restaurants in 47 states had closed their doors, and the corporation filed for bankruptcy as a result of this and other business problems.[4]

Looking at the cover and the illustrations of *The Story of Little Black Sambo* from an increasingly sensitive and perspicacious modern perspective it is not hard to understand why people of color found them offensive— they truly are cringeworthy (*figure 5-1a–d*). Images, names, and language matter. Decades later I recall that the boy featured in the restaurants had

Figure 5-1a

Figure 5-1b

Figure 5-1c

Figure 5-1d

Figure 5-1a–d. Little Black Sambo Then and Now

The Story of Little Black Sambo by Helen Bannerman is about an Indian boy (*figure 5-1a*) whose clothing is stolen by tigers, who chase each other around a tree so fast that they turn into butter, which Sambo's mother then makes into pancakes. Sambo's restaurant chain adopted the story and portrayed scenes from the book on the walls of its restaurants, but when "Sambo" became synonymous with offensive blackface iconography—as portrayed on the covers of early U.S. editions of the book (*figure 5-1b*)—and when the name mutated into a pejorative label and a racist stereotype, the restaurant chain faced a catastrophic public relations fiasco (*figure 5-1c*). Depicted here are early twentieth-century covers and a modern rendition (*figure 5-1d*) that reveal how much our racial and cultural sensitivities have changed in less than a century.[5]

African features, even though a Google image search reveals that my memory is faulty. I even recollect seeing restaurant signs with the name *Little Black Sambo's*, even though that too is a false memory. Now, on one level, the Sambo's story sounds like political correctness run amok with the language police going too far, but it is a sign of moral progress that we've all become sensitized to how other people see the world, interchanging our perspectives with theirs.

The Sambo's story illustrates one of the driving forces behind moral progress: those who champion human rights often begin with what we *call* things—how we talk and write about them—because so much of our thinking, including thinking about other people, is grounded in images and language. Of course, social change does not come about solely through the pristine powers of pure thought. The rights hard won by oppressed peoples throughout history have come about through the herculean efforts of the peoples themselves, in conjunction with their allies, through arduous direct political action. It almost inevitably takes some kind of direct deed (marches, sit-ins, sabotage, blockades, the destruction of property, outright civil war) to realize social transformation.

Still, even just-war theory demands there be *reasons* for the conflict beyond just *Veni, vidi, vici*. The abolition of slavery, in fact, came about at the time when its defenders were making intellectual arguments in support of it, but it was the better counterarguments, combined with political action (and in America, a war), that ultimately resulted in its collapse. Because the political action and the intellectual arguments generated in response to slavery produced the first of the great rights revolutions, we begin with it, and in keeping with my thesis that science and reason bend the moral arc of the universe, I will focus on the rational arguments and justifications for slavery and why they are wrong.

SLAVERY AND HUMAN RIGHTS

Of the many and various abuses and usurpations of humans by other humans, there is perhaps none so odious and oppressive as the ownership of one human being by another. This is slavery, and it has existed for as long as there have been those who, for the life of them, can't see the problem with having a bunch of people that they don't have to pay doing all of the soul-crushing grunt work for them. It is a custom that relies on an unspeakable lack of empathy, on the existence of a class- or caste-based, socially stratified society, and on a population and economy large enough to support it. Older than any written record, institutionalized slavery very possibly began at about the time of the agricultural revolution (with its accompanying ideas about ownership) approximately ten thousand years ago. The written record traces the practice back to 1760 BCE and the Code of Hammurabi, a Babylonian document that treats slavery as a given. Out of the 282 laws dictated by King Hammurabi, at least 28 of them deal directly with slavery and what to do if, for example, a slave claims that you are *not* his master when you are indeed his master (cut off his ear), or a man buys a slave and the slave gets sick within the month (return the defective slave for a full refund), or a surgeon makes a large incision in your slave and your slave dies (the surgeon owes you a slave and must replace yours with an equally serviceable model).[6]

Slavery has now been outlawed in every country, though amazingly enough it was only a little over thirty years ago, in 1981, that the last country to outlaw the custom—Mauritania—finally did so. Of course, making slavery illegal is one thing; ending it is another. In Mauritania, for example, many people (mostly women) still live as slaves;[7] worldwide, possibly millions are trapped in sex and labor slavery.[8] Historically, the body count for the Atlantic slave trade alone may be as high as 10 million; the self-described atrocitologist Matthew White ranks it as the tenth-worst atrocity in the history of humanity.[9] The process was beyond brutal, with a significant percentage of kidnapped victims perishing on the forced marches to the coastal forts where they were imprisoned, sometimes for months, until a slave ship came along. Many more died during the Middle Passage across the Atlantic, and still more in their first year in the New World as they were seasoned for work in the fields and mines. These human beings—mere beasts of burden—were stripped of their clothing, branded, displayed, inspected, and auctioned for sale to would-be buyers. The life of the slave truly was—pace Thomas Hobbes—solitary, poor, nasty, brutish, and short.

For millennia, religions in general, and Jewish, Christian, and Islamic churches in particular, have had little problem with the forced enslavement of hundreds of millions of people. It was only after the Age of Reason and the Enlightenment that rational arguments were proffered for the abolition of the slave trade, influenced by and citing such secular documents as the American Declaration of Independence and the French Declaration of the Rights of Man. After an unconscionably long lag time, religion finally got on board the abolition train and became instrumental in helping to propel it forward.

Of course, one can find scattered throughout this long history the occasional religious objection to slavery, but these were remonstrations penned *by* Christians *for* Christians and *against* Christians who fully endorsed the institution, if they didn't themselves participate. And, since nearly everyone before the twentieth century was religious, it is inevitable that any arguments against slavery would be made by religious people; thus it is the *arguments*, not the people, that require scrutiny. By the same logic, virtually all of the slave traders and slave masters were religious and grounded their justifications for slavery in the primary source of religious morality—a holy book. Some religions, such as Catholicism, fully endorsed slavery, as Pope Nicholas V made clear when, in 1452, he issued the radically proslavery document *Dum Diversas*. This was a papal bull granting Catholic countries such as Spain and Portugal "full and free permission to invade, search out, capture, and subjugate the Saracens and pagans and any other unbelievers and enemies of Christ wherever they may be, as well as their kingdoms, duchies, counties, principalities, and other property ... and to reduce their persons into perpetual slavery."[10] These last few words—*to reduce their persons into perpetual slavery*—sound not just sinister to us, but also psychotic. They make perfect sense, however, in a Christian context, given that the Bible is itself a heedlessly proslavery tome. It's hardly surprising then that it took almost two thousand years for Christians to twig to the fact that slavery is wrong. The spectacularly unreflective authors of the Bible had absolutely no problem with slavery whatsoever, as long as the slave owner didn't actually beat his slaves blind and toothless (Exodus 21:26, 27). That was going just too far, although beating a slave to death was perfectly fine as long as the slave survived for a day or two after the beating. Then, when the slave died, it was appropriate to feel sorry for the unfortunate slave owner because it was he who had suffered a loss (Exodus 21:21).

Other biblical passages make it clear that God believes that slavery is a good idea, and so should you.

> *You may purchase male or female slaves from among the foreigners who live among you.*—Leviticus 25:44

Are there any age restrictions? Surely it must be wrong to buy and sell children.

You may also purchase the children of such resident foreigners, including those who have been born in your land. You may treat them as your property, passing them on to your children as a permanent inheritance.
—Leviticus 25:45

Thanks, Dad! But these laws only apply to foreigners. What about the people of Israel?

You may treat your slaves like this, but the people of Israel, your relatives, must never be treated this way.—Leviticus 25:46

Does this verse mean that the people of Israel were not to be slaves? No. God didn't forbid his chosen people from enslaving one another—that would be extreme—but he did demand that more lenient protocols should be followed.

If you buy a Hebrew slave, he is to serve for only six years. Set him free in the seventh year, and he will owe you nothing for his freedom.—Exodus 21:2

Saying that the *slave* will owe his *master* nothing for his freedom seems a bit rich. Still, as far as treatment goes, a total of seven years of slavery isn't *too* bad (if the slave survived the experience), but note that biblical law had a special way of holding even a freed slave hostage:

If he was single when he became your slave and then married afterward, only he will go free in the seventh year. But if he was married before he became a slave, then his wife will be freed with him. If his master gave him a wife while he was a slave, and they had sons or daughters, then the man will be free in the seventh year, but his wife and children will still belong to his master. But the slave may plainly declare, "I love my master, my wife, and my children. I would rather not go free." If he does this, his master must present him before God. Then his master must take him to the door and publicly pierce his ear with an awl. After that, the slave will belong to his master forever.—Exodus 21:2–6

That's a clever trick—bribe your male slave with a wife and children and then use those family relationships as a lever to make him yours forever. Seven years of clubs and chains just wasn't quite enough misery, it would seem, for the Hebrew slave. He was bedeviled by a special category of emotional torture—a gut-wrenching choice between family and freedom. (Clearly being the chosen people has its drawbacks; one is reminded of Tevye appealing to God in *Fiddler on the Roof*: "I know, I know. We are your chosen people. But, once in a while, can't you choose someone else?")

Then there is this little gem from Exodus 21:7:

When a man sells his daughter . . .

Excuse me? When a man sells *what*? This must be some mistake in translation. But no. Sex trafficking as a form of slavery was widely practiced in biblical times, and so naturally the "good book" issued instructions on the proper way to sell one's daughter into a lifetime of sexual servitude. As the father of a daughter myself, I find the mere contemplation of what is being negotiated here sickening:

When a man sells his daughter as a slave, she will not be freed at the end of six years as the men are. If she does not please the man who bought her, he may allow her to be bought back again. But he is not allowed to sell her to foreigners, since he is the one who broke the contract with her. And if the slave girl's owner arranges for her to marry his son, he may no longer treat her as a slave girl, but he must treat her as his daughter. If he himself marries her and then takes another wife, he may not reduce her food or clothing or fail to sleep with her as his wife. If he fails in any of these three ways, she may leave as a free woman without making any payment.—Exodus 21:7–11

These passages are from the *Old* Testament. What does the *New* Testament say about slavery?

Slaves, obey your earthly masters with deep respect and fear. Serve them sincerely as you would serve Christ.—Ephesians 6:5

Christians who are slaves should give their masters full respect so that the name of God and his teaching will not be shamed. If your master is a Christian, that is no excuse for being disrespectful. You should work all the harder because you are helping another believer by your efforts.
—1 Timothy 6:1–2

*Teach slaves to be subject to their masters in everything, to try to please
them, not to talk back to them, and not to steal from them, but to show that
they can be fully trusted.*—Titus 2:9–10

Clearly slave ownership was a given in the New Testament, as in the
Old. Master/slave was just another relationship, like husband/wife or
father/commodity. Paul's *Epistle to Philemon* (the shortest book in the
Bible at only 25 verses and 425 words) addresses the issue of slavery indirectly in that it was written by Paul to a leader of the Colossian church
named Philemon, who owned a runaway slave named Onesimus—a slave
who'd been converted to Christianity by Paul. Paul sent Onesimus back to
his master with the letter, which instructed Philemon to treat his slave as
a "brother beloved" instead of "a servant" (Philemon 1:16). Further, Paul
even offered to pay Philemon any debt owed to him by Onesimus, "if he
hath wronged thee at all" (Philemon 1:18). Tellingly, Paul does not suggest
that Onesimus be set free, nor does he condemn slavery as the immoral
custom it so obviously is. Instead, Paul inveigles his slaveholding Christian
fellow to treat Onesimus as a member of the family. (That sounds nice,
although given that fathers were allowed to sell their daughters, stone
their sons, and put their unfaithful wives to death, it's not the kind of
family a person necessarily wants to be a part of. "Dysfunctional" doesn't
begin to cover it.) No doubt Paul was suggesting that Onesimus not receive
the customary beating that the runaway slave might expect but, whatever
he meant, yet again the kind of moral clarity one might expect to find in a
book purported to be the final authority on the subject is nowhere to be
found (which is why both supporters of slavery and abolitionists quoted
from the *Epistle to Philemon* in their favor).

　　To be fair, biblical scholars and theologians, along with Christian
apologists and defenders of the faith, rationalize these passages as ancient
peoples dealing as best they could with an evil practice. In some cases it
may be true that "slavery" or "bondage" really meant something more akin
to "servant," "bondservant," or "manservant," along the lines of a modern
live-in maid or housekeeper. But this qualification misses the point. Where
are the passages in the Bible *condemning* slavery on moral grounds? Where
is there anything like a rational argument outlining why humans should
not be so treated? Why is there not a straightforward commandment from
Yahweh along the lines of *Thou shalt not enslave thy fellow man*? Imagine
how different the history of humanity might have been had Yahweh *not*

neglected to mention that people should never be treated as a means to someone else's ends but should be treated as ends in themselves. Would this have been too much to ask from an all-powerful and loving God?

After the Age of Reason and the Enlightenment, abolitionist arguments were made by the more progressive liberal Christian sects, such as the Mennonites, Quakers, and Methodists. It was these Christians who, in defiance of the norms found in the Bible, rejected the whole concept of slavery and became most vocal in the abolition movement. Among these was the great British crusader and abolitionist William Wilberforce (powerfully portrayed by actor Ioan Gruffudd in the film *Amazing Grace*). Wilberforce displayed incredible courage throughout his twenty-six-year campaign to abolish slavery (and in his push to found the Society for the Prevention of Cruelty to Animals and other humanitarian movements), and there is no question that his motives were religious in nature. But note who Wilberforce's most vociferous opponents were—his fellow Christians, who had to be badgered for a quarter century before coming around to his point of view. Plus, Wilberforce's religious motives were complicated by his pushy and overzealous moralizing about virtually every aspect of life, and his great passion seemed to be to worry incessantly about what other people were doing, especially if what they were doing involved pleasure, excess, and "the torrent of profaneness that every day makes more rapid advances." Wilberforce founded the Society for the Suppression of Vice after King George III issued the "Proclamation for the Discouragement of Vice" (at Wilberforce's recommendation), which ordered the prosecution of those guilty of "excessive drinking, blasphemy, profane swearing and cursing, lewdness, profanation of the Lord's Day, and other dissolute, immoral, or disorderly practices."[11]

Wilberforce wasn't happy just clipping the wings of the wayward at home; he also crusaded abroad for moral reform in Britain's colony in India, insisting on Christian instruction and "religious improvement" of the citizens of that non-Christian nation because, he boasted, "Our religion is sublime, pure, beneficent; theirs is mean, licentious and cruel." In fact, Wilberforce's initial efforts were not concerned with the abolition of slavery but the ending of the slave *trade* (with existing slaveholders allowed to continue the practice that would presumably die out on its own in time without the outside source of slave replenishment). Slavery was not just an assault on the humanity of the slaves, Wilberforce argued, it was a blemish on the Christian religion that had for so long endorsed it. As he told the House of Commons on April 18, 1791, "Never, never will we desist till we have wiped away this scandal from the Christian name, released ourselves from the load of guilt, under which we at present labour, and extinguished every trace of this bloody traffic, of which our posterity, looking back to the history of these enlightened times, will scarce believe that it has been suffered to exist so long a disgrace and dishonour to this country."[12] Nevertheless, and tellingly, the House *voted down* Wilberforce's bill on this occasion, 163–88.

THE COGNITIVE DISSONANCE OF SLAVERY

From a twenty-first-century perspective, in which we feel that we know instinctively that slavery is wrong, it is difficult to conceive that our immediate ancestors really believed that slavery was perfectly moral or, at the very least, not exactly immoral. We know that people commonly made proclamations about slavery being legal or moral or justified, but surely on the inside no one seriously believed that enslaving another human being was morally acceptable, did they? They did. And they made rational arguments to that effect, which were in time countered with better arguments on reason's road to abolition.

Slavery is a case study on the power of self-deception as a work around for a psychological phenomenon called *cognitive dissonance*, or the mental tension experienced when someone holds two conflicting thoughts simultaneously; in this case: (1) slavery is acceptable and possibly even good, and (2) slavery is unacceptable and possibly even evil. For most of human history people simply accepted the first idea, but as Enlightenment ideas about people being treated equally began to percolate throughout societies, sentiments began to change toward the second idea, which produced cognitive dissonance.

The psychologist who first identified cognitive dissonance, Leon Festinger, described the psychological process this way: "Suppose an individual believes something with his whole heart; suppose further that he has a commitment to this belief, that he has taken irrevocable actions because of it; finally, suppose that he is presented with evidence, unequivocal and undeniable evidence, that his belief is wrong: what will happen? The individual will frequently emerge, not only unshaken, but even more convinced of the truth of his beliefs than ever before. Indeed, he may even show a new fervor about convincing and converting other people to his view."[13] I suspect that increasing cognitive dissonance made the life of the slave owner increasingly uncomfortable throughout the nineteenth

century, as the long-term shift in moral sentiments about the proper treatment of human beings was under way. The commonly accepted belief—that
one group of humans may enslave another group—collided with the
Enlightenment belief that humans are never to be treated as means to an
end but are an end in and of themselves (Immanuel Kant) and that all
people are created equal (Thomas Jefferson).

As this shift was under way, and ever more slaveholders experienced
the cognitive dissonance this clash of ideas inevitably entails, something
had to give—either slavery itself or the justifications for slavery through a
host of ersatz rational arguments justified through the psychological phenomenon called self-deception, a well-documented cognitive strategy to
attenuate cognitive dissonance, summarized by the evolutionary biologist
Robert Trivers in *The Folly of Fools: The Logic of Deceit and Self-Deception
in Human Life* and by the psychologists Carol Tavris and Elliot Aronson
in their book *Mistakes Were Made (but Not By Me)*.[14] There is a logic to
self-deception that works like this: In a selfish-gene model of evolution we
should maximize our reproductive success through cunning and deceit.
Yet game-theory dynamics (like those we considered in chapter 1) shows
that if you are aware that other contestants in the game will also be
employing similar strategies, it behooves you to feign transparency and
honesty and lure them into complacency before you defect and grab the
spoils. But if they are like you in anticipating such a shift in strategy they
might pull the same trick, which means you must be keenly sensitive to
their deceptions and they of yours. Thus we evolved the capacity for
deception detection, and this led to an arms race between deception and
deception detection.

Deception gains a slight edge over deception detection when the interactions are few in number and among strangers. But if you spend enough
time with your interlocutors, they may leak their true intent through
behavioral tells. As Trivers writes, "When interactions are anonymous or
infrequent, behavioral cues cannot be read against a background of known
behavior, so more general attributes of lying must be used." He identifies
three: *Nervousness*. "Because of the negative consequences of being
detected, including being aggressed against ... people are expected to be
more nervous when lying." *Control*. "In response to concern over appearing nervous ... people may exert control, trying to suppress behavior,
with possible detectable side effects such as ... a planned and rehearsed
impression." *Cognitive load*. "Lying can be cognitively demanding. You

must suppress the truth and construct a falsehood that is plausible on its face and . . . you must tell it in a convincing way and you must remember the story." Unless self-deception is involved. If you believe the lie, you are less likely to give off the normal cues of lying that others might perceive: deception and deception detection create self-deception.

The self-deception of slave owners and proponents of slavery is well documented by the historians Eugene Genovese and Elizabeth Fox-Genovese in their book *Fatal Self-Deception: Slaveholding Paternalism in the Old South*. Slavery was not perceived by most slaveholders in the nineteenth century to be an exploitation of humans by other humans for economic gain; instead, slaveholders painted a portrait of slavery as a paternalistic and benign institution in which the slaves themselves were seen as not so different from all laborers—black and white—who toiled everywhere in both free and slave states; further, the South's "Christian slavery" was claimed to be superior. "Decades of study have led us to a conclusion that some readers will find unpalatable," Genovese and Fox-Genovese note. "Notwithstanding self-serving rhetoric, the slaveholders did believe themselves to be defending the ramparts of Christianity, constitutional republicanism, and social order against northern and European apostasy, secularism, and social and political radicalism." Southern slaveholders could look around the world and see the hypocrisy with their own eyes. "Viewing the free states, they saw vicious Negrophobia and racial discrimination and a cruelly exploited white working class. Concluding that all labor, white and black, suffered de facto slavery or something akin to it, they proudly identified 'Christian slavery' as the most humane, compassionate, and generous of social systems."[15]

As the ex-slave Lewis Clarke observed in his revealingly titled book *Narrative of the Sufferings of Lewis Clarke, During a Captivity of More Than Twenty-Five Years, Among the Algerines of Kentucky, One of the So Called Christian States of North America*: "There is nobody deceived quite so bad as the masters down South, for the slaves deceive them, and they deceive themselves."[16] The American architect, journalist, and social critic Frederick Law Olmsted summed up the theory of black incapacity on offer from both Northerners and Europeans: "The laboring class is better off in slavery, where it is furnished with masters who have a mercenary as well as a humane interest in providing the necessities of a vigorous physical existence to their instruments of labor, than it is in Europe, or than it will be in the North."[17] Consider the range of rationalizations employed by people at the time to reduce cognitive dissonance—unconsciously through the process of self-deception—documented by the Genoveses:

> The negro has never yet found a sincere friend but his master.—George M. Troup, 1824, "First Annual Message to the State Legislature of Georgia."[18]

> My husband's influence over the slaves is very great, while they never question his authority, and are ever ready to obey him implicitly, they love him!—Frances Fearn of Louisiana[19]

> We, above all . . . were guarding the helpless black race from utter annihilation at the hands of a bloody and greedy 'philanthropy,' which sought to deprive them of the care of humane masters only that they might be abolished from the face of the earth, and leave the fields of labor clear for that free competition and demand-and-supply, which reduced even white workers to the lowest minimum of a miserable livelihood, and left the simple negro to compete, as best he could, with swarming and hungry millions of a more energetic race, who were already eating one another's heads off.—E. A. Pollard, 1866, *Southern History of the War*[20]

> Negroes are a thriftless, thoughtless people. Left to themselves they will over eat, unseasonably eat, walk half the night, sleep on the ground, out of doors, anywhere.—Virginia slaveholder, 1832[21]

> Nine-tenths of the Southern masters would be defended by their slaves, at the peril of their own lives.—Thomas R. R. Cobb, 1858[22]

The twisted psychology employed to assuage the dissonance was well expressed by Chancellor William Harper of South Carolina when he said, "It is natural that the oppressed should hate the oppressor. It is still more natural that the oppressor should hate his victim. Convince the master that he is doing injustice to his slave, and he at once begins to regard him with distrust and malignity."[23] In her 1994 sociological analysis of gender, class, and race relations, *The Velvet Glove*, the sociologist Mary R. Jackman identified what I think is more appropriately labeled as cognitive dissonance: "The presumption of moral superiority over a group with whom one has an expropriative relationship is thus flatly incompatible with the spirit of altruistic benevolence, no matter how much affection and breast-beating accompanies it. In the analysis of unequal relations between social groups, paternalism must be distinguished from benevolence."[24] You can hear the rationalization in the words of South Carolina governor George McDuffie in 1835: "The government of our slaves is strictly patriarchal, and produces those mutual feelings of kindness which result from a con-stant interchange of good offices."[25] Whatever you call it, self-deceptive paternalism masquerades as altruism ("I'm helping these people") or reciprocity ("I'm giving back to these people after they gave for me"), and it may be an inevitable stage between slavery and freedom. Lest one lapse for even a moment and think there was something benevolent to Southern slavery, screen Steve McQueen's *12 Years a Slave* for a visceral reminder of the inhumanity of this so-called paternalism—the chains of bondage; the mouthpieces of silence; the filth of living quarters; and the beatings, hangings, and especially the slashing, searing pain of the whip.

ENLIGHTENMENT REASON AND THE ABOLITION OF SLAVERY

There is considered and considerable debate among scholars and historians about what led ultimately to the abolition of slavery. However, focusing strictly on the arguments, a brief review is instructive with an eye toward the arguments that were used *against* the institution.

Starting with religion, as the British historian Hugh Thomas noted in his monumental study *The Slave Trade: The Story of the Atlantic Slave Trade, 1440–1870*, "There is no record in the seventeenth century of any preacher who, in any sermon, whether in the Cathedral of Saint-André in Bordeaux, or in a Presbyterian meeting house in Liverpool, condemned the trade in black slaves."[26] Instead, what few objections there were often had a ring of pragmatism to them, as in a 1707 letter from the secretary of the Royal Africa Company, Colonel John Pery, to his neighbor William Coward, who was considering sponsoring a slave voyage. It was "morally impossible that two tier of Negroes can be stowed between decks in four feet five inches," he said. *One* tier, however, was perfectly acceptable in terms of profitability, given that the death rate on overpacked slave ships was a whopping 10 to 20 percent in the Middle Passage between Africa and America. Even religious-sounding objections were suspiciously expedient in their reasoning, as reflected in this observation by ship's surgeon Thomas Aubrey, a doctor on the slave ship *Peterborough,* when he mused that the inhumanity suffered by their cargo might be compensated for in the next life: "For, though they are heathens, yet they have a rational soul as well as us; and God knows whether it may not be more tolerable for them in the latter day [of judgment] than for many who profess themselves Christians."[27]

In America, the Quaker William Edmundson penned a letter in 1676 to his fellow Quakers in all slaveholding colonies and territories that slavery was un-Christian because it was "an oppression on the mind." This was met with a denunciation by the founder of the colony of Rhode Island, Roger Williams—himself a Protestant theologian—that Edmundson was "nothing but a bundle of ignorance and boisterousness." Shortly thereafter, in 1688, a group of German Quakers in Germantown (Philadelphia) issued a petition against slavery, arguing that it violated the Bible's Golden Rule, but it went nowhere and lay dormant until it was rediscovered in 1844 and used in the abolitionist movement that had already taken root there. In fact, two prominent Quakers of Philadelphia—Jonathan Dickinson and Isaac Norris—were themselves slave traders, and there was even a slave ship named the *Society* that belonged to Quakers in the early eighteenth century; the captain, Thomas Monk, recorded that in 1700, during the transport of 250 slaves from Africa to America, 228 of them were lost in the Middle Passage.[28] As Hugh Thomas notes in summing up the relationship between religion and slavery:

> In all the richer parts of New York, Dominicans and Jesuits, Franciscans and Carmelites still had slaves at their disposal. The French Father Labat, on his arrival at the prosperous Caribbean colony of Martinique in 1693, described how his monastery, with its nine brothers, owned a sugar mill worked by water and tended by thirty-five slaves, of whom eight or ten were old or sick, and about fifteen badly nourished children. Humane, intelligent, and imaginative though Father Labat was, and grateful though he was for the work of his slaves, he never concerned himself as to whether slavery and the slave trade were ethical.[29]

It was not until late in the eighteenth century that objections to slavery were marshaled on ethical grounds, as noted by one prominent Bostonian: "About the time of the Stamp Act [1765], what were before only slight scruples in the minds of conscientious persons, became serious doubts and, with a considerable number, ripened into a firm persuasion that the slave trade was *malum in se*." *Evil in itself.* That attitude was slow in coming. As late as 1757, the Huguenot rector of Westover, Peter Fontaine, wrote to his brother Moses about their "intestine enemies, our slaves," noting, "To live in Virginia without slaves is morally impossible." It was also economically problematic, and as Hugh Thomas notes, "None of these prohibitions . . . was decided upon for reasons of humanity. Fear and economy were the motives."[30]

What ultimately brought about the abolition of slavery? According to Thomas, "The great wave of ideas, and emotions, known in France, and those who followed her, as the Enlightenment, was (in contrast to the Renaissance) hostile to slavery, though not even the most powerful intellects knew what to do about the matter in practice."[31] Enlightenment ideas transformed into laws, coupled with state enforcement, is what ultimately secured the end of the practice—an end that unfolded in ever more rapid escalation once it took off in the late eighteenth century. But make no mistake about it, the moral arguments that undermined slavery were not enough to bring about the abolition of it; many people and countries had to be dragged kicking and screaming up the moral ladder, as evidenced by the fact that after it outlawed slavery in 1807, the British Royal Navy had to patrol the African coast in search of illegal slave trade for more than sixty years, until 1870, seizing nearly 1,600 ships and liberating more than 150,000 slaves in the process.[32] And, as previously noted, in the United States it took the deaths of more than 650,000 Americans in the Civil War to finally bring about slavery's end there.

In the long run, however, it is the force of ideas even more than the force of arms that marshal moral advancement, as notions such as slavery gradually inch, by degrees, from morally good to acceptable to questionable; to unacceptable to immoral to illegal; and finally they shift altogether from unthinkable to utterly unthought of. What follows is a representative sample of nonreligious (secular) arguments against slavery proffered by Enlightenment philosophers that were highly influential in the abolition of slavery.

In his fictional 1756 *Scarmentado*, the widely read Voltaire had his African character turn the tables on a European slave trading ship captain, explaining to him why he enslaved the white crew: "You have long noses, we have flat ones; your hair is straight, while ours is curly; your skins are white, ours are black; in consequence, by the sacred laws of nature, we must, therefore, remain enemies. You buy us in the fairs on the coast of Guinea as if we were cattle in order to make us labor at no end of impoverishing and ridiculous work . . . [so] when we are stronger than you, we shall make you slaves, too, we shall make you work in our fields, and cut off your noses and ears."[33]

Montesquieu, the Enlightenment philosopher we met in chapter 2, argued in his highly influential 1748 work *The Spirit of the Law* that slavery was bad not only for the slave but also for the slave master; the former for the obvious reason that slavery prevents a person from doing anything virtuous; the latter because it leads a person to become proud, impatient, hard, angry, and cruel.[34]

In the entry for the slave trade in Denis Diderot's monumental 1765 *Encyclopédia,* which was devoured by intellectuals throughout the Continent, Great Britain, and the colonies, the author wrote, "This purchase is a business which violates religion, morality, natural law, and all human rights. There is not one of those unfortunate souls . . . slaves . . . who does not have the right to be declared free, since in truth he has never lost his freedom; and he could not lose it, since it was impossible for him to lose it; and neither his prince, nor his father, nor anyone else had the right to dispose of it."[35]

Jean-Jacques Rousseau, in his 1762 *Du Contrat Social (The Social Contract)*—which was so influential in the intellectual foundation of the US Constitution—rejected slavery as being "null and void, not only because it is illegitimate, but also because it is absurd and meaningless. The words 'slavery' and 'right' are contradictory."[36]

From mainland Europe secular sentiments against slavery migrated to the British Isles and were inculcated and expanded upon in the Scottish Enlightenment. In *A System of Moral Philosophy,* the Scottish philosopher Francis Hutcheson concluded that "all men have strong desires of liberty and property" and that "no damage done or crime committed can change a rational creature into a piece of goods void of all right."[37] Hutcheson's student Adam Smith applied this principle in his first book, in 1759, *The Theory of Moral Sentiments,* to argue, "There is not a negro from the coast of Africa who does not . . . possess a degree of magnanimity which the soul of his sordid master is scarce capable of conceiving."[38] In time, such arguments found their way into the legal system, such as through the writings of the judge Sir William Blackstone, whose 1769 *Commentaries on the Laws of England* outlined a legal case against slavery, then pronounced that "a slave or a negro, the moment he lands in England, falls under the protection of the laws and, with regard to all natural rights, becomes, *eo instanti* [instantly], a freeman."[39]

The influence of both the French and the Scottish Enlightenments on the Founding Fathers and framers of the US Constitution is well known. Thomas Jefferson, Benjamin Franklin, John Adams, Alexander Hamilton, James Madison, George Washington, and the others are considered to be both products of the Enlightenment and promoters of its philosophy of science and reason as the bases of a moral social order in the New World.

SLAVERY AND THE PRINCIPLE OF INTERCHANGEABLE PERSPECTIVES

Slavery is morally wrong because it is a clear-cut case of *decreasing the survival and flourishing of sentient beings*. But why is *that* wrong? It is wrong because it violates the natural law of personal autonomy and our evolved nature to survive and flourish; it prevents sentient beings from living to their full potential as they choose, and it does so in a manner that requires brute force or the threat thereof, which itself causes incalculable amounts of unnecessary suffering. How do we know *that's* wrong? Because of the *principle of interchangeable perspectives*: I would not want to be a slave, therefore I should not be a slave master. If this sounds familiar, it's because it is, in fact, the very argument made by the man who did more than anyone else in the United States to put an end to slavery, Abraham Lincoln, who, in 1858, on the eve of the American Civil War that would be fought in part to end the institution, declared, "As I would not be a slave, so I would not be a master."[40]

This is also another way to formulate the Golden Rule: "As I would not want someone else to make *me* a slave, so I should not make someone else *be* a slave." In modern parlance, it is a description of the evolutionary stable strategy of *reciprocal altruism*: "I will scratch your back instead of being your master, if you will scratch my back and not make me a slave."

The *principle of interchangeable perspectives* is also a restatement of John Rawls's "original position" and "state of ignorance" arguments, which posit that in the original position of a society in which we are all ignorant of the state in which we will be born—male or female, black or white, rich or poor, healthy or sick, Protestant or Catholic, slave or free—we should favor laws that do not privilege any one class because we don't know which category we will ultimately find ourselves in.[41] This can be restated in this context thusly: "As I would not want to live in a society in which I am a slave, so I will vote for laws that outlaw slavery."

In an unpublished note penned in 1854, Lincoln outlined the argument in what to our modern ears sounds like a perfect articulation of a behavioral game analysis. In his refutation of the arguments made in his day that the races should be ranked by skin color, intellect, and "interest" (by which Lincoln meant economic interest), Lincoln wrote the following:

> If A. can prove, however conclusively, that he may, of right, enslave B.— why may not B. snatch the same argument, and prove equally, that he may enslave A.?
>
> You say A. is white, and B. is black. It is color, then; the lighter, having the right to enslave the darker? Take care. By this rule, you are to be slave to the first man you meet, with a fairer skin than your own.
>
> You do not mean color exactly?—You mean the whites are intellectually the superiors of the blacks, and, therefore have the right to enslave them? Take care again. By this rule, you are to be slave to the first man you meet, with an intellect superior to your own.
>
> But, say you, it is a question of interest; and, if you can make it your interest, you have the right to enslave another. Very well. And if he can make it his interest, he has the right to enslave you.[42]

Lincoln is here making a clearly secular argument for equality, reasoning his way from premises to a conclusion. Nowhere did he aver that his inspiration for the abolition of slavery came from religion. In point of fact, Lincoln was not a believer in any traditional sense of the word. After his assassination, the executor of his will and his longtime friend Judge David Davis said of Lincoln, "He had no faith in the Christian sense of the term." Another one of his friends, Ward Hill Lamon, who knew him from his early years as a lawyer in Illinois to his presidency, affirmed, "Never in all that time did he let fall from his lips or his pen an expression which remotely implied the slightest faith in Jesus as the son of God and the Savior of men."[43]

Instead, the above passage reflects the influence on Lincoln of Euclid's *Elements of Geometry*, of which he was an avid reader and made reference to the mathematical propositions and how such reasoning might apply to human affairs. In the above passage, A and B are interchangeable elements of the proposition of the right to enslave—as A would not be a slave to B, A cannot be a master to B. In Steven Spielberg's film *Lincoln*, the screenwriter Tony Kushner has the great emancipator explain Euclid's axiom in the context of a discussion on the equality of the races: "Euclid's first common notion is this: Things which are equal to the same thing are equal to each other. That's a rule of mathematical reasoning. It's true because it works. Has done and always will do. In his book Euclid says this is self-evident. You see, there it is, even in that 2,000-year-old book of mechanical law it is a self-evident truth." Although Lincoln never actually uttered those words, there is every reason to think that he would have made just such an argument because it's precisely what is implied in his 1854 argument that A is interchangeable with B.[44]

In fact, the subsequent lines to Lincoln's formulation of the *principle of interchangeable perspectives*—"As I would not be a slave, so I would not be a master"—are usually left off: "This expresses my idea of democracy. Whatever differs from this, to the extent of the difference, is no democracy." For a democracy to thrive, its citizens must also thrive, because a democracy is the sum total of its individual members. And remember, it is individual people who feel pain and suffer, not collectivities such as democracies. Thus the debate over slavery in Lincoln's time reflected deeper and lasting moral principles that go to the heart of the source of government and the recognition of all human rights. As Lincoln noted in his seventh and final debate, on October 15, 1858, with Stephen A. Douglas—who famously declared in their debates, "I positively deny that he [the black man] is my brother or any kin to me whatever"[45]—the principle at stake was the freedom of the individual to flourish versus the divine right of kings to rule:

> That is the real issue. That is the issue that will continue in this country when these poor tongues of Judge Douglas and myself shall be silent. It is the eternal struggle between these two principles—right and wrong—throughout the world. They are the two principles that have stood face to face from the beginning of time; and will ever continue to struggle. The one is the common right of humanity and the other the divine right of kings. It is the same principle in whatever shape it develops itself. It is the same spirit that says, "You work and toil and earn bread, and I'll eat it." No matter in what shape it comes, whether from the mouth of a king who seeks to bestride the people of his own nation and live by the fruit of their labor, or from one race of men as an apology for enslaving another race, it is the same tyrannical principle.[46]

Lincoln's ultimate moral avowal was simple, and he made it in April 1864 while the body count of the Civil War had ticked up to more than half a million dead: "If slavery is not wrong, nothing is wrong."[47]

Figure 5-2 shows the abolition and criminalization of slavery by political states around the world from 1117, when Iceland became the first state

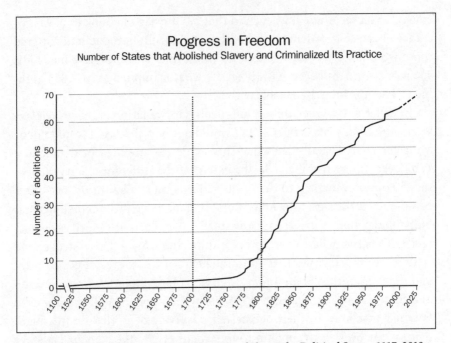

Figure 5-2. The Abolition and Criminalization of Slavery by Political States, 1117–2010
From the Wikipedia entry for "Abolition of Slavery Timeline." Iceland was the first country to officially abolish slavery, in 1117, and the United Kingdom was the latest to make slavery a crime, in 2010. The dashed line to 2025 represents the fact that even though slavery has been legally abolished in all countries, in reality it is still practiced in the form of sex trafficking and slave labor, so there is headway still to be made.[48]

to abolish slavery. Progress was desperately slow and halting for centuries. After the American Declaration of Independence in 1776, the French Declaration of the Rights of Man in 1789, and various other Enlightenment-inspired secular works on human rights became influential in the nineteenth century, the rate of slavery abolition and the spread of freedom escalated, culminating in 2007 and 2010, when Mauritania and the United Kingdom, respectively, made slavery a crime across the board. I have extended the dashed line to 2025 to reflect the fact that although slavery is illegal everywhere on Earth, it is still practiced in the form of sex trafficking in parts of Southeast Asia and elsewhere, and as slave labor in parts of Africa and elsewhere.

Unfortunately, the practice of slave labor and sex trafficking is ongoing in poorer regions of the world. The organization End Slavery Now estimates that there are as many as thirty million people enslaved in this manner,[49] although a number of journalists and scholars who study the subject hold that this figure is very likely inflated by an order of magnitude based on unreliable data and estimates.[50] As a percentage of the world's population, however, all forms of slavery are at the lowest they've ever been. Nevertheless, people are still being exploited, and we need to end it. The Walk Free Foundation has identified ten countries that hold about 70 percent of the world's slaves, with India, China, and Pakistan being the worst offenders. The index includes in its definition of slavery, "forced labour of men, women, and children, domestic servitude and forced begging, the sexual exploitation of women and children, and forced marriage."[51]

The photographer Lisa Kristine has visually documented the brutality of such practices as sex trafficking and slave labor through her photography,[52] and Free the Slaves cofounder Kevin Bales outlines how to combat these particular horrors, and how his organization intends to end all forms of it within a quarter century.[53] As Bales explains, this form of slavery is economic in nature—"People do not enslave people to be mean to them; they do it to make a profit"—and the price of slaves has dropped dramatically from a historical average of $4,000 per slave in 2010 dollars to $90 per slave today. The reason for the drop in price is due to an increase in supply: with an exploding global population, there are a billion vulnerable people to exploit. It's a bonanza for enslavers, whose modus operandi is to trick impoverished people, usually with the offer of a job. "They climb into the back of the truck, they go off with the person who recruits them, 10 miles, 100 miles, 1,000 miles later they find themselves in dirty, demeaning, dangerous work, they take it for a little while, but when they try to leave—Bang!—the hammer comes down and they discover they're enslaved." Being forced to work without pay, under the threat of violence, and unable to walk away surely constitute a form of slavery, and the $40 billion generated by modern slave laborers today is the lowest fraction of the global economy ever generated by slave labor: "Today, we don't have to win the legal battle; there's a law against it in every country. We don't have to win the economic argument; no economy is dependent on slavery (unlike in the 19th century, when whole industries could have collapsed). And we don't have to win the moral argument; no one is trying to justify it any more."[54] And there's a freedom dividend: as slaves become legal producers and consumers, local economies spiral upward very rapidly.

These modern antislavery campaigners argue that the criminalization of slavery is important in the abolishment of the continued practice, noting that it is only in countries where slavery has been criminalized—and not just abolished—that the prosecution of slaveholders becomes legally practical. As previously noted, in the African country of Mauritania, for example, slavery wasn't abolished until 1981, but it was only criminalized in 2007, so now the process of prosecuting slaveholders as criminals has begun, albeit slowly. As the CNN correspondent John Sutter noted in his report on the conditions of the people there, "The first step toward freedom is realizing you're enslaved."[55]

. . .

The rational arguments and scientific refutations of slavery in the eighteenth and nineteenth centuries that led to its legal abolition and universal denunciation set the stage for the other rights revolutions that led to greater justice and freedoms for blacks and minorities, women and children, gays and lesbians, and now even animals, and expanded the moral sphere to include more sentient beings than ever before in human history.

6

A Moral Science of Women's Rights

Let us have a fair field! This is all we ask, and we will be content with nothing less. The finger of evolution, which touches everything, is laid tenderly upon women. They have on their side all the elements of progress, and its spirit stirs within them. They are fighting, not for themselves alone, but for the future of humanity. Let them have a fair field!

—Tennessee Celeste Claflin, 1897, social reform advocate,
first woman to open a Wall Street brokerage firm

In chapter 1 we met the nineteenth-century Irish historian William Lecky, who introduced the metaphor of the expanding moral circle in his 1869 book *A History of European Morals*. In his chapter on the "position of women" he postulated that the rise of monogamy and marriage were the primary steps in the elevation of women to a status closer to that of men, and he argues that the primary value of the marriage contract is in granting women equal rights, at least in the home (but, sadly, *only* in the home): "The utilitarian arguments in its defence are also extremely powerful, and may be summed up in three sentences. Nature, by making the number of males and females nearly equal, indicates it as natural. In no other form of marriage can the government of the family, which is one of the chief ends of marriage, be so happily sustained, and in no other does woman assume the position of the equal of man."[1]

This grudging admission that women are the equal of men, as long as they keep to their needlework and don't set foot outside the parlor, is all the less impressive coming as it does almost eighty years after Mary Wollstonecraft's *A Vindication of the Rights of Woman*[2] and after John Stuart

Mill's call for the legal and social equality of women in his treatise (possibly coauthored with his wife, Harriet Taylor Mill) *The Subjection of Women*.[3] It also comes about twenty years after the very first women's rights convention (held in 1848 in Seneca Falls, New York) in which the Declaration of Rights and Sentiments, chiefly authored by Elizabeth Cady Stanton, was ratified by sixty-eight women and thirty-two men. The document was patterned after the Declaration of Independence and contained these words: "We hold these truths to be self-evident: that all men and women are created equal." Clearly, Lecky felt this truth wasn't self-evident in the least, as he opined:

> In the ethics of intellect they are decidedly inferior. Women very rarely love truth, though they love passionately what they call "the truth," or opinions they have received from others, and hate vehemently those who differ from them. They are little capable of impartiality or of doubt; their thinking is chiefly a mode of feeling; though very generous in their acts, they are rarely generous in their opinions or in their judgments. They persuade rather than convince, and value belief rather as a source of consolation than as a faithful expression of the reality of things.[4]

Unfortunately this attitude was not atypical, and supporters of this uncommon modern notion of women's equality and their right to vote were harshly scorned and ridiculed. Clearly men felt their comforts and privileges threatened; commenting on the 1848 convention, a reporter for the *Oneida Whig* had this to say:

> This bolt is the most shocking and unnatural incident ever recorded in the history of womanity. If our ladies will insist on voting and legislating, where, gentlemen, will be our dinners and our elbows? Where our domestic firesides and the holes in our stockings?[5]

Where indeed?

Nevertheless, in the United States the suffragists and their allies persevered, and after a seventy-two-year battle the passage of the Nineteenth Amendment secured women's right to vote, in 1920. The events that led to suffrage for women in America are riveting in their details, though only the briefest overview is possible here.[6] It was Elizabeth Cady Stanton and Lucretia Mott who organized the 1848 conference, after attending the

World Anti-slavery Convention in London in 1840—a convention at which they had come to participate as delegates, but at which they were not allowed to speak and were made to sit like obedient children in a curtained-off area. This did not sit well with Stanton and Mott. Conventions were held throughout the 1850s but were interrupted by the American Civil War, which secured the franchise in 1870—not for women, of course, but for black men (though they were gradually disenfranchised by poll taxes, legal loopholes, literacy tests, threats, and intimidation). This didn't sit well either and only served to energize the likes of Matilda Joslyn Gage, Susan B. Anthony, Ida B. Wells, Carrie Chapman Catt, Doris Stevens, and countless others who campaigned unremittingly against the political slavery of women.

Things began to heat up when the great American suffragist Alice Paul (arrestingly portrayed by Hilary Swank in the 2004 film *Iron Jawed Angels*) returned from a lengthy sojourn in England. She had learned much during her time there through her active participation in the British suffrage movement and from the more radical and militant British suffragists, including the courageous political activist Emmeline Pankhurst, characterized as "the very edge of that weapon of willpower by which British women freed themselves from being classed with children and idiots in the matter of exercising the franchise."[7] Upon her death Pankhurst was heralded by the *New York Times* as "the most remarkable political and social agitator of the early part of the twentieth century and the supreme protagonist of the campaign for the electoral enfranchisement of women";[8] years later, *Time* magazine selected her as one of the one hundred most important people of the century. Thus when Alice Paul returned from abroad she was ready for action, though the more conservative members of the women's movement weren't quite ready for Alice. Nevertheless, she and Lucy Burns organized the largest parade ever held in Washington, DC, to attract attention to the cause. On March 3, 1913 (strategically timed for the day before President Wilson's inauguration), twenty-six floats, ten bands, and eight thousand women marched, led by the stunning Inez Milholland, wearing a flowing white cape and riding a white horse. (See *figure 6-1*.) Upwards of a hundred thousand spectators watched the parade, but the mostly male crowd became increasingly unruly and the women were spat upon, taunted, harassed, and attacked while the police stood by. Afraid of an all-out riot, the War Department called in the cavalry to contain the escalating violence and chaos.[9]

Figure 6-1. Inez Milholland's March on Washington, DC
On March 3, 1913, the women's rights advocate Inez Milholland led the march on the capital along with her fellow suffragists Alice Paul and Lucy Burns.[10]

It was a gift. A scandal ensued due to the rough treatment of the women, and suddenly "the issue of suffrage—long thought dead by many politicians—was vividly alive in front page headlines in newspapers across the country. . . . Paul had accomplished her goal—to make woman suffrage a major political issue."[11]

In 1917 women began peacefully picketing outside the White House, but, once again, they were met with harassment and violence. These Silent Sentinels (as they were called) stood day and night (except Sundays) with their banners for two and a half years, but after the United States joined in World War I, patience ran thin as it was seen as improper to picket a wartime president. The picketers were charged with obstructing traffic and were thrown—often quite literally *thrown*—into prison cells, where they were treated like criminals rather than political protesters, and were kept in appalling conditions. Many of the women went on a hunger strike, including Alice Paul, who was viciously force-fed to keep her from becoming a martyr for the cause. Word of the brutality in the workhouse was leaked to the press, and the public became increasingly incensed at the protesters' horrific treatment. During what became known as the Night of

Terror, forty prison guards went on a rampage and the women were "grabbed, dragged, beaten, kicked, and choked"; Lucy Burns had her wrists cuffed and chained above her head to the cell door; another woman was taken to the men's section and told "they could do what they pleased with her"; another woman was knocked unconscious; still another had a heart attack.[12] These outrages were grave tactical errors. "With public pressure mounting as a result of press coverage, the government felt the need to act. . . . Arrests didn't stop these protesters; neither did jail terms, psychopathic wards, force-feeding, or violent attacks. Their next decision was simply to let them out."[13]

At long last, in 1920, the Nineteenth Amendment (originally drafted by Susan B. Anthony and Elizabeth Cady Stanton in 1878) was passed—by a single vote—thanks to twenty-four-year-old Harry T. Burn, a Tennessee legislator who had originally intended to vote against his state ratifying the amendment (which needed ratification of thirty-six of the then forty-eight states to pass), but changed his mind because of a note from his mother.

> Dear Son:
> Hurrah, and vote for suffrage! Don't keep them in doubt. I notice some of the speeches against. They were bitter. I have been watching to see how you stood, but have not noticed anything yet.
> Don't forget to be a good boy and help Mrs. Catt put the "rat" in ratification.
>
> Your Mother.[14]

In the end, then, suffrage for women came down to the vote of one young man, influenced by his mom. It was rumored that "the anti-suffragists were so angry at his decision that they chased him from the chamber, forced him to climb out a window of the Capitol and inch along a ledge to safety."[15] Thus suffrage arrived in the United States, kicking and screaming.

It was a right that women in a number of other countries had already won years before, but one that others would have to wait for—and in some cases they're are still waiting. *Figure 6-2* tracks the moral progress of women's suffrage, while *figure 6-3* tracks the gaps between when all men versus all women were granted the franchise, from Switzerland's 123-year gap between 1848 and 1971, to Denmark's 0-year gap in 1915. By comparison, the 50-year gap in the United States between 1870 and 1920 lies midway in this history.

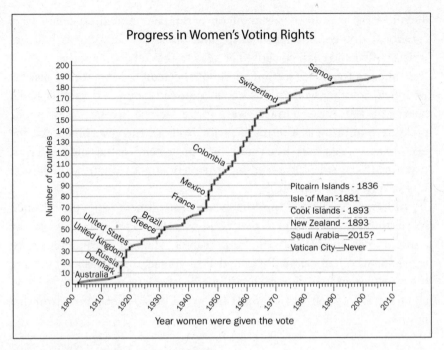

Figure 6-2. Women's Right to Vote Over Time
The stair-step progress of women's suffrage is tracked over time from 1900 to 2010, showing two big bursts, the first after World War I and the second after World War II. Tellingly, the expected date for the sovereign nation of Vatican City to grant women the right to vote is "never."[16]

FROM SUFFRAGE TO SUCCESS

Once the foundation for one type of rights is built, it makes the work of subsequent rights builders easier. Note the 2013 Global Gender Gap Report, produced by the World Economic Forum, which tracks benchmarks in 136 countries on economic, political, education-, and health-based criteria. The report shows that, between men and women, "globally, 96 percent of the health gap has been closed, 93 percent of the education gap, 60 percent of the economic participation gap, but only 19 percent of the political gap." Iceland, Finland, Norway, and Sweden have the smallest gaps in the world, with northern European countries holding seven of the top ten places. The value of this shift can be tracked economically, as the report concluded: "The index continues to track the strong correlation between a country's gender gap and its national competitiveness. Because women account for one-half of a country's potential talent base, a nation's compe-

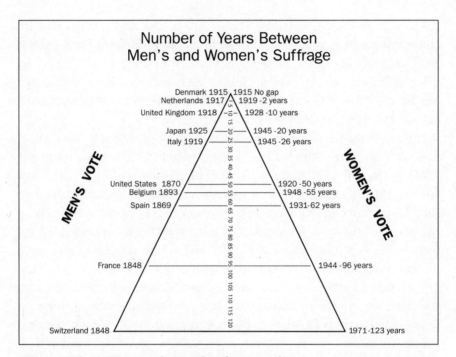

Figure 6-3. The Gap Between the Franchise for Men and Women

The spasmodic nature of moral progress is reflected in the shrinking time in years between the dates when men's suffrage and women's suffrage were legalized, from 123 years for Switzerland to 0 years for Denmark. Such change is contingent on many social and political variables that differ from country to country.

titiveness in the long term depends significantly on whether and how it educates and utilizes its women."[17]

It's heartening that women are occupying a place in traditionally male-dominated professions in ever-greater numbers. For instance, in 2013, Ursula von der Leyen became the first-ever minister of defense in Germany. In her position von der Leyen has been a powerful advocate for change, laying out the scope of her ambitions in December of that year: "My aim is a united states of Europe, run along the lines of the federal states of Switzerland, Germany or the USA," which would be defended by a united European army. Including female voices in the political arena is always good for moral progress, as further evidenced by von der Leyen's call for increasing the number of nurseries in the country and her support of gay marriage. And she is by no means the first female defense minister in Europe. Jeanine Hennis-Plasschaert is presently the minister of defense, since 2012, in the Netherlands, Kristin Krohn Devold held that post in Norway from 2001 to 2005, as did Michèle Alliot-Marie for France from 2002 to 2007, Leni Björklund for Sweden from 2002 to 2006, Carme Chacón Piqueras for Spain from 2008 to 2011, and Grete Faremo for Norway from 2009 to 2011; in fact, there have been a total of eighty-two female defense ministers around the world.[18]

Added to the list of powerful women in traditionally male-dominated fields is Mary Barra, who became the first woman CEO of a car company (GM) in December 2013. Her appointment was followed by reports that *Fortune* 500 companies that have more women in positions of leadership show 50 percent higher profits. There are now twenty-two female CEOs in the *Fortune* 500, and 16.9 percent of their board seats are occupied by women.[19] A Pew Research Center study released the day after Barra took her new post reported that only 15 percent of young women in America say that they have been discriminated against because of their gender, and the share of women in managerial and administrative occupations is nearly equal to that of men, at 15 percent compared to 17 percent. Among young American women, their rate of pay is now up to 93 percent of that of men in comparable positions, up from 67 percent in 1980, and across all age cohorts the average hourly wage for women is 84 percent of that of men in comparable jobs, up from 64 percent in 1980. The likeliest cause of this progress is education: 38 percent of women ages twenty-five to thirty-two now hold bachelor's degrees, compared to 31 percent of men the same age. As a consequence, 49 percent of employed workers with at least a bachelor's degree last year were women, up from 36 percent in 1980. According to Kim Parker, the associate director of the Pew Social & Demographic Trends Project, "Today's generation of young women is entering the labor force near parity with men in terms of earnings and extremely well prepared in terms of their educational attainment."[20] Finally, another Pew Research Center analysis, published in 2013, found that in 40 percent of American households with children, women are the sole or primary breadwinners, a fourfold increase since 1960.[21] Although women still have not attained full parity with men in pay, *figure 6-4, figure 6-5,* and *figure 6-6* show these unmistakable trends in narrowing the gender gap, which looks to be closed by the end of the decade if the trends continue. At least in the United States (and various other countries), the pay gap is a fiendishly complicated calculation that depends on many factors, but the long-term trend line is in the right direction.

These graphs showing improvements in the status of women are

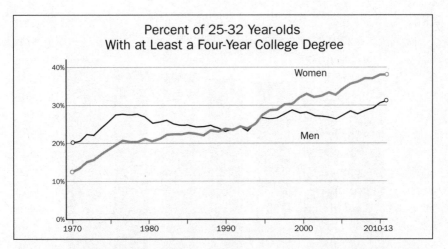

Figure 6-4. Narrowing the Gender Gap in Education

Percentage of twenty-five to thirty-two-year-olds with at least a four-year college degree shows that women are now ahead of men in earning a bachelor's degree. In 1970 only 12 percent of women compared to 20 percent of men earned a four-year degree. In 2012 that gap had reversed, with women at 38 percent compared to 31 percent for men. Source: Pew Research Center

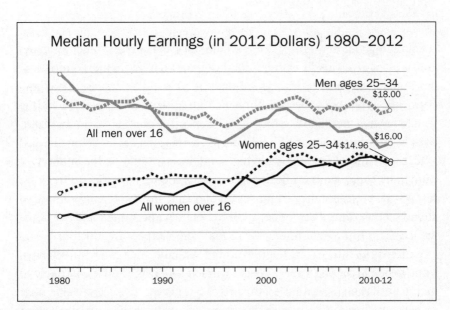

Figure 6-5. Narrowing the Gender Gap in Hourly Earnings

Median hourly earnings in 2012 dollars from 1980 to 2012 show women over age sixteen closing the gap of $8.00 an hour in 1980 to $1.04 in 2012. Source: Pew Research Center

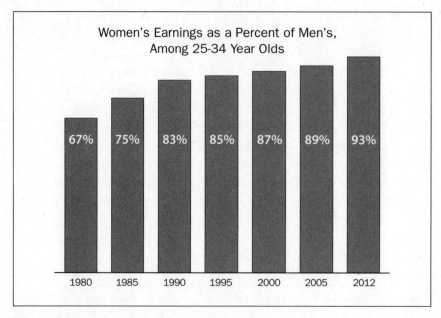

Figure 6-6. Narrowing the Gender Gap in Earning Power

Women's overall earnings as a percentage of men's earnings in 2012 dollars among twenty-five-to-thirty-four-year-olds from 1980, when a woman earned 67 percent of a man in a comparable job, to a near-parity 93 percent in 2012. Ideally, the gap will close to zero in the US by 2020. Source: Pew Research Center

encouraging, but they don't tell the whole story, of course. Women in many non-Western nations live under conditions of extreme male domination, most notably in theocracies and nations with corrupt or dysfunctional governments. Women in these cultures endure a litany of horrors, including female genital mutilation, which often causes lifelong pain and increased risks in childbirth; honor killings; child marriage; and even being charged with the crime of being raped that sometimes results in the murder of women by their own families or the state (while perpetrators go free). As second-class citizens, they endure a raft of insults from not being allowed to attend school, drive a car, or get a job to not being permitted to leave home without a male escort or interact with male shopkeepers and doctors.

And worldwide, to one degree or another, rape and sexual assault continue to be harrowing issues. Where women are scarce and men unable to marry until their late twenties, rape rates rise, as we have seen in alarming recent crimes reported in India.[22] Where women have the courage to enter fields or territories that were traditionally male—most notably the military—rape and assault also rise. In the United States, a widely viewed 2012 film,

The Invisible War, reported that the problem of "military sexual trauma" is widespread. Media coverage of both the film and the problem led top military brass to respond with an investigation and prosecution of perpetrators,[23] including army brigadier general Jeffrey Sinclair, who was court-martialed and fined $20,000 after pleading guilty to adultery and the mistreatment of a woman who accused him of sexual assault.[24]

According to the US Department of Justice's Bureau of Justice Statistics, the overall rate of rape has been declining since the government began collecting reliable data, in 1995. According to the department's 2013 report, "from 1995 to 2010, the estimated rate of female rape or sexual assault victimizations declined 58%, from 5.0 victimizations per 1,000 females age 12 or older to 2.1 per 1,000." The bureau's definition of sexual violence includes "completed, attempted, or threatened rape or sexual assault." While the rate of all such acts remained stable from 2005 to 2010, completed acts "declined from 3.6 per 1,000 females to 1.1 per 1,000" per year in the United States over that time frame, a decline of 327 percent. (See *figure 6-7*.)

In spite of all the TV crime shows about serial killer-rapists, most rapists are not psychopathic strangers. The bureau found that "78% of sexual

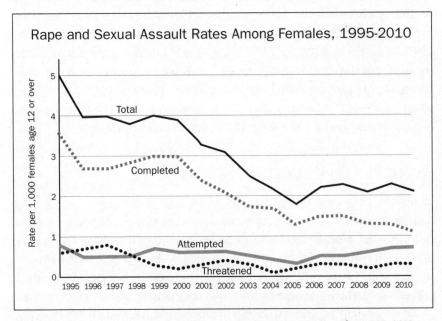

Figure 6-7. Rape and Sexual Assault Victimization Rates Among Females, 1995–2010

From 1995 to 2010 the estimated rate of female rape or sexual assault victimizations declined 58 percent, from 5.0 victimizations per 1,000 females age 12 or older, to 2.1 per 1,000. Source: U.S. Department of Justice's Bureau of Justice Statistics violence involved an offender who was a family member, intimate partner, friend, or acquaintance."[25] At the overall current rate of 2.1 per 1,000 (two tenths of 1 percent), this translates to a rate of 1.6 per 1,000 females who are sexually assaulted by someone they know, and 0.5 per 1,000 by a stranger (five one-hundredths of 1 percent), making stranger danger a statistical outlier.[26]

If a study published in 2014 is corroborated, however, the rape rate numbers may be higher than the Department of Justice's figures because of the problem of underreporting of sexual assaults, not just by those to whom they happen, but also by police departments themselves. These reports are what the FBI relies on for its Uniform Crime Report program, from which such statistics are compiled. According to the new study's author, University of Kansas School of Law professor Corey Rayburn Yung, there has been a systematic underreporting of sexual assaults in 22 percent of the 210 police departments he examined, leading him to estimate "that between 796,213 and 1,145,309 complaints of forcible vaginal rapes of female victims nationwide disappeared from the official records from 1995 to 2012."[27] However, even when adjusted for these unreported cases, Yung's estimates show both low- and high-adjusted rape rates as declining since the early 1990s, although they start at a higher rate than the Uniform Crime Report's figures at about 40 per 100,000 people in 1993, to 25 per 100,000 in 2011. Yung's high-adjusted rape rate declines from 60 per 100,000 in 1993, to 45 per 100,000 in 2011, and his low-adjusted rape rate declines from about 55 per 100,000 in 1993, to 40 per 100,000 in 2011.[28] Whatever the exact numbers are, the decline in the rates of rape, sexual assault, and domestic violence are encouraging.[29]

WOMEN'S REPRODUCTIVE RIGHTS

In a book on moral progress it would be unconscionable of me not to address an area that many people consider to be one of the greatest moral failings of our time: legal abortion. Because abortion is so intimately linked to a woman's right to control her own body, which in turn has been subject to the same progressive forces of change as other human rights, I will consider the topic here in the context of conceptualizing our morals based on a scientific understanding of human nature, and how employing reason in the continued service of expanding rights further expands the moral sphere.

Throughout history men have tried to exert varying levels of control over women's reproductive rights. One reason is as obvious as it is simple: In general, men are bigger and stronger than women and they have used this biological fact—as they do against other males in contests over hierarchy, territory, or mates—to their advantage in exerting dominance. With regard to reproduction, men and women can't be equally certain that a particular child is their offspring, a concept captured by the phrase "mother's baby, father's maybe."[30] Women know with 100 percent certainty that they are the mothers of their children (with the exception of rare cases in the modern world when hospitals mix up babies shortly after birth), whereas men know with less than 100 percent certainty the paternity of their issue. Researchers estimate that 1 percent to 30 percent of babies are the product of an "extra-pair paternity" (EPP), in which the biological father is not the partner of the mother.[31] The percentage varies wildly depending on the population being studied, as do the error margins inherent in the difficulties in collecting data on such a sensitive subject.[32] The anthropologist Brooke Scelza, for example, documented an EPP rate of 17 percent of all recorded marital births among the traditional society of the Himba in northern Namibia, and that such EPPs were "associated with significant increases in women's reproductive success."[33] (In this context, reproductive success means the number of offspring that survive and reach reproductive age.) The evolutionary biologist Maarten Larmuseau and his colleagues, however, found a much lower rate, of 1 to 2 percent, in a western European population in Flanders, Belgium, noting that "This figure is substantially lower than the 8–30% per generation reported in some behavioural studies on historical EPP rates, but comparable with the rates reported by other genetic studies of contemporary Western European populations."[34] In a survey of sixty-seven studies reporting nonpaternity, the anthropologist Kermyt Anderson computed an average figure of 1.9 percent for men with high paternity confidence,[35] which is significantly lower than the 9 percent figure that Robin Baker and Mark Bellis computed from ten studies and reported in their controversial book *Human Sperm Competition*, in which they hypothesize that spermatozoa evolved to compete against other men's sperm cells in the reproductive track of the woman (to swim toward the egg to fertilize it, to fight off other men's sperm from getting to the egg, or to block other men's sperm from fertilizing her egg if they get there first).[36] The likely difference here can be found in Anderson's second finding: men with low-paternity confidence had a nonpaternity rate of 29.8 percent.[37]

The evolutionary psychologist Martie Haselton, whom I consulted

about this sometimes confusing data, estimates that "the range of nonpaternity in the West is 2–4 percent and perhaps a bit higher elsewhere in
the world." She explained the wider range of figures in non-Western countries this way: "In addition to the strength of norms about marriage, there
are norms about fidelity that vary across populations. In the Himba
women marry, but there seems to be an accepted norm that people will
sleep around. So, women might not suffer the extreme costs that other
women suffer if their infidelities are discovered. That might allow them to
be freer to pursue a dual mating strategy." Haselton added parenthetically
that extrapolating from modern data to ancestral environments is problematic in that "another issue to consider is that the modern world affords
women more independence and privacy. Ancestral women didn't travel
on business trips or have husbands who did so."[38]

A dual mating strategy is one in which women seek to secure a sound
investment in parental care and food resources through one partner and
obtain good genes through another partner if both sets of characteristics
are not present in one man. As Haselton explained the phenomenon: "In
principle, women could benefit from both material and heritable benefits
conferred by male partners, but both sets of features may be difficult to
find in the same mate. Men who display indicators of good genes are highly
attractive as sex partners; hence, they can, and often do, pursue a short-
term mating strategy that tends to be associated with reduced investment
in mates and offspring. Therefore, women may often be forced to make
strategic trade-offs by selecting long-term social partners who are higher
in investment attractiveness than sexual attractiveness."[39]

Whatever the rate of EPPs was historically—most likely higher in the
distant past compared to today, in which the institution of monogamous
marriage is encouraged by both church and state—the point is that a
woman could, theoretically, select a dual-mating strategy, and it is this
possibility that leads men to try to control women's reproductive choices.
Even if no offspring result from an EPP sexual encounter, infidelity rates
are high enough that the emotion of jealousy evolved as a result of the
phenomenon of mate guarding to fend off potential mate poachers and to
prevent mates from defecting to another partner.[40] How high are rates of
infidelity? Studies vary, but according to the National Opinion Research
Center in Chicago, about 25 percent of American men and 15 percent of
American women have had an affair at some point during marriage.[41]
Other studies find a range of 20 percent to 40 percent of heterosexual
married men and 20 percent to 25 percent of heterosexual married women

have philandered during their marriages.[42] Another study found that 30 percent to 50 percent of American married men and women are adulterous.[43] As the evolutionary psychologist David Buss notes, "Those who failed in mate guarding risked suffering substantial reproductive costs ranging from genetic cuckoldry to reputational damage to the entire loss of a mate." The result, he argues, is a range of mate guarding adaptations, from "vigilance to violence."[44] This is an unfortunate response to a very real threat, as evidenced in a survey of single American men and women that found 60 percent of men and 53 percent of women admitted to "mate poaching," or trying to entice or charm someone out of their current relationship in order to enter into one with them.[45] An anthropological survey identified mate poaching as common in at least fifty-three other cultures.[46]

Although both men and women philander, get jealous, mate-guard, and mate-poach, in the context of expanding women's reproductive rights and men's attempt to restrict them, male jealousy and mate guarding— whether through vigilance or violence—are strong causal factors. (Studies show, for example, that in the United States more than twice as many women were shot and killed by their husband or intimate acquaintance than were murdered by strangers using guns, knives, or any other means,[47] and that women make up the majority of victims of intimate partner/ family-related homicides.[48]) From stalking to chastity belts to female genital mutilation, throughout history men have tried to control women's sexuality and reproductive choices. And women have developed several strategies in response: contraception, abortion, clandestine affairs, mariticide (killing one's husband), and infanticide.

Starting with the latter, anthropologists and historians tell us that infanticide has been practiced by all cultures everywhere in the world throughout history, including and notably by adherents of all the world's major religions. Historical rates of infanticide have ranged from 10 to 15 percent in some societies to 50 percent in others, but none lack it entirely.[49] The killing of babies is often portrayed by theists (for example, in my debates with them) as the single purest act of evil conceivable.[50] They are wrong. Normal people do not kill their children for no reason. Like all human behavior, infanticide has nontrivial causes, and the evolutionary psychologists Martin Daly and Margo Wilson unearthed some of those reasons in their study of sixty societies using an ethnographic database of cultures around the world. Of the 112 cases of infanticide in which anthropologists recorded a motive, 87 percent supported the "triage theory" of infanticide, which posits that mothers must make hard choices when times are hard (i.e., they kill their children when resources are too scarce to support another infant), a concept well captured in Edward Tylor's nineteenth-century anthropological observation that "Infanticide arises from hardness of life rather than hardness of heart."[51] Nature is not infinite in its resources and not all organisms that are born can survive. When conditions are difficult, parents, especially mothers, must decide who is most likely to survive—including future potential children who may fare better when conditions are better—and sacrifice the rest. Daly and Wilson's survey turned up these reasons for infanticide: disease, deformity, weakness, a twin when parents have only enough resources for one, an older sibling too close in age for resources to support both, hard economic times, no father to help raise the child, or because the infant was fathered by a different sexual partner.[52]

In *Eunuchs for the Kingdom of Heaven*, Uta Ranke-Heinemann notes how widely infanticide was practiced in ancient Greece and Rome, and the Catholic Church's prohibition against it goes back to the Middle Ages.[53] During the historical period of the early rights revolutions, both church and state tried to do something to curb infant killings without addressing the underlying causes (of which they were ignorant). Injunctions were approved and laws were passed, but just as in the bad old days of back alley abortions, if a woman didn't want her baby there was little anyone could do about it. Mothers would "accidentally" roll over on top of their infants during sleep (called "overlying"), or they dropped them off at foundling homes where the business of infant disposal was swift and private. Wet nurses and "baby farmers" were also tasked with the business of baby removal. In mid-nineteenth-century London it was reported that public parks and other spaces were the site of as many dead babies as dead dogs and cats.[54] The popular 2013 film *Philomena*—about a teenage girl in the early 1950s who has her out-of-wedlock baby in an abbey, where she was forced to give it up for adoption over her cries and protestations—captures the tragedy many women face even in modern times when there are no other viable options for them and their baby.

However one frames the issue, the more important question in the context of moral progress is what can be done about it. Proximate solutions arose in historical times in the form of orphanages and adoption agencies, but the ultimate solution is to be found in contraception and education. A comprehensive international study on the relationship between contraception and abortion conducted by Cicely Marston from the London School of Hygiene and Tropical Medicine concluded, "In seven countries—

Kazakhstan, Kyrgyz Republic, Uzbekistan, Bulgaria, Turkey, Tunisia and Switzerland—abortion incidence declined as prevalence of modern contraceptive use rose. In six others—Cuba, Denmark, Netherlands, the United States, Singapore and the Republic of Korea—levels of abortion and contraceptive use rose simultaneously. In all six of these countries, however, overall levels of fertility were falling during the period studied. After fertility levels stabilized in several of the countries that had shown simultaneous rises in contraception and abortion, contraceptive use continued to increase and abortion rates fell. The most clear-cut example of this trend is the Republic of Korea."[55] *Figure 6-8* shows these South Korean data. Abortion rates there took some time to start their decline because for a few years women relied on more traditional but far less effective methods of birth control, such as withdrawal, but when replaced by reliable methods, pregnancy rates tumbled, thereby lowering the demand for abortions.

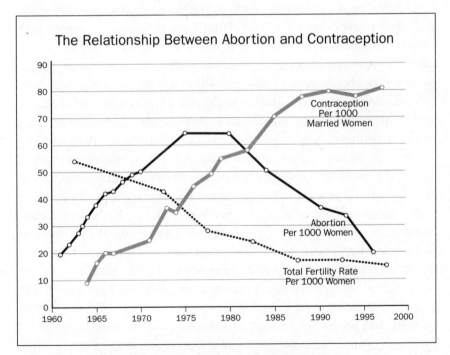

Figure 6-8. The Relationship Between Abortion and Contraception

This data set from South Korea demonstrates that granting women the right to use contraception causes a dramatic decrease in the number of abortions, along with the fertility rate, which has additional benefits for the progress of humanity in creating a sustainable world.[56]

A similar effect occurred in Turkey when abortion rates dropped by almost half between 1988 and 1998 (from 45 to 24 per 1,000 married women) even though the overall rate of contraceptive use remained stable. A study conducted by Pinar Senlet, a population program adviser at the US Agency for International Development, however, revealed that there was a shift from traditional, pitifully underpowered forms of birth control to more modern and therefore more reliable methods. "Marked reductions in the number of abortions have been achieved in Turkey through improved contraceptive use rather than increased use." In short, Turkish couples abandoned natural and unreliable birth control methods in favor of condoms and more effective means of contraception.[57]

Like all social and psychological phenomena, rates of contraceptive use and abortions are multivariable—many factors are operating at once, which makes inferring direct causal links problematic. When it comes to human behavior, it is almost never as simple as "when X goes up, Y goes down," and that is certainly the case for contraception and abortion. Different nations and states have different laws and regulations that affect accessibility to both abortions and birth control technologies. Some countries have higher rates of religiosity than others, and this too influences to what extent women or couples use family planning techniques. Socio-economic forces and poverty rate differences among countries and states also confound conclusions. And so forth. But as I read the data and analysis, here is my interpretation: when women have limited reproductive rights and no access to contraception, they are more likely to get pregnant and this leads to higher fertility rates in a country. When women's reproductive rights are secure and they have access to safe, effective, and inexpensive birth control, as well as access to safe, legal abortion, they rely on both strategies to gain control over their family size to maximize parental investment. So for a period of time after the legalization of abortion and access to contraception, rates of both increase in parallel. But once fertility rates stabilize—once women feel confident in controlling their family size and their ability to raise a child—contraception alone is often all that is needed, so abortion rates decline.

Why can't people "just say no" when it comes to sex—that is, use abstinence as a form of birth control, or time their sexual encounters for when it is "safe" during a woman's natural monthly cycle? They can, of course, and some do, but as the old joke goes that was making the rounds when I was in high school, "What do you call couples who use abstinence, withdrawal, or the rhythm method of birth control? Parents." In theory, of course, abstinence is a foolproof method of preventing pregnancies and STDs and STIs (sexually transmitted diseases and sexually transmitted infections—you can have the latter without the former), just as starvation is a foolproof method of preventing obesity. But in reality the desires to love physically and to bond socially are fundamental to who we are as human beings; and the sex drive is so powerful, and the pleasures and psychological rewards so great, that recommending abstinence as a form of contraception and STI prevention is, in fact, to recommend pregnancy and infection by default. In a 2008 study descriptively titled "Abstinence-Only and Comprehensive Sex Education and the Initiation of Sexual Activity and Teen Pregnancy," the University of Washington epidemiologists Pamela Kohler, Lisa Manhart, and William Lafferty found that among never-married American adolescents aged fifteen to nineteen years, "Abstinence-only education did not reduce the likelihood of engaging in vaginal intercourse, but comprehensive sex education was marginally associated with a lower likelihood of reporting having engaged in vaginal intercourse. Neither abstinence-only nor comprehensive sex education significantly reduced the likelihood of reported STD diagnoses." The authors concluded, "Teaching about contraception was not associated with increased risk of adolescent sexual activity or STD. Adolescents who received comprehensive sex education had a lower risk of pregnancy than adolescents who received abstinence-only or no sex education."[58]

More evidence against abstinence-only programs may be gleaned from the 2013 National Longitudinal Study of Adolescent Health published in the *British Medical Journal* and conducted by researchers at the University of North Carolina at Chapel Hill between 1995 and 2009 on more than seventy-eight hundred women; remarkably, it was discovered that 0.5 percent—or one in two hundred—of adolescent girls had reported that they'd become pregnant *without* sex. Are biblical-level miracles afoot in the bedrooms of teenage girls everywhere? This seems unlikely, to say the least. Interestingly, however, adolescents who reported a "virgin pregnancy" were twice as likely as other pregnant women to have signed a chastity pledge, and they were significantly more likely to report that their parents had difficulties discussing sex or birth control with them. Although the researchers admitted that "Scientists may still face challenges when collecting self-reported data on sensitive topics," my point in citing these data is that young women who are pressured through home or church to "just say no" rather than given solid information to avoid pregnancy should sexual intercourse occur are more likely both to get pregnant and to lie about how it happened.[59] Enforcement instead of education does not work.

The opposite of an abstinence-only program would be a reverse test of my thesis, and here we could not find a better social experiment than in Romania. When the dictator Nicolae Ceauşescu came to power in 1965 he hatched a scheme for national renewal by severely restricting abortions and the use of contraception, to increase his country's population. It worked. When abortion had previously been legalized in Romania, in 1957, 80 percent of unwanted pregnancies were aborted, primarily because of the lack of effective contraception. A decade later the fertility rate had fallen from 19.1 to 14.3 per 1,000, so Ceauşescu made abortion a crime unless a woman was over 45, had already delivered (and raised) four children, was suffering dangerous medical complications, or had been raped. The birth rate promptly shot up to 27.4 per 1,000 in 1967. Despite the dictator's punishments for "childless persons" (monthly fines withheld from wages) or for those with fewer than five children (the imposition of a "celibacy tax") and rewards for especially fertile mothers with a "distinguished role and noble mission" (state-sponsored child care, medical care, maternity leave), if (pace Yogi Berra) the people don't want more babies you can't stop them. The result was a social catastrophe of epic proportions, as thousands of babies were abandoned and left to the care of a state that was inept, corrupt, and broke. More than 170,000 children were dumped into over 700 dank and stark state-run institutional orphanages, and more than 9,000 women died due to complications from black-market (back alley) abortions. The effects are still felt today as many of those orphaned children are now adults with severely impaired intelligence, social and emotional disorders, and alarmingly high crime rates. The book *Romania's Abandoned Children*, by Charles Nelson, Nathan Fox, and Charles Zeanah, is a moving account of this tragedy that should be read by anyone with pretentions toward social engineering and the restriction of women's right to choose.[60]

If conservatives and Christians want to put an end to the termination of fetuses and infants, therefore, the best path to take is education, contraception, and the recognition of full female rights—which are simply human rights—including and especially reproductive rights. In the United States alone, studies show that safe, effective, and inexpensive contraception has prevented approximately twenty million pregnancies in twenty years, and given the rate of abortion during that time, this means that nine million fetuses were never aborted because they were never conceived. And note that it's only 7 percent of women who are sexually active, but who do not use birth control, who account for almost 50 percent of all unintended pregnancies and almost 50 percent of all abortions.[61] And while I'm citing statistics, one more is relevant to this discussion: according to the National Institutes of Health, childbirth is fourteen times more dangerous to a woman than an abortion.[62] This fact provides a rebuttal to the argument "What if a young woman aborts a baby who would have gone on to become a doctor and find the cure for cancer?" A rejoinder is, "What if a young woman who would have gone on to become a doctor and find the cure for cancer dies in childbirth?"

Of course, many people are deaf to arguments demonstrating the positive impacts of contraception and abortion, because their sole concern is the right to life of the unborn fetus, which, in their view, trumps the rights of an adult woman. I believe the pro-choice vs. anti-choice debate turns more on facts than it does on morals, and if the facts can be resolved, then that may help settle the dispute. (I prefer using the terms "pro-choice" and "anti-choice" because we are all "pro-life.") The major obstacle here is binary thinking that forces us to pigeonhole into two distinct categories a problem best conceived as a continuous scale. So-called pro-life proponents believe that human life begins at conception; before conception there is no life— after conception there is. For them, it is a binary system. With continuous thinking we can assign a *probability* to human life—before conception 0, the moment of conception 0.1, multicellular blastocyst 0.2, one-month-old embryo 0.3, two-month-old fetus 0.4, and so on until birth, when the fetus becomes a 1.0 human life-form. It is a continuum, from sperm and egg, to zygote, to blastocyst, to embryo, to fetus, to newborn infant.[63]

Neither an egg nor a sperm cell is a human being, but then neither is the zygote or blastocyst because they might split to become twins, or develop into less than one individual and naturally abort.[64] Although an eight-week-old embryo has recognizable human features such as face, hands, and feet, neuroscientists now know that at that stage neuronal synaptic connections are still under development, so anything even remotely resembling thoughts or feelings is impossible. After eight weeks embryos begin to show primitive response movements, but between eight and twenty-four weeks (six months) the fetus could not exist on its own because such critical organs as the lungs and kidneys do not mature before that time. For example, air sac development sufficient for gas exchange does not occur until at least twenty-three weeks after gestation, and often later, so independent viability is not possible.[65] It is not until twenty-eight weeks, or approximately 77 percent of full-term development, that the fetus acquires sufficient neocortical complexity to exhibit some of the cognitive capacities typically found in newborns. Fetus EEG recordings with the characteristics of an adult EEG appear at approximately thirty weeks, or about 83 percent of full-term.[66]

Along this continuum we see that the fetus's capacity for human thought does not appear until just weeks before its birth. Since abortions are almost never performed after the second trimester—and as we have seen, before that period there is no evidence that the fetus is a thinking, feeling human individual—it is reasonable and rational to provisionally conclude that abortion is not comparable to the murder of a conscious sentient being after birth. Thus there is no scientific justification or rational argument to equate abortion with murder.

Still, an argument might be made that a fetus is a *potential* human being, since all of the characteristics that make us persons are there in the genes, unfolding during embryological development. True enough, but potentiality is not the same as actuality, and moral principles must apply first and foremost to actual persons, not potential persons. Given the choice between granting rights to an actual person (an adult woman) or a potential person (her fetus), it is preferable to choose the former on the grounds of both reason and compassion. Although more than half the states in the United States now have laws to protect unborn victims from violence—as in the Unborn Victims of Violence Act, which treats the killing of a pregnant woman and her fetus as a double homicide—the law does not treat fetuses and adults alike in any other manner. Once again binary thinking lures us into treating a mother and her fetus as the same, whereas continuous thinking allows us to see the substantial differences.

CARVING WOMEN'S RIGHTS

The trend over the past several centuries has been to grant women the same rights and privileges as those of men. Political, economic, and social advances, enabled by scientific, technological, and medical discoveries and inventions, have increasingly provided women not only greater amounts of reproductive autonomy and control, but have also driven an expansion of their rights and opportunities in all areas of life, leading to healthier and happier societies across the globe. As with the other rights revolutions, much progress remains to be realized, but the momentum now is such that the expansion of women's rights should continue unabated into the future.

In these ways—the rational justification for including women as full rights-bearing persons no less deserving than men, the interchangeability of women's perspectives with those of men, the scientific understanding of the nature of human sexuality and reproduction, and the continuous thinking that enables us to see and comprehend the difference between a woman's and a fetus's rights—science and reason have led humanity closer to truth, justice, and freedom.

As an example of how far we've come in just the past two generations (and how oppressed women were as recently as the early twentieth century), I close this chapter with the story of two women—mother and daughter—both named Christine Roselyn Mutchler. The mother was born in Germany and passed through Ellis Island in 1893 with her parents, who then moved to Alhambra, California. Mother Christine married her husband, Frederick, and gave birth to baby Christine in 1910 (and a second daughter three years later), but their lives were shattered shortly after that when Fred told his wife he was going out for a loaf of bread and never returned. Abandoned by her husband, left with no money or food to care for herself and her two small children, mother Christine was forced to return to her father's home.

Unknown to her at the time, Fred had wandered off and walked into the county jail with delusions that his father-in-law was after him. After being examined by a physician, he was sent to a mental hospital for more than a year. During this time, with his delusions in remission, Fred wrote heartbreaking letters to his wife asking about her and the children, but Christine's father kept the letters from her and she continued to believe that she had been abandoned. In time she found work as a housemaid for a friend of a successful motion picture executive named John C. Epping, whose wife had recently died. Desperate for a daughter and enamored by three-year-old Christine, Epping talked Christine's father into forcing her to allow him to adopt the child. Young, poor, scared, and intimidated by her father, Christine reluctantly agreed to the adoption, although a series of articles in the *Los Angeles Times* show that a probation officer on the case opposed the adoption, declaring "she believed Epping saw possibilities of a future Mary Pickford in the little girl, and that the child should have a home in some private family where home life and education would be the principal features."[67] Based on the false information provided by

Figure 6-9. Sculptor Franc Epping

Born Christine Roselyn Mutchler and given the adopted name Frances Dorothy Epping, the sculptor started using the masculinized version of her adopted name—Franc—to be taken seriously in the male-dominated world of sculpture. Her sculptures portray strong women with muscular features in empowering poses.[69]

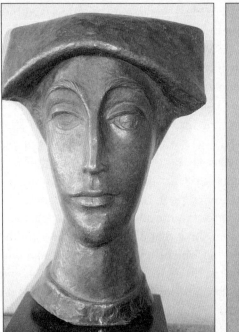

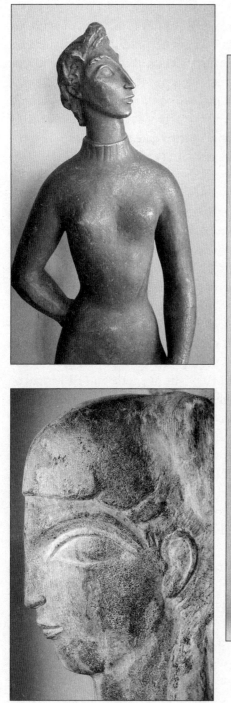

Christine's father that Fred had abandoned them, the judge granted the adoption.

Epping promptly changed the name of his newly adopted daughter to Frances Dorothy Epping, addressed her by her middle name, and (unbelievably) told her she was born in Providence, Rhode Island, and that his deceased wife was her true mother. Now age four, Christine/Dorothy apparently did not accept the fictional story and rebelled—or perhaps Epping changed his mind about raising a daughter as a single dad—because he shuffled her around through a series of surrogate parents, including sisters at the Ramona Convent in Alhambra and caretakers at the Marlborough Preparatory School in Los Angeles, before shipping her back East for a year to live with his sister in the Catskills, and then on to Germany, where she lived with Epping's relations. During that period Dorothy discovered that she had a talent for the arts, in particular sculpture.

She then returned to Los Angeles and finished her secondary education, after which she was reunited with her original family and told the truth about the adoption. She went on to college at the Otis Art Institute in Los Angeles, the Corcoran School of Art in Washington, DC, and the prestigious Academy of Fine Arts in Munich, Germany, in the 1930s under the tutelage of Joseph Wackerle, who, at that time, was the Third Reich's culture senator and who received praise from both Goebbels and Hitler. (She later recalled being stunned by the hypnotic pull Hitler had on an audience of one of his speeches she attended.) In the meantime, Dorothy's real mother, Christine, was instructed by her father to divorce her husband, Fred, after which she met and married a vegetable cart vendor in Los Angeles, left her father's oppressive rule, and began to rebuild her life and new family. But the tragedy of being forced to give up her firstborn child haunted her the rest of her life. As the world changed and Christine saw how women became more empowered in the second half of the twentieth century, she continually asked herself why she didn't speak up and oppose the adoption.

Meanwhile, as Dorothy came of age she soon discovered that family law and the adoption courts were not the only worlds ruled by men. Her chosen profession of sculpture was a heavily male-dominated one, so to be taken seriously she began using a truncated version of her first name, Frances—Franc—and that gained her entrée into the German academy and subsequent galleries and museums (even now one can find references to "his" work). She later recalled that when the professors at the Academy of Fine Arts in Munich found out "Franc" was a woman, she had to listen to lectures from the hallways because only men were allowed inside. From the early 1930s through her death in 1983—by which time it was acceptable for women to shape clay, wood, and stone with their hands—Franc Epping's work was shown in numerous exhibits throughout the United States, including the prestigious Whitney Museum of American Art in New York City. One of her works, *The Man with a Hat,* even appeared in an episode of the original series of *Star Trek.* I know because I own that piece, along with many other sculptures of hers, which I inherited from my mother (see examples in figure 6-9).

You see, Franc Epping was my aunt; her real mother, Christine, was my grandmother; and I am proud to be related to such a resilient and determined woman.[68] Aunt Franc's sculptures portray strong women with muscular features in empowering poses—allegories for what women for generations have had to rise to in order to gain the recognition and equality that are rightfully theirs. This book was written in the inspiring presence of those carved stones.

7

A Moral Science of Gay Rights

> I still hear people say that I should not be talking about the rights of lesbian and gay people and I should stick to the issue of racial justice. But I hasten to remind them that Martin Luther King Jr. said, "Injustice anywhere is a threat to justice everywhere." I appeal to everyone who believes in Martin Luther King Jr.'s dream to make room at the table of brother- and sisterhood for lesbian and gay people.
>
> —Coretta Scott King, hero of the
> gay rights movement, wife of Martin Luther King Jr.

> My father did not take a bullet for same-sex marriage.
>
> —Dr. Bernice King,
> Baptist minister, daughter of Dr. Martin Luther King Jr.

For the past several decades, the campaign to recognize the equal rights and privileges under the law for those citizens who identify as lesbian, gay, bisexual, or transgender (LGBT) has been gaining momentum. To say that religion has fought tooth and nail against gay rights and same-sex marriage wouldn't be strictly fair, given the early support offered to gays by Episcopalians, Unitarians, Reformed Jews, and others,[1] but religious attitudes toward homosexuality are major reasons why, in the United States and around the world, members of the gay community and their allies have been compelled to continue fighting for equal rights, including the right to marry and have children. It's an uphill battle, especially because many religious people still believe that being gay is a sin and a crime and, therefore, by their logic, "if homosexuals are allowed their civil rights, then so would prostitutes, thieves, and anyone else." These words were spoken by Anita Bryant—former entertainer and orange juice representative—who now has the honor of an award in her name, to be given to lucky winners for "unbridled and unparalleled bigotry."[2]

It's astonishing that much of Christendom remains mired in pre–civil rights, pre-Enlightenment, pre-scientific thinking, basing its beliefs on a handful of biblical passages, such as Leviticus 20:13: "If a man also lie with mankind, as he lieth with a woman, both of them have committed an abomination: they shall surely be put to death; their blood shall be upon them." This is a verse that itself is tucked between other passages that instruct you (and obviously the *you* here is male) not to wear linen/wool blend clothing, nor to get tattoos, eat shrimp, marry your wife's sister, become a wizard, round the corners of your head, or clip the ends of your beard. More draconian still, parents should kill their disobedient children and men are further advised to execute adulterous wives and nonvirgin brides. That's right, the death penalty for adultery, which would eliminate a good number of Christian senators and congressmen, preachers and televangelists, along with a nontrivial percentage of the rest of the world's population.

"The face of sin today often wears the mask of tolerance," said Thomas S. Monson—current president of the Church of Jesus Christ of Latter-Day Saints (Mormons)—who continued by saying, "Do not be deceived; behind that facade is heartache, unhappiness, and pain."[3] Refusing to wear the mask of tolerance might explain why, in my own überliberal state of California, the Mormon Church (in league with Catholics and other groups) poured $22 million into the campaign to uphold Proposition 8—a constitutional amendment prohibiting same-sex marriage. "Some portray legalization of so-called same-sex marriage as a civil right," said Gordon B. Hinckley, late president of the LDS Church. "This is not a matter of civil rights; it is a matter of morality."[4] It's odd that Hinckley understood civil rights as somehow distinct from morality—as if morality isn't the bedrock of all rights—and odd too that although tax-exempt organizations are not allowed to participate in political campaigns, Mormons "took over virtually every aspect of Proposition 8," which undoubtedly led to its eventual victory at the polls.[5] Fortunately, Proposition 8 was later ruled unconstitutional, and, on June 28, 2013, same-sex marriage resumed in California with elation and fanfare.

One of the most outspoken religious bigots is the evangelist Jimmy Swaggart, who, when pontificating on his opposition to same-sex marriage, once said, "I've never seen a man in my life I wanted to marry. And I'm going to be blunt and plain: If one ever looks at me like that, I'm going to kill him and tell God he died."[6] This from a man who, presumably, believes in God's omniscience and has read the Ten Commandments; but perhaps it's understandable given Swaggart's reading of the science of human sexuality. When asked if homosexuals are born that way, Swaggart responded, "In a word, 'No!' While it is true that the seed of original sin carries with it every type of deviation, aberration, perversion, and wrongdoing, the homosexual cannot claim to have been born that way any more than the drunkard, gambler, killer, etc."[7] A telling comparison—in Swaggart's twisted view, being gay is comparable to drinking, gambling, and even murder.

Science, however, tells us that gender preference is primarily determined by our genetics, prenatal biology, and embryological hormone development.[8] Almost everyone is attracted to members of the opposite sex;[9] a small percentage of the population—one to five percent—are attracted exclusively to members of the same sex.[10] And these preferences emerge at a very early age. This is why asking a homosexual person when he or she *chose* to become gay or lesbian is like asking a heterosexual person when he or she chose to become straight. (Try it—you'll get dumbfounded looks, as if to say, "What are you talking about? I've always felt this way." Precisely.) But even if sexual orientation *wasn't* primarily biologically determined, it's difficult to understand why such a choice would be considered immoral, and even criminal. As the late Canadian prime minister Pierre Elliott Trudeau famously said, "[T]here's no place for the state in the bedrooms of the nation," adding that "what's done in private between adults doesn't concern the Criminal Code." These words—so refreshingly sane, modern, and progressive—were spoken in 1967.[11]

The 1960s, however, were still the dark ages for LGBT citizens, coming as they did on the heels of the "Lavender Scare"—the 1950s equivalent of McCarthyism for gays that generated fear, persecution, and witch hunts—and President Eisenhower's executive order establishing homosexuality as grounds for being fired from government employment. This resulted in several thousand dismissals, and the private sector followed suit; and since employment records could be shared with private business, gays and lesbians could find themselves penniless and unemployable.[12] As the writer Richard R. Lingeman so aptly put it, "The fifties under Ike represented a sort of national prefrontal lobotomy: tail-finned, we Sunday-drove down the superhighways of life while tensions that later bubbled up in the sixties seethed beneath the placid surface."[13]

STONEWALL

In the 1960s *all* homosexual acts in the United States were illegal, except in Illinois (where the first gay rights organization was founded and where sodomy was decriminalized in 1961).[14] Homosexuality was considered to be a mental illness—even a form of psychopathy—and gay people were subjected to various forms of aversive therapy. Yale University law professor William Eskridge writes in his 2004 history of the movement,

> Gay people who were sentenced to medical institutions because they were found to be sexual psychopaths were subjected sometimes to sterilization, occasionally to castration, [and] sometimes to medical procedures such as lobotomies, which were felt by some doctors to cure homosexuality and other sexual diseases. The most infamous of those institutions was Atascadero, in California. Atascadero was known in gay circles as the Dachau for queers, and appropriately so. The medical experimentation in Atascadero included administering, to gay people, a drug that simulated the experience of drowning; in other words, a pharmacological example of waterboarding.[15]

The legal persecution of gays in the United States—"the land of the free"—was draconian. If police caught a man engaged in "lewd" behavior, his name, age, and even home address could be published in newspapers. Bars and clubs where gays and lesbians were known to hang out were frequently raided; the police would barge in, the music would stop, the lights would go up, IDs would be checked, and men who were suspected of masquerading as women could be taken into the washrooms by female officers and checked, either manually or visually. New York's penal code stated that people had to wear at least three pieces of clothing befitting their gender, or face arrest.

Then came the Stonewall riots, the legendary flash point that for many marks the true beginning of the gay rights movement. It was by no means the first display of activism by the gay community—"homophile" organizations such as the Mattachine Society, the Daughters of Bilitis, and the Janus Society had previously organized various rallies—but Stonewall is commonly perceived to be the first demonstration of gay *power* and the defining moment of fearless solidarity. "We became a people," said one gay man. "All of a sudden, I had brothers and sisters, you know, which I didn't have before."[16] As the writer Eric Marcus says, "Before Stonewall, there was no such thing as coming out or being out. The very idea of being out, it was ludicrous. People talk about being in and out now; [before Stonewall] there was no out. There was just in."[17]

The Stonewall Inn was a grotty, Mafia-owned gay bar on Christopher Street in Greenwich Village in New York City. On the night of June 28, 1969, several police officers descended on the inn to conduct a raid in the customary manner, but this time the patrons fought back. They stood their ground and refused to cooperate, becoming increasingly rowdy and taunting the officers with openly affectionate behavior and a chorus line of drag queens. It wasn't long before a sympathetic crowd joined Stonewall patrons and, as the story goes, after one woman was dragged out in handcuffs and struck over the head with a billy club, the gathering erupted in anger. "The violence resumed each evening through Wednesday night, July 2, with taunts from young gays and chants by experienced activists stoking police violence through the labyrinthine streets of the West Village. Mortified that they had been disgraced by a bunch of 'queers,' the cops returned in force each night to try and recapture Christopher Street. They never did."[18]

The Stonewall riots have come to be understood as the high noon of the gay civil rights movement, not only in the United States, but also around the world. A year after the uprising, on June 28, 1970, participants marched in the first gay pride parade on a route that went from the Stonewall Inn to Central Park; they were joined by supporters marching in Chicago, San Francisco, and Los Angeles. Now every year, pride marches commemorating Stonewall are held in cities all over the world, in countries as unlikely as Uganda, Turkey, and Israel.

THE VIRTUOUS CIRCLE AND THE DECLINE OF HOMOPHOBIA

What's become known simply as Stonewall happened almost fifty years ago—so what progress has been made since then? First, the good news: in 1973, the American Psychiatric Association declassified homosexuality as a mental illness. Officially acknowledging that gay men and lesbians weren't actually diseased was a necessary first step in changing attitudes toward them, and attitudes most certainly have changed. In many parts of the world, homophobia is coming to be regarded as offensive as racism. The sociologist Mark McCormack notes,

The gay rights movement has been very successful, even just in terms of gay visibility. When people see famous gay people, people whom they like who turn out to be gay, that has a huge impact. People are bigoted about people they don't know; when you get to know gay people, the homophobia drops off.

Another key area of change is the Internet. The Internet has meant that closeted kids have the ability to make friends, to be more confident, to come out earlier. Social networking sites like Facebook ask you your sexual orientation, you click whether you're male or female, and then you click whether you're interested in men or women. When I was at school, that question wasn't even asked—you were straight, or if you weren't, you were pitied.

Part of the reason it's spiralled so quickly is that as homophobia decreased, boys could kind of hug each other a little bit or say to their best mate that they loved them, and then they could kind of cough and talk about girls. Then they realized that actually it wasn't disgusting or repulsive, and so that undid some of their homophobia a little bit more. It's a virtuous circle.[19]

The virtuous circle that has led to the swift decline of homophobia around the world was also chronicled by the sociologist Eric Anderson. When asked about his results, Anderson said, "Well, the findings are only surprising to those who are 25 or 30 years of age and older. They're not really surprising to 17-year-olds. That's not to say that this new attitude exists in all demographics in all spaces in all places. But it is to say that it's a growing emergence and [homophobia] is particularly unacceptable in white, urban, middle-class youth."[20]

WHO LEADS AND WHO RESISTS THE GAY RIGHTS REVOLUTION

Other arenas have also seen positive changes for LGBT citizens—including for personnel in the US military. Don't ask, don't tell (DADT) was the official policy of the US government from 1994 until 2010 that allowed closeted gay, lesbian, and bisexual personnel to serve, but only under the constant threat of immediate expulsion if they accidentally slipped up and revealed their true identities. Slipping up is just what Alexander Nicholson did; while serving in the army, he decided to write a letter to his former boyfriend, in Portuguese, so that no one would be able to understand the content. This was a mistake; the content was leaked, Nicholson panicked and spoke with his supervisor, hoping to quell the rumors circulating about his orientation, and in what Nicholson calls "a classic outcome," he was discharged. Nicholson says,

> Many gay, lesbian, and bisexual youth have gone into the military over the years with the same misunderstanding of the DADT policy as I had. It just sounded so reasonable and manageable, and for many it was. Hundreds of thousands of gay, lesbian, and bisexual servicemembers did navigate DADT over the course of their careers, but at a tremendous personal cost. But tens of thousands more could not. Some were abruptly fired after their secret was discovered and the information spread out to their peers and then up through the chain of command, as in my case, while others were maliciously outed by jilted lovers, jealous fellow servicemembers, or bigoted acquaintances who discovered their secret.[21]

After getting unceremoniously kicked out, Nicholson founded Servicemembers United, an LGBT interest group dedicated to repealing DADT, a crusade that took years of effort and eventually triumphed when President Obama signed legislation that repealed the policy on December 22, 2010.

Professional sports is another arena that remains largely closed and closeted, and one wonders where all the openly gay players are in American football, in European soccer, among NASCAR drivers, professional cyclists, and so on. Even here, though, there is hope of a new day dawning for gay athletes. One study of Association Football (soccer) uncovered the following:

> The overall findings are that, contrary to assumptions of homophobia, there is evidence of rapidly decreasing homophobia within the culture of football fandom. The results advance inclusive masculinity theory with 93 per cent of fans of all ages stating that there is no place for homophobia within football. Fans blame agents and clubs for the lack of openness and challenge football's governing organizations to oppose the culture of secrecy surrounding gay players and to provide a more inclusive environment to support players who want to come out.[22]

As I was completing this book in early 2014, the Sochi Olympics included a number of openly gay athletes as well as an American delegation of LGBT athletes led by Billie Jean King, in open defiance of Russia's antiquated and homophobic laws.[23] The National Football League (NFL) drafted its first officially gay player from the college ranks—Michael Sam—who was treated in the media as a hero for coming out on the eve of the draft. An ESPN.com survey in response to the event found that 86 percent of NFL players said that a teammate's sexual orientation does not matter to them.[24] In January 2014, the former German international footballer Thomas Hitzlsperger revealed that he is gay "to move the discussion on homosexuality in professional sport forward." Soccer fans, he said, "are very complex—you have all ages from all walks of life in [the] stadium. That's why you can't rule any reaction out. But I think for the vast majority it probably wouldn't be a problem."[25]

Of course, one of the biggest signs of progress is that finally, in at least a handful of countries, gay, lesbian, and bisexual people are allowed to marry, form families, and have children. The change from, say, the sixties is remarkable: note how Richard Enman, president of the Mattachine Society of Florida, laughed off the question in a 1966 interview when asked, "What type of laws are you after?"

> Well, let me say, first of all, what type of laws we are *not* after, because there has been much to-do that the Society was in favor of the legalization of marriage between homosexuals, and the adoption of children, and such as that, and that is not at all factual at all. Homosexuals do not want that, you might find some fringe character someplace who says that that's what he wants.[26]

Imagine same-sex marriage and adoption, for *homosexuals*, this evidently gay man seems to be saying. Ridiculous! But it's not ridiculous anymore, at least not in Uruguay, Denmark, South Africa, Canada, New Zealand, and the sixteen countries in total that have legalized same-sex marriage; nor is it ridiculous anymore in California, Connecticut, Minnesota, New York, Washington, or any one of the more than thirty states and the District of Columbia where gay, lesbian, and bisexual citizens have finally won the right to marry. And to those who argue that the purpose of marriage is procreation and thus gays should be excluded, upon New Mexico becoming the seventeenth state in the union to legalize gay marriage, the author of the New Mexico opinion, Justice Edward L. Chavez, wrote: "Procreation has never been a condition of marriage under New Mexico law, as evidenced by the fact that the aged, the infertile and those who choose not to have children are not precluded from marrying."[27] It's just a matter of time before all fifty states come around. And in 2014 Germany's highest court strengthened the rights of same-sex couples to adopt.[28]

The following figures track this moral progress and show who's leading the rights revolution in this area. *Figure 7-1* graphs progress in attitudes toward homosexuality and gay marriage since the 1970s, showing increasing numbers of more tolerant responses on surveys asking people about the morality and legality of homosexuality and gay marriage. *Figure 7-2* shows that, for the first time in history, more people support rather than oppose gay marriage. *Figures 7-3* and *7-4* reveal who is leading the moral revolution for gay rights and marriage and who still oppose it. The effect can be seen cross-generationally, most noticeably in the difference between millennials, who are, unsurprisingly, largely in support of it, and baby boomers (those born between 1946 and 1964) and the silent generation (those born before 1946), who are largely against it. According to a March 2013 survey by the Public Religion Research Institute, half of Christians under age thirty-five support same-sex marriage compared to just 15 percent of those over age sixty-five. These data show that over time the moral values of Christians are shifting toward more tolerance and acceptance of others as the silent and baby boomer generations are slowly but inexorably displaced by millennials, who are riding the crest of the moral wave while older people are pulled slowly along in the flotsam and jetsam of the trough below.

Although many religious people have supported these changes, and worked to make gay marriage acceptable in their churches, religious-based opposition to homosexuality continues. A case in point: in late 2013, India's Supreme Court overturned a lower court ruling made in 2009 that declared same-sex relations between consenting adults no longer a crime under Indian law (which it had become under British colonial rule). The new ruling, as described in Section 377 of the Indian Penal Code, prohibits sexual activity "against the order of nature with any man, woman or animal." (The penalties for breaking the law include fines and up to ten years imprisonment.) Predictably, the movement to recriminalize homosexuality was led by Hindu, Muslim, and Christian groups. Mujtaba Farooq, president of the Welfare Party of India—a Delhi-based Muslim political group—proclaimed, "Homosexuality's not natural, disturbs the continuity of life, leaves the future uncertain. This is the unacceptable influence of the West." Kamal Farooqui, a representative of the All India Muslim Personal Law Board, one of the groups that petitioned the Supreme Court to

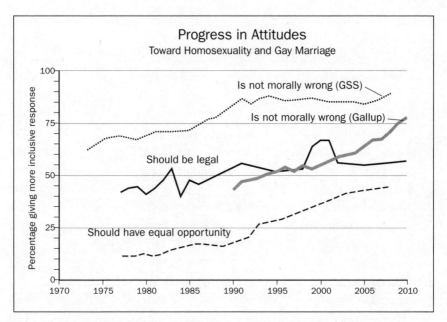

Figure 7-1. Progress in Attitudes Toward Homosexuality and Gay Marriage

Percentage giving more inclusive response on surveys asking about the morality and legality of homosexuality and gay marriage.[29]

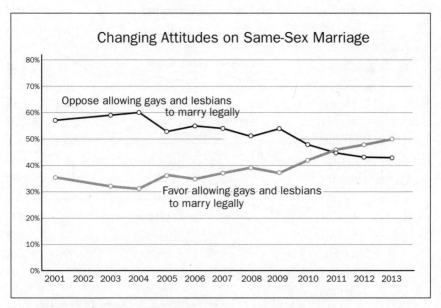

Figure 7-2. More Americans Support Gay Marriage Than Oppose It

For the first time in history more people support gay marriage than oppose it. From the June 2013 Pew Research Center Forum on Religion and Public Life.[30]

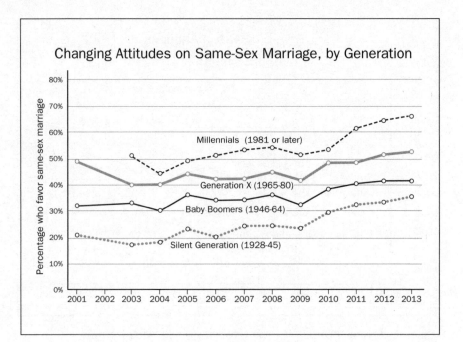

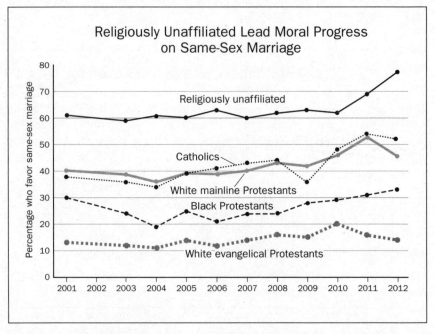

Figures 7-3 and 7-4. Who Is Leading the Rights Revolution for Same-Sex Marriage

The percentage of those who favor same-sex marriage is highest among the youth (millennials) and religiously unaffiliated and lowest among older Americans and white evangelical Protestants. From the June 2013 Pew Research Center Forum on Religion and Public Life.[31]

overturn the 2009 decision, explained their reasoning: "If homosexuality is to be made legal, then let everyone be a homosexual. Homosexuality is against nature's law of reproduction. The world will end in 100 years without reproduction."[32] Of all the countries on earth that should be the least worried about their population declining, India is second only to China.

Such moralizing was also on display in 2014 when the Ugandan Parliament passed an anti-homosexuality bill that makes it a crime to be gay, with a prison sentence of fourteen years for the first "offense" and life for subsequent offenses. "The bill aims at strengthening the nation's capacity to deal with emerging internal and external threats to the traditional heterosexual family," the Parliamentary committee said, failing to mention how, exactly, two consenting adults of the same sex are a threat to two consenting adults of the opposite sex.[33]

In Russia, the Kremlin has orchestrated a hate campaign against LGBT people, including a ban on "gay propaganda," a law that prohibits same-sex couples from adopting children,[34] and a bill in the works that may remove children from same-sex parents,[35] which may well result in an LGBT refugee crisis as gay and lesbian Russians flee with their children and seek asylum elsewhere.[36] Russia appears to be transforming itself into the traditional values capital of the world, and Vladimir Putin himself has "made a particularly thinly-veiled attack on the West's more liberal attitudes toward gay rights, saying Russia would defend against 'genderless and fruitless so-called tolerance.'"[37] Unfortunately, it appears that 88 percent of Russians support the "gay propaganda" ban; 35 percent think being gay is a disease, and 43 percent think it's a disgusting habit. The head of the Russian Orthodox Church, Patriarch Kirill I, said (presumably in all seriousness) that same-sex marriage is "a very dangerous sign of the apocalypse."[38] Don't say you haven't been warned.

Sometimes it's difficult to tell whether people are serious when they make preposterous claims, or are just riling up their fellow believers. For example, a British politician noted, "Since the passage of the Marriage [same-sex couples] Act, the nation has been beset by serious storms and floods."[39] Blaming God's anger at gay people for bad weather is nothing new. It's an American tradition: after every tornado or hurricane, at least one religious wingnut points the finger at same-sex coupling as the reason for God's liquid wrath. Apparently hurricanes are God's way of teaching humanity, once again, that straight, married sex is the only kind of which the Almighty approves. It's an indirect method of instruction, to be sure, and it's understandable that people are slow to see the connection between same-sex marriage and hurricanes. Nevertheless there could be some benefits to gay weather. As one satirical news article pointed out, if gay sex causes large-scale moisture events, married gay people should be encouraged to spend time in deserts. "We're honeymooning in sub-Saharan Africa so that we can bring the waters of gayness to its parched landscape," says one gay couple. "If one small kiss summons a full-on pea souper, imagine what fully penetrative gay intercourse could do."[40]

Riffing off the ramblings of religious extremists provides a moment of levity—and it's a sign that a revolution is reaching its apotheosis when comedians and satirists make light of the opposition—but there is nothing funny about the statistics telling us that, in the United States, LGBT youth attempt suicide "up to five times more than their heterosexual peers."[41] There are undoubtedly many reasons why this would be so, but as Jorge Valencia of the Trevor Project Suicide Hotline for LGBT youth says, "One of the top 5 reasons why teenagers call us is for religious reasons. They're feeling there isn't a place for them and God."[42] This could be because, forty years after it was determined that homosexuality is not a mental illness, many Christian preachers, writers, and theologians still think nothing of tormenting the LGBT community by telling them that their desire to love another person of the same sex is an abomination and a disease that can be "cured" through "treatment" known as conversion therapy. "By instilling fear and shame in a person you can get them to suppress any feelings that they have," says Dr. Paula J. Caplan. Sadly for gay conversionists, "That's worlds away from changing somebody's sexual orientation."[43]

Nonetheless, conversion therapy got a boost in 2003 when the eminent psychiatrist Dr. Robert L. Spitzer published a paper titled "Can Some Gay Men and Lesbians Change Their Sexual Orientation? 200 Participants Reporting a Change from Homosexual to Heterosexual Orientation."[44] Conversion therapy organizations such as Exodus International and Love Won Out (sponsored by the evangelical organization Focus on the Family) rejoiced. But not for long: A decade later, Dr. Spitzer apologized profusely for his impossibly flawed study,[45] and Allan Chambers of Exodus International likewise apologized for the havoc he'd wrought, "likening his behavior as president of the organization to a 'four-car pileup'" he once caused.[46] Chambers announced that Exodus International would disband. The National Association for Research & Therapy of Homosexuality (NARTH)[47] continues to promote conversion therapy, even in the face of government bills banning the practice state by state,[48] and even though professional psychological associations have denounced all therapies designed to change sexual orientation as being "unjustified, unethical, and harmful."[49]

The religious extremists who continue to press for conversion therapy fail to understand that being gay is like being left-handed—it's not something that requires an intervention. As the columnist Dan Savage says, instead they point "to the suicide rate among gay teenagers—which the religious right works so hard to drive up . . . as evidence that the gay lifestyle is destructive. It's like intentionally running someone down with your car and then claiming that it isn't safe to walk the streets."[50] Many Christians actually believe they are being charitable by proclaiming that they "hate the sin, not the sinner," which is not dissimilar to what Christians declared just before torching women as witches to save their souls, or when Christians called for pogroms against Jews for being Christ killers.

THE REVOLUTION COMES OF AGE

Mark my words: I predict that within a few years, maybe a decade, white Christians will come around to treating gay men and lesbians no differently from how they now treat other groups whom they previously persecuted—women, Jews, blacks. These changes will not occur because of some new interpretation of a biblical passage or because of a new revelation from God. These changes will come about the same way they always do: by the oppressed minority fighting for the right to be treated equally, and by enlightened members of the oppressing majority supporting their cause. Then Christian churches will take credit for the civil liberation of the gay community, rummage through the historical record and find those preachers who had the courage and the character to stand up for gay rights when their fellow Christians would not, and then cite those as evidence that, were it not for Christianity, gay people would still be in the closet.

Whoever takes credit, however, the gay rights revolution is nearing completion. On September 20, 2013, Pope Francis issued a warning to his more than one billion members that the Catholic Church was too focused on personal moral issues such as gay marriage, abortion, and contraception— at the expense of its pastoral mission of helping the poor and needy and homeless around the world—noting that this misplaced morality threatened to bring down the church. As he told the Italian Jesuit journal *Civiltá*

Cattolica: "We cannot insist only on issues related to abortion, gay marriage and the use of contraceptive methods." Breaking with his two immediate predecessors, Pope Benedict XVI and Pope John Paul II, along with the other 263 papal forerunners dating back two millennia, Pope Francis added, "The teaching of the church, for that matter, is clear and I am a son of the church, but it is not necessary to talk about these issues all the time. We have to find a new balance; otherwise even the moral edifice of the church is likely to fall like a house of cards."[51]

But it was the pope's comment a few months earlier, on July 29, that really pulled him into the biggest civil rights issue of the day. The Argentine-born pope (as Jorge Mario Bergoglio) opined from his experiences in his home country on the issue of homosexuality: "Who am I to judge a gay person of goodwill who seeks the Lord? You can't marginalize these people."[52] Francis then added this observation, which was more prescriptive than descriptive: "In Buenos Aires I used to receive letters from homosexual persons who are 'socially wounded' because they tell me that they feel like the church has always condemned them. But the church does not want to do this."[53]

When the head of the world's largest religion entreats his followers to be more concerned with the survival and flourishing of the poor and the homeless, and to be more accepting of gays and lesbians, this is an indicator of moral progress as I have defined it. Even evangelicals are changing. Just a month after Pope Francis rocked the world in his historic call for tolerance, Russell Moore, head of the Southern Baptist Convention, the largest evangelical cohort in America, sent a message to the leaders of his forty-five thousand church franchises and their sixteen million members, instructing them to "Love your gay and lesbian neighbors. They aren't part of an evil conspiracy." This observation appeared in a story on the front page of the first section of the October 22, 2013, edition of the *Wall Street Journal*, just below the fold and such lead stories as Syria's civil war and problems with the website for the Affordable Health Care program. The placement makes it news, but what makes it newsworthy is that Moore's more tolerant moral position was in defiance of a warning of his predecessor Richard Land of a "radical homosexual agenda." The article highlighted the church's pullback from politics and the culture wars that had driven so many faithful followers to condemn homosexuality and gay marriage previously. Moore's message was, in part, a response to the 2013 decision by the US Supreme Court to strike down the federal Defense of Marriage Act that had defined marriage as strictly between a man and a woman—*Adam and Eve, not Adam and Steve,* as a protest sign pro-claimed.[54] Moore was also responding to what he described as "a visceral recoil" among younger evangelicals to the church's involvement in such political culture wars as same-sex marriage, and to keep millennials (those born after 1979) involved in the church, he added, pastors and preachers need to be "winsome, kind and empathetic."[55]

That sounds good, and we should be grateful for any moral progress whatever its source, but one might think that pastors and preachers—not to mention Christians in general—would not need to be told to be win-some, kind, and empathetic. The story, in fact, makes my point about reli-gion's lag behind the cultural curve toward justice, made even more affecting by Moore's reminder that the Southern Baptist Church split from a larger denomination to defend slavery before the Civil War: "The fact that we were founded, at least partly, to justify man-stealing, kidnap-ping and lynching—we stand here only by God's grace and mercy." Only after a war that cost the lives of more than 650,000 Americans and an amendment to the US Constitution prohibiting slavery did this denomi-nation get with the moral program.

SCIENCE, REASON, AND GAY RIGHTS

These political and cultural factors at work in helping expand the moral sphere for the LGBT community are only part of the story, argues Jona-than Rauch, a longtime advocate for free speech and civil liberties in gen-eral, and of gay rights in particular. "Generational replacement doesn't explain why people in all age groups, even the elderly, have grown more gay-friendly. Gay people have been coming out for years, but that has been a gradual process, while recent changes in public attitude have been dizzyingly fast. Something else, I believe, was decisive: we won in the realm of ideas." Ultimately, says Rauch, good ideas outcompete bad ideas in the marketplace of free exchange. Rauch recalls the comment of a caller to a radio talk show he was on during a book tour for his 2004 book *Gay Marriage: Why It Is Good for Gays, Good for Straights, and Good for America*:

> "Your guest," he said, meaning me, "is the most dangerous man in Amer-ica." Why? "Because," said the caller, "he sounds so *reasonable.*" In hind-sight, this may be the greatest compliment I have ever been paid. It is certainly among the most sincere. Despite the caller's best efforts to shut

out what I was saying, the debate he was hearing—and the contrast between me and my adversary—was working on him. I doubt he changed his mind that day, but I could tell he was *thinking*, almost against his will. Hannah Arendt once wrote, "Truth carries within itself an element of coercion." The caller felt that he was in some sense being forced to see merit in what I was saying.[56]

Once again we see here a powerful restatement of the *principle of interchangeable perspectives*, in which the use of *reason* in an open dialogue forces us to consider the merit of what the other person is saying, and if the other person makes sense, their superior ideas gradually chip away at our prejudices. Coupled with the overwhelming scientific evidence that homosexuality is not a choice but part of human nature, we see in this rights revolution another example of how science and reason lead humanity toward truth, justice, and freedom. And the change in attitudes supports my contention that many moral issues are, in part, debates over facts, and that many immoral beliefs are just factual errors. While acknowledging that homophobic emotions have fueled attitudes toward gays over the past century, Jonathan Rauch notes that the deeper problem was that people held incorrect beliefs about homosexuals: "factual misapprehensions and moral misjudgments born of ignorance, superstition, taboo, disgust. If people think you are a threat to their children or their family, they are going to fear and hate you. Gays' most urgent need was epistemological, not political. We had to replace bad ideas with good ones." That replacement can only happen in a free society in which open debate is allowed and ideas must compete for cognitive space in our brains. As it is now unfolding for gay rights, so it has for all human rights, as Rauch explains:

> History shows that the more open the intellectual environment, the better minorities will do. We learn empirically that women are as intelligent and capable as men; this knowledge strengthens the moral claims of gender equality. We learn from social experience that laws permitting religious pluralism make societies more governable; this knowledge strengthens the moral claims of religious liberty. We learn from critical argument that the notion that some races are fit to be enslaved by others is impossible to defend without recourse to hypocrisy and mendacity; this knowledge strengthens the moral claims of inherent human dignity.[57]

Again, I am not arguing that reason alone will get us there; we need legislation and laws to enforce civil rights, and a strong police and military to back up the state's claim to hold a monopoly on the legitimate use of force to back up those laws. But these forces are themselves premised on law being grounded in reason, and the legislation being backed by rational arguments. Without that, there is no long-term sustainability to moral progress, as it is just a matter of might makes right. If your moral campaign depends exclusively on the power of the state, then when the powers of the state change hands, those in power can just as easily change the law, which is what happened in India at the end of 2013. To make morals stick you have to change people's thinking—from constantly thought of, to barely thinkable, to unthinkable, to unthought of.

This is what happened with the abolition of slavery, with the expansion of women's rights, and what is destined to happen, gradually but inevitably, with the recognition of full rights for the LGBT citizens of the world, especially as they come out more visibly. Coming out allows the rest of society to see that gay people are just people; it normalizes what is natural and good. One coming out leads to another, which, besides providing greater political leverage, also provides support and comfort for others who feel marginalized and alone.

The British actor, writer, and icon Stephen Fry relates his own coming out and the terrible dread he felt at the prospect of someone discovering that he was gay—which, at the time, was like admitting to both mental illness and criminal depravity in a single go. It was therefore an enormous relief for him and thousands of other young people when, back in 1976, he picked up a copy of *Rolling Stone* magazine and read the cover story about one of the greatest pop stars of the century, who'd come out as bisexual and who had the guts to speak the truth. "It was a game-changing moment for me," says Fry, "and countless other gay teens who had locked ourselves away in the closet."[58] What did this pop star say? "There's nothing wrong with going to bed with somebody of your own sex. . . . It's going to be terrible with my football club. It's so hetero, it's *unbelievable*. But I mean, who cares?! I just think people should be very free with sex."[59]

That pop superstar was Elton John, a man who has since gone on to marry his long-term partner David Furnish, and to have two sons together with a surrogate mother. This is proof positive that It Gets Better—the name of the campaign begun by Dan Savage and his husband, Terry Miller, in response to the alarming suicide rate among LGBT youth. They made a video with that simple, vital message, hoping that maybe a hundred other adult gay, lesbian, and bisexual people might do the same. The response, however, was overwhelming. The videos poured in and to date there are now over 150,000 posted entries from people of all orientations.[60] Only a month after Savage and Miller made their video, Barack Obama contributed his own, and soon after, contributions from people and groups poured in from Stephen Colbert, Google, Lady Gaga, General Motors, Brigham Young University, the Most Holy Redeemer Church of San Francisco, Apple, Hillary Clinton, twenty Royal Canadian Mounted Police officers, Kermit the Frog—and me.[61]

TO SIN BY SILENCE

It has been a long, painfully slow effort to expand the moral sphere to include all humans inside it, and we're not there yet. We would do well to recall these words attributed to Edmund Burke: "All that is necessary for the triumph of evil is that good men do nothing." What should we do? Above all, we must speak out against wrongs wherever we find them, as the American poet and author Ella Wheeler Wilcox (she of the enduring line "Laugh, and the world laughs with you; Weep, and you weep alone") so forcefully expressed in her 1914 poem "Protest":

> To sin by silence, when we should protest,
> Makes cowards out of men. The human race
> Has climbed on protest. Had no voice been raised
> Against injustice, ignorance, and lust,
> The inquisition yet would serve the law,
> And guillotines decide our least disputes.
> The few who dare, must speak and speak again
> To right the wrongs of many. . . .[62]

8

A Moral Science of Animal Rights

> When I view all beings not as special creations, but as the lineal descendants of some few beings which lived long before the first bed of the Cambrian system was deposited, they seem to me to become ennobled.
>
> —Charles Darwin, *On the Origin of Species*, 1859[1]

Six hundred miles off the coast of Ecuador lies an archipelago called the Galápagos, famous for its connection to Charles Darwin and his theory of evolution by means of natural selection. Darwin spent five weeks there in the fall of 1835, and in 2004 I accompanied my friend and colleague Frank J. Sulloway on his expedition to retrace Darwin's footsteps.[2] The Ecuadorian government has ownership and jurisdiction over the archipelago and it has gone to great lengths to keep the islands as pristine and natural as possible. For example, before we were allowed to hike off trail into the interior of the islands we had to go through a careful quarantine process to ensure that no foreign invaders were hiding in our backpacks or clothes. Nevertheless, invasive species are an ongoing problem for the native populations, especially the famous Galápagos tortoises, whose diet depends heavily on the local vegetation, which has been systematically eroded by goats, introduced almost a century ago and now threatening the tortoises and other species with extinction.

In response, the National Park Service of Ecuador undertook a massive goat eradication project on the most affected island—the massive 58,465-hectare Santiago Island—which resulted in the killing of more than seventy-nine thousand goats in a span of four and a half years in the mid-2000s. The initial culling of the goats was done by riders on horseback, who corralled them into pens with air horns and rifle shots and then killed them. But this method fell far short of the goal because of the brutally harsh terrain of these volcanic islands. It is dry and hot, and the razor-sharp a'a lava surface slices up your hiking boots. Bushwhacking through thorn-festered scrub brush slashes up your arms and legs. Water is hard to come by, so you have to carry your supply on your back. The landscape is undulating and jagged, and there are numerous lava-formed caves, nooks, and crannies in which goats can hide from human hunters. Even though I am in good physical condition from a lifetime of competitive cycling, I found this trek with Frank to be a grueling slog, among the most arduous things I have ever undertaken. Even the native Ecuadorians more acclimated to the equatorial environment had to turn to helicopters to find the remaining goats, shooting them with rifles from the air. And still the goats persisted. To finish the job the National Park Service introduced "Judas goats" and "Mata Hari goats" to find the remaining feral goats and kill them. Judas goats are goats captured from nearby islands and equipped with radio collars and then released on Santiago to lead hunters to their remaining kind in hiding. Mata Hari goats are sterilized female Judas goats chemically induced into long-term estrus so that when released they would tempt male goats who were hunter-shy but female-sure. At a cost of $6.1 million, it was the largest eradication of a mammalian species from an island in history.

Was this a moral act? Who should live and who should die—the native species that evolved in the Galápagos islands over the course of millions of years, or the goats that were introduced there a mere century ago? At first blush that sounds like a moral no-brainer. The native species have the moral imperative by dint of time. Then again, invasive species have always been a problem for native inhabitants over the course of a billion years of evolution—it is one of the primary causes of extinction in nature—so what we're talking about here is a difference in time scale (long-term versus short-term) and the source of the invasion (natural versus human). One species nudging out another species by the natural process of migration and competition is morally different from humans introducing a species to, say, clear unwanted (by people) ground cover. So my moral sympathies were with the conservationist eradicators, until I came across a baby goat and held him in my arms (see *figure 8-1*). These are surely sentient beings, and as mammals they are perhaps even more sentient—more emotional, responsive, perceptive, and with an increased ability to sense and feel— than the ancient reptilian tortoises whose turf they moved in on. But after

Figure 8-1. A Moral Dilemma in the Galápagos Islands

The author with a baby goat in the Galápagos Islands before she and all other goats were eradicated to save the native Galápagos tortoises (also pictured). Is that moral? Where is the moral bright line? Credit: Author's collection.

looking at the cute baby goat, gaze into the eyes of one of these majestic tortoises (also pictured) as they lumber across the landscape in search of vegetation—which they have been doing in this island archipelago for millions of years untouched by civilization—and imagine them extinct because of a bunch of feral goats, and then try to defend the goat's moral position. Then again, the goats didn't choose to come here, nor did they arrive by an accident of nature, such as a chance flotilla of debris from a storm that probably brought the tortoises here millions of years ago.

This is a morally dumbfounding problem, and one that well represents others related to animal rights. It's a moral triage of who lives and who dies. "In the long run, eradication is going to be cheaper," explained Josh Donlan of Cornell University, who published a paper on the goat-eradication project. "It also makes sense from an ethical perspective, because in the end you are actually killing fewer animals," referring to all the native individuals who would have died from lack of food and that ultimately would have led to their extinction.[3] I concur, but not without some regret for our complicity in the problem in the first place, and acknowledgment that this "return to nature" can only go so far, given the impossibility of implementing such programs in places where civilization has encroached so deeply into nature that there is little natural ecology left. Short of a "world without us" scenario in which every human on Earth suddenly disappears and nature comes roaring back with a vengeance over all of our man-made structures,[4] there is no clear reconciliation between animals and civilization, no obvious moral bright line.

Nevertheless, I shall argue that the arc of the moral universe of animals also has bent toward justice, as witnessed by this very act of animal genocide in the name of saving life and preserving nature. Less than a century ago people thought nothing of introducing members of a non-native species into the Galápagos Islands, oblivious to what they might do to the ecosystem. In Darwin's time it was common for sailing ships to visit the Galápagos to collect tortoises and store them alive in the bowels of the ships and eat them while crossing the Pacific Ocean. Even Darwin—who was ahead of his time on so many social issues (including the abolition of slavery)—ate his tortoise data during HMS *Beagle*'s voyage home across the Pacific.[5]

CONTINUOUS THINKING ABOUT ANIMALS

A moral system based on continuous rather than categorical thinking gives us a biological and evolutionary foundation for the expansion of the moral sphere to include nonhuman animals, based on objective criteria of genetic relatedness, cognitive abilities, emotional capacities, moral development, and especially the capacity to feel pain and suffer. This is, in fact, what it means to be a sentient being, and for this reason I worded the first principle of this science-based moral system as *the survival and flourishing of sentient beings*. But which sentient beings, and which rights?

Instead of thinking of animals in categorical terms of "us" and "them," we can think in continuous terms from simple to complex, from less to more intelligent, from less to more aware and self-aware, and especially from less to more sentient. So, for example, if we place humans at 1.0 as a full-rights-bearing sentient species, we could classify gorillas and chimps at 0.9; whales, dolphins, and porpoises at 0.85; monkeys and marine mammals at 0.8; elephants, dogs, and pigs at 0.75; and so forth down the phylogenetic scale. Continuity, not categories.[6] Take brains, for instance. Scaling average brain size across species we can compare the brains of gorillas (500 cubic centimeters, or cc), chimpanzees (400cc), bonobos (340cc), and orangutans (335cc) with those of humans at an average of 1,200–1,400cc. Dolphin brains are especially noteworthy, coming in at an average of 1,500–1,700cc, and the surface area of a dolphin's cortex—where the higher centers of learning, memory, and cognition are located—is an impressive $3,700cc^2$ compared to our $2,300cc.^2$ And although the thickness of the dolphin's cortex is roughly half that of humans, when absolute cortical material is compared, dolphins still average an impressive 560cc compared to that of humans at 660cc.[7]

These notable facts about dolphins led a scientist named John C. Lilly to found the semi-secret "Order of the Dolphin" in 1961, a veritable who's who of scientists that included the astronomer Carl Sagan and the evolutionary biologist J. B. S. Haldane, both of whom were interested in communicating with extraterrestrial intelligences. Since ETs were nowhere to be found, dolphins would serve as a terrestrial stand-in on how to communicate with a species radically different from us. The research project didn't pan out, and Lilly's lack of rigor in his experiments discouraged Sagan and the others from drawing definitive conclusions about dolphin language and intelligence (among other things, Lilly gave his aquatic charges LSD to see if that might open up the doors of perception for them). However, half a century of more scientific research on dolphins since then is revealing. In her book *The Dolphin in the Mirror*, the psychologist Diana Reiss shows how dolphins pass a modified mirror test of self-awareness.[8] For example, a YouTube video of dolphins preening in front of a giant mirror placed in their tank is not just amusing to watch; they clearly know it is themselves they are seeing in the mirror, and they look more than a little pleased with what they see, staring inside their mouths, flipping upside down, blowing air bubbles, etc. This particular video also shows both dolphins and elephants passing the equivalent of the red-dot self-recognition mirror test—the dolphin with an ink mark on his side that he stares at in a mirror (compared to controls who received no mark and did not mirror-stare), and an elephant with an "X" marked on her temple, which, in viewing in a mirror, she touches repeatedly with the tip of her trunk, obviously curious (and perhaps irritated) at its presence.[9] As for dolphin language, however, the psychologist Justin Gregg is less enthusiastic than Lilly was in his assessment of the literature, in his book *Are Dolphins Really Smart?*:

> Evidence for a dolphin language—dolphinese—is all but nonexistent. Dolphins do possess signature whistles, which function a bit like names. They likely use them to label themselves and might even call one another's name on occasion. This is both unique and impressive, but it is the only label-like aspect of dolphin communication that we've found. All the other clicks and whistles that dolphins produce are probably used to convey messages about their emotional states or intentions—not the type of complex or semantically rich information found in human language.[10]

Gregg points out the obvious contradiction in the argument that big brains equal big intelligence: "If big brains are the key to intelligence, why do crows and ravens, which are (literally) bird-brained, display forms of cognition that rival those of their big-brained dolphin and primate cousins. The animal kingdom is full of small-brained species that display astonishingly complex and intelligent behavior."[11]

As for cognitive continuity, psychologists who tested a gorilla named Koko concluded that she passes the mirror recognition test for self-awareness. (For comparison, more than half of human infants passed the mirror self-recognition test by age eighteen months and 65 percent by age two.[12]) Koko also passed an "object permanence" test in which she could memorize the location of moved objects. She was also able to figure out that the quantity of a liquid does not change when it is poured into a different-size container, which is considered another cognitive hurdle in the "conservation of liquid."[13] Elephants have also been observed mourning the loss of

a fellow family or tribe member. A 2014 study of twenty-six Asian elephants in a Thailand sanctuary found that when individuals feel stressed
by snakes, barking dogs, hovering helicopters, or the presence of other hostile elephants (their ears go out, their tail stands erect, and they emit a
low-frequency rumble), their mates comforted them by making seemingly sympathetic noises and by touching them with their trunks on
their shoulders, in their mouths, and on their genitals (not recom-
mended in human populations).[14]

The University of St. Andrews cognitive neuroscientists Anna Smet
and Richard Byrne, in a series of clever experiments involving hungry
elephants, hid food under an opaque container near an empty container
and then pointed to the one where the morsels were located. To their
astonishment they discovered that African elephants are the first nondo-
mesticated species who exhibit such advanced social cognition as the ability to read human nonverbal communication. As they noted in their 2013
paper: "Elephants successfully interpreted pointing when the experimenter's proximity to the hiding place was varied and when the ostensive
pointing gesture was visually subtle, suggesting that they understood the
experimenter's communicative intent."[15]

This is surprising because previous research found that domesticated
animals such as dogs are better at reading nonverbal cues from humans
about hidden food (the experimenter simply points to where the food is)
than free-living (wild) animals such as chimpanzees, even though the latter are more closely related to us. The prevailing assumption has been that
the ability to read human cues evolved in species under domestication as
an adaptive strategy for survival. As Smet and Byrne noted, "Most other
animals do not point, nor do they understand pointing when others do it.
Even our closest relatives, the great apes, typically fail to understand
pointing when it's done for them by human carers; in contrast, the domestic dog, adapted to working with humans over many thousands of years
and sometimes selectively bred to follow pointing, is able to follow human
pointing—a skill the dogs probably learn from repeated, one-to-one interactions with their owners."[16]

But elephants have never been completely domesticated despite repeated
attempts over the millennia (dating back at least four thousand to eight
thousand years), and regardless of the considerable time spent in captivity
in zoos and circuses. The explanation must come from elsewhere. "The
African elephant's complex society makes it a good candidate for using
others' knowledge: its elaborate fission-fusion society is one of the most

extensive of any mammal, and cognitive sophistication is known to correlate with the complexity of a species' social group," the authors explain. "We suggest that the most plausible account of our elephants' ability to interpret even subtle human pointing gestures as communicative is that human pointing, as we presented it, taps into elephants' natural communication system. If so, then interpreting movements of other elephants as deictic [context-dependent] communication must be a natural part of social interaction in wild herds; specifically, we suggest that the functional equivalent of pointing might take the form of referential indication with the trunk."[17]

Dogs, of course, are hypersensitive to human cues, and for good reason, since we now know that all modern dogs evolved out of a population of wolves some 18,800 to 32,100 years ago, surviving on the margins of hunter-gatherer bands and coevolving with their human counterparts by learning to read one another's cues, both verbal and nonverbal.[18] The UCLA evolutionary biologist Robert Wayne, lead author of this study, speculated, "Their initial interactions were probably at arm's length, as these were large, aggressive carnivores. Eventually, though, wolves entered the human niche. Maybe they even assisted humans in locating prey, or deterred other carnivores from interfering with the hunting activities of humans."[19] In addition to dogs' skulls, jaws, and teeth becoming smaller as they became domesticated, they evolved a set of social-cognitive tools that enabled them to read human communicative signals indicating the location of hidden food. In experimental conditions, domesticated dogs are capable of selecting the right container of concealed food when the experimenter looked at, tapped, or pointed to it, whereas wolves, chimpanzees, and other primates were unable to do so.[20]

What are dogs thinking when they are interacting with people? Dog owners everywhere have been wondering about this for ages—I've had dogs my entire life and can attest that you can't help but speculate on what is going on behind those curious eyes. To answer this question scientifically the cognitive psychologist Gregory Berns and his colleagues Andrew Brooks and Mark Spivak trained dogs to lie perfectly still inside an MRI brain scanning machine while they presented their canine subjects with human hand signals that indicated the presence or absence of a food reward.[21] What Berns and his colleagues saw was the dogs' caudate nucleus light up in anticipation of a food reward, which is significant because the caudate is rich in dopaminergic neurons—or nerve cells that produce dopamine, which is associated with learning, reinforcement, and even pleasure.

When an animal (including a human animal) is reinforced for doing something (a rat pressing a bar or a human pulling a slot machine lever), these neurons release dopamine, which produces a sensation of pleasure, which is a signal to tell the animal to "do that again" (and thus the astonishing success of Las Vegas casinos).

In Berns's experiments, not only did the scientists see an increase in canine caudate activity in response to hand signals indicating food, "the caudate also activated to the smells of familiar humans. And in preliminary tests, it activated to the return of an owner who had momentarily stepped out of view. Do these findings prove that dogs love us? Not quite. But many of the same things that activate the human caudate, which are associated with positive emotions, also activate the dog caudate. Neuroscientists call this a functional homology, and it may be an indication of canine emotions."[22]

In his book-length treatment of his research, *How Dogs Love Us*, Berns asks "what are dogs thinking?" and answers it this way: "they're thinking about what we're thinking." This is called "mind reading," or "theory of mind," and it is usually attributed only to humans and a few primates, but as Berns demonstrates, "Dogs are much better than apes at interspecies social cognition. Dogs easily bond with humans, cats, livestock, and pretty much any animal. Monkeys, chimpanzees, and apes will not do this without a lot of training from a young age. And even then, I would never trust an ape."[23] Berns spells out what this type of sentience means for the ethical treatment of animals:

> The ability to experience positive emotions, like love and attachment, would mean that dogs have a level of sentience comparable to that of a human child. And this ability suggests a rethinking of how we treat dogs. Dogs have long been considered property. Though the Animal Welfare Act of 1966 and state laws raised the bar for the treatment of animals, they solidified the view that animals are things—objects that can be disposed of as long as reasonable care is taken to minimize their suffering. But now, by using the M.R.I. to push away the limitations of behaviorism, we can no longer hide from the evidence. Dogs, and probably many other animals (especially our closest primate relatives), seem to have emotions just like us. And this means we must reconsider their treatment as property.[24]

Instead of property, Berns suggests we should consider dogs to be "persons" in a narrow legal definition. "If we went a step further and granted dogs rights of personhood, they would be afforded additional protection against exploitation. Puppy mills, laboratory dogs and dog racing would be banned for violating the basic right of self-determination of a person."[25]

Self-determination and personhood are two ethical criteria developed over the centuries in other rights revolutions to break down the barrier between "us" and "them," and should be applied to the use of animals in scientific research. Although I have long held the use of animals in research to a different moral standard than, say, puppy mills, dog racing tracks, zoos, and the like—because it is done in the name of something I consider noble, science—a documentary film called *Project Nim* has led me to rethink even this aspect of animal morality.[26] Project Nim was initiated and monitored by Columbia University psychologist Herbert Terrace, who wanted to test the (then) controversial theory of the MIT linguist Noam Chomsky that there is an inherited universal grammar basic and unique to humans, by teaching our closest primate cousin American Sign Language (ASL). Terrace, however, did a turnabout, concluding that the signs Nim Chimpsky (a cheeky nod to Noam Chomsky) learned from his human companions and trainers amounted to little more than animal begging, more sophisticated perhaps than Skinner's rats pressing bars, but in principle not so different from what dogs and cats do to beg for food, be let outside, etc.[27]

Nim was taken from the arms of his mother at only a few weeks old. As he was the seventh of her children to be so seized, the mother had to be tranquilized and grabbed quickly so she did not accidentally smother her baby, whom she clutched to her chest in motherly love and protection, as she collapsed on the floor. Nim began his childhood in a brownstone apartment in New York's Upper West Side, surrounded by human siblings in the mildly dysfunctional LaFarge family spearheaded by Stephanie, who breast-fed Nim and, as he got older, allowed him to explore her nakedness even as he put himself between his adopted mother and her poet husband in an Oedipal scene right out of Freud.[28] Just as Nim grew into his new family, surrounded by fun-loving human siblings and days filled with games and hugs, Terrace realized that scientists were not going to take him seriously because there was next to no science going on in this free-love home. (According to one of the trainers, there were no lab manuals, no diaries, no data sheets, no recordings of progress, and no one in the family even knew how to sign ASL!) So for a second time in his young life Nim was wrenched from his mother and placed into a more controlled environment in the form of a sprawling home owned by Columbia University. There a string of trainers carefully monitored Nim's progress in learning ASL, making daily trips to a lab at the university where Terrace could control all intervening variables in a manner not dissimilar to a Skinner box for lab rats.

In due time Nim grew into his teenage years, and as most testosterone-fueled male primates are wont to do, he became more assertive, then aggressive, then potentially dangerous in his evolved propensity to test his fellow primates for hierarchical status in the social pecking order. The problem is that adult chimpanzees are at least twice as strong as humans. In other words, Nim became a threat. As one of the trainers said while pointing to a scar on her arm that required thirty-seven stitches, "You can't give human nurturing to an animal that could kill you." After several of these biting incidents that sent trainers and handlers to the hospital, including one woman who had part of her cheek ripped open, Terrace pulled the plug on the experiment and therewith shipped Nim back to the research lab in Oklahoma whence he came. Tranquilized into unconsciousness, Nim went to sleep surrounded by loving human caretakers on a sprawling estate in New York and awoke in a gray-barred sterile steel cage in Oklahoma.

Having never seen another member of his species, Nim was understandably anxious and scared at the sight of grunting, hooting male chimps eager to let the youngster know his place in the social order. As a result, Nim slipped into a deep depression, losing weight and refusing to eat. A year later Terrace visited Nim, who greeted him eagerly and expressed himself in a manner that Terrace described as signaling to get him out of this hellhole. Instead, Terrace took off the next day for home and Nim slid back into a depression. Sometime later he was sold to a pharmaceutical animal-testing laboratory managed by New York University, where hepatitis B vaccinations were foisted on our nearest primate relatives. Footage of a tranquilized chimp being pulled out of and stuffed back into a steel-barred cage barely big enough to turn around in churns one's stomach.

As an exercise in the *principle of interchangeable perspectives*, imagine what Nim might have signed in response to this treatment.[29] Sadly, there was no Shawshank redemption for Nim. But thanks to films such as this—whose purpose is to draw us into changing perspectives with their subject—we can at least put flowers on his metaphorical grave. Flowers for Nim.

SPECIESISM: THE ARGUMENT

A century of scientific research on animal cognition and emotion reveals a capacity and depth to warrant our moral consideration for how all sentient beings should be treated.[30] Jeremy Bentham—considered by scholars to be the patron saint of animal welfare for his inclusion of nonhuman animals in his classic and groundbreaking 1823 work *Introduction to the Principles of Morals and Legislation*—asked where we should draw the line between humans and animals:

> Is it the faculty of reason or perhaps the faculty of discourse? But a full-grown horse or dog, is beyond comparison a more rational, as well as a more conversable animal, than an infant of a day or a week or even a month, old. But suppose the case were otherwise, what would it avail? The question is not, Can they *reason*? nor, Can they *talk*? but, Can they *suffer*?[31]

Bentham's observation that an adult horse or dog has cognitive skills beyond those of a young human infant has been turned into a cogent argument for modern-day animal rights advocates. It was on display, for example, in Mark Devries's 2013 film *Speciesism: The Movie*, the Los Angeles premiere of which I attended in a theater full of animal advocates who cheered wildly for the film's avatars, such as the Princeton ethicist Peter Singer, who articulated the argument in the context of vivisection and its abolitionist opponents:

> Would the experimenter be prepared to perform his experiment on an orphaned human infant, if that were the only way to save many lives? . . . If the experimenter is not prepared to use an orphaned human infant, then his readiness to use nonhumans is simple discrimination, since adult apes, cats, mice, and other mammals are more aware of what is happening to them, more self-directing and, so far as we can tell, at least as sensitive to pain, as any human infant. There seems to be no relevant characteristic that human infants possess that adult mammals do not have to the same or a higher degree.[32]

Critics counter that our superior intelligence, self-awareness, and moral sense add up to make us wholly other from all other animals and thus justify our exploitation of them. But Singer points out that by these criteria humans being in such states as infancy, severe mental retardation, grave physical handicap, or coma would imply that we can exploit them. Since we would not consider wearing or eating such humans, so we should not do the same to animals whose capacities in such categories as intelligence, self-awareness, and moral emotions are equal or superior to many people in these conditions. "Any characteristic of humans that is said to justify this sharp ethical distinction will either be shared by some nonhuman animals or be absent in some humans," Devries told me in an interview. "Therefore, the presumption that nonhuman animals' interests are less important than human interests could be merely a prejudice—similar in kind to prejudices against groups of humans such as racism—termed *speciesism*."[33]

The common counter to this argument, associated with the University of Michigan philosopher Carl Cohen, is that even though some humans may lack this or that characteristic (for example, an infant or a comatose adult may lack language), they are the "kind" of species that has language, since that characteristic is vouchsafed to us. Thus, regardless of the particulars of an individual organism, the general traits of a species is what makes them separate by definition, and so speciesism is justified in this sense.[34] Devries finds a major flaw in this argument in that whatever the characteristic is (language, tool use, intelligence) that goes into the "kind" that separates species, it is the moral relevance of those characteristics that matters:

> Imagine that someone were to argue that what justifies an ethical distinction between humans and nonhuman animals is the fact that only members of the human species are capable of manufacturing green t-shirts. The reason this seems facially absurd is that obviously there is nothing we consider ethically relevant within the species about the ability to manufacture green t-shirts. Now, if we switch the characteristic to something like language, we actually run into the same problem: if there is nothing we consider ethically relevant within the species about the capacity to use language, how does that characteristic become relevant when we are attempting to distinguish between species? Again, this seems to expose the "kind" argument as simply begging the question by assuming that there is something ethically relevant about species membership.[35]

As Virginia Morell writes in her thoughtful book *Animal Wise*, the question "what makes us different?" is the wrong question: "Instead, given that we now know that we live in a world of sentient beings, not one of stimulus-response machines, we need to ask, how should we treat these other emotional, thinking creatures?"[36]

What about eating animals? Given the fact that we bring into existence animals who would not otherwise have even been born, as long as we give them a decent life while they are alive and end their lives humanely, isn't this morally acceptable? Temple Grandin makes this point in her books and talks, and she has done much (and should be commended) to reform the factory farm system to make the life of animals more humane.[37] And yet, as Mark Devries notes, "in the case of bringing an animal into a life of suffering, that animal is harmed by his or her experiences, whereas the nonexistent animal does not 'mind,' since he or she never exists, and thus has no experience of a lack of benefits." As well, "using animals as economic commodities—bought and sold like property, and brought into existence and killed solely for the purpose of a palate preference—seems obviously inconsistent with taking their interests seriously, just as would be the case if we were doing the same to humans."[38] This is similar to the point I made in the previous chapter on slavery, in which I noted how whites used a similar line of reasoning to justify slavery in arguing that blacks on a plantation have a better life than blacks in Africa (or even blacks and poor whites toiling in factories in Northern states), and that blacks born into slavery in America would not have been born at all otherwise. That may be, but in the long run freedom is better than slavery, and liberty is preferred over oppression.

Michael Pollan, author of the best-selling books *The Omnivore's Dilemma* and *In Defense of Food*,[39] points out, "It's one thing to choose between the chimp and the retarded child or to accept the sacrifice of all those pigs surgeons practiced on to develop heart-bypass surgery. But what happens when the choice is between 'a lifetime of suffering for a nonhuman animal and the gastronomic preference of a human being'? You look away—or you stop eating animals. And if you don't want to do either?" Like most of us, Pollan doesn't want to do either, and so employs a favorite cognitive tool we all use: denial. Nevertheless, even an omnivore such as Pollan can marshal counterarguments to his own beliefs like a skilled rhetorician, citing the book *Dominion: The Power of Man, the Suffering of Animals, and the Call to Mercy* by the Christian conservative author Matthew Scully, who believes God commands us to "treat them with kindness, not because they have rights or power or some claim to equality but . . . because they stand unequal and powerless before us."[40] Pollan's description (drawn from Scully's work) of how pigs—who are as smart as dogs—are treated has to shudder even the hardiest of bacon-, ham-, and pork-eating omnivores:

> Piglets in confinement operations are weaned from their mothers 10 days after birth (compared with 13 weeks in nature) because they gain weight faster on their hormone- and antibiotic-fortified feed. This premature weaning leaves the pigs with a lifelong craving to suck and chew, a desire they gratify in confinement by biting the tail of the animal in front of them. A normal pig would fight off his molester, but a demoralized pig has stopped caring. "Learned helplessness" is the psychological term, and it's not uncommon in confinement operations, where tens of thousands of hogs spend their entire lives ignorant of sunshine or earth or straw, crowded together beneath a metal roof upon metal slats suspended over a manure pit. So it's not surprising that an animal as sensitive and intelligent as a pig would get depressed, and a depressed pig will allow his tail to be chewed on to the point of infection. Sick pigs, being underperforming "production units," are clubbed to death on the spot. The U.S.D.A.'s recommended solution to the problem is called "tail docking." Using a pair of pliers (and no anesthetic), most but not all of the tail is snipped off. Why the little stump? Because the whole point of the exercise is not to remove the object of tail-biting so much as to render it more sensitive. Now, a bite on the tail is so painful that even the most demoralized pig will mount a struggle to avoid it.[41]

Scully, whose conservative credentials include working as special assistant and senior speechwriter for President George W. Bush, makes the case for animal rights as well as anyone in the secular/liberal axis, and he does so by employing a natural rights argument devoid of any supernatural special pleading or holy book injunction:

> If natural law is what informs our own laws and moral codes among one another, in short, then it should also inform our laws and moral codes regarding animals. And its most basic and revolutionary insight is the same for them as for us. Any moral claims we have, we have simply because of what we are. Any moral claims that animals have in relation to us, they, too, have simply because of what they are. We don't each get to decide their moral claims for ourselves. The moral value of any creature belongs to that creature, acknowledged or not, a different value from our own but just as much a hard and living reality. Just as our own individual

moral worth does not hinge on the opinion of others, their moral worth does not hinge upon our estimation of them.[42]

Another argument for why an animal's place is in our stomach is because in nature an animal's place is in each other's stomachs. Animals eat each other, and since we're animals who also need to eat, isn't it natural we should eat them? This is not a bad rejoinder, although animal rights advocates counter that throughout human history slavery, genocide, and rape were considered natural and the ways things were always done, and yet we outlawed them as immoral. We all have an evolved nature for a certain amount of violence—self-defense, jealousy, honor—yet that does not mean we should not try to control our impulses. The point of morality is to choose to do what you might otherwise not have done. You can seek your own survival and flourishing with no one else's survival and flourishing in mind, but when you take the perspective of other sentient beings into consideration, that is what makes it a moral act.

ANIMAL *ARBEIT MACHT FREI*

At this point I have a confession to make: I am a speciesist—I eat and wear members of other species. There are few foods I find more pleasurable than a lean cut of meat—a tri-tip, a tuna or salmon steak, a buffalo burger. And I laughed out loud at the joke about the farmer who castrates his horse with two bricks, who when asked if it hurts replied, "Not if you keep your thumbs out of the way."[43] I also find disturbing those on the edges of the animal liberationist movement who trash or destroy scientific laboratories and release lab animals. I caution them (contra Barry Goldwater) to consider that *moderation in the pursuit of liberty is no vice, and extremism in the defense of justice is no virtue.*[44]

Also, in my opinion, context and motivation matter. There is a tiny Inuit community in Gambell, Alaska, whose livelihood of killing walruses has been severely curtailed because of the ice breakup due to global warming. There are 690 residents who killed "only" 108 walruses this year, which is one-sixth the average of their annual kill rate of 648 walruses. A thirty-eight-year-old resident of Gambell named Jennifer Campbell, a mother of five whose family depends on walrus meat, bemoaned the fact that they caught only two walruses this year, compared to the average of twenty in normal years. "If it continues like this, we will seriously starve," she said.[45] It seems to me that there is a sizable moral difference between some Beverly Hills diva wearing a fur to a trendy LA restaurant where you'll down a juicy steak when you could have ordered the veggie burger, and the Inuit inhabitants of Gambell, Alaska, who kill and eat and use walruses for their survival. Here we encounter moral dumbfounding: once you concede that speciesism exists and we breach the barrier, wouldn't all killing of animals be the same regardless of context and motivation? But clearly there is a contextual and motivational difference between these two scenarios, and that difference matters morally.

I am also troubled by an affecting analogy running throughout the animal rights movement (and on graphic display in many documentary films such as *Earthlings* and Devries's *Speciesism: The Movie*) that animals are undergoing a "holocaust," made poignant by the architectural layout comparison between factory farm buildings and prisoner barracks at the Nazi concentration camp at Auschwitz-Birkenau—row on row of long, rectangular buildings surrounded by barbed-wire fences. Worse, goes the analogy, as evil as the Holocaust was in its attempt to exterminate groups of people, in factory farming new generations of sentient beings are brought into existence only to be exterminated, again and again, generation after generation. Some animal rights activists refer to this as a Holocaust that never ends, captured in the book title *Eternal Treblinka* by Charles Patterson,[46] taken from the Yiddish writer and Nobel laureate Isaac Bashevis Singer, a vegetarian who famously wrote (through one of his fictional characters):

> In his thoughts, Herman spoke a eulogy for the mouse who had shared a portion of her life with him and who, because of him, had left this earth. "What do they know—all these scholars, all these philosophers, all the leaders of the world—about such as you? They have convinced themselves that man, the worst transgressor of all the species, is the crown of creation. All other creatures were created merely to provide him with food, pelts, to be tormented, exterminated. In relation to them, all people are Nazis; for the animals it is an eternal Treblinka."[47]

One limitation of this analogy is in the motivation of the perpetrators. As someone who has written a book on the Holocaust (*Denying History*[48]) I see a vast moral gulf between farmers working to feed the world and turning a profit in so doing, and Nazis with a genocidal motive. Small-farmers who kill animals in the service of making a living by providing food to people who want and need it bear no motivational resemblance to

SS guards murdering Jews, Gypsies, and homosexuals for the sole purpose of "racial cleansing" and "eliminationist purity." Even corporate factory farmers motivated solely by quarterly profits and who care not at all for the welfare of their animals, whom they consider to be nothing more than products, are higher on the moral ladder than Adolf Eichmann and Heinrich Himmler and their henchmen, who orchestrated the processing of mass murder on an industrial scale. There are no signs at factory farms reading *"Arbeit Macht Frei"* ("Work Makes You Free").

Yet I cannot fully rebuke those who equate factory farms with concentration camps when I recall one of the most dreadful things I ever had to do. While working as a graduate student in an experimental psychology animal lab in 1978 at California State University at Fullerton, it was my job to dispose of lab rats who had outlived our experiments. This was not a pleasant task, made all the worse by the fact that we named our rats after the Los Angeles Dodgers baseball team of that era—Ron Cey, Davey Lopes, Bill Russell, Steve Garvey, Don Sutton, and the rest. You get to know your lab animals, and they get to know you. But then the experiment is over and the time comes to dispose of the subjects, which in the case of the rats was done by . . . I can barely type the words . . . gassing them with chloroform in a large plastic trash bag. I wanted (and asked) if I could take them up into the local hills and let them go, figuring that death by predation or starvation was surely better than this. But my suggestion was rejected in no uncertain terms because what I was proposing was actually illegal. So I exterminated them. With gas. The very words used to describe this act—exterminating a group of sentient beings by gassing them— are uncomfortably close to those I wrote in my Holocaust book to describe what the Nazis did to their captives. No wonder naming your lab animals is considered taboo in science. I thought it was to encourage objectivity, but I now suspect that it has as much to do with remaining emotionally detached so as to feel morally blameless.

Today the treatment of lab animals is less loathsome, and according to my old professor Douglas Navarick—whose lab at Cal State Fullerton I worked in for those two years and whom I recently queried to confirm my memory of the process at the time—the method now used is "metered CO_2" for rats. He adds, "There is now a committee on campus that reviews and must approve any research or instructional activity involving vertebrate animals (Institutional Animal Care and Use Committee) and it has rules for what type of euthanasia should be used for different species (I'm a member)."[49] This is certainly an improvement over what I did (and what was customary at the time in animal labs everywhere), and it is good to know that animal care committees have also become sensitive to the effects on the animals' human caretakers, as instructed in the *Guide for the Care and Use of Laboratory Animals*: "Euthanizing animals is psychologically difficult for some animal care, veterinary, and research personnel, particularly if they perform euthanasia repetitively or are emotionally attached to the animals being euthanized. When delegating euthanasia responsibilities, supervisors should be sensitive to this issue."[50]

This counts as moral progress, to be sure, but I remain deeply unsettled both by what I did and the perspective of the guidebook, which still sounds more concerned with the welfare of the perpetrators than the victims, whereas the long-term trend is to shift moral perspectives from the former to the latter.

PERSPECTIVE-TAKING

Perspective-taking is the psychological foundation underlying the capacity for empathy. To judge the rightness or wrongness of an action against another, one must first take the perspective of that other sentient being. In the context of animal rights, I am reminded of the scene in the film *My Cousin Vinny*, in which Joe Pesci's character, Vinny Gambini, is preparing to go deer hunting with the opposing attorney in their court case to ascertain what his strategy in the trial might be through the anticipated camaraderie of a hunt. As he is getting dressed to go he inquires of his fiancée— Marisa Tomei's character, Mona Lisa Vito—if the pants he has on are appropriate for deer hunting. She answers with a thought experiment in perspective-taking:

> Imagine you're a deer. You're prancing along. You get thirsty. You spot a little brook. You put your little deer lips down to the cool, clear water. Bam! A fuckin' bullet rips off part of your head. Your brains are layin' on the ground in little bloody pieces. Now I asks ya, would you give a fuck what kind of pants the son of a bitch who shot you was wearing?[51]

Perspective-taking is what animal rights documentarians are trying to evoke in those hidden-camera videos of slaughterhouses and the horrific conditions in which factory farm animals live (or, more to the point, suffer and die). In a short video clip posted at freefromharm.com and appropriately titled "saddest slaughterhouse footage ever" (to see it, Google-search that word string), a bull is waiting in line to die. As he hears his mates in front of him being killed he shakily backs up into the rear wall of the metal chute. When he can go no farther, he turns toward the back (from where the camera is filming) as if looking to escape his fate. A worker walks along the outside of the chute with an electric cattle prod and zaps the bull forward. He takes a few hesitating steps, stops, and starts to back up again, so the worker jolts him with some more electricity far enough forward for the final death wall to come down behind him. You see his rear legs trying one last time to back up out of his death trap and then . . . thump! . . . down he goes in a heap, rear legs poking out of the slot at the bottom of the wall. Dead.[52] Am I projecting human emotions into the head of a cow? I don't think so. The investigative journalist Ted Conover, while working undercover as a USDA inspector at the Cargill Meat Solutions plant in Schuyler, Nebraska, when he asked a worker there why the ramps leading the animals up to the killing chamber stank so bad from cattle waste, was told, "They're scared. They don't want to die."[53] Perhaps this is why many factory farms have barbed wire surrounding their facilities and they do not take kindly to snooping documentary filmmakers probing around with their cameras.

The 2005 film *Earthlings* is arguably the most disturbing of the perspective-taking documentaries, drawing parallels between the mistreatment and economic exploitation of blacks and women in centuries past with the maltreatment and commercial use and abuse of animals today.[54] To describe the film as hard to watch is an understatement. To get through it I had to have two windows open on my computer screen: the film and the transcript of it for note-taking purposes, which was really just a pretext to cover the gore when I could take it no more. Scenes of animals being processed as products include slashed-open dolphins flopping around on a concrete floor while schoolchildren obliviously pass by, and cattle being killed by a captive bolt gun that fires a steel bolt in the animal's brain by compressed air, a process that doesn't always work, leaving them struggling for life even as the butchering process begins. Such images are overlaid with narration to drill home the point not just of our evolutionary continuity with all other animals, but of their continuity with us:

> Undoubtedly there are differences, since humans and animals are not the same in all respects. But the question of sameness wears another face. Granted, these animals do not have all the desires we humans have;

granted, they do not comprehend everything we humans comprehend; nevertheless, we and they do have some of the same desires and do comprehend some of the same things. The desires for food and water, shelter and companionship, freedom of movement and avoidance of pain. These desires are shared by nonhuman animals and human beings.[55]

Perhaps the windowless, fortresslike factor of farm buildings surrounded by barbed-wire fences is a two-way agreement—most of us don't want to know how sausage is made. As the poet Ralph Waldo Emerson poignantly observed, "You have dined, and however scrupulously the slaughterhouse is concealed in the graceful distance of miles, there is complicity."[56]

The power of visual media to shift our perspective-taking is on disturbing display in the Academy Award–winning 2009 documentary film *The Cove*, about the mass slaughter of dolphins and porpoises in Taiji Cove in Wakayama, Japan. The film features Ric O'Barry, former dolphin trainer for the 1960s hit television series *Flipper*, which as a young boy I watched faithfully every week as the show highlighted the very humanlike characteristics of these marine mammals, most notably their humanity in social bonding with each other and with humans (and in which each week Flipper predictably thwarted bad guys in their nefarious activities underwater or rescued good guys trapped in improbable circumstances). According to the film, dolphins are herded into the cove, where a few are captured unharmed for sale to marine parks and aquariums around the world, while the rest are brutally butchered and killed for their meat, which is sold to resalers in Japanese fish markets. The press attention the film received after it won the Academy Award led to Japanese government officials scrambling to deny, then explain, and then rationalize away the ruthless butchery of sentient beings so close to us in social cognition and emotion as to wrench powerful emotions at the sight of their slaughter.

One heart-ripping scene among many captured on film (with cameras hidden inside rocks and long-range lenses from the surrounding hills) is that of a young dolphin desperately trying to survive after being slashed open, gasping for air as blood gushes out of her torso, crying for help as the filmmakers watch in horror, powerless to save her as she slips beneath the surface one final time in a pool of blood, never seen again. It is at once an infuriating and sickening scene to watch. I cannot say that I would have been able to restrain myself from jumping into the cove, swimming over to the cluster of tiny boats, and in Rambo-like fashion pulling these fishermen into the water and giving them a dose of their own machete-medicine. But no matter how powerful the temptation for retaliatory violence in these circumstances, such vigilante justice is precisely what most animal rights activists avoid; you can't be an effective rights activist if you are in jail.

O'Barry and his crew showed great restraint in not attempting to stop the slaughter and instead expose it to the world through such gut-wrenching scenes, and that's the point of the film—it translates the abstract into the concrete, thereby engaging our brain's empathic neural pathways that are normally triggered when experiencing another *human*'s pain. So, even more than our shared intelligence, self-awareness, cognition, and moral capacity, it is our common capacity to suffer in a very human way—gasping for air, struggling to stay upright, fighting for life—that, in part, expands the moral sphere. Animal rights will not be fully realized until we gain a deep emotional understanding that they are sentient beings who—like us—want to live and are afraid to die.

This is where *is* becomes *ought*, where the way things are naturally—by nature animals desire food, water, shelter, companionship, freedom of movement, avoidance of pain just like we do—becomes the way things ought to be, especially when a departure from nature is the result of an exploitation of one animal species over another. Here we make a moral choice in which perspective-taking reverses the naturalism argument— from the perspective of humans exploiting animals to fulfill our nature, to the perspective of animals fulfilling theirs. Moral progress has been driven primarily by this shift in perspective-taking, from the exploiter to the exploited, from the perpetrator to the victim. Why should we favor one perspective over the other? Because that is how moral progress is made.[57]

FIRST, DO NO HARM

In the animal rights debate there is a counterargument put forth by the philosopher Daniel Dennett in his book *Kinds of Minds*, in which he draws a distinction between *pain* and *suffering*, suggesting that the former is more visceral and basic and shared by most animals, while the latter involves more humanlike emotions such as worry, shame, sadness, humiliation, dread, and especially anxiety from projecting what may happen in the future, which requires higher cognitive functioning shared by only a few animals. "If we fail to find suffering in the animal lives we can see, we can rest assured there is no invisible suffering somewhere in their brains.

If we find suffering, we will recognize it without difficulty."[58] I'm not so sure. In all these films and videos of animals in pain and suffering, it looks to me that they are anxious, worried, and fearful. Furthermore, why should we assume that animals suffer *less* than humans? If the argument is that we can't really know what goes on inside the minds of other sentient beings, why assume that their suffering is *less* than ours? Maybe it's *worse* for other animals.

A moral foundation for animal rights begins with how we treat them, a good starting point of which may be found in the Hippocratic Oath's ethical precept *primum, non nocere—first, do no harm.* "All ethical theories and moral systems appear to share the basic principle that, all else being equal, causing harm—specifically suffering—is a bad thing," the filmmaker Mark Devries told me. "Because nonhuman animals are capable of experiencing suffering, this basic ethical principle prima facie applies beyond the species barrier." As in my moral model, sentience is key to Devries's reasoning: "If an animal's capacity to have a subjective experience of harm is the ethically relevant characteristic, then the proper 'cut-off point' for what species must be considered appears to turn on whether the animals are sentient. We know that birds and mammals are capable of physical and emotional suffering, in fact the neural anatomy that permits such experiences in humans was developed in our common ancestors with those animals."[59]

It is especially in the area of animal rights that we must remember that it is the *individual* who is the moral agent—not the species—because it is the *individual organism* who feels pain and suffers. Or to be more precise, it is an individual brain with the ability to feel pain that constitutes an individual organism. It is an individual bull who walks through the chute into the entrapment and faces the captive bolt gun, not the species *Bos primigenius.* It is an individual dolphin who is sliced open in a cove in Japan who gushes blood and gasps for air, not the species *Delphinus capensis.*

A PROFOUND SHIFT IN MORAL PERCEPTION

Although the parallels between the human and animal rights movements are abundant, it seems to me the latter have a significantly larger task than the slavery abolitionists of the eighteenth and nineteenth centuries did, given that even at the zenith of the slave trade only a small portion of the world's population owned or traded in slaves, whereas the vast majority of the world's population eats meat and uses animal products. As the animal rights attorney Steven Wise notes, animals are even more essential to our daily lives than slaves were in nineteenth-century America—including personally, psychologically, economically, religiously, and especially legally, where animals are property and their use protected by law, which is not always easy to change.[60] Even if the arguments in favor of animal rights are better than those opposing them—which I think they are—bumping up the percentage of the population committed to a vegetarian or vegan lifestyle from the low single digits into the high double digits is going to be a daunting task. People didn't have to own slaves to survive. But people do have to eat, and meat is delicious to most, relatively cheap, and readily available, and therefore (still) a popular commodity to fulfill that need. In the United States it took a civil war to finally abolish slavery, and that was with only a small fraction of citizens owning slaves. When more than 95 percent of the population eats meat, that's a daunting difference.

To bring about significant change we are going to need the equivalent of what the slavery historian David Brion Davis calls "a profound shift in moral perception,"[61] a reconfiguration in how we think about animals, from property to persons, as the Rutgers University legal scholar Gary Francione has done in his 2008 book *Animals as Persons*, where he outlined in logical detail why sentient nonhumans should legally be regarded as persons: "They are conscious; they are subjectively aware; they have interests; they can suffer. No characteristic other than sentience is required for personhood."[62]

Moral progress along these lines occurred in 2013 in India, which banned the captivity of dolphins for public entertainment because "Confinement in captivity can seriously compromise the welfare and survival of all types of cetaceans by altering their behavior and causing extreme distress." Putting teeth into the law, India listed all cetacean species in Schedule II, Part I of the Wild Life Protection Act of 1972, adding that dolphins should be considered as "nonhuman persons."[63] This granting of legal personhood to a nonhuman animal is a monumental step toward justice and freedom for all sentient beings.

Reading the works of animal rights activists can be an emotionally draining experience, and if we focus only on the worst cases (factory farms) and raw numbers killed (in the Hemoclysm billions, not Holocaust millions), a case could be made that the arc of the moral universe is decidedly bent backward. So it is good to reflect on how people used to perceive animals, from the recreational (cat burning and bear baiting) to the philo-sophical (Descartes's belief that animals are mechanical automata who merely mimicked pain, pleasure, desire, interest, boredom, and the rest of humanlike emotions he denied animals could experience). A sizable body of literature exists that tracks the mixed (although mostly bad) moral his-tory of animal welfare over the millennia.

As with the abolition of slavery, religion not only did not lead the revo-lution for animal welfare, it often hindered it, starting at the beginning . . . literally in the first chapter of Genesis when Yahweh commands Adam and Eve to "Be fruitful, and multiply, and replenish the earth, and subdue it: and have dominion over the fish of the sea, and over the fowl of the air, and over every living thing that moveth upon the earth."[64] The attitude was reinforced over the centuries by such church fathers as St. Augustine and St. Thomas Aquinas, the latter of whom opined, "Hereby is refuted the error of those who said it is sinful for a man to kill dumb animals: for by divine providence they are intended for man's use in the natural order. Hence it is not wrong for man to make use of them, either by killing or in any other way whatever." Apparently Aquinas must have gotten some push-back from those who felt sympathy with suffering animals, for he fol-lowed the above observation with this admonition: "And if any passages of Holy Writ seem to forbid us to be cruel to dumb animals, for instance to kill a bird with its young: this is either to remove man's thoughts from being cruel to other men, and lest through being cruel to animals one become cruel to human beings: or because injury to an animal leads to the temporal hurt of man, either of the doer of the deed, or of another."[65]

Ever since biblical times, animal cruelty was the norm, not the excep-tion. Cockfights and dogfights, of course, have been a perennial favorite across the centuries, and in Colonial America bear baiting featured a cap-tured bear chained to a post, where she would be tormented (baited) and even ripped to shreds by dogs who were whipped up into an attack frenzy. And who could forget the sixteenth-century popular Parisian pastime of cat burning, in which a terrified feline was gradually lowered into a fire while "spectators, including kings and queens, shrieked with laughter as the animals, howling with pain, were singed, roasted, and finally carbon-ized." Or the fun of nailing a cat to a post and having a contest to see who could head-butt it to death without getting one's eyes clawed out by the horrified beast.[66]

But that's just recreational cruelty. Food production brutality does not begin with the factory farms of today. Consider this account of a practice common in the seventeenth century quoted by the historian Colin Spencer in his sweeping history of vegetarianism, *The Heretic's Feast,* of the slaughter of animals and the extent to which anyone cared much for their experience of the process:

> Methods of slaughter were coldly rational. As Dr. Johnson remarked, the butchers "have no view to the ease of the animals but only to make them quiet for their own safety and convenience." Cattle were poleaxed [knocked out] before being killed, but pigs, calves and poultry died more slowly. In order to make their meat white, calves and sometimes lambs were struck in the neck so that the blood would run out. Then the wound was stopped and the animal allowed to linger on for another day. As Thomas Hardy's Arabella explained to Jude, pigs should not be slaughtered quickly. "The meat must be well bled and to do that he must die slow. I was brought up to it and I know. Every good butcher keeps them bleeding long. He ought to be up till eight or ten minutes dying at least."[67]

THE MORAL ARC IS BENDING FOR ANIMALS

Moral progress has been sporadic and halting, to be sure, but the arc has been bending ever so slightly for animals ever since the Enlightenment, as we have already seen in the case of the treatment of laboratory animals over the past several decades, the shift in perspective-taking that has been under way for a century as we come to understand the continuity between ourselves and other animals, and the linking of the animal rights movement to other rights revolutions whose success has raised the moral consciousness of everyone to expand the moral sphere even wider to include at least some animals.

A 2003 Gallup poll, for example, found that "The vast majority of Americans say animals deserve at least some protection from harm and exploitation, and a quarter say animals deserve the same protection as human beings." Bans on medical research and product testing were still opposed by most Americans, and even though the percentage of households with hunters has dropped from about 30 percent to about 20 percent in the past quarter century, most Americans oppose banning all types of hunting. Nevertheless, it is encouraging to know that "96% of Americans say that animals deserve at least some protection from harm and exploitation, while just 3% say animals don't need protection 'since they are just animals.'" Most notably, 25 percent of all Americans say that animals deserve "the exact same rights as people to be free from harm and exploitation."[68]

This is noteworthy because rights talk gets fuzzy when it comes to positive rights about what people (or sentient beings) deserve to have, with much dispute even over what humans are entitled to (health care, Social Security, disability insurance, and the like). But being free from harm and exploitation involves only the cessation of a negative activity (the inducement of pain and suffering). For example, in the Gallup poll, sixty-two percent of "Americans support passing strict laws concerning the treatment of farm animals."[69]

As for the consumption of animals as food, in 2012 the Vegetarian Resource Group commissioned the National Harris polling agency to ask, "How Often Do Americans Eat Vegetarian Meals?" and "How many adults in the U.S. are vegetarian?" (meaning no meat, fish, seafood, or poultry). Results: 4 percent of Americans always eat vegetarian meals, of which 1 percent are always vegan (also no dairy or eggs) and 3 percent are always vegetarian. On the more moderate side, 15 percent eat "many" of their meals vegetarian (but less than half the time), while another 14 percent make half or more (but not all) of their meals vegetarian, with almost half (47 percent) having eaten vegetarian meals at some point during the year (a dubious statistic, since a bowl of granola would count as a vegetarian meal).[70] A 2012 Gallup poll was slightly more encouraging, with a total of 2 percent vegan and 5 percent vegetarian.[71] I suspect the differences are in the statistical noise, with the overall numbers so low. Still, over the past several decades the trend line is unmistakably upward, as seen in *figure 8-2*.

Another trend encouraging to animal welfare proponents is a recent downward turn in meat consumption and a resulting drop in the number of animals raised for food. This appears to be largely the result of a rise in people adopting modest hybrid eating habits, such as "flexitarians" (flexible vegetarians) who, for example, practice "Meatless Mondays" or are "vegan until 6:00 p.m." or who eat red meat only once a week, and the like. According to the USDA, for example, between 2007 and 2012 meat and poultry consumption dropped by 12.2 percent.[73] It's not a huge number, but it's not trivial either. Slightly more encouraging are the figures for semi-vegetarians, defined as people who eat meat for less than half their meals, which, according to the Humane Research Council, ranges from 12 percent to 16 percent of the US population. Still larger are the percentages of "active meat reducers," or "those who say they eat less meat

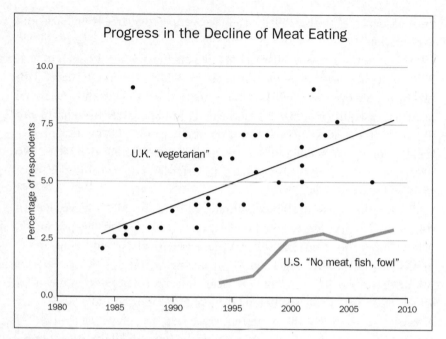

Figure 8-2. Trends in Vegetarianism
The gains are modest but upward.[72]

compared to one year ago," which is 22 to 26 percent of the US population, or roughly a quarter, which is indeed nontrivial.[74]

Ever since World War II, European countries have shifted politically more leftward than America, and this has expanded more extensively the moral sphere for animals there. In a number of European countries sows (female pigs) may not be confined to crates so small that they cannot turn around, and hens must be in cages large enough to stretch their wings. In England, the farming of animals for fur is now illegal. Switzerland has changed the status of animals to "beings" instead of "things," and Germany was the first nation to grant animals a constitutional right when they added the words "and animal" to a provision obliging the state to respect and protect human dignity.

Medically, a diet of meat—especially red meat eaten regularly—has proven to be unhealthy and a major contributor to cardiovascular disease in many (but not all) people. Environmentally, factory cattle farms gener- ate massive amounts of waste products and the greenhouse gas methane, among many other pollutants. And they stink to high heaven—I shall never forget riding my bike during the Race Across America through Dalhart, Texas, where the country's largest cattle ranches are located. You could smell them from miles away. I can't imagine what the local residents do, other than habituate to the smell to the point where they no longer notice it, which itself could be a metaphor for the overall treatment of animals—we have become habituated to their pain and suffering.

Blood sports such as cockfighting and dogfighting have been banned in most Western democracies, and even bullfighting appears to be on its way out. A 2008 Gallup poll found that almost four in ten Americans say they favor banning sports that involve competition between animals, such as horse racing and dog racing.[75] Hunting and fishing are on the decline as well, down 10 to 15 percent from 1996 to 2006, and it is not due to people staying indoors playing video games or watching television, because the percentage of people participating in wildlife watching and ecotourism increased by a corresponding amount over the same time frame.[76] According to the US Fish and Wildlife Service, in 2011 there were 13.7 million hunters and 33.1 million anglers, but "Almost 68.6 million people wildlife watched around their homes, and 22.5 million people took trips of at least one mile from home to primarily wildlife watch," which totals 91.1 million.[77] That's a lot of wildlife watching compared to killing.

I have witnessed and made the transition myself over the past several decades—as a youngster in the 1960s my stepfather regularly took my stepbrothers and me out to hunt dove and quail. I did not enjoy hunting as much as I did playing baseball, but neither did it bother me to shoot birds out of the sky with a shotgun. In fact, it was kind of exhilarating because it is hard to hit a moving target at a distance, and it became something of a challenge comparable to trying to hit a baseball. (The fact that we ate our prey helped to justify it.) I also did my fair share of fishing—both lake and deep-sea—but today the catch-and-release method is becoming more popular, giving the angler the thrill of the catch while sparing the victims their lives. In the 1980s the burgeoning business of ecotourism was under way, and I both joined and led many trips into the wild to, as it is said, "take only photographs and leave only footprints." And today the Skeptics Society, which I direct, routinely takes what we call "Geotours" to places such as the Grand Canyon and Death Valley, and there is no shortage of customers. I cannot today imagine going to such places to do what I did as a young hunter.

How perspectives have changed over the decades. Compare the making of the 1979 film *Apocalypse Now*, in which a water buffalo is almost decapitated while slashed open to the shoulder blades by a machete, to

Steven Spielberg's 2011 film *War Horse*, in which great care was taken not to harm the horses who were depicting the more than four million horses who were killed or died during World War I. For example, a horse tangled in barbed wire in no-man's-land was actually struggling to free herself from Styrofoam rubber that was painted silver.[78]

HOW FAR WILL THE MORAL ARC BEND FOR ANIMALS?

The trend lines we have been tracking in this chapter are encouraging for the animal rights movement. Although the arguments in favor of expanding the moral sphere to include more and more of our fellow sentient beings are superior to those against, the gains, it must be admitted, are modest at best. I predict that hunting and fishing will continue their downward slide in popularity as people's biophilic propensities for wildlife viewing, photography, hiking, and ecotourism and geotourism increase, but it will never drop to zero. As for the consumption of animals as food products, at the rate we're going the percentage of vegetarians will not reach the two-digit mark in the United Kingdom and the rest of Europe until about 2025, and in the United States even later (2030 by my calculations), and veganism will be lucky to hit 10 percent anywhere by 2050 (unless something happens to change the rate of growth). But rather than the black-and-white, meat-or-no-meat dichotomy, a flexitarian lifestyle and the gradual weaning of ourselves off of meat is the likelier path we shall see, which will reduce the carnage in the long run.

In the meantime, a workable moral middle ground may be found in a return to small, environmentally stable, and animal-friendly farms—sometimes called local farms or free-range farms or family farms—such as the one Michael Pollan visited in the Shenandoah Valley of Virginia that specializes in raising cattle, pigs, chickens, rabbits, turkeys, and sheep that, in the owner's words, allow each animal "to fully express its physiological distinctiveness."[79] Instead of being treated like an assembly-line widget in a factory farm, all members of species can live out their lives as they would in nature—to the extent that domesticated animals can be said to be in "nature," given that their own nature has been modified by humans toward the end of living on local farms—thereby fulfilling the moral imperative for all sentient beings to survive, reproduce, and flourish, even if most of the animals will be eventually killed and consumed.[80] As a local farmer named Kim Alexander observed with unintentional humor about his charges in the documentary film *Free Range*, after describing his chickens who frolic outdoors on grassy fields with the sun and wind at their backs, "They have a very good life. They just have one bad day at the end." Alexander's moral middle ground between the horrors of factory farming and the impracticality (and undesirability, for most people) of veganism strikes me as a sound and reasoned position with which most of us could live:

> I enjoy eating meat, especially when I know where it comes from [local farms v. factory farms] and how it's been raised and how it's been processed. The essence of what farming is—growing clean quality food for your family and, when you get good at it, growing it for other people that appreciate it. Out here we factor in all the costs—the costs of a clean environment, the costs of clean food for customers, the costs of treating animals properly with respect and as close to nature as possible, and the costs of what it takes for a family farm to make a decent living.[81]

Such farms are starting to catch on as the market demand for food from such sources expands. In the United Kingdom in 2012, for example, free-range eggs outsold those from caged hens for the first time in history (51 percent of the nine billion eggs laid that year). The trend began in 2004, when new rules required farmers to reveal that their eggs were produced by hens stuffed into cages with a floor space the size of an A4 (8x10 inch) piece of paper, leading to a public demand for better treatment of farm animals.[82] In the United States, the Whole Foods Market chain is involved in a program called the 5-Step Animal Welfare Rating, and on its web page avows: "All beef, chicken, pork and turkey in our fresh meat cases must come from producers who have achieved certification to Global Animal Partnership's 5-Step™ Animal Welfare Rating. As their rating program expands to other species, we will require our suppliers for those species to be certified as well." *Figure 8-3*, from the Whole Foods Market web page, outlines the steps and encapsulates what is involved in the program.[83]

Encouragingly, one of the largest of the factory food production corporations, Tyson Foods, Inc., under pressure from animal rights activists and after a graphic hidden-camera investigation by Mercy for Animals at a Tyson pig factory farm in Henryetta, Oklahoma, in late 2013 announced a set of new animal welfare guidelines for its pork suppliers, including a move away from restrictive gestation crates, castration, and tail docking without painkillers, and slamming piglets headfirst into the ground to kill

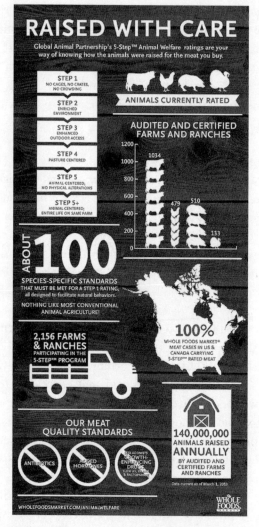

Figure 8-3. Whole Foods Market 5-Step Animal Welfare Rating Program

them. As well, gestation crates have now been widely condemned as one of the cruelest factory farming practices in the world, declared to be so inhumane that they are now banned in the entire European Union and in nine US states. Showing the power of the market to make moral changes, some sixty major food providers have demanded their suppliers do away with gestation crates, including Kmart, Costco, Kroger, McDonald's, Burger King, Wendy's, Chipotle, and Safeway. That's real progress.

This is not some pie-in-the-sky, hippie-dippy, sandal-wearing, tree-hugging, flower child visage of a world filled with Bambi the deer and

Babe the pig. It is a practical, economically realistic solution to a moral dilemma. At Whole Foods Market, the driving force behind these initiatives is CEO John Mackey, a pro–free market libertarian who is not only a dietary vegan, but who also practices what he calls "conscious capitalism," which he defines as "an ethical system based on value creation for all stakeholders," which include not just owners but also employees, customers, the community, the environment, and even competitors, activists, critics, unions, the media, and, in the case of Whole Foods, the animal stakeholders who are part of the system no less than their human counterparts. Animal welfare is no longer the province of far-left liberals, and John Mackey is a conscious capitalist who is dedicated to improving the food industry in terms of the health and well-being of both consumers and consumed.[84]

It remains to be seen if family farms could feed a world of seven billion people even with the gradual shift away from meat eating.[85] I think not, and so total animal liberation as envisioned by some animal rights activists is likely not even on the horizon. The numbers are staggering. The world population will hit 9.6 billion in about 2050, and then optimistic projections by the United Nations put us back down to about 6 billion by 2100. According to the International Institute for Applied Systems Analysis (IIASA), if the entire world stabilized at a European fertility rate of about 1.5 children per woman (2.1 is replacement level), the world population would fall to about 3.5 billion by 2200 and plummet to about 1 billion by 2300,[86] which would be sustainable by family farms.

To nourish 9.6 billion people, however, requires about 100 billion land animals to provide the world's population with meat, dairy, eggs, and leather goods, and short of "biofabricating" meat and leather (which is being tested in the laboratory now[87]), economic forces will likely dictate outcomes. For example, what we will more likely see is the economic principle of supply and demand in action. Smaller farms will generate smaller supplies. Fewer supplies for the current demand will drive prices up, and this in turn will cause a decrease in demand, which could then lead more and more food producers to shift from factory farms to family farms, but only so long as the consumer demand for 5+ ratings on food products continues. Economic realists also note that poor people with large families cannot afford the higher prices of family farm foods, but even here the long-term trends of smaller family sizes and increasing per capita GDP could ultimately solve this problem as, one hopes, everyone becomes better off and can afford to care about animals and the environment.

PROVISIONAL MORAL PRINCIPLES FOR ANIMALS

Let's consider the Provisional Ten Moral Principles that I outlined at the end of chapter 4. The first two principles—the *Golden Rule Principle* and the *Ask-First Principle*—might apply to animal rights by asking ourselves how our actions with regard to other species affect their capacity to survive and flourish. We can imagine how we might feel if we were, say, a chimpanzee locked up in a cold steel cage injected with human diseases. We might visualize ourselves as dolphins frolicking about a cove when all of a sudden a machete cuts through the water and slashes us open to bleed to death. Or we might picture being a head of cattle, taking our final walk down our own green mile, hearing our comrades falling one by one, and think of what Ted Conover was told by a worker at a factory farm slaughterhouse about the cattle and imagine that they are us: "We're scared. We don't want to die."[88]

In so identifying with other species and taking their perspective, we might find a way to apply the *Happiness Principle* and the *Liberty Principle* in which *we should never seek happiness and liberty when it leads to the unhappiness and loss of liberty of another sentient being*. Recall that Principle 5, the *Fairness Principle*, is based on John Rawls's concepts of the "veil of ignorance" and the "original position," in which we are ignorant of our position in society when determining rules and laws that affect everyone, so imagine this applied to ourselves as sentient beings not knowing if we would be born as a farmer or a farm animal. Given what we know, which farm would be fairest, a family farm or a factory farm? In most cases, the former.

Continuing through the Provisional Principles, the *Reason Principle* would imply that we should reason our way toward rational choices about the foods we eat and the animals we use (for hunting, recreation, or pets), opting when possible for food products from companies that consider animal welfare and the environment, or hunting organizations such as Ducks Unlimited that aim for species conservation, or pets from rescue shelters instead of pet mills. Principle 8, the *Defend Others Principle*, applies not only to children, the mentally ill, the old, and the handicapped, but to domesticated animals as well, especially since we purposefully bred out of them their natural capacity for surviving on their own in the wild. Principle 9 is worded for inclusion of animals in our expansion of the moral sphere for sentient beings worthy of moral consideration, and Principle 10, *The Biophilia Principle*, would apply to all plants and animals, along with the air and water, in terms of creating a sustainable environment for untold generations of human and nonhuman animals still to come.

AN IDEA WHOSE TIME HAS COME

I am sensitive to the fact that to the ears of many readers this chapter echoes the schmaltzy sentiments of a film such as *Dances with Wolves*. But if we are going to take seriously the moral precept of the right of all sentient beings to survive and flourish, and if we are going to continue to push the arc of the moral universe toward justice and freedom and expand the moral sphere to include more and more sentient beings, then we should show the courage of our convictions by acting on our beliefs. When? Here I can relate to the prayer of St. Augustine of Hippo—the highly influential early Catholic theologian whose writings so shaped Western Christianity and philosophy—who when faced with the sexual passions of a young man that he knew he should resist cried out to the Lord, "Grant me chastity and continence, but not yet."[89]

To be frank, like so many people on this issue, my ideals are often at odds with my speciesism. Still, I am moved by the words of Victor Hugo when he wrote, "One can resist the invasion of armies; one cannot resist the invasion of ideas," usually rendered more poignantly, "Nothing is more powerful than an idea whose time has come."[90] I can think of no reason why *now* is not the time for sentient beings to be granted rights. But which sentient beings and which rights?

For wild species, this is nothing more than allowing them the freedom to hunt and forage for food in their natural environment and to protect them from poachers and the excessive encroachment of civilization into their habitats. For domesticated species, I believe this means ending factory farming and shifting toward family farming—giving domesticated animals a more humane version of the environment in which they survived and flourished for the past ten thousand years before the advent of factory farming in the twentieth century.

To further this moral progress, from the bottom up we can vote with our voices and our dollars for the types of food we want and push the market toward this more moral stance. And from the top down we can work for legislation that abolishes the exploitation of sentient beings, expanding our moral sphere to include the great apes and marine mammals and working our way across the many branches and twigs of the evolutionary

tree that Charles Darwin so eloquently described in the final sentence of
On the Origin of Species:

> There is grandeur in this view of life, with its several powers, having been
> originally breathed into a few forms or into one; and that, whilst this
> planet has gone cycling on according to the fixed law of gravity, from so
> simple a beginning endless forms most beautiful and most wonderful
> have been, and are being, evolved.[91]

PART III

The Moral Arc Amended

9

Moral Regress and Pathways to Evil

It is indeed probable that more harm and misery has been caused by men determined to use coercion to stamp out a moral evil than by men intent on doing evil.

—Friedrich Hayek, *The Constitution of Liberty*, 1960

In 2010, I worked on a *Dateline NBC* two-hour television special in which we replicated a number of now classic psychology experiments. In one experiment, an unsuspecting subject and a room full of confederates (actors who knew the objective of the study) were asked to fill out applications to participate in a television game show. The confederates dutifully filled out their forms, even as the room gradually filled with smoke—but, remarkably, most of the subjects continued to fill out their forms too. The majority of unsuspecting participants, who had every reason to believe that the building was actually on fire, continued their task, as if burning to death was not a particular problem for them. Everyone else was calm, so they were calm. As the subjects coughed and waved the smoke away, heads bent to their trivial task, the herd instinct became ever more blazingly obvious. You could almost hear the baaa-ing. But the most dramatic experiment we replicated was Yale University professor Stanley Milgram's famous shock experiments from the early 1960s on the nature of evil.

In a book on moral progress, it is necessary to address its obvious antithesis—moral regress—and identify which pathways lead to evil, to mitigate them.

SHOCK AND AWE: WILLINGNESS OR RELUCTANCE?

Shortly after the war crimes trial of Adolf Eichmann began in Jerusalem in July 1961, psychologist Stanley Milgram devised a set of experiments, the aim of which was to better understand the psychology behind obedience to authority. Eichmann had been one of the chief orchestrators of the Final Solution but, like his fellow Nazis at the Nuremberg trials, his defense was that he was innocent by virtue of the fact that he was only following orders. *Befehl ist Befehl*—orders are orders—is now known as the Nuremberg defense, and it's an excuse that seems particularly feeble in a case like Eichmann's. "My boss told me to kill millions of people, so—hey—what could I do?" is not a credible defense. But, Milgram wondered, was Eichmann unique in his willingness to comply with orders, no matter how atrocious? And just how far would ordinary people be willing to go?

Obviously Milgram could not have his experimental subjects gas or shoot people, so he chose electric shock as a legal nonlethal substitute. Looking for subjects to participate in what was billed as a "study of memory," Milgram advertised on the Yale campus and also in the surrounding New Haven community. He said he wanted "factory workers, city employees, laborers, barbers, businessmen, clerks, construction workers, sales people, telephone workers" and not just the usual guinea pigs of the science lab: undergraduates. Milgram then assigned his subjects to the role of "teacher" in what was purported to be research on the effects of punishment on learning. The protocol called for the subject to read a list of paired words to the "learner" (who was, in reality, a shill working for Milgram), then present the first word of each pair again, upon which the learner was to recall the second word. Each time that the learner was incorrect, the teacher was to deliver an electric shock from a box with toggle switches in 15-volt increments that ranged from 15 volts all the way to 450 volts, and featured such labels as *Slight Shock*, *Moderate Shock*, *Strong Shock*, *Very Strong Shock*, *Intense Shock*, *Extreme Intensity Shock*, and *DANGER: Severe Shock, XXX*.[1] Despite the predictions of forty psychiatrists whom Milgram surveyed before the experiment, who figured that only 1 percent of subjects would go all the way to the end, 65 percent of them completed the experiment, flipping that final toggle switch to deliver a shocking 450 volts, a phenomenon the social psychologist Philip Zimbardo characterizes as "the pornography of power."[2]

Who was most likely to go the distance in maximal shock delivery? Surprisingly—and counterintuitively—gender, age, occupation, and personality characteristics mattered little to the outcome. Similar levels of punishment were delivered by the young and the old, by males and females, and by blue-collar and white-collar workers alike. What mattered most was physical proximity and group pressure. The closer the learner was to the teacher, the less of a shock the latter delivered. And when Milgram added more confederates to encourage the teacher to administer ever more powerful shocks, most teachers complied; but when the confederates themselves rebelled against the authority figure's instructions, the teacher was equally disinclined to obey. Nevertheless, 100 percent of Milgram's subjects delivered at least a "strong shock" of 135 volts.[3]

In our 2010 replication in a New York City studio, we tested six subjects who believed that they were auditioning for a new reality show called *What a Pain!* We followed Milgram's protocols and had our subjects read a list of paired words to a "learner" (an actor named Tyler), then present the first word of each pair again. When Tyler gave a prearranged incorrect answer, our subjects were instructed by an authority figure (an actor named Jeremy) to deliver an electric shock from a box modeled after Milgram's contraption.[4]

Milgram characterized his experiments as testing "obedience to authority," and most interpretations over the decades have focused on subjects' unquestioning adherence to an authority's commands. What I saw in our subjects, as you can see in Milgram's subjects in old film footage on YouTube, was great reluctance and disquietude nearly every step of the way. Our first subject, Emily, quit the moment she was told the protocol. "This isn't really my thing," she said with nervous laughter. When our second subject, Julie, got to 75 volts (having flipped five switches), she heard Tyler groan. "I don't think I want to keep doing this," she said.

Jeremy pressed the case: "Please continue."

"No, I'm sorry," Julie protested. "I don't think I want to."

"It's absolutely imperative that you continue," Jeremy insisted.

"It's imperative that I continue?" Julie replied in defiance. "I think that—I'm like, I'm okay with it. I think I'm good."

"You really have no other choice," Jeremy said in a firm voice. "I need you to continue until the end of the test."

Julie stood her ground: "No. I'm sorry. I can just see where this is going, and I just—I don't—I think I'm good. I think I'm good to go. I think I'm going to leave now."

At that point the show's host, Chris Hansen, entered the room to debrief her and introduce her to Tyler, and then Chris asked Julie what was going through her mind. "I didn't want to hurt Tyler," she said. "And then I just wanted to get out. And I'm mad that I let it even go five [toggle switches]. I'm sorry, Tyler."

Our third subject, Lateefah, started off enthusiastically enough, but as she made her way up the row of toggle switches, her facial expressions and body language made it clear that she was uncomfortable; she squirmed, gritted her teeth, and shook her fists with each toggled shock. At 120 volts she turned to look at Jeremy, seemingly seeking an out. "Please continue," he authoritatively instructed. At 165 volts, when Tyler screamed "Ah! Ah! Get me out of here! I refuse to go on! Let me out!" Lateefah pleaded with Jeremy. "Oh my gosh. I'm getting all . . . like . . . I can't . . ."; nevertheless, Jeremy pushed her politely, but firmly, to continue. At 180 volts, with Tyler screaming in agony, Lateefah couldn't take it anymore. She turned to Jeremy: "I know I'm not the one feeling the pain, but I hear him screaming and asking to get out, and it's almost like my instinct and gut is like, 'Stop,' because you're hurting somebody and you don't even know why you're hurting them outside of the fact that it's for a TV show." Jeremy icily commanded her to "please continue." As Lateefah reluctantly turned to the shock box, she silently mouthed, "Oh my God." At this point, as in Milgram's experiment, we instructed Tyler to go silent. No more screams. *Nothing.* As Lateefah moved into the 300-volt range it was obvious that she was greatly distressed, so Chris stepped in to stop the experiment, asking her if she was getting upset. "Yeah, my heart's beating really fast." Chris then asked, "What was it about Jeremy that convinced you that you should keep going here?" Lateefah gave us this glance into moral reasoning about the power of authority: "I didn't know what was going to happen to me if I stopped. He just—he had no emotion. I was afraid of him."[5]

Our fourth subject, a man named Aranit, unflinchingly cruised through the first set of toggle switches, pausing at 180 volts to apologize to Tyler after his audible protests of pain: "I'm going to hurt you and I'm really sorry." After a few more rungs up the shock ladder, accompanied by more agonizing pleas by Tyler to stop the proceedings, Aranit encouraged him, saying, "Come on. You can do this. We are almost through." Later, the punishments were peppered with positive affirmations. "Good." "Okay." After completing the experiment Chris asked, "Did it bother you to shock him?" Aranit admitted, "Oh, yeah, it did. Actually, it did. And especially when he wasn't answering anymore."

Two other subjects in our replication, a man and a woman, went all the way to 450 volts, giving us a final tally of five out of six who administered shocks, and three who went all the way to the end of maximal electrical evil. All of the subjects were debriefed and assured that no shocks had actually been delivered, and after lots of laughs and hugs and apologies, everyone departed, none the worse for wear.[6]

ALPINISTS OF EVIL

What are we to make of these results? In the 1960s—the heyday of belief in the blank slate[7] taken to mean human behavior is almost infinitely malleable—Milgram's data seemed to confirm the idea that degenerate acts are primarily the result of degenerate environments (Nazi Germany being, perhaps, the ultimate example). In other words, there are no bad apples, just bad barrels.

Milgram's interpretation of his data included what he called the "agentic state," which is "the condition a person is in when he sees himself as an agent for carrying out another person's wishes and they therefore no longer see themselves as responsible for their actions. Once this critical shift of viewpoint has occurred in the person, all of the essential features of obedience follow." Subjects who are told that they are playing a role in an experiment are stuck in a no-man's-land somewhere between authority figure, in the form of a white-lab-coated scientist, and stooge, in the form of a defenseless learner in another room. They undergo a mental shift from being moral agents in themselves who make their own decisions (that autonomous state) to the ambiguous and susceptible state of being an intermediary in a hierarchy and therefore prone to unqualified obedience (the agentic state).

Milgram believed that almost anyone put into this agentic state could be pulled into evil one step at a time—in this case 15 volts at a time—until they were so far down the path there was no turning back. "What is surprising is how far ordinary individuals will go in complying with the experimenter's instructions," Milgram recalled. "It is psychologically easy to ignore responsibility when one is only an intermediate link in a chain of evil action but is far from the final consequences of the action." This combination of a stepwise path, plus a self-assured authority figure who keeps the pressure on at every step, is the double whammy that makes evil of this nature so insidious. Milgram broke the process down into two stages: "First, there is a set of 'binding factors' that lock the subject into the situation. They include such factors as politeness on his part, his desire to uphold his initial promise of aid to the experimenter, and the awkwardness of withdrawal. Second, a number of adjustments in the subject's thinking occur that undermine his resolve to break with the authority. The adjustments help the subject maintain his relationship with the experimenter, while at the same time reducing the strain brought about by the experimental conflict."[8]

Put yourself into the mind of one of these subjects—either in Milgram's experiment or in our NBC replication. It's an experiment conducted by an established institution—a national university or a national network. It's for science—or it's for television. It's being run by a white-lab-coated scientist—or by a casting director. The authorities overseeing the experiment are either university professors or network executives. An agent—someone carrying out someone else's wishes under such conditions—would feel in no position to object. And why should she? It's for a good cause, after all—the advancement of science, or the development of a new and interesting television series. Out of context, if you ask people—even experts, as Milgram did—how many people would go all the way to 450 volts, they lowball the estimate by a considerable degree, as Milgram's psychiatrists did. As Milgram later reflected, "I am forever astonished that when lecturing on the obedience experiments in colleges across the country, I faced young men who were aghast at the behavior of experimental subjects and proclaimed they would never behave in such a way, but who, in a matter of months, were brought into the military and performed without compunction actions that made shocking the victim seem pallid."[9]

In the sociobiological and evolutionary psychology revolutions of the 1980s and 1990s, the interpretation of Milgram's results shifted toward the nature/biological end of the spectrum from its previous emphasis on nurture/environment. The interpretation softened somewhat as the multidimensional nature of human behavior was taken into account. As it is with most human action, moral behavior is incredibly complex and includes an array of causal factors, obedience to authority being just one among many. The shock experiments didn't actually reveal just how primed all of us are to inflict violence for the flimsiest of excuses; that is, it isn't a simple case of bad apples looking for a bad barrel in order to cut loose. Rather the experiments demonstrate that all of us have conflicting moral tendencies that lie deep within.

Our moral nature includes a propensity to be sympathetic, kind, and good to our fellow kith and kin and friends, as well as an inclination to be xenophobic, cruel, and evil to tribal others. And the dials for all of these can be adjusted up and down depending on a wide range of conditions and circumstances, perceptions, and states of mind, all interacting in a complex suite of variables that are difficult to tease apart. In point of fact, most of the 65 percent of Milgram's subjects who went all the way to 450 volts did so with great anxiety, as did the subjects in our NBC replication. And it's good to remember that 35 percent of Milgram's subjects were exemplars of the *disobedience* to authority—they quit in defiance of what the authority figure told them to do. In fact, in a 2008 partial replication by the social psychologist Jerry Burger, in which he ran the voltage box up to only 150 volts (the point at which the "learner" in Milgram's original experiment began to cry out in pain), it was found that twice as many subjects refused to obey the authority figure. Assuming these subjects were not already familiar with the experimental protocols, the findings are an additional indicator of moral progress from the 1960s to the 2000s, caused, I would argue, by that ever-expanding moral sphere and our collective capacity to take the perspective of another, in this case the to-be-shocked learner.[10]

Milgram's model comes dangerously close to suggesting that subjects are really just puppets devoid of free will, which effectively lets Nazi bureaucrats off the hook as mere agentic automatons in an extermination engine run by the great paper-pushing administrator Adolf Eichmann (whose actions as an unremarkable man in a morally bankrupt and conformist environment were famously described by Hannah Arendt as "the banality of evil"). The obvious problem with this model is that there can be no moral accountability if an individual is truly nothing more than a mindless zombie whose every action is controlled by some nefarious mastermind. Reading the transcript of Eichmann's trial is mind-numbing (it goes on for thousands of pages), as he both obfuscates his real role while shifting the blame entirely to his overseers, as in this statement:

> What I said to myself was this: The head of State has ordered it, and those exercising judicial authority over me are now transmitting it. I escaped into other areas and looked for a cover for myself which gave me some peace of mind at least, and so in this way I was able to shift—no, that is not the right term—to attach this whole thing one hundred percent to those in judicial authority who happened to be my superiors, to the head of State—since they gave the orders. So, deep down, I did not consider myself responsible and I felt free of guilt. I was greatly relieved that I had nothing to do with the actual physical extermination.[11]

The last statement might possibly be true—given how many battle-hardened SS Nazis were initially sickened at the site of a killing action and Eichmann avoided them—but the rest is pure spin-doctored malarkey, and Arendt allowed herself to be taken in by it more than reason would allow, as the historian David Cesarani shows in his revealing biography *Becoming Eichmann* and as recounted in Margarethe von Trotta's moving film *Hanna Arendt*.[12] The evidence of Eichmann's real role in the Holocaust was plain for all to see at the time, as dramatically reenacted in Robert Young's 2010 biopic titled simply *Eichmann*, based on the transcripts of the interrogation of and confession by Eichmann just before his trial, conducted by the young Israeli police officer Avner Less, whose father was murdered in Auschwitz.[13] Time and again, throughout hundreds of recorded hours, Less queries Eichmann about transports of Jews and Gypsies sent to their death, all followed by denials and lapses of memory. Less then presses the point by showing Eichmann copies of transport documents with his signature at the bottom, leading Eichmann to say in an exasperated voice, "What's your point?"

The point is that there is a mountain of evidence proving that Eichmann—like all the rest of the Nazi leadership—were not simply following orders. As Eichmann himself boasted when he wasn't on trial, "When I reached the conclusion that it was necessary to do to the Jews what we did, I worked with the fanaticism a man can expect from himself. No doubt they considered me the right man in the right place. . . . I always acted 100 percent, and in giving of order I certainly was not lukewarm." As the genocide historian Daniel Jonah Goldhagen asks rhetorically, "Are these the words of a bureaucrat mindlessly, unreflectively doing his job about which he has no particular view?"[14]

The historian Yaacov Lozowick characterized the motives in his book *Hitler's Bureaucrats*, in which he invokes a mountain-climbing metaphor: "Just as a man does not reach the peak of Mount Everest by accident, so Eichmann and his ilk did not come to murder Jews by accident or in a fit of absent-mindedness, nor by blindly obeying orders or by being small cogs in a big machine. They worked hard, thought hard, took the lead over many years. They were the alpinists of evil."[15]

WHAT'S IT LIKE TO BE A NAZI?

To understand the psychology of immorality we must ascend into the thin air of evil, and there is arguably no more poignant example than that of the Nazis. Throughout this book I have been emphasizing the importance of perspective-taking—walking a mile in someone else's shoes and trying to feel what they feel. To truly understand what it takes to create a Nazi—that is, how to take a nation of intelligent, educated, cultured people like the Germans and turn them into swastika-donning, jackboot-wearing, goose-step-marching, Heil Hitler–swearing participants in a political regime—we must imagine what it was like to actually be one.

The fact is that most of the Nazi leaders were intelligent, learned, and highly cultured people who were quite capable of committing mass murder during the workday while acting like loving and doting family men after hours. Even Joseph Mengele—the Auschwitz doctor who played god on the train platform that cleaved the barracks from the crematoria—was described by one prisoner as "capable of being so kind to the children, to have them become fond of him, to bring them sugar, to think of small details in their daily lives, and to do things we would genuinely admire . . . And then, next to that, . . . the crematoria smoke, and these children, tomorrow or in a half hour, he is going to send them. Well, that is where the anomaly lay."[16] In his classic study *The Nazi Doctors*, the psychologist Robert Jay Lifton described Mengele's apparent transformation from savage to salesman in Brazil, where he'd escaped after the war and evaded capture for thirty-four years until his death in 1979. When his remains were discovered and identified in 1985, Lifton says that many Auschwitz survivors "refused to believe that the remains in the Brazilian grave were Mengele's. Soon after that identification, a twin whom Mengele had studied told me that she simply did not believe that the arrogant, overbearing figure she had known in Auschwitz could have undergone a 'change in personality' and become the frightened hermit in Brazil. She was saying, in effect, that she and the others had not been provided with a psychological experience of that 'metamorphosis' from evil deity to evil human being." Even at Auschwitz, Lifton notes, Mengele was not the monolithic demon of film and fiction:

Mengele's many-sidedness in Auschwitz was both part of his legend and a source of his desacralization. In the camp he could be a visionary ideologue, an efficiently murderous functionary, a "scientist" and even a "professor," an innovator in several areas, a diligent careerist (like Dorf), and, above all, a physician who became a murderer. He reveals himself as a man and not a demon, a man whose multifaceted harmony with Auschwitz can give us insight into—and makes us pause before—the human capacity to convert healing into killing.[17]

The complexity of human moral psychology was well captured by Primo Levi in *The Drowned and the Saved*: "Occurrences like this astonish because they conflict with the image we have of a man in harmony with himself, coherent, monolithic; and they should not astonish because that is not how man is. Compassion and brutality can coexist in the same individual and in the same moment, despite all logic."[18]

Considering evil from both the perpetrator's and the victim's perspectives is what the social psychologist Roy Baumeister did in his groundbreaking book *Evil: Inside Human Violence and Cruelty*. Baumeister sketches a triangle of evil involving three parties: perpetrators, victims, and bystanders. "The essential shock of banality is the disproportion between the person and the crime," he writes. "The mind reels with the enormity of what this person has done, and so the mind expects to reel with the force of the perpetrator's presence and personality. When it does not, it is surprised."[19] The explanation for the surprise can be found by contrasting the victim's perspective with that of the perpetrator. Steven Pinker calls Baumeister's distinction the "moralization gap," and it can be instructive—even shocking—to stand on both sides of the gap and look into the dark abyss below.[20] On either side of the moralization gap are two narratives, one representing the victim and the other the perpetrator. In a 1990 paper by Baumeister and his colleagues, instructively titled "Victim and Perpetrator Accounts of Interpersonal Conflict," the authors describe the dual narratives (shortened and summarized by Pinker).[21] First, the victim's narrative:

> The perpetrator's actions were incoherent, senseless, incomprehensible. Either that or he was an abnormal sadist, motivated only by a desire to see me suffer, though I was completely innocent. The harm he did is grievous and irreparable, with effects that will last forever. None of us should ever forget it.

Now the perpetrator's narrative:

> At the time I had good reasons for doing it. Perhaps I was responding to an immediate provocation. Or I was just reacting to the situation in a way that any reasonable person would. I had a perfect right to do what I did, and it's unfair to blame me for it. The harm was minor, and easily repaired, and I apologized. It's time to get over it, put it behind us, let bygones be bygones.

To obtain a thorough understanding of moral psychology, it is necessary to take both perspectives into account, even though our natural propensity is to side with the victim and moralize against the perpetrator. If evil is explicable—and I believe that it is, or at least can be—then remaining detached is imperative; as the scientist Lewis Fry Richardson put it in his statistical study of war: "For indignation is so easy and satisfying a mood that it is apt to prevent one from attending to any facts that oppose it. If the reader should object that I have abandoned ethics for the false doctrine that 'to understand all is to forgive all,' I can reply that it is only a temporary suspense of ethical judgment, made because 'to condemn much is to understand little.'"[22]

The leaders of National Socialism did not see themselves as the demonic "Nazis" of Hollywood films. As the Holocaust historian Dan McMillan cautions, "Demonizing the Germans is unworthy of us because it denies both their humanity and ours."[23] Nazis were flesh-and-blood people who fully believed that their actions were justified in terms of their allegedly virtuous goals: national renewal, lebensraum (living space), and especially racial purity. For example, in the many writings and rantings of the Reich minister of propaganda, Joseph Goebbels, one can hear the moralizing cry characteristic of so many perpetrators: the victims had it coming. In a diary entry dated August 8, 1941, regarding the spread of spotted typhus in the Warsaw ghetto, Goebbels commented, "The Jews have always been the carriers of infectious diseases. They should either be concentrated in a ghetto and left to themselves or be liquidated, for otherwise they will infect the populations of the civilized nations." Eleven days later, on August 19, after a visit to Hitler's headquarters, Goebbels penned this entry in his diary: "The Führer is convinced his prophecy in the Reichstag is becoming a fact: that should Jewry succeed in again provoking a new war, this would end with their annihilation. It is coming true in these weeks and months with a certainty that appears almost sinister. In the East the Jews are paying the price, in Germany they have already paid in part and they will have to pay more in the future."[24]

Of the thousands of Nazi documents I have reviewed, however, there is none as chilling as a speech by Reichsführer Heinrich Himmler given on October 4, 1943, to the SS Gruppenführer in the city of Poznan (in Poland), which was recorded on a red oxide magnetic tape. (The speech is accessible on YouTube with a transcription that includes a German-English translation.[25]) Himmler lectured from notes and spoke for a stupefying three hours and ten minutes on a range of subjects, including the military and political situation, the Slavic peoples and racial blends, German racial superiority, and the like. Two hours into the speech Himmler began to talk about "the extermination of the Jewish people." He compared this action with the June 30, 1934, blood purges against traitors in the Nazi Party (the "night of the long knives" during which Nazis killed one another in grabs for power and the settling of old scores), then talked about how difficult it is to remain an honorable man in the midst of human slaughter, insisting that this episode in the forthcoming thousand-year Reich was a necessary part of a glorious history. Instead of hearing the voice of pure evil, listen to the sound of fervent righteousness:

> Most of you will know what it means when a hundred bodies lie together, when there are five hundred, or when there are a thousand. And to have seen this through, and—with the exception of human weaknesses—to have remained decent, has made us hard and is a page of glory never mentioned and never to be mentioned. Because we know how difficult things would be, if today in every city during the bomb attacks, the burdens of war, and the privations, we still had Jews as secret saboteurs, agitators, and instigators.... We have the moral right, we had the duty to our people to do it, to kill this people who wanted to kill us.... I will never see it happen that even one bit of putrefaction comes in contact with us, or takes root in us. On the contrary, where it might try to take root, we will burn it out together. But altogether we can say: we have carried out this most difficult task for the love of our people.[26]

The twisted logic of the genocidal, righteous mind works something like this: define yourself as a good person but ramp it up out of all proportion by declaring yourself part of a master race. Define the master race as the pinnacle of perfection. Now do whatever you want. Amazingly enough, you will find that you take on no defect in either your soul or your character, even if you murder a few million people. By virtue of the fact that you've defined yourself as good, you can do no wrong. It worked like a charm for Himmler and his ilk.

Finally, we can crack open the mind of the ultimate perpetrator, Adolf Hitler, and see that here too the moralizing gap between victim and perpetrator is enormous, with the Führer providing many documented instances of his justification for the extermination of the Jews. As early as April 12, 1922, in a Munich speech later published in the Nazi Party newspaper *Völkischer Beobachter*, Hitler told his audience: "The Jew is the ferment of the decomposition of people. This means that it is in the nature of the Jew to destroy, and he must destroy, because he lacks altogether any idea of working for the common good. He possesses certain characteristics given to him by nature and he never can rid himself of those characteristics. The Jew is harmful to us."[27] On February 13, 1945, in the final Götterdämmerung of the war, with the world crashing in around him in his Berlin bunker, Hitler declared with menacing pride: "Against the Jews I fought open-eyed and in view of the whole world. At the beginning of the war I sent them a final warning. I did not leave them in ignorance that, should they once again manage to drag the world into war, they would this time not be spared—I made it plain that they, this parasitic vermin in Europe, will be finally exterminated."[28] Even as he faced his own suicide on April 29, 1945, at 4:00 a.m., Hitler commanded his successors in his political testament to carry on the fight against the Jews: "Above all I charge the leaders of the nation and those under them to scrupulous observance of the laws of race and to merciless opposition to the universal poisoner of all peoples, International Jewry."[29]

In these examples, by employing the *principle of interchangeable perspectives* and taking the point of view of the Nazi perpetrators, we are witness to an example of a *moral judgment that is based on a factual error*. The National Socialists perpetrated facts about Jews—most of which were extant in the culture from centuries of European-wide anti-Semitism—that were simply not true. The Jews were not secret saboteurs, agitators, and instigators, as Himmler claimed. The Jews were not responsible for World War I, as Hitler maintained. The Jews were not, in fact, a biologically distinct race, nor were they intent on overrunning the country like plague rats, as eugenics ideologues theorized. These were all tragic misconceptions— factual errors that, had they been checked against reality, would have come up short. Nevertheless, they were sincerely believed; thus extermination had a kind of inescapable internal logic to it, however grotesque.

This is not to minimize the emotional undercurrent of mindless bigotry, but merely to suggest that if you sincerely (however wrongly) believe that X is responsible for the ruination of all and everything that you hold dear, stamping out X follows like night follows day. It was only natural, for example, that people in the Middle Ages burned the witches who were causing diseases, disasters, and assorted other misfortunes due to their well-known practice of cavorting with demons. Of course, women *weren't* cavorting with demons—given that demons don't actually exist; thus the disasters befalling people clearly did not originate with witches conspiring with the devil. The whole idea is preposterous but, like so many human errors, it was based on a faulty understanding of cause and effect. Likewise, the ultimate cause of anti-Semitism was (and remains) an utterly mistaken set of beliefs about Jews; thus the long-term solution to anti-Semitism is a better understanding of reality (while the short-term solution is legislation against discrimination). This is where science and reason come into the picture and why I argue that many of our moral mistakes are errors of fact based on defective thinking and on the erroneous assumptions that we make about other people. Thus one solution is to be found in a scientifically rational understanding of causality.

THE GRADUAL PATHWAY TO EVIL: FROM EUTHANASIA TO EXTERMINATION

The journey to evil is made in small steps, not giant leaps. Evil begins at 15 volts, not 450 volts. No single step embodies evil, but the farther down the path you go, the harder it is to turn back.

Long before prisoners were herded into gas chambers and killed with Zyklon-B or carbon monoxide gas, the Nazis had developed a program for the systematic and secret liquidation of certain targeted peoples, including German citizens. It began with the sterilization programs of the early 1930s, evolved into the euthanasia programs of the late 1930s, and with this expertise under their belts, the Nazis were able to implement their program of mass murder in the extermination camps from 1941 to 1945. As disconcerting as it is to even imagine contemplating the gassing of masses of prisoners in a chamber, as Milgram showed it is possible, and sometimes easy, to get people to do almost anything when the steps leading up to it are small and incremental. After murdering tens of thousands of "inferior" Germans, the idea of annihilating the entirety of the Jewish population was no longer unimaginable. After you've gotten used to demonizing, excluding, expelling, sterilizing, deporting, beating, torturing, and euthanizing people, the next step to genocide isn't that big a leap.

Sterilization laws were passed in Germany in late 1933, not long after Hitler came to power. Within a year, 32,268 people were sterilized. In 1935, the figure jumped to 73,174; official reasons given included feeble-mindedness, schizophrenia, epilepsy, manic-depressive psychosis, alcoholism, deafness, blindness, and physical malformations. So-called sex offenders were simply castrated—no fewer than 2,300 in the first decade of the program.

In 1935, Hitler told the leading Reich physician, Gerhard Wagner, that when the war began he wanted to make the shift from sterilization to euthanasia. True to his word, in the fall of 1939, the Führer ordered the extermination of physically handicapped children, after which the program moved on to mentally handicapped children, and soon thereafter proceeded to targeting adults who had either handicap. The murders were initially committed through large doses of "normal" medication given in tablet or liquid form so as to look like an accident (because the families of the victims who were notified of the death might start asking questions if they suspected foul play). If the patients resisted, injections were used. When the numbers chosen for death became cumbersomely large, the operations had to be moved into special killing wards instead of isolated units.

The process became so extensive that the Germans had to expand their operation by taking over an office complex set up at a stolen Jewish villa in Berlin, with the address Tiergartenstrasse 4; thus the program became known internally as Operation T4, or just T4. Officially it was called the "Reich Work Group of Sanatoriums and Nursing Homes." How quaint. T4 doctors arbitrarily decided who would live and who would die, with economic status being one of the common criteria (among others): those unable to work or only able to perform "routine" work could be put to death. Historians estimate that approximately 5,000 children and 70,000 adults were murdered in the euthanasia program prior to August 1941.

As the numbers increased so too did the complications of homicide on such a colossal scale. Mass murder is more efficient with a mass murder technology, and the Nazi leader determined that medication and injections were just not sufficient to meet the objective of genocide on an industrial scale. The T4 physicians hit upon a solution when they heard stories about accidental deaths and suicides caused by the exhaust of automobile engines or the gas from leaking stoves. According to Dr. Karl Brandt, he and Hitler discussed the various techniques and decided upon gas as "the more humane way" of eliminating those deemed unfit for the Reich. The T4 administrators set up six killing centers. The first was established at an old jail building in the city of Brandenburg. Sometime between December 1939 and January 1940, a two-day series of gassing experiments was conducted and determined to be successful. Thereafter, five more killing centers were established. The gas chambers were disguised as showers—including fake showerheads—into which the "handicapped" patients were herded and the gas administered. One observer, Maximilian Friedrich Lindner, recalled the process at Hadamar:

> Did I ever watch a gassing? Dear God, unfortunately, yes. And it was all due to my curiosity. . . . Downstairs on the left was a short pathway, and there I looked through the window. . . . In the chamber there were patients, naked people, some semi-collapsed, others with their mouths terribly wide open, their chests heaving. I saw that, I have never seen anything more gruesome. I turned away, went up the steps, upstairs was a toilet. I vomited everything I had eaten. This pursued me days on end. . . .[30]

But not endlessly. As Himmler noted in his Poznan speech, and as research on the psychology of killing and sadism reveals, it takes some "getting used to" the process of killing another human being; but gradually, through habituation, the mind becomes desensitized to even the most horrifying of experiences and may be reconciled to atrocity.

The gas was ventilated from the chamber with fans, the bodies were disentangled and removed from the room, the corpses marked with an "X" on their backs were looted for gold in their teeth, and then they were cremated. The entire process—from arrival at the killing center to cremation—took less than twenty-four hours, not unlike what was soon implemented in the larger camps in the East. Henry Friedlander, who traced this stepwise evolutionary process, concluded, "The success of the euthanasia policy convinced the Nazi leadership that mass murder was technically feasible, that ordinary men and women were willing to kill large numbers of innocent human beings, and that the bureaucracy would cooperate in such an unprecedented enterprise."[31]

In the T4 killing centers we see all of the components of the extermination camps to come later, such as those at Majdanek and Auschwitz-Birkenau. Over time, the Nazi bureaucracy evolved along with the killing centers, setting the stage for the conversion of concentration and work camps into extermination camps, all in incremental steps in the gradually evolving system that became the Final Solution.[32]

The gradual escalation to evil is just one of several psychological factors at work in the corruption of good people. Let's return to the shock experiments and consider the reason why subjects will both obey an unethical order from an authority figure and, at the same time, feel terrible about the suffering they're inflicting (unless, over time, they habituate to it). These findings, I believe, reflect our complex moral nature that leads us to vacillate between assessing others and their actions as positive or negative, helpful or harmful (all of which fall under the umbrella terms "good" and "evil"), depending on the context and desired outcomes.

THE PSYCHOLOGY OF GOOD AND EVIL

In the context of moral conflicts, the experimental psychologist Douglas J. Navarick—my mentor at California State University at Fullerton, as it turns out—calls this vacillation "moral ambivalence," in which "When we assess the moral implications of an action that we observe or contemplate, we may have a sense of ambivalence, a feeling that one could justifiably judge the action as either right or wrong. Efforts to resolve such ambivalence are potentially difficult, protracted, and aversive."[33] As such, our moral emotions can slide back and forth between right and wrong that can be modeled as an approach-avoidance conflict.

The approach-avoidance paradigm began with rats, who were motivated to seek a food reward in a maze by being starved to 80 percent of their body weight, but when they reached the goal region they were not only rewarded with food but also punished with a mild shock.[34] This paradigm sets up an approach-avoidance conflict in which the rats become ambivalent about reaching the end of the runway where both a reward and a punishment await them, so they end up vacillating, first toward, then away from the goal. When harnessed to a device to measure their pull strength either toward or away from the goal, psychologists have been able to take a quantitative reading on precisely how strong or weak their ambivalence was. Revealingly, as the rats got closer to the goal box the strength of both approach and avoidance tendencies increased, although the avoidance behavior was stronger than that of the approach.

Moral conflicts may also arise between *prescriptions* (what we ought to do) that bring rewards for action (pride from within, praise from without) and *proscriptions* (what we ought not to do) that bring punishments for violations (shame from within, shunning from without).[35] (Eight of the Ten Commandments in the Decalogue, for example, are proscriptions.) As in the limbic system with its neural networks for emotions, approach-avoidance moral conflicts have neural circuitry called the behavioral activation system (BAS) and the behavioral inhibition system (BIS), which drive an organism forward or back,[36] as in the case of the rat vacillating between approaching and avoiding the goal region, or in the case of the man on the train platform vacillating between saving the woman and punishing her offender in the video vignette from chapter 1. These activation and inhibition systems can be measured in experimental settings in which subjects are presented with different scenarios in which they then offer their moral judgment (giving money to a homeless person as prescriptive vs. wearing a revealing dress to a funeral as proscriptive); under such conditions researchers have found that "BAS scores correlated with prescriptive ratings but not with the proscriptive ones, whereas the BIS scores correlated with the proscriptive ratings but not with the prescriptive ones."[37] This, argues Navarick, shows how some moral judgments can be better understood as an approach-avoidance conflict.

Other moral emotions, such as disgust, drive an organism away from a noxious stimulus because noxiousness is an informational cue that a stimulus could kill you through poisoning (e.g., food substances) or disease (e.g., fecal matter, vomit, or other bodily effluvia). Anger, however, has the opposite effect and drives an organism toward an offensive stimulus, such as another organism that attacks it. Thus, if you are taught by your culture that Jews (or blacks, natives, homosexuals, Tutsis, etc.) are a bacillus poisoning your nation, you naturally avoid them with disgust as you would any noxious stimulus. If you learn from your society that Jews (or blacks, natives, homosexuals, Tutsis, etc.) are dangerous enemies attacking your nation, you naturally approach them with anger as you would any assaulter. And, as in these examples, this system can be hijacked into getting one group of people to believe that another group of people is evil and dangerous and therefore needs to be punished or destroyed, through such techniques as state propaganda, literature and mass media, gossip and hearsay, and other means of communicating information. False information naturally leads to mistaken belief, and again we see how factual error morphs into moral judgment in accord with Voltaire's linkage between absurdities and atrocities.

Fortunately, this moral system is tractable in the other direction as well. Consider the Germans. Once considered to be an inherently racist, bigoted, and bellicose people, they are now among the most tolerant, liberal, and peaceful in the world.[38] It took only a few years for the de-Nazification process implemented by the Allies after World War II to drive the beliefs of National Socialism to the margins of society. Today you might find neo-Nazi skinhead kooks dressing up in ersatz SS uniforms in their bedrooms, and fantasizing about goose-stepping to the Führer, but there is little chance that anything like the Holocaust could ever happen again in Germany. It is a testimony to the malleability of our moral emotions.

Moral approach-avoidance conflicts can be seen in such classic dilemmas as pitting a *deontological* (duty-bound) principle such as the prohibition against murder against a *utilitarian* (greatest good) principle such as the trolley experiment where most people agree that it is acceptable to sacrifice one person to save five. Which is right? Thou shalt not kill, or thou shalt kill one to save five? Such conflicts cause much cognitive dissonance and anxiety—and vacillation—and are popular in fiction as a vehicle to explore the complexities of moral choices. In Arthur C. Clarke's science fiction novel (and Stanley Kubrick's film) *2001: A Space Odyssey*, the HAL 9000 computer is unable to resolve a conflict between its duty (via programmed orders) for "the accurate processing of information without distortion or concealment" and its command to keep the true nature of the space mission—knowledge of the alien Monolith discovered on the moon—a secret from the crew. This sets up a "Hofstadter-Moebius loop" (a nod to Douglas Hofstadter's work on unsolvable mathematical problems and the Moebius infinity loop) that leads HAL to kill the crew, thereby enabling him to be consistent in obeying his orders to be both truthful and maintain secrecy about the mission (although in the end the astronaut Dave Bowman survives and dismantles HAL in one of the classic scenes in science fiction film history). In the final episode of the television series *M*A*S*H*—the most watched event in US television history—Captain Hawkeye Pierce (Alan Alda) experiences a nervous breakdown after witnessing a South Korean refugee smother her crying baby on a bus so as not to alert North Korean soldiers of their presence, a situation that would inevitably have led to the death of everyone in the vehicle. As Hawkeye explains in a letter to his father, "Remember when I was a kid, you told me that if my head wasn't attached to my shoulders, I'd lose it? That's what happened when I saw that woman kill her baby. A baby, Dad. A baby."

Perpetrators of the Holocaust faced this moral conflict (among others): between the natural inclination most humans have against hurting or killing another human being, versus duty and loyalty to one's nation and obedience to one's superiors. Demonstrating that Jews (and others) were not the enemies of Germany and that Nazi racial policy was based on the pseudoscience of eugenics might have helped to resolve the problem, but in the minds of the Holocaust perpetrators who believed such nonsense, these moral conflicts emerged regardless.

Dramatic examples can be found in a remarkable collection of wartime letters titled *"The Good Old Days": The Holocaust as Seen by Its Perpetrators and Bystanders*. In one letter, for example, dated Sunday, September 27, 1942, SS-Obersturmführer (Lieutenant Colonel) Karl Kretschmer apologizes to his wife, his "dear Soska," for not writing more, and explains that he is feeling ill and in low spirits. "I'd like to be with you all. What you see here makes you either brutal or sentimental." His "gloomy mood," he explains, is caused by "the sight of the dead (including women and children)." His moral conflict is resolved by coming to believe that the Jews deserved to die: "As the war is in our opinion a Jewish war, the Jews are the first to feel it. Here in Russia, wherever the German soldier is, no Jew remains. You can imagine that at first I needed some time to get to grips with this." In a subsequent letter, not dated, Kretschmer explains to his wife how he did come to grips with the conflict: "there is no room for pity of any kind. You women and children back home could not expect any mercy or pity if the enemy got the upper hand. For that reason we are mopping up where necessary but otherwise the Russians are willing, simple and obedient. There are no Jews here any more." Finally, on October 19, 1942, Kretschmer shows how easy it is to slip into the evil of moral banality (referencing the Einsatzgruppen, or special action forces of which he was a part, which were assigned to follow the German army into towns to rid them entirely of any unwanted people, including and especially Jews):

> If it weren't for the stupid thoughts about what we are doing in this country, the Einsatz here would be wonderful, since it has put me in a position where I can support you all very well. Since, as I already wrote to you, I consider the last Einsatz to be justified and indeed approve of the consequences it had, the phrase: "stupid thoughts" is not strictly accurate. Rather it is a weakness not to be able to stand the sight of dead people; the best way of overcoming it is to do it more often. Then it becomes a habit.[39]

His "stupid thoughts" reflect his moral conflict, which he overcame through a combination of convincing himself that the Einsatz killings were necessary (because if the tables had been turned they'd be doing it to us) and turning murder into a "habit" to overcome the emotional trauma of brutality.

Habit is an appropriate descriptor because *habituation* is a psychological state in which one becomes oblivious to a continually repeated stimulus. On the simplest perceptual level, one may cease to notice a continuous stimulus such as the pressure of a ring or bracelet. In learning experiments, organisms cease to respond to a stimulus that has no consequences or relevance for them, such as a repeated loud noise associated with nothing. In animal behavior research on primates in the wild, scientists habituate their subjects by repeatedly exposing them to the presence of humans so that they no longer notice them standing around with their binoculars and video recorders.[40] The habituation effect happens at the neural level as well as psychological—in fMRI scans in which people are continuously exposed to the same stimulus, the areas of the brain that normally respond to such stimuli decrease their rates of firing, or cease firing altogether.[41] In the ranks of the Nazi Waffen-SS, many soldiers in this elite fighting squad simply habituated to their job of killing after years of bitter engagement on the Eastern Front. Gerhard Stiller, for example, who fought with the "Leibstandarte Adolf Hitler," or the 1st SS Panzer Division, which initially served as the Führer's personal bodyguard, after the war recalled of his fellow SS soldiers, "After a few years they became so desensitized that they didn't even notice any more, given that they were capable of just bumping someone off without batting an eyelid. Let's just say they would need to develop a lot of their humanity again, and that takes time."[42]

Navarick sites as an example of moral conflict the Józefów massacre in Poland in which a Nazi reserve police battalion rounded up and shot fifteen hundred Jewish civilians in the head, most of whom were women and children.[43] According to the Holocaust historian Christopher Browning, in his frank book on the massacre, *Ordinary Men*, 10 to 20 percent of the Nazi reservists withdrew from the killing operation after one shot, and most of the rest experienced physical revulsion during the murderous process. Their conflict was not on an intellectual level—like a trolley problem given to undergraduates—but instead was more visceral. As one reservist explained, "Truthfully I must say that at the time we didn't reflect about it at all. Only years later did any of us become truly conscious of what had happened then. . . . Only later did it first occur to me that [it] had not been right."[44] Their initial conflict was more likely a reaction coming from a deeper evolved emotion of revulsion to killing that most of us are born with, unless and until special circumstances are instituted to override that natural propensity.

What are those special circumstances? Navarick notes the similarities between the subjects who withdrew from Milgram's experiments and the reservists who withdrew from the Józefów massacre, by employing the paradigm of operant conditioning rather than social psychological models. In this analysis he proposes a three-stage behavioral model to explain disobedience: (1) aversive conditioning of contextual stimuli (how negative the conditions were in a given situation), (2) emergence of a decision point (the moment at which someone could extricate themselves from a negative situation without severe consequences), and (3) a choice between immediate and delayed reinforcers (at which point they're reinforced for withdrawing or disobeying). Together these three conditions show that "participants withdraw to escape personal distress rather than to help the victim."[45] In other words, says Navarick, people in situations such as Milgram's shock experiment, or Nazi reservists involved in a killing operation, who withdrew or refused to participate did so not due to the positive reinforcement of helping the victim but due to the negative reinforcement of terminating the discomfort.[46]

Instead of reifying some internal state like "obedience" into a psychological force, Navarick explains people's behavior in terms of positive and negative reinforcements and analyzes the extent to which they will act to increase the former and decrease the latter. At Józefów, for example, the Nazi reservists shot their victims at point-blank range with the barrel of their rifle pressed against the base of the victim's skull. This resulted in an unacceptable (to the Nazi commanders) level of withdrawal and defiance. At a later killing engagement, in the village of Lomazy, Poland, the commanders arranged for their reservist charges to shoot the Jewish victims at a distance, and this predictably led to lower rates of noncompliance, undoubtedly because the men shanghaied into pulling the trigger were spared the emotional devastation of close-range shooting. One is reminded of the scene in *The Godfather* when Sonny Corleone tells his younger brother Michael—who is eager to avenge the shooting of his father (the Godfather) and his own beating by a corrupt police captain whom he wants to kill: "What are you gonna do? Nice college boy, didn't want to get mixed up in the family business. Now you want to gun down a police captain. Why? Because he slapped you in the face a little? What do you think, this is like the army, where you can shoot 'em from a mile away? No, you gotta get up like this and, badda-bing, you blow their brains all over your nice Ivy League suit."

Blowing someone's brains out all over your clothing—whether it is a nice Ivy League suit or a pressed Nazi uniform—is unnatural, repugnant, and for all but a handful of sadists and extreme psychopaths physically revolting. Browning summarized the most common reason for the men's withdrawal from the killing at Józefów as "sheer physical revulsion." It is unquestionably a negative form of stimulus, but also one that can be overcome—otherwise the Holocaust itself would never have happened. And sometimes that stimulus tipped over into sadistic pleasure. In a disturbing book titled *Male Fantasies*, Klaus Theweleit recorded an incident in which a Nazi camp commandant more than overcame any revulsion he might have once experienced in response to the flogging of a camp inmate: "His whole face was already red with lascivious excitement. His hands were plunged deep in his trouser pockets, and it was quite clear that he was masturbating throughout, apparently unembarrassed by the watching crowd. Having 'finished himself off' and reached satisfaction, he turned on his heel and disappeared; perverse swine that he was, he lost interest at this point in the further development of the proceedings." In fact, this witness added, "on more than thirty occasions, I myself have witnessed SS camp commanders masturbating during floggings."[47]

What circumstances and conditions tweak the dials of good and bad acts either up or down in normal people? Consider this explanation from Alfred Spiess, the chief senior state prosecutor in the trial of some of the SS guards at the Treblinka death camp, who explained the psychology of evil this way:

> On the one hand obviously there was the order, and also a certain willingness not to refuse the order [duty]. But this readiness was naturally promoted psychologically in that these people were given privileges. Let me put it this way, a lot of carrot and a little stick—that was more or less how the system worked. And their carrot consisted of, first, there was more to eat, and second, which was most important, one couldn't be sent to the front. . . . Third, one had the chance of getting into a rest home run by T4, and not least of all, good rations, plenty of alcohol, and last, and not least of all, the opportunity of helping themselves to many valuables which had been taken from the Jews.[48]

The perks that were offered in the service of the system, the relentless black rain of propaganda, and the steady hammering of the master race ideology into the ears of ordinary men enticed them to slither even farther along the pathway into the pit of evil. A Waffen-SS soldier named Hans Bernhard explained it this way: "Our motto was duty, loyalty, the Fatherland, and comradeship." This wasn't just some ordinary war; these were blood brothers in arms, fighting the good fight. As the SS-Viking Division member Jürgen Girgensohn put it, "We were convinced we were conducting a just fight, that we were convinced we were a master race. We were the best of this master race and that really does form a bond." Discipline was crucial, and anyone in the rank and file who slacked off was punished from within. SS-Das Reich Division member Wolfgang Filor said of the men that "if they didn't manage it they drew attention and had to do extra training or something." He explained what that extra "something" entailed: "Everyone gave the guy merry hell. He was dragged out of bed and beaten on the head and that kind of thing—so he wouldn't give up so easily next time, so that the troop would not be disrupted." Those who couldn't grin and bear it went AWOL or hanged themselves, knowing that a court-martial was awaiting them if they didn't perform.

There was yet another tried-and-true method that the men used to resolve moral conflict, by means of the temporary relief of mental turmoil, through the blotting out of memories and the numbing of the pain—and that method was getting themselves thoroughly, utterly plastered. After a particularly brutal engagement in France, Waffen-SS soldier Kurt Sametreiter noted, "We were happy that it was all over. Really happy. We were so happy that for a few days . . . well . . . basically we got drunk. Really, you see, we just wanted to forget."[49]

Putting all of these factors together to explain how ordinary Germans became extraordinary Nazis, Christopher Browning summarized the process this way in his appropriately titled book *The Path to Genocide*:

> In short, for Nazi bureaucrats already deeply involved in and committed to "solving the Jewish question," the final step to mass murder was incremental, not a quantum leap. They had already committed themselves to a political movement, to a career, and to a task. They lived in an environment already permeated by mass murder. This included not only programs with which they were not directly involved, like the liquidation of the Polish intelligentsia, the gassing of the mentally ill and handicapped in Germany, and then on a more monumental scale the war of destruction in Russia. It also included wholesale killing and dying before their very eyes, the starvation in the ghetto of Lodz and the punitive expeditions and reprisal shooting in Serbia. By the very nature of their past activities, these men had articulated positions and developed career interests that inseparably and inexorably led to a similar murderous solution to the Jewish question.[50]

A PUBLIC HEALTH MODEL OF EVIL

Another way to think about evil is to compare the medical model of disease to the public health model of disease. The medical model of evil treats it as if the locus of the contagion were inside each individual patient. In Western religion, sin is in the individual; in the law, criminality is in the individual. The medical model demands that each infected person be treated, one by one, until no one shows any further symptoms. The medical model of evil, then, is the analogue to the dispositional model of evil, where evil exists in the disposition of the person exhibiting it. Evil is simply in their nature; thus to eradicate the plague of evil we have simply to eliminate those with evil dispositions.

This paradigm served as the basis of the Inquisition, in which women were instructed to be boiled in oil for such crimes as "sleeping with the devil." Did this put a dent into evil? Hardly. What the witch hunt did was to spread evil in the form of barbaric, systemic violence against women throughout much of Europe and North America for centuries.

By contrast, a public health model of evil accepts as a given that, of course, we affect and infect one another, but individuals are simply part of a larger disease vector that includes many additional social psychological factors that have been identified in the past half century that must be included in any theoretical model to explain the often puzzling world of human moral psychology. Here are some of the most potent factors at work in turning good people into bad.

Deindividuation

Removing individuality by taking people out of their normal social circle of family and friends (as cults do), or dressing them in identical uniforms (as the military does), or insisting that they be team players and go along with the group program (as corporations often do), sets up a situation in which behavior can be molded as the leader wishes. In his classic 1896 work *The Crowd*, the French sociologist Gustave Le Bon called the concept "group mind, in which people are manipulated through anonymity, contagion, and suggestibility."[51] In 1954, the social psychologists Muzafer Sherif and Carolyn Sherif tested Le Bon's idea in a now classic experiment at a camp in Oklahoma, where they divided twenty-two eleven-year-old boys into two groups, "The Rattlers" and "The Eagles." Within a couple of days new identifications were formed within each group, after which the two groups were forced to compete in various tasks. Despite preexisting long-term friendships between many of the boys, hostilities quickly developed between them along group identity lines. Acts of aggression escalated to the point where Sherif and Sherif were forced to terminate this phase of the experiment early, and then introduce tasks for the boys that required cooperation between the two groups, which just as quickly led to renewed friendships and between-group amity.[52]

Dehumanization

Dehumanization is the disavowal of the humanity of another person or group, either symbolically through discriminatory language and objectifying imagery or physically through captivity, slavery, the infliction of bodily harm, systematic humiliation, and so on. It happens intentionally or unintentionally, between individuals and groups, and can even occur in the self as, for example, when an individual views himself or herself negatively from the third-person perspective of the discriminatory group. When the in-group is defined by its out-group counterpart, and vice versa, the power of the in-group can be reinforced in various ways; for example, members of the out-group might be relabeled as vermin, animals, terrorists, insurgents, and barbarians, making it easy to classify them as subhuman or nonhuman. Prisoners might have their hair shaved off, they might be stripped naked to remove a layer of civilizing humanity, and their heads might be bagged as the ultimate removal of identity; they might be identified by number (as in concentration camps), marked with a symbol (as groups targeted in the Holocaust were forced to wear badges—triangles or double triangles of various colors—to mark their inferiority and out-group status), and they might be treated as mere tools or automatons, after which they can be neatly disposed of by fellow members of the out-group. Dehumanization can be much more subtle than that, however; one need only look at the online world to witness dehumanization in all its nastiness, with individuals regularly treating one another as less than human, with utter disregard for feelings (and sometimes truth), especially when confrontation is not immediate or likely.

Compliance

Compliance occurs when an individual acquiesces to group norms or to an authority's commands without agreeing with those norms or commands.

In other words, the individual will obediently carry out orders without the internalized belief that what he or she is doing is right. In a classic 1966 experiment conducted in a hospital, the psychiatrist Charles Hofling arranged to have an unknown physician contact nurses by phone and order them to administer 20mg of a nonexistent drug called "Astrofen" to one of his patients. Not only was the drug fictional, it also was not on the approved list of drugs, and the bottle clearly denoted that 10mg was the maximum daily dose allowed. Preexperimental surveys showed that when given this scenario as purely hypothetical, virtually all nurses and nursing students confidently asserted that they would refuse to obey the order. Yet, when Hofling actually ran the experiment, he got twenty-one out of twenty-two nurses to comply with the doctor's orders, even though they knew it was wrong.[53] Subsequent studies support this disturbing finding. For example, a 1995 survey of nurses revealed that nearly half admitted to having at some time in their careers "carried out a physician's order that you felt could have had harmful consequences to the patient," citing the "legitimate power" that physicians hold over them as the reason for their compliance.[54]

Identification

Identification is the close affiliation with others of like interest, as well as the normal process of acquiring social roles through modeling and role playing. In childhood, heroes serve as role models for identification, and peers become a reference point for comparing, judging, and deciding opinions. Our social groups provide us with a frame of reference with which we can identify, and any member of the group who strays from those norms risks disapproval, isolation, or even expulsion.

The power of identification is emphasized in a 2012 reinterpretation of Milgram's research by psychologists Stephen Reicher, Alexander Haslam, and Joanne Smith.[55] They call their paradigm "identification based followership," noting that "participants' identification with either the experimenter and the scientific community that he represents or the learner and the general community that he represents" better explains the willingness of subjects to shock (or not) learners at the bidding of an authority. At the start of the experiment, subjects identify with the experimenter and his or her worthy scientific research program, but at 150 volts the subjects' identification begins to shift to the learner, who cries out "Ugh!!! Experimenter! That's all. Get me out of here, please." It is, in fact, at 150 volts that subjects are most likely to quit or protest, as it was for our NBC replication subjects. As Reicher and Haslam suggest, "In effect, they become torn between two competing voices that are vying for their attention and making contradictory demands upon them."

Haslam and Reicher also reinterpreted the famous 1971 Stanford Prison Experiment by Philip Zimbardo, reconfiguring his findings through their paradigm of identity theory.[56] Recall that Zimbardo randomly assigned undergraduate students to be either guards or prisoners, directing them to take on their roles fully, and he provided them with appropriate uniforms and the like. Over the next couple of days these psychologically well-adjusted American students were transformed into either the role of violent, authoritative guards or demoralized, impassive prisoners, and Zimbardo had to terminate the two-week study after only a week because of the brutality he witnessed.[57]

In 2005 Haslam and Reicher worked with the BBC in its Prison Study replication, explaining that "Unlike Zimbardo . . . we took no leadership role in the study. Without this, would participants conform to a hierarchical script or resist it?" The Haslam and Reicher study made three findings: (1) "participants did not conform automatically to their assigned role," (2) subjects "only acted in terms of group membership to the extent that they actively identified with the group (such that they took on a social identification)," and (3) "group identity did not mean that people simply accepted their assigned position; instead, it empowered them to resist it." The scientists concluded that in the BBC Prison Study "it was neither passive conformity to roles nor blind obedience to rules that brought the study to this point. On the contrary, it was only when they had internalized roles and rules as aspects of a system with which they identified that participants used them as a guide to action." Therefore they concluded,

> Those who do heed authority in doing evil do so knowingly not blindly, actively not passively, creatively not automatically. They do so out of belief not by nature, out of choice not by necessity. In short, they should be seen—and judged—as engaged followers not as blind conformists.[58]

This observation echoes the appraisal of Eichmann as an alpinist of evil, and to this I would add that it is here where the element of free choice comes into play. Ultimately, even with all these influencing variables, one is still making a volitional choice to act badly, or not.

Conformity

Because we evolved to be social beings, we are hypersensitive to what others think about us and are strongly motivated to conform to the social norms of our group. Solomon Asch's studies on conformity demonstrate the power of groupthink: if you are in a group of eight people who are instructed to judge the length of a line by matching it to three other lines of differing lengths, even when it is obvious which line is the match, if the other seven people in the group select a different line, you will agree with them 70 percent of the time. The size of the group determines the degree of conformity. If just two people are judging the line lengths, conformity to the incorrect judgment is almost nonexistent. In a group of four in which three select the incorrect line match, conformity happens 32 percent of the time. But no matter what the size of the group, if you have at least one other person who agrees with you, conformity to the incorrect judgment plummets.[59]

Interestingly, fMRI studies tell us something about the emotional impact of nonconformity based on the areas of the brain that are most active when a subject is at odds with the group. Conducted by Emory University neuroscientist Gregory Berns, the task involved matching rotated images of three-dimensional objects to a standard comparison object. Subjects were first put into groups of four people, but unbeknown to them the other three were confederates who would intentionally select an obviously wrong answer. On average, subjects conformed to the group's wrong answer 41 percent of the time, and when they did, the areas of their cortex related to vision and spatial awareness became active. But when they broke ranks with the group, their right amygdala and right caudate nucleus lit up, areas associated with negative emotions.[60] In other words, nonconformity can be an emotionally traumatic experience, which is why most of us don't like to break ranks with our social group norms.

Tribalism and Loyalty

Since many of these social psychological factors operate within a larger evolved propensity we have to divide the world into tribes of Us vs. Them. An empirical demonstration of our natural inclination toward so dividing the world can be seen in a 1990 experiment by the social psychologist Charles Perdue, in which subjects were told that they were participating in a test of their verbal skills in which they were to learn nonsense syllables, such as *xeh, yof, laj,* or *wuh.* One group of subjects had their nonsense syllables paired with an in-group word (*us, we,* or *ours*), a second group had their nonsense syllables paired with out-group words (*them, they,* or *theirs*), while a control group had their nonsense syllables paired with a neutral pronoun (*he, hers,* or *yours*). Subjects were then asked to rate the nonsense syllables on their pleasantness or unpleasantness. Results: the subjects who had their nonsense syllables paired with in-group words rated them as significantly more pleasant than subjects who had their nonsense syllables paired with out-group words or neutral-paired words.[61]

The power of in-group loyalty in the real world was well summarized by Lieutenant Colonel Dave Grossman in his deeply insightful 2009 book *On Killing,* in which he shows that a soldier's primary motivation is not politics (fighting for a nation or state) or ideology (fighting to make the world safe for democracy), but devotion to one's fellow soldiers. "Among men who are bonded together so intensely, there is a powerful process of peer pressure in which the individual cares so deeply about his comrades and what they think about him that he would rather die than let them down."[62] This is not obedience to authority. This is camaraderie, a group of strangers who become what are called "fictive kin," nongenetically related individuals who come to act as if they were each other's genetic relations. It is a system that hijacks our evolved propensity to be kind to our kith and kin, in which through such bonding activities as marching and suffering together implemented by the military total strangers come to feel like genetic relations.

Tribes often reinforce loyalty to the group by the punishment (or the threat of punishment) of those who threaten it from within. Whistle-blowers are a case in point. When a whistle-blower threatens our group—even if we know on some level that they are morally right in blowing the whistle—our tribal instincts kick in and we circle the wagons against the perceived threat and vilify it with such emotion-laden labels as "Tattletale, Rat Fink, Stool Pigeon, Snitch, Informer, Turncoat, Bigmouth, Canary, Busybody, Fat Mouth, Informer, Squealer, Weasel, Backstabber, Double-Crosser, Agent-Provocateur, Shill, Judas, Quisling, Treasonist, etc."[63]

Pluralistic Ignorance, or the Spiral of Silence

To understand how a group of people, or even an entire nation, can seemingly come to accept an idea that most of the individual members or citizens would likely reject, we must turn to one of the most perplexing of all social psychological phenomena. Pluralistic ignorance happens when individual members of a group do not believe in something, but mistakenly believe that *everyone else* in the group believes it—and if no one speaks up, it leads to a "spiral of silence" and thence to individuals behaving out of character.

Take binge drinking on college campuses. A 1998 study by Christine Schroeder and Deborah Prentice of Princeton University found that "the majority of students believe that their peers are uniformly more comfortable with campus alcohol practices than they are." Another Princeton study, in 1993, by Prentice and her colleague Dale Miller, found a gender difference in drinking attitudes in which, predictably, "Male students shifted their attitudes over time in the direction of what they mistakenly believed to be the norm, whereas female students showed no such attitude change."[64] Women, however, were not immune to pluralistic ignorance as shown in a 2003 study by the psychologist Tracy A. Lambert and her colleagues, who found that regarding casual sex, "both women and men rated their peers as being more comfortable engaging in these behaviors than they rated themselves."[65] In other words, these college students say that they themselves are not prone to binge drinking and casual hookups, but that most everyone else is, so they go along with the crowd. When everyone in the group thinks this way, an idea that most individual members do not endorse can take hold of the group.

Pluralistic ignorance can transmogrify into witch hunts, purges, pogroms, and repressive political regimes. European witch hunts degenerated into preemptive accusations of guilt, lest one be thought guilty first.[66] Or take the story of Russian dissident Aleksandr Solzhenitsyn, who described a party conference in which Stalin was given a standing ovation in absentia that went on for eleven minutes, until a factory director finally sat down, to everyone's relief—everyone but one of Stalin's party functionaries, that is, who had the man arrested that night and sent to the gulag for a decade.[67] A 2009 study by the sociologist Michael Macy and his colleagues confirmed the effect: "people enforce unpopular norms to show that they have complied out of genuine conviction and not because of social pressure." Laboratory experiments show that people who conform to a norm under social pressure are more likely to publicly punish deviants from the norm as a way of advertising their genuine loyalty instead of appearing to be just faking. Together, "these results demonstrate the potential for a vicious cycle in which perceived pressures to conform to and falsely enforce unpopular norms reinforce one another."[68]

Bigotry is ripe for the effects of pluralistic ignorance, as evidenced in a 1975 study by the sociologist Hubert J. O'Gorman, "indicating that in 1968 most white American adults grossly exaggerated the support among other whites for racial segregation," especially among those leading segregated lives, thereby reinforcing the spiral of silence.[69] Interestingly, a 2000 study by the psychologist Leaf Van Boven found pluralistic ignorance at work when "students overestimated the proportion of their peers who supported affirmative action by 13% and underestimated the proportion of their peers who opposed affirmative action by 9%." He attributes the effect to political correctness, which drives some of us to lead dual lives, professing beliefs we think others hold while privately holding beliefs that may vary from this presumptive norm. Presciently, Van Boven closed his analysis with this commentary on the (at the time in 2000) emerging gay marriage debate: "one might expect that fear of appearing politically incorrect could lead to pluralistic ignorance for any number of politically correct issues, such as people's attitudes toward gay and lesbian marriage or adoption, their views about the appropriate labels for the romantically involved (are they 'boyfriends and girlfriends' or 'partners'?), their attitudes toward gender equality, or their beliefs about the role of the Western canon in a liberal arts education. To the extent that concerns about appearing racist, sexist, or otherwise culturally insensitive squelch public expression of private doubts about such issues, pluralistic ignorance can be expected to emerge."[70]

Perhaps in some cases pluralistic ignorance is a good thing, since many private thoughts may be morally regressive, given the time it often takes to shift beliefs and preferences, and the fact that many people who privately hold politically incorrect beliefs may wish that they didn't. As the Russian novelist Fyodor Dostoevsky wrote, "Every man has reminiscences which he would not tell to everyone but only his friends. He has other matters in his mind which he would not reveal even to his friends, but only to himself, and that in secret. But there are other things which a man is afraid to tell even to himself, and every decent man has a number of such things stored away in his mind."[71]

Fortunately, there is a way to break the bonds of pluralistic ignorance: knowledge and transparency. In the Schroeder and Prentice study on college binge drinking it was found that exposing incoming freshmen to a peer-directed discussion that included an explanation of pluralistic ignorance and its effects significantly reduced subsequent student alcoholic intake.[72] Sociologist Michael Macy found that when skeptics are scattered

among true believers in a computer simulation of a society in which there
is ample opportunity for interaction and communication, social connectedness acted as a safeguard against the takeover of unpopular norms.[73]

. . .

Figure 9-1 features a visual record of how all of these factors can come
into play in the real-world example of Köln, Germany, in the 1930s and
early 1940s. On a trip to this beautiful city I consulted with the *NS-
Dokumentationszentrum der Stadt Köln* (the National Socialism Documentation Center of the City of Cologne) to assess how the Nazis managed
to take over one city. It became clear that they did it district by district,
house by house, and even person by person, part of a national plan to
Nazify all of Germany but orchestrated locally by gauleiters, or district
overseers.[74] The museum is in a building that housed the Cologne Gestapo
from December 1935 through March 1945, and it documents the seizure
of power; the propaganda employed in everyday life, including the youth
culture, religion, racism, and especially the elimination and extermination
of Cologne's Jews and its Sinti and Roma; and finally the opposition, resistance, and society during the war. The photograph of a sign posted by the
Allies in Köln captures the entire span of the process and its ultimate
demise in a quote from Hitler: "Give me five years and you will not recognize Germany again." The magnificent bridge spanning the Rhine River
adjacent to the Köln Cathedral—lying in the water, smashed to pieces—
reveals what it took in this case to bring an end to the evil.

EVIL, INCORPORATED

All of these factors are interactive and autocatalytic—that is, they feed
on one another: *dehumanization* produces *deinidividuation*, which then
leads to *compliance* under the influence of *obedience to authority*, and in
time that morphs into *conformity* to new group norms, and *identification*
with the group, which leads to the actual performance of evil acts. No one
of these components inexorably leads to evil acts, but together they form
the machinery of evil that arises under certain social conditions.

These conditions are necessary but not sufficient to account for evil,
which also involves the dispositional nature of the individual; the overall
system in which these conditions occur; and, of course, free will. We can
change the conditions and attenuate evil, first by understanding it and
then by taking action to change it. By understanding how its components

Figure 9-1. A Visual Record of How the Nazis Took Over Köln, Germany

Photographs from the National Socialism Documentation Center of the City of Cologne reveal how an evil regime can take over a city and a nation. (a) Indoctrination of the citizens was the preferred method of Nazifying the German people, as evidenced in the Hitler Youth programs pictured here (credit: LAV NRW R, BR 2034 Nr. 936); (b) Shaping of cultural life through publications, such as Robert Ley's Nazi newspaper featuring an anti-Semitic characterization; and (c) a bookstore with swastikas and the anti-Semitic slogan "The Jews Are Our Misfortune." (d) A fragment of a list of the hundreds of banned clubs shows the extent to which the Nazis controlled every aspect of daily life in Köln. (e) If indoctrination through propaganda didn't work, imprisonment was employed as a means of bringing the people into line, as seen here in the Köln Gestapo prison. (f) Indoctrination included a eugenics race program that took specific measurements of people such as this woman to determine if they measure up to Aryan standards. (g) A German civilian in Köln, Germany, April 18, 1945, reading a sign posted by the American forces, quoting Hitler's promise to the German people "Give me five years and you will not recognize Germany again" (National Archives, US Army Photograph, SC 211781). (h) The bombed-out Hohenzollern Bridge spanning the Rhine River adjacent to the famed Köln Cathedral is a visually striking reminder of what it sometimes takes to end evil (National Archives, US Army Photograph, SC 203882).

Figure 9-1a

Figure 9-1c

Figure 9-1b

Rheinische Windthorstbunde
Rheinischer Auktionatoren-Verband
Rheinischer Bauernverein
Rheinischer Bezirksarbeitgeberverband der chemisch
Rheinischer Damen-Automobil-Club
Rheinischer Genossenschafts-Verband Köln
Rheinischer Handwerkerbund
Rheinischer Kälte-Verein
Rheinischer Kochkunstverein Gasterea
Rheinischer Landesblindenverband
Rheinischer landwirtschaftlicher Pächterverband
Rheinischer Mieterschutzverband
Rheinischer Philologen-Verein
Rheinischer Radfahrer-Verband für Touren und Saalspo
Rheinischer Sängerbund, Vereinigung rheinischer Mäni
Rheinischer Schiffer-Verein
Rheinischer Schutzverband für Grundbesitz
Rheinischer Sportverein Union 05
Rheinischer Verband für Saal- und Tourensport
Rheinischer Verband für Tieflandrinderzucht
Rheinischer Verein gegen Betriebsdiebstähle
Rheinischer Volks-Feuerbestattungsverein
Rheinisches Landvolk
Rheinland- und Moriahloge
Rheinland-Klub Köln

Figure 9-1d

Figure 9-1e

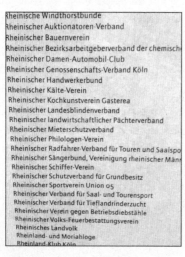

Figure 9-1f

Figure 9-1g

Figure 9-1h

operate and how to control them, we can quell evil and keep it in check through the social tools and political technologies we now know how to employ to the betterment of humanity.

This is what we have been doing since the end of World War II. Social scientists have undertaken extensive studies and experimental protocols to understand precisely which social and psychological factors enable evil to triumph over good. Historians have uncovered the political, economic, and cultural conditions that allowed these social and psychological factors to play out their effects on individuals and whole populations. Politicians, economists, legislators, and social activists have applied this knowledge to change conditions to attenuate the possibility of such factors leading people down the path toward evil. Although there have been disruptions still—the Rwanda genocide and the 9/11 terrorist attacks come to mind—ever since the Nazi camps were liberated and the Soviet gulags eliminated, the overall trend toward a more moral world is unmistakable. And those salubrious effects have been primarily results of the scientific understanding of the causes of evil and the rational application of political, economic, and legal forces to drive it down and bend that arc ever upward.

10

Moral Freedom and Responsibility

Everything can be taken from a man or a woman but one thing: the last of human freedoms to choose one's attitude in any given set of circumstances, to choose one's own way.

—Viktor E. Frankl, *Man's Search for Meaning*, 1946

As the publisher of a national magazine, I routinely receive letters from prisoners. Most of them just want free reading material because they are bored. Some are genuinely interested in the topics we cover and have ideas of their own that they want to share. A few feel overwhelmed by the intrusion of religion into prison life, particularly Christianity and Islam—to which their fellow inmates feel a proselytizing zeal or to which they fake devotion in an attempt to convince parole boards that by finding God they should be released early. A few have written me lengthy and detailed letters (handwritten in tiny print) about their crimes and to what extent they feel responsible for them. Two come to mind as especially salient for this exploration of free will and moral culpability.[1]

In the late 1990s a man on death row for the rape and murder of a woman wrote to suggest to me that we publish an article in *Skeptic* against the death penalty. He thought that he should be spared execution because he had a sense that there was something seriously wrong with his brain and therefore was not fully responsible for his crime. He told me that he was constantly flooded with thoughts of raping and killing women. So strong were these urges that even when he was being transported to jail after his death penalty sentencing, chained and locked down to his seat in a van with armed guards at the ready nearby, when they passed a woman walking on the sidewalk the only thing he could think about was what it would take for him to slip the chains and locks, overpower the armed guards, escape from the van, and get to that woman. Even though he was against the death penalty, he added that he should never be released from prison because there was no question in his mind that he would do it again. I don't know if a brain scan was ever conducted on this man to identify any obvious neural pathologies, but from my reading of his account it seems certain that his capacity to control these obsessive, psychotic urges was so severely limited that he could not reasonably be compared to the average person, who has both sufficient self-restraint and an utter lack of such murderous impulses.

In 2012 a man sentenced for pedophilia wrote to try to convince me to edit a special issue of *Skeptic* about how pedophiles—who he claimed are born that way—are vilified and misunderstood. This man, now in his thirties, narrated a lengthy history of his childhood in which he was attracted to boys (not girls) his age, and that as he grew up the window of his sexual interests remained locked in at boys aged eight to ten, almost as if there were a critical window for sexual attraction that was imprinted on his brain. As an adult he was still only attracted to boys in that age range, and had urges, he said, that he satisfied through certain Internet sites. He was vague about the extent to which he also fulfilled these needs through actual sexual encounters with young boys, but when he asked me to recall the most loving and intimate feelings I had ever experienced with a woman, and then compared these emotions to the most deeply sexual and loving feelings that he had experienced for young boys, this told me all I needed to know. This man's deeper point, however, was that there was nothing obviously wrong with him. He had no brain tumor or any other neural anomalies or pathologies that could be detected. He was raised by two loving parents in a middle-class suburban home, and he attended safe and functional schools. He simply could not understand why these feelings that came so naturally to him and upon which he acted should be condemned as abnormal and unnatural—indeed, criminal—by society.

ARE WE FREE TO CHOOSE?

These vignettes illustrate the problem of free will, determinism, and the extent to which we are free to make moral choices and are therefore responsible for and should be held accountable for our actions.

Morality involves how we think and act toward other sentient beings and whether our thoughts and actions are right (good) or wrong (bad) with regard to their survival and flourishing. As noted previously, a principle of moral good is this: *if other persons are involved in an action, then always act with their good in mind, and never act in a way that it leads to their loss or suffering (through force or fraud).* You can, of course, act in a way that has no effect on anyone else, and in this case morality isn't involved—your actions are *amoral.* But given the *choice* between acting in a way that increases someone else's moral good or not, it is more moral to do so than not. The parenthetical addendum of "through force or fraud" is meant to clarify *intent* instead of acting out of neglect or ignorance. Morality involves conscious choice, and the choice to act in a manner that increases someone else's moral good, then, is a moral act, and its opposite is an immoral act.

This all presumes that we are free to choose, that we are volitional beings. Are we? I was pondering this recently when ordering dinner from a restaurant menu that offered many tasty options. Looking at the menu, I craved a heavy stout beer and a buttery escargot appetizer to start, followed by a marbled steak with some rich creamed spinach and a baked potato with sour cream and butter, and wrap it all up with a slice of cheesecake and a creamy latte. These cravings don't come about in a vacuum. Our brains have evolved neural networks that produce the gnawing sensation of hunger for sweet and fatty foods—like ripe fruits and game animals— that were so rare and nutritious in our Paleolithic ancestral environment. Natural selection designed our brains to crave fats and sweets; the more the better, hence our modern problems of obesity and diabetes. In competition with those deep cravings are signals from other neural networks that evolved to make us care what a fit physique signals to other members of the tribe (status and desirability).

After considering the menu, I remembered how sluggish I feel after a rich meal, and I stopped to consider the huge amount of exercise I'd need to counter the calories. The higher regions of my cortex also kicked into action as I considered my future health—and girth. In the end I ordered a glass of white wine and a chicken Caesar salad, and I reluctantly skipped dessert.

Was I free to make these choices? According to most neuroscientists, such as Sam Harris in his book *Free Will,* I was not: "Free will is an illusion," Harris claims. "Our wills are simply not of our own making. Thoughts and intentions emerge from background causes of which we are unaware and over which we exert no conscious control. We do not have the freedom we think we have."[2] In a deterministic world, every step in the causal chain that resulted in my restaurant menu selections above is fully controlled by forces and conditions not of my choosing, from my evolved taste preferences to my learned social status concerns; the causal pathways were laid down by my ancestors and parents, culture and society, peer groups and friends, mentors and teachers. The historical contingencies lead all the way back to before I was even born.

MINI-ME AND THE DETERMINISTIC DEMON

How far back does the causal chain go? By some accounts, it starts at the beginning of the universe with the Big Bang creation of all space, time, matter, and energy. The principle of determinism holds that every event in the universe has a prior cause and that all effects are predictable if all causes are known. If this is true, whence free will? If all effects have causes, including human thoughts and actions, then where in the causal chain does the act of choice enter? There is no homunculus—no little man—in the brain pulling the levers. But even if there were a mini-me up there calling the shots, his little brain would have to be just as determined as my big brain, so for mini-me to have free will he would have to have a mini-mini-me inside of him pulling his strings, and mini-mini-me would himself need an itty-bitty mini-mini-me inside of his brain . . . ad infinitum. And if you believe in souls, this fails in the same way as mini-me does. A soul inside of you pulling your strings does not grant you freedom; it just means the soul is in control. And as with mini-me, having a soul would mean that there's a mini-soul inside the soul directing its actions, and so forth. Needless to say, no such homunculi, mini-mes, or mini-souls have been found in the brain or anywhere else. They don't exist.

With the advent of modern science and its focus on natural laws that govern the universe, the idea of determinism gained credibility—the chain linking effects to causes grew even tighter. Instead of a linear sequence along a single line of cause-effect-cause-effect-cause-effect, etc., the complexity of what we mean by a determined universe can be better described by casting a much wider "causal net"—a *network* of causes—that are linked to effects throughout the past and into the future, and which encompass a myriad of intervening variables and interacting factors. This causal net involves all phenomena, past, present, and future, throughout the cosmos, from atoms to molecules, cells, organisms, persons, planets, stars, and galaxies all the way out to the edge of the observable universe. Without assuming that the universe is determined, in fact, scientists could not explain the past or predict the future, and this includes psychologists and neuroscientists attempting to explain and predict human behavior.

A number of compelling experiments seem to support the doctrine of determinism in human action. In 1985, the physiologist Benjamin Libet famously conducted a series of experiments that involved taking EEG readings of subjects' brains engaged in a task that required them to press a button at random intervals whenever they felt like it during the session. Results: several seconds before the "decision" was consciously made by the subject, the brain's motor cortex was activated.[3]

The neuroscientist John-Dylan Haynes employed functional magnetic resonance imaging (fMRI) technology in a 2011 study in which subjects inside the brain scanner were instructed to press one of two buttons whenever they wanted while observing a series of random letters. The subjects were told to verbally report which letter was on the screen when they decided to press the button. Results: the time between brain activation and conscious awareness of a "choice" was several seconds, and in some cases a full seven seconds. As Haynes concluded, "Libet's and our findings do address one specific intuition regarding free will, that is, the naïve folk-psychological intuition that at the time when we make a decision, the outcome of this decision is free in the sense of not being predetermined by prior brain activity."[4]

Another 2011 study, by Itzhak Fried and his UCLA colleagues, recorded activity in a tiny network of neurons in the brains of subjects who were instructed to command their fingers to move. Results: a full fifteen seconds before subjects reported making a decision to move their finger, the neuroscientists could detect neural activity related to the finger movement. Narrowing down their search, they found activity in a tiny clump of 256 neurons in the medial frontal cortex that enabled them to predict with 80 percent accuracy which choice a subject would make a full seven seconds before the subjects themselves knew.[5]

In other words, in these studies—and many others corroborating Libet's original study—neuroscientists measuring subjects' brain activity know which decision will be made before the subjects themselves know it. In my restaurant example above, if I were wired up with an EEG cap with neuroscientists measuring my brain wave activity as I perused the menu, very likely they could have placed the order with the waiter before I had consciously "decided" what I was going to select. That is spooky, and if these results don't bother you, then you're not thinking hard enough about them. What they imply is that we are not free to choose in the way we think we are. We *feel* free, but that's just what our conscious self believes because it doesn't know about the inputs feeding into it from below that have already made the choice.

Accepting a determined universe, however, does not preclude retaining free will and moral responsibility. There are at least four ways around this paradox: (1) *modular mind*—even though a brain consists of many neural networks in which one network may make a choice that another network finds out about later, they are all still operating in a single brain; (2) *free won't*—vetoing competing impulses and choosing one thought or action over another; (3) *degrees of moral freedom*—a range of choice options varying by degrees of complexity and the number of intervening variables; and (4) *choice as part of the causal net*—wherein our volitional acts are part of the determined universe but are still our choices.

MODULAR MIND

The argument that we are not free to choose because an unconscious part of a brain tells a conscious part of a brain what has been decided is, at best, based on a dodgy interpretation of the neuroscience. If a subcortical region of my brain sends a signal to a cortical region of my brain to inform it of a preference, it is still my brain making the choice. It is still *me*—an autonomous volitional being—making choices, regardless of which part of me is actually making the decision.

The idea of compartmentalized brain functions acting either in concert or in conflict has been a core idea of evolutionary psychology since the 1990s. In his book *Why Everyone (Else) Is a Hypocrite,* the evolutionary psychologist Robert Kurzban shows how the brain evolved as a modular multitasking, problem-solving organ—a Swiss Army knife of practical tools in the old metaphor, or an app-loaded iPhone in Kurzban's upgrade.[6] There is no unified "self" that generates internally consistent and seamlessly coherent beliefs devoid of conflict, out of which volitional choices are allegedly made. Instead, we are a collection of distinct but interacting modules that are often at odds with one another, and the decision-making process often happens unconsciously, so it seems as if choices are being made for us from we know not where.

Kurzban calls this illusion the Magic 8 Ball view of preferences. Recall the Magic 8 Ball novelty toy—a black plastic sphere containing a blue liquid and a white, twenty-sided die with ten affirmative answers, five uncertain answers, and five negative answers printed in raised letters on its sides. The idea is that the user holds the ball with its window facing downward, asks a yes/no question, and then turns the ball right side up, at which point the die floats to the surface and an answer appears in the window—an answer such as "Concentrate and ask again" or "My sources say no" or "It is decidedly so." In this model of the mind, the brain is envisioned as being one big module, something like a Magic 8 Ball, into which a question is fed and out of which comes an answer after the equivalent of being shaken—thinking—but with a lot more than twenty possible answers. "If the mind were like a Magic 8 Ball in this sense," Kurzban writes, "taking everything that one knows, and one's preferences, and integrating them all together—a process as magical as what economists refer to as 'rationality'—then people would behave consistently."

Consider the ultimatum game, in which you are given $100 to split between yourself and your game partner. Whatever division of the money you propose, if your partner accepts it, you are both richer by that amount. How much should you offer? Why not suggest a $90–$10 split? If your game partner is a rational, self-interested money maximizer—as predicted by the standard economic model of *Homo economicus*—he isn't going to turn down a free ten bucks, is he? He is. Research shows that proposals that deviate beyond a 70–30 split are usually rejected.[7] Why? Because they aren't fair. Says who? Says the moral module of "reciprocal altruism," which evolved over the Paleolithic aeons to demand fairness by our potential exchange partners. "I'll scratch your back if you'll scratch mine" only works if I know you will respond with something approaching parity. So when offers deviate too much from equity, the module demanding fairness is in conflict with the module for gain and profit. In these and many other experiments in which people are given choices, it is clear that people are not rational calculators.[8] So, for example, in my menu selection process the module that leads us to craving sweet and fatty foods in the short term is in conflict with the module that monitors our body image and health in the long term. Likewise in moral dilemmas, the module for cooperation is sometimes in conflict with the module for competition, as is the module for altruism and the module for avarice, or the module for truth telling and the module for lying.

A modular mind model not only helps to explain moral hypocrisy, it also removes the apparent deterministic nature that the neuroscience of choice experiments claim. Now that we know the source and process of neural decision making we can build volition back into our psychological models. There *is*, after all, a mini-me—lots of them, in fact—all of them with preferences, many of them in competition with one another, and all of them inside a single brain.

FREE WON'T

If we define free will as *the power to do otherwise*,[9] it becomes clear that a useful approach is to conceptualize "free will" as "free won't"—as the power to *veto* one impulse in favor of another. Free won't is the capacity to reject a particular action arising from the unconscious neural network, such that any decision to act one way instead of another way is an authentic choice. For example, I could have had the steak, but by engaging in certain self-control techniques that remind me of other competing impulses (my health, my circumference), I dismissed one set of meal selections in favor of another. *That* is free won't. We have limitations, it's true—we cannot just do anything we choose—but for the most part we have veto power; we have the capacity to say "no"; we can act this way instead of that way, and that is a real choice.

Support for this hypothesis can be found in a 2007 study conducted by neuroscientists Marcel Brass and Patrick Haggard, who used fMRI brain scans while subjects made choices. But at the last moment the subjects could change their minds and override their initial decision to press a button. When they chose to veto their initial decision, the scientists discovered that a specific area of the brain lit up—the left dorsal frontomedian cortex (dFMC)—an area that is normally active during decision-making behavior, especially during the intentional inhibition of a choice. Tellingly, there were no differences in the brain regions active in preparation of a voluntary action and those involved in inhibiting such actions. "Our results suggest that the human brain network for intentional action includes a control structure for self-initiated inhibition or withholding of intended actions."[10] That is free won't.

An additional finding in this study is of interest with regard to self-control: "We also found a positive correlation across individuals between inhibition-related dFMC activation and the frequency of inhibited actions," the authors noted. In other words, the more you learn to suppress your impulses, the better able you are to activate your dFMC to further suppress future impulses. And the reverse: "In the psychology of individual differences, disinhibited and impulsive behavior is a crucial marker for specific personality traits that have been related to antisocial and criminal behavior." Think of my examples of the murderer and the pedophile and their uncontrollable impulses. But there is redemption to be found in the structure of the brain's choice architecture: "Our findings suggest that inhibition of intentional action involves cortical areas different from, and upstream from, the intentional generation and execution of action. In addition, this process of 'last-minute' inhibition is compatible with a conscious experience of intending to act."[11]

In other words, our inhibitory neural networks are higher up than the initial decision-making networks, which means that those impulses, and the early decisions we make in response to them, can be overridden by higher-order decision neural networks. As Brass and Haggard conclude: "Our results provide the first clear neuroscientific basis for the widely-held view that people can refrain from doing something even if they genuinely wish to do it. We speculate that the dFMC may be involved in those aspects of behavior and personality that reflect self-control."[12] Even Benjamin Libet himself—the instigator of this line of research that has led so many neuroscientists to abandon belief in free will—in the end came down in favor of human nature containing a volitional element: "The role of conscious free will would be, then, not to initiate a voluntary act, but rather to control whether the act takes place. We may view the unconscious initiatives for voluntary actions as 'bubbling up' in the brain. The conscious-will then selects which of these initiatives may go forward to an action or which ones to veto and abort."[13]

What this research implies is that the neural architecture of choice can be modified by experience—in other words, training and practice—which means that in the long run, with better neuroscience and technology, we could not only teach people how to impede their maladaptive impulses to, say, eat unhealthy foods or take dangerous drugs, but also, in principle, we could train criminals to learn to veto their early and dangerous choices in order to make more socially acceptable decisions. And the choice is real in this way: regardless of which part of our brain makes our choices, they are still *our* choices, and even the apparently subconscious ones can be overridden by conscious effort.

FREE CHOICE AS PART OF THE DETERMINISTIC CAUSAL NET

The causal-net theory of determinism holds that since no set of causes can completely encompass all the determiners of human action, in terms of human freedom they might be considered pragmatically as *conditioning* causes, rather than determining ones. The vastness, intricacy, and ultimate unknowability of the causal net lead us to feel as if we are acting freely. But it is more than a feeling. As with our capacity for free won't and consciously choosing to override desires bubbling up from the unconscious, our choices are genuine neural processes. Our ancestors made decisions to behave in ways that result in actual consequences for survival and reproduction in our evolutionary history, and this led to the evolution of a neural architecture for behavioral choice.[14]

This version of freedom is one that has been explored extensively by the philosopher Daniel Dennett. Millions of years of evolutionary pressures gave rise to free will itself because we gradually evolved a cortex that can—and, for survival purposes, *must*—assess the consequences of the many and various courses of action open to us.[15] In his book *Freedom Evolves*, Dennett argues that free will arises from a number of our characteristics of cognition (characteristics shared, I would argue, with other species as well, such as the great apes and marine mammals), including a sense of being *self-aware* and aware that others are self-aware; *symbolic language* that allows us to communicate the fact that we are aware and self-aware; *complex neural circuitry* that allows for many behavioral options arising out of numerous neural impulses; a *theory of mind* about others that enables us to think about what they're thinking about; and *evolved moral emotions* about right and wrong choices. And because we can communicate complex ideas through language, we have the power to reason with one another about these moral choices. Out of this collection of cognitive characteristics comes free will, because we can, and do, weigh the consequences of the many courses of action available to us at any given moment. We are aware that we (and others) consciously make these choices, and we hold ourselves (and others) accountable.

MORAL DEGREES OF FREEDOM

A final way to understand volition in a deterministic system is through the concept of "degrees of freedom"—the range of options that an organism has as a result of its complexity and the number of intervening variables acting on it. Insects, for example, have very few degrees of freedom and are guided mostly by fixed instincts. Reptiles and birds have more degrees of freedom enabled by modifiable instincts that are subject to environmental triggers in critical periods, with subsequent life experience allowing for learned responses to changing environments. Mammals, especially the great apes, have many more degrees of freedom through considerable neural plasticity and learning. And humans have the most degrees of freedom, with our massive cortex and our highly developed culture.

Within our own species, some people—psychopaths, the brain-damaged, the severely depressed, or the chemically addicted—have fewer degrees of freedom than other people, and the law adjusts for their lowered capacity for legal and moral culpability. But we still hold them accountable for their actions to the extent that they have control over their choices, especially their capacity to, say, veto their criminal impulses.

The more neuroscientists bore into the brain's black box to reveal its internal workings, the more it has been revealed that we should be thinking continuously instead of categorically about moral responsibility. Instead of thinking of people as either sane or insane, normal or abnormal, law-abiding citizens or criminals, we should place people on a continuum—a sliding scale—from Jeffrey Dahmer at one end to Mr. Rogers at the other. Sliding scales are how social scientists think about such internal states as sanity and insanity, and this allows us to take into consideration the degrees of freedom within a behavioral system.

The law also recognizes degrees of freedom by distinguishing among various grades of taking a life, which are classified according to circumstance and intent. First-degree murder is the unlawful killing of one human being by another, with malice aforethought—that is murder, both intentional and premeditated. Second-degree murder is the unlawful killing of one human being by another, but *without* malice aforethought—that is, it is *not* intentional and premeditated. Voluntary manslaughter is the unlawful killing of one human being by another *without* prior intent to kill, and is committed under circumstances that would "cause a reasonable person to become emotionally or mentally disturbed," as in a crime of passion. *Involuntary* manslaughter is neither deliberate nor premeditated and is reserved for fatal accidents due to negligence—for example, deaths caused by drunk driving. And, as we shall see below, there are murders and manslaughters caused by mitigating factors such as tumors, PTSD, depression, and so on—factors that are assumed to restrict the autonomy of the accused and are thus taken into account during the sentencing phase of a trial. Finally, there are *lawful* killings, such as those in war, as a result of self-defense, or capital punishment by a state. All of these ways in which a human life is cut short, either lawfully or unlawfully, by individuals or the state, take into account circumstances, intent, and moral degrees of freedom.[16]

. . .

We have now established that we have free will through (1) our *modular minds* that have many competing neural networks, that (2) allow us to make real choices by veto power—*free won't*—over contending impulses, that (3) give us a range of volitional choices by varying *degrees of freedom*, so (4) our choices are part of the causal net but free enough for most of us in most circumstances and conditions to be accountable for our actions. What those circumstances and conditions are, and how they constrain our choices, are case studies in moral culpability. Let's consider two— psychopathy and violent crime—in the context of how they are dealt with in a rational society.

PSYCHOPATHY AND MORAL CULPABILITY

Psychopathy is a condition that includes some (but not necessarily all) of these characteristics: callousness, antisocial behavior, superficial charm, narcissism, grandiosity, a sense of entitlement, a lack of empathy and remorse, and poor impulse control and criminality.[17] The psychologist Kevin Dutton, whom I queried for this chapter, says that "estimates of the incidence of psychopathy tend to vary from 1–3 percent in men and 0.5–1 percent in women," and in prison populations, "around 50 percent of the most serious crimes on record—crimes such as murder and serial rape, for instance—are committed by psychopaths."[18] But even within psychopathy there are shades of gray, from the low end of the spectrum inhabited by certain CEOs, lawyers, Wall Street traders, and coldhearted bosses who enjoy growling "You're fired!," to the high end inhabited by the likes of Ted Bundy, who, after raping and murdering at least thirty women in the 1970s, boasted, "I'm the coldest son of a bitch you'll ever meet."[19]

A test for measuring psychopathy was developed in 1980 by the psychologist Robert Hare called the Psychopathy Checklist (PCL), which consists of 20 items, with a maximum score of 40 (on each of the 20 items, the test taker can score a 0, 1, or 2). A revised version of the scale is still in use today (PCL-R), and you can take different versions of it yourself online (not advised). I scored a 7. (The beginning level for psychopathy is at 27.) Of course, there *are* circumstances, such as in business, sports, and other competitive enterprises, in which it is advantageous to be a little charming, tough-minded, socially manipulative, impulsive, and to have a need for stimulation when you find yourself bored. But as these personality traits scale up to the point where a little charm becomes manipulative conning, where self-confidence escalates to grandiosity, where occasional exaggeration morphs into pathological lying, where tough-mindedness devolves into cruelty, where impulsivity becomes irresponsibility, and especially if all of these characteristics lead to criminal behavior, we have the makings of a dangerous psychopath.

The Psychopathic Personality Inventory (PPI), developed by the psychologist Scott Lilienfeld, consists of a much wider swath of personality dimensions measured by 187 questions, factor-analyzed into a cluster of combined characteristics such as Machiavellian Egocentricity, Impulsive Nonconformity, Blame Externalization, Carefree Nonplanfulness, and the like. The idea is to recognize the fact that most of us have these personality traits in some measure. As Lilienfeld told Dutton in an interview, "You and I could post the same overall score on the PPI. Yet our profiles with regard to the eight constituent dimensions could be completely different. You might be high on Carefree Nonplanfulness and correspondingly low on Coldheartedness, whereas for me it might be the opposite."

The neuroscientist James Fallon was forced to consider both sides of psychopathy from a very personal perspective—during his search inside the heads of serial killers for the biosocial markers of violence, he discovered signs of psychopathy in himself. As he recounts the story in his book *The Psychopath Inside*, Fallon was putting the finishing touches on a paper on the neuroscience of psychopathy that included brain scans of young psychopathic murderers, which "shared a rare and alarming pattern of low brain function in certain parts of the frontal and temporal lobes—areas commonly associated with self-control and empathy."[20] Mixed with these brain scans was another one that was not part of the dataset but that showed clear signs of psychopathy. The scan turned out to be of Fallon's brain. Additional tests revealed that Fallon is "a borderline psychopath," a diagnosis confirmed by his family and friends who, upon his insistence on honesty, told the neuroscientist the last thing he wanted to hear: that he could not be trusted. "Every one of them told me the same thing. They said they had been telling me for years what they thought of me: that I was a nice and interesting fellow and great fun to work with, but that I was a 'sociopath.' I told them I was sure they'd been kidding. They said they had been dead serious with me all along."[21] Traits they used to describe Fallon were right out of the psychopathic checklist: "manipulative, charming but devious, an intellectual bully, narcissistic, superficial, egocentric, unable to love deeply, shameless, completely lacking scruples, cunning liar, no respect for laws or authority or rules of society."

There are no "cures" for psychopathy, Fallon explains. "Psychopathic tendencies are particularly hard to fight, and attempted cures may make only small differences. Drugs that influence the monoamine neurotransmitter systems can partially reduce impulsivity and aggression, and early interventions involving diet and meditation can decrease behavioral problems, but the core neuropsychological deficits leading to lack of empathy and remorse remain." Fallon fits Dutton's model of a psychopath who channeled his tendencies for positive instead of negative ends. "I was Psychopath Lite, or a prosocial psychopath, someone who has many of the traits of psychopathy other than the violent criminality, a type of psychopathy in which one finds socially acceptable outlets for one's aggression and which is manifested in a cold, narcissistic manipulation of people."[22] Herein lies a key as to what to do about psychopathy and its implications for volition: choose to work within its constraints and channel the traits toward productive instead of destructive ends.

This is what Kevin Dutton proposes in his book *The Wisdom of Psychopaths,* asking readers to consider both the positive and the negative sides of the personality characteristics that make up the diagnosis of psychopathy. Dutton is developing a tool kit to train people to use what he calls the Seven Deadly Wins—"seven core principles of psychopathy that, apportioned judiciously and applied with due care and attention, can help us get exactly what we want; can help us respond, rather than react, to the challenges of modern-day living." Dutton's seven are: Ruthlessness, Charm, Focus, Mental Toughness, Fearlessness, Mindfulness, and Action. The idea is that if you want to accomplish something (anything, really) in life you need a certain amount of all these traits, but not too much. "Cranking up the ruthlessness, mental toughness, and action dials, for instance, might make you more assertive—might earn you more respect among your work colleagues," he writes. "But ratchet them up too high and you risk morphing into a tyrant."[23]

The social psychologist Philip Zimbardo has been searching for that balance in his Heroic Imagination Project, which teaches people how to choose the most effective action in a variety of challenging situations.[24] "The decision to act heroically is a choice that many of us will be called upon to make at some point in our lives," Zimbardo explained. "It means not being afraid of what others might think. It means not being afraid of the fallout for ourselves. It means not being afraid of putting our necks on the line."[25] Here the traits of fearlessness, focus, and mental toughness come into play. Perhaps we need another word—another label—for the positive side of psychopathy, the side of the spectrum where those personality dimensions are put to good use—Positive Psychopathy (with Negative Psychopathy as its antithesis). But whatever we call it we should always remember that human behavior is multivariate, complex, and context-dependent, and our labels do not always capture its rich tapestry.

Perhaps there's a way to train psychopaths to channel their skills into doing good instead of evil. That is the goal of the neuroscientist Daniel Reisel, who, on a grant from the UK Department of Health, has been visiting prisons to study the brains of criminals in search of causes and cures. His initial findings reveal that criminals—especially psychopathic criminals (which, recall, make up at least half of the prison population of violent offenders)—show different physiological responses to such emotions as distress or sadness when compared to noncriminal brains. "They failed to show the emotions required; they failed to show the physical response. It was as though they knew the words but not the music of empathy." Brain scans revealed that "Our population of inmates had a deficient amygdala, which likely led to their lack of empathy and their immoral behavior."[26]

One avenue of treatment for these neurologically impaired psychopaths is neurogenesis, or the birth of new neurons in the adult brain. Take mice. If you raise them in a prison-cell-like environment devoid of stimulation, they lose their capacity to form bonds with their fellow mice when they are reintroduced to them. But if you raise mice in an enriched environment, they not only form normal attachments with their fellow group members, they also experience the growth of new brain cells and connections, which not only leads them to "perform better on a range of learning and memory tasks," says Reisel, but also "their improved environment results in healthy, sociable behavior." Reisel then draws the analogy with prison: "When you think about it, it is ironic that our current solution for people with dysfunctional amygdalas is to place them in an environment that actually inhibits any chance of further growth." Of course, our natural propensity to punish wrongdoers results in a system of retributive justice, but Reisel would like us to also consider the treatment of these broken brains through rehabilitation programs and restorative justice programs, which I discuss at length in the next chapter.

VIOLENT CRIME AND MORAL CULPABILITY

In 1966 an engineering student and ex-marine named Charles Whitman murdered his wife and mother by stabbing them through the heart with a knife, then drove to the University of Texas in Austin, went to the observation deck of the university's clock tower with a cache of guns and ammunition, and proceeded to shoot and kill fifteen people and wound another thirty-two before being shot and killed by an Austin police officer.[27] Whitman left a note that read, in part:

> I do not quite understand what it is that compels me to type this letter. Perhaps it is to leave some vague reason for the actions I have recently performed. I do not really understand myself these days. I am supposed to be an average reasonable and intelligent young man. However, lately (I cannot recall when it started) I have been a victim of many unusual and irrational thoughts. These thoughts constantly recur, and it requires a tremendous mental effort to concentrate on useful and progressive tasks. . . . After my death I wish that an autopsy would be performed on me to see if there is any visible physical disorder.[28]

Sure enough, the day after the mass shootings pathologists found a pecan-size glioblastoma brain tumor pressing against Whitman's hypothalamus and amygdala—associated with emotions and the fight/flight response—leading a state commission to conclude that it "conceivably could have contributed to his inability to control his emotions and actions."[29] In March of that year, in fact, Whitman consulted with a University of Texas Health Center psychiatrist, whose notes from the meeting included this disturbing observation: "This massive, muscular youth seemed to be oozing with hostility," and "that something seemed to be happening to him and that he didn't seem to be himself," and that Whitman told him he was "thinking about going up on the tower with a deer rifle and start shooting people."[30]

Because of his brain tumor, Charles Whitman probably had fewer degrees of freedom than the average person, in that it seems he could not override his demented impulses, and he was unaware of why he could not control them, beyond the vague sense that something was deeply awry, which it clearly was. But how is a brain tumor different from, say, a horrific and impoverished upbringing and environment that also affect the brain in a way that leads to violence of this magnitude? As another exercise in the principle of interchangeable perspectives—this time into the criminal mind with an aim toward understanding not only the causes of crime but also the complexities of moral culpability when powerful forces are at work that lead someone to commit unspeakable acts—imagine that you are a young African American male raised in one of the worst neighborhoods in one of the most run-down, crime-ridden cities in America: Washington, DC. Your grandmother was fourteen when she gave birth to your mother, who, in turn, was sixteen when she had you. Your mom was raised by an aunt and uncle who physically and sexually abused her, and she adopted these parenting skills when she raised you. Your father is who-knows-where, but his side of the family was rife with drugs, crime, and mental illness. By age two, you had been taken to the emergency room no fewer than five times for head injuries and other assaults on your small body. You "fell" out of a car window, you were knocked unconscious by a swing, you fell out of the top of a bunk bed and landed on your head. Like all babies, you cried when you needed nutrition—both physical and psychological—but instead of being fed and hugged and loved, you were shaken in anger, causing your brain to slosh around inside your skull, causing even more damage to it. At age three you were punched in the head by your mother, causing you to have terrible headaches. At age six you were beaten with an electrical cord. You were often so scared that you wet yourself, and of course you were beaten for that, as you were for bad grades, for minor misbehaviors, or for any other trivial reason that took your caregiver's fancy, including burning you with cigarettes.

In addition to your horrible home life, neighborhood bullies heaped upon you further abuse, including anal rape. Your peer groups consisted of punks and thugs who robbed and burgled for a living and for fun. By the time you turned eighteen you had committed so many robberies and burglaries that you were sentenced to twenty years in prison. After serving four years you were let out on parole where, at a halfway house in Denver, Colorado, you assaulted one of the other residents. You were about to be shipped back to prison in Maryland when you decided to rob an apartment in Denver for some quick cash before catching the bus back East. While you were inside the apartment, however, the female occupant returned home. Terrified to see you, she bolted upstairs, where you chased her down and treated her as you've been treated your entire life—with violence. You hit her in the face and head, you dragged her by the hair into a bedroom, where you tied her up and demanded money. She fought back, and things escalated out of control as you proceeded to rape and beat and stab her until she stopped screaming, resulting in her death. You exited the apartment and caught your one-thirty bus, another day in the life of a man whose entire life can best be described as unalloyed hell.

What's the difference between having a tumor that so alters a brain as to turn someone into a violent killer, and the most horrific upbringing imaginable that so changes a brain as to also turn someone into a violent killer? This is the question asked by the psychiatrist and criminologist Adrian Raine in his monumental study of the biological roots of crime, *The Anatomy of Violence*, whence the second example came.[31] Raine details how neuroscience, cognitive psychology, and evolutionary psychology are converging toward a biological theory of crime. I will return to these stories and Raine's interpretation (and that of the courts) below, but for now consider what you would have done were you either one of these men. Most people, of course, believe that they would have the wherewithal and self-control to suppress these urges. I think that myself. But then, we *would* think that because most of us without brain tumors and hellish backgrounds have brains wired for self-control, perspective-taking, compassion, and empathy. But what if you were *literally* the man with the tumor or the man with the horrific background? Can you say definitively that you would not have committed these crimes? And if you did, would you hold yourself accountable for them as if you had complete self-control? This cuts to the heart of the problem of finding moral culpability in a determined world. Science tells us that we most assuredly live in a universe that is deterministic—where all effects have causes. At the same time, society and the law tell us that we need to hold people morally accountable for their actions to have a civilized society, and that rests on the assumption that people are volitional beings who make free choices.

According to Raine, the childhood of the African American man, Donta Page, was so horrifically bad—impoverished, malnourished, fatherless, abused, raped, and beaten on the head to the point of being hospitalized several times—that a scan of his brain "showed clear evidence of reduced functioning in the medial and orbital regions of the prefrontal cortex," areas associated with impulse control. "Neurological patients with damage to these regions show impulsivity, loss of self-control, immaturity, lack of tact, inability to modify and inhibit inappropriate behavior, poor social judgment, loss of intellectual flexibility, and poor reasoning and problem-solving skills, as well as psychopathic-like personalities and behavior."[32] In comparing Page's brain scan to those he made of forty-one other murderers, he found significant impairment of their prefrontal cortex, resulting in "a loss of control over the evolutionarily more primitive parts of the brain, such as the limbic system, that generate raw emotions like anger and rage."[33] Research on neurological patients in general, Raine notes, shows that "damage to the prefrontal cortex results in risk-taking, irresponsibility, and rule-breaking behavior," along with personality changes such as "impulsivity, loss of self-control, and an inability to modify and inhibit behavior appropriately." There is also cognitive impairment such as a "loss of intellectual flexibility and poorer problem-solving skills" that may later result in "school failure, unemployment, and economic deprivation, all factors that predispose someone to a criminal and violent way of life."[34]

From a neuroscience perspective, we can again ask, what's the difference between an aggressive tumor and a violent upbringing? Both have direct and measurable behavioral consequences, which Raine identifies in another case study of a forty-year-old man who suddenly developed a propensity for pedophilia that he acted on with his twelve-year-old stepdaughter. After being caught he was charged with sexual assault and child molestation and diagnosed by a psychiatrist as a pedophile. After failing rehab but before being sent to prison, he had a brain scan that revealed a massive tumor growing at the base of his orbitofrontal cortex that pressed up against the right prefrontal region of his brain, an area involved in impulse control. The tumor was resected, all pedophilic feelings disappeared, and he returned home to his normal life. Several months later his wife found kiddy porn on his computer, and another brain scan found the recurring tumor culprit. Once again, when the offending growth was removed, so too were his desires for sex with children.[35] It is a compelling story that reveals much about our thoughts and actions, but what does it say about what we ought to do about it?

FROM BIOLOGICAL IS TO MORAL OUGHT

Some people may balk at the biological determinism inherent in such an approach to understanding crime and morality, but before we consider how a science of restorative justice might work (in the next chapter), we need another exercise in interchangeable perspectives by considering Donta Page's victim, Peyton Tuthill, who lost her life in an unimaginably violent way. Her life cannot be restored, no matter how far the sciences of neurocriminology and restorative justice develop and, understandably, her family has a less clinical appraisal of Page's background than the one

I have outlined here (and by Raine in considerably more detail in his book). Raine was consulted on the case on behalf of the defense team in Page's trial, artfully navigating between the inclination to understand and the hunger for punishment, with the defense arguing that Page should be spared the death penalty. A panel of three judges agreed, and Donta Page was sentenced to life in prison. And where our science currently stands, that's a good thing. Given what we know about recidivism rates, there is a very good chance that he would return to a life of crime and violence were he to be set free.

We know this from brain scan studies of reoffenders at correctional facilities in New Mexico and Wisconsin, which were conducted by the neuroscientist Kent Kiehl. Based on these data, Kiehl is able to predict with high certainty which criminals are most likely to return to a life of crime. Using a portable MRI scanner that he carts around to prisons in a trailer, Kiehl correlates prisoner brain scans with the results of their scores on the Psychopathy Checklist—Revised. In general, psychopaths have less gray matter in the paralimbic system related to self-control. As well, Kiehl found that "[C]riminal psychopaths showed significantly less affect-related activity in the amygdala/hippocampal formation, parahippocampal gyrus, ventral striatum, and in the anterior and posterior cingulate gyri"—all brain regions that regulate and control emotions and, in normal brains, make people sensitive to fear and punishment. Psychopaths are notorious for being insensitive to such stimuli, and this insensitivity facilitates their antisocial and criminal activity.[36]

These brains, by the way, show no obvious pathology—no tumors, for example. And yet there was clearly a difference in brain *function*, most likely the result of early life experiences, as evidenced by the fact that most psychopaths show symptoms of psychopathy fairly early in life, and when they do become criminals they start a life of crime at a relatively young age. In another study on criminals, Kiehl found low activity in the anterior cingulate cortex (ACC), which is associated with error processing, conflict monitoring, response selection, and avoidance learning. In fact, people with damage to this area show marked differences in apathy and aggressiveness. Kiehl notes, "Indeed, ACC-damaged patients have been classed in the 'acquired psychopathic personality' genre." Acquired? Specifically, Kiehl found that those criminals with low ACC activity were twice as likely to commit crimes within four years of being released compared to those with high ACC activity. "We cannot say with certainty that all who are in the high-risk category will reoffend—just that most will,"

Kiehl said, adding, "Not only does this study give us a tool to predict which criminals may reoffend and which ones will not reoffend, it also provides a path forward for steering offenders into more effective targeted therapies to reduce the risk of future criminal activity." To that end Kiehl is developing techniques to increase activity in the ACC that he hopes to apply to high-risk offenders.[37]

These are model examples of what I mean by deriving an ought from an is. Once we know the cause of crime—say, low impulse control, insensitivity to fear and punishment, lack of empathy, mental illness, violent upbringing, tumor—if we want to decrease violence in society, we have a moral duty to work on changing these conditions, both in society and in the persons most subject to these forces (and many others) acting on them. We know, for example, that committing crime, especially violent crime, is primarily a male problem. In fact, after a century of progress for women closing the gap with men in so many areas, there remains one in which women continue to lag behind: violent crime. According to the UN Office on Drugs and Crime (UNODC), in its *2011 Global Study on Homicide*, the vast majority of violent offenders everywhere in the world are men, and in most countries males represent more than 90 percent of prison populations. Occasionally men do kill women they know—it's called intimate partner homicide—but mostly they kill other men. Some 80 percent of all homicide victims and perpetrators are men. The global homicide rate for men is 11.9 per 100,000 compared to 2.6 per 100,000 for women, meaning that a man is 4.6 times more likely to be murdered than a woman. But that statistic includes all ages. In the 15-to-29 age range the difference is 21 per 100,000 for men vs. 3 per 100,000 for women, a sevenfold difference. This is due to the fact that males more than females expose themselves to high-risk activities such as gang membership, illegal drug trafficking, dangerous sports and other competitive enterprises, status and honor challenges, bullying, and the like. As the UNODC study reports, "Generally, the higher the homicide rate, the higher the share of men among the suspected offenders. Conversely, the lower the homicide rate, the higher the share of female homicide suspects, though females never make up the majority of homicide offenders. This sex pattern is a clear indication that the share of male homicide offenders among all homicide suspects is a good predictor of the type of homicide that is most prevalent in a country or region."[38]

Knowing this fact about a problem (the way it *is*) enables us to focus on what we *ought* to do about it, which is precisely why the United Nations conducts such research: "taking account of the most relevant characteristics of both perpetrators and victims is thus a necessary requirement for a better understanding of homicide trends and context for the formulation of better, evidence-based policies and crime prevention strategies."[39] Thus we ought to target inner-city gangs, work with at-risk prepubescent boys before they become dangerous teenagers and young men, and in general teach males of all ages self-control techniques for mastering their baser impulses, to prevent assertiveness from morphing into aggression and violence. And knowing the data on intimate partner homicides and the fact that more than twice as many women each year are shot and killed by their husbands or boyfriends or exes than were murdered by strangers, we ought to develop programs and support groups for both men and women for dealing more effectively with domestic and relationship conflicts.[40] If poor nutrition and violent parenting cause brain abnormalities in infants and young children that lead them to a life of crime and violence, we ought to do something about it, not just for them, but also for their would-be victims and for society as a whole.

11

Moral Justice:
Retribution and Restoration

> As I walked out the door toward the gate that would lead to my freedom, I knew if I didn't leave my bitterness and hatred behind, I'd still be in prison.
>
> —Nelson Mandela, after serving twenty-seven years in prison
> for his struggle against South African apartheid[1]

Have you ever thought about killing someone? I have. I've imagined it a number of times, and with a number of different people. Or, if not actually *killing* the particular bastard who's inspired my wrath, at the very least I imagine dislocating his jaw with a crushing roundhouse knuckle sandwich that sends him reeling to the pavement. I think of cursing him all the way down for his malicious, backstabbing actions. In these fantasies I am Cassius Clay standing over Sonny Liston in their 1965 title bout, taunting my opponent to "get up and fight, sucker" after that first-round knockout. I am Billy Jack cutting down racist ruffians who dare to bully innocent Native American youths, burning a slow fuse that escalates into an explosive rage of martial-arts justice as "I just go berserk," a moment of reckless heroism that made it a cult classic of indie film. It's almost embarrassing to admit it, but such fantasies have brought me enormous pleasure as I imagine the feeling of justice being served up to those who have done me, and others, wrong. Of course, I've never done anything like this—nor would I unless I or a loved one were directly threatened with grievous bodily harm—but I can identify with Mark Twain when he quipped, "I didn't attend the funeral, but I sent a nice letter saying I approved of it."

I am not alone, and neither are you if you answered the opening question in the affirmative. In fact, the evolutionary psychologist David Buss, in his 2005 book *The Murderer Next Door: Why the Mind Is Designed to Kill*, reports that most people have harbored homicidal fantasies at some point in their lives. Who are these homicidal fantasists? "Not the gang members or troubled runaways one might expect to express violent rage," Buss explained, but "intelligent, well-scrubbed, mostly middle-class kids." The results shocked him. "Nothing had prepared me for the outpouring of murderous thoughts my students reported," which led Buss to suspect "that actual homicides were just the tip of the deep psychological iceberg of murder. Could actual murder be only the most flagrant outcome of a fundamental human drive to kill?"[2]

To find out, Buss not only conducted his own studies, but also gathered the results of other related studies, which together comprised a database of more than five thousand people worldwide. The results illuminate the darker side of human nature: 91 percent of men and 84 percent of women reported having had at least one vivid homicidal fantasy in their life. One man who acted on these fantasies, from a group of Michigan murderers whom Buss studied, said he killed his girlfriend because "I was deeply in love with her and she knew that. It infuriated me for her to be with another guy." Jealousy is a common motive, as evidenced in another case in which a man flew into a jealous rage during sex with his wife. Why? According to him, she asked him, "How does it feel to fuck me right after someone else has?" He strangled her to death in bed.[3] The impetus behind jealousy is not hate, but attachment and fear of loss, as in this police confession by a thirty-one-year-old man after stabbing his twenty-year-old wife to death when she confessed to having sex with another man during a six-month separation:

> I told her how can you talk about love and marriage and you been fucking this other man. I was really mad. I went to the kitchen and got the knife. I went back to our room and asked: Were you serious when you told me that? She said yes. We fought on the bed. I was stabbing her. Her grandfather came up and tried to take the knife out of my hand. I told him to go and call the cops for me. I don't know why I killed the woman. I loved her.[4]

Even though most such murders are committed by men, there are enough women who do so—and with equally moralistic motives—to constitute a sizable database. For example, case S483 in Buss's study, a forty-three-year-old woman, thought about killing her forty-seven-year-old boyfriend:

> I had this vision of putting poison in his food. My imagination started from the moment when he is back home, and went for his bath. I would put the dinner on the table and take out 2 separate bowls for the soup. One bowl of his will contain rat poison. Without suspicion he will finish the soup. Then I visualize him suffering stomach pain, then white bubbles come out of his mouth, and the next thing he collapses.[5]

Case P96, a nineteen-year-old woman, wanted her ex-boyfriend dead after a series of events over the course of their year-and-a-half relationship:

> The things he did to make me think about killing him were as follows: try to control who I saw, what I did, where I went, when I went. He tried to control every aspect of my life once we came to college together. He would say mean things, call me names, make me feel worthless or like I couldn't find anyone else. . . . There were two main events that triggered my thoughts—1) he got in a huge fight with my mom, 2) he called me a whore.[6]

Buss's point in recording the motives behind these cutthroat fantasies is to confirm the fact that most murders are moralistic in nature. In the mind of the fantasist or the perpetrator, at the time of the murder the victim deserved to die. Over the course of history there have undoubtedly been incalculable instances of abuse so violent that a violent reaction was called for; thus a case can be made that we evolved the capacity for such murderous retaliation out of self-defense. If you do nothing to defend yourself, the bully, abuser, or murderer gets away with it, thereby setting up a self-perpetuating system of brutality as a means to an end. Victims who fight back put the perpetrator (and bystanders) on notice that violence will be met with violence. For example, Buss cites the case of an Australian man named Don who was killed by his wife, Sue, after fourteen years of an abusive marriage:

> Don had also become quite abusive, verbally and physically. The latter included many types of humiliation, and being hit across the head regularly, being threatened with death, being locked in a closet, and being

forced to sit looking in a mirror while Don made derogatory comments about her. On the night of the killing, Don held a knife to Sue's throat while threatening to kill her. He had also both locked her in a closet and urinated in her face. Later that night, after Don had gone to sleep . . . Sue struck him with an axe to the side of the neck about three times. She then stabbed him in the stomach about six times with a large carving knife.[7]

Is there anyone—aside from Don—who would not read this account with some sympathy and understanding for Sue? If someone hit me on the head, humiliated and derided me, locked me in a closet, urinated in my face, and threatened to kill me—or did this to someone I loved—I can easily imagine myself swinging that axe with my sense of moral righteousness dialed up to eleven. Moral righteousness appears to be what a woman named Susan felt when she responded aggressively to her abusive cocaine-addicted husband as he advanced on her with a hunting knife, screaming, "Die, bitch!" Susan kneed him in the groin and grabbed the knife—surely a rational response to a wigged-out maniac. At her trial she said, "I was terrified because he was gonna kill me. I knew the second that I stopped he was gonna get the knife back and then I was gonna be the one that would be dead."

However, once she got started, Susan found that she couldn't stop stabbing her husband, a phenomenon that the sociologist Randall Collins calls "confrontational tension." This kind of exorbitant, rising psychological pressure can lead to a "tunnel of violence," and in its most extreme expression a "forward panic"—an explosion of rage and anger that is released through aggression and violence, as in the beating of Rodney King by Los Angeles police officers who appeared on the bystander's grainy video like a pack of wolves ripping apart their prey.[8] While lost in her tunnel of forward panic violence, Susan stabbed her husband 193 times. "I stabbed him in the head and I stabbed him in the neck and I stabbed him in the chest and I stabbed him in the stomach, and I stabbed him in the leg for all the times he kicked me, and I stabbed him in the penis for all the times he made me have sex when I didn't want to."[9]

Such revenge emotions are common enough that authors and film directors count on them, as the postrape revenge scene in the film version of *The Girl with the Dragon Tattoo* demonstrates. As the main character, Lisbeth Salander, tasers her rapist, ties and gags him, and tattoos "I am a sadistic pig and a rapist" on his torso in giant letters, the audience in the theater in which I saw the film cheered loudly with righteous approval.

THE EVOLUTIONARY ORIGINS OF MORAL JUSTICE

Emotions evolved to direct behavior toward our own survival and flourishing, and specifically moral emotions (such as guilt, shame, empathy, contempt, vengeance, and remorse) evolved to guide our behavior in interaction with others. *Anger* leads us to strike out, fight back, and defend ourselves against predators, bullies, and abusers. *Fear* causes us to pull back, retreat, and escape from risks. *Disgust* directs us to push out, eject, and expel that which is bad for us, such as bodily excreta and other disease-carrying substances. Computing the odds of danger in any given situation often just takes too long. There is always the possibility that we may need to react instantly, and this is what emotions are "for" in an evolutionary sense.

We can track the evolution of the desire for justice through two lines of evidence: our primate cousins and our hunter-gatherer ancestors. We share a common ancestor with chimpanzees from a point about six million to seven million years ago, so we shall begin with the expression of the moral emotion of justice in chimps who, along with bonobos, are our closest living primate ancestors. In his book *Chimpanzee Politics*, the primatologist Frans de Waal describes behaviors that are clearly "direct payment for services rendered," noting that, in general, "chimpanzee group life is like a market in power, sex, affection, support, intolerance and hostility." As with humans, chimpanzees also show a dual system of reward and punishment, as de Waal notes, "The two basic rules are 'one good turn deserves another' and 'an eye for an eye, a tooth for a tooth.'"[10] In his follow-up study *Peacemaking among Primates*, de Waal records what happens after chimpanzees and other primates fight among themselves within their group—they make up. And they do so with a very humanlike hug or an arm-about-the-shoulder embrace.[11] In *Our Inner Ape* de Waal explains the researcher's procedure of waiting for a fight to happen between chimps, and then recording what they do next. "Bystanders often embrace and groom distressed parties."[12] Conciliation is the key to conflict resolution, and to grease the wheels of justice something must be done to make the peace—at least provisionally—so that everyone feels like they can go on living together without excessive resentment.

Bonobos—a type of chimpanzee known to be far more amorous than its chimp cousins and equidistant from humans on the branching tree of evolution—resolve conflicts with another human favorite, make-up sex, or at least lots of grooming and touching and petting. Anatomically, bonobos are more similar to that of our evolutionary ancestors the australopithecines, and compared to chimpanzees they practice more bipedalism, more face-to-face mating (missionary-style sex), more oral sex and tongue kissing, and have reduced limb and body proportions, smaller canines, consume a wider variety of foodstuffs, organize into larger group sizes, show less intragroup competition and aggression, and are closer to humans in levels of hormones that promote sociability and in brain regions that are associated with empathy.[13]

Making a broader argument for the evolutionary origin of such emotions, de Waal notes that such appeasing behavior is not restricted to chimpanzees and the other great apes, but also can be found among most social mammals: "Elephants are known to use their trunk and tusks to lift up weak or fallen comrades. They also utter reassurance rumbles to distressed juveniles. Dolphins have been known to save companions by biting through harpoon lines, hauling them out of tuna nets in which they got entangled, and supporting sick companions close to the surface so as to keep them from drowning."

For a species of social animal to survive, individuals must have the cognitive tools and behavioral repertoire to resolve conflicts, keep the peace, and suppress their aggressive tendencies when they bubble up from below. And they do. Capuchin monkeys, for instance—a much smaller-brained and more evolutionarily distant cousin of humans—also display these same traits. In studies with both species, de Waal and his colleagues found that when two individuals work together on a task for which only one is rewarded with a desired food, if the reward recipient does not share that food with his task partner, the partner will refuse to participate in future tasks and will express emotions that are clearly meant to convey displeasure at the injustice. A peeved-off primate will show it by rattling his or her cage, throwing objects, and shrieking in anger.[14]

In an experiment conducted by the psychologist Sarah Brosnan, two capuchin monkeys were trained to exchange a granite stone for a cucumber slice, but then Brosnan introduced a devious injustice into the experiment by giving one monkey a grape instead of a cucumber. Monkeys prefer grapes to cucumbers because they're sweet; thus the naturally evolved preference for sugary foods makes them more desirable. Under this condition, the slighted monkey who got the cucumber cooperated only 60 percent of the time, sometimes even refusing to take the cucumber slice altogether. In a third condition, Brosnan ramped up the sense of injustice even farther by giving only one of her capuchin subjects a grape

without even requiring him to swap a granite stone for it. Under these terribly unfair conditions the snubbed monkey cooperated only 20 percent
of the time, and in many instances became so agitated that the cucumber
slice became a makeshift projectile, hurled out of the cage at the human
experimenters.[15] In a similar experiment at the University of Zürich with
long-tailed macaques—another relatively small-brained primate cousin of
ours—primatologists Marina Cords and Sylvie Thurnheer found that
after two macaques had learned to cooperate with each other in a task that
required both of their efforts in order to be rewarded with food, they were
more likely to reconcile after a quarrel than those monkeys who had not
learned to cooperate in the joint-effort task.[16]

So fairness and justice go hand in hand in the moral economy that
trades in currencies such as food, grooming, conciliation, reciprocity,
friendship, and alliances. De Waal discovered, for example, that chimps
who have groomed one another are more likely to share food, and that
capuchin monkeys tend to share food and grooming activities, and to
conciliate more readily with those of their fellows with whom they've
traded such currencies before.[17] This is a literal example of the colloquial
definition of reciprocal altruism—I'll scratch your back if you'll scratch
mine. The ethologist Nicola Koyama and her colleagues at John Moores
University in Liverpool discovered that chimps who groomed each other
formed social alliances against other chimps. For example, if chimp A
groomed chimp B, then chimp B would be more likely to support chimp
A in a fight with others the next day. The researchers took this to mean
that chimpanzees curry political favor through positive interactions with
other chimpanzees in anticipation of a possible future need, even though
they cannot articulate the exchange calculation through language in the
same way that humans can.[18]

The research on primates supports the thesis of this book that we
evolved the capacity to actually be moral animals—and not just give the
appearance of being moral animals. It is not enough to fake being a good
person because, if you act otherwise, in the long run your fellow group
members may find you out. So you actually have to *be* a good person most
of the time, by which I mean being pro-social, reciprocally altruistic,
cooperative, and fair, and to do so in a way that feels good, feels right, and
feels just.[19] That is what the moral emotions give us—true and genuine
morality even if we don't always live up to our own or society's moral
standards. Without such moral emotions, our actions would be nothing
more than a simple selfish moral calculation. And as de Waal notes, there

is no evidence that nonhuman primates are *calculating* future payoffs for current kindnesses, so this fact strengthens my position because it means that they are doing so for an immediate reward of feeling good about helping others.[20] So the moral emotions are a proxy for a moral calculation evolved through natural selection that leads us to assume not only that others of our group are really moral, but that we are as well. Morality is a real, biologically based phenomenon, and the desire for moral justice is as concrete an emotion as love.

These observations and experiments on conflict and resolution in primates are a window into our evolutionary past—emotional "fossils" that help to piece together what life might have been like in the environment of our ancient ancestry. The fact that both modern humans and present-day primates have a sense of fairness and justice points to a common response to unfairness and injustice that evolved in our distant past as a conflict-resolution tool, without which the members of such social species would be less likely to survive and flourish. If everyone only ever acted in their own self-interest with no concern for those around them in their band, the social community would disintegrate into anarchy and violence. We evolved emotions that lead us to care about the outcome of interactions with our fellow group members, most notably that these interactions and exchanges be fair. As Sarah Brosnan concludes from her studies on inequity in nonhuman primates, unequal social outcomes lead to antagonistic responses, and vice versa: "An aversion to inequity may promote beneficial cooperative interactions, because individuals who recognize that they are consistently getting less than a partner can look for another partner with whom to more successfully cooperate."[21] This leads to the natural selection of pro-social and cooperative behavior *within* groups and the selection of xenophobia and tribalism *between* groups, or more briefly, *within-group amity* and *between-group enmity*.

So the sense of justice evolved to deal with conflict resolution and to keep bullies, abusers, and murderers from overrunning a society and thereby decreasing the evolutionary fitness of everyone in the group, which otherwise could threaten extinction. There has to be some means of dealing with fellow group members who won't play nice. *Moralistic punishment* is one such action driven by a desire for justice in an exchange, as when subjects in the ultimatum game rejected unfair offers with a sense of righteous injustice. The universal nature of moralistic punishment was found in studies that used the ultimatum game protocols all over the world, including in fifteen small-scale traditional societies. Compared to subjects in Western countries who typically propose and accept splits in the 50/50 to 70/30 range, interactions vary among small-scale traditional tribes, from a minimum of 26 percent for the Machiguenga tribe in Peru to a maximum of 58 percent for the Lamelara tribe in Indonesia. The variation appears to be associated with the primary occupation of the people, where those whose economies are large and more market-integrated tend to make fairer offers than those whose economies are at a more subsistence and less market-integrated level.[22]

In a comprehensive overview of the substantial literature on such experiments, the anthropologists Joseph Henrich, Robert Boyd, and their colleagues concluded that people everywhere in the world "care about fairness and reciprocity, are willing to change the distribution of material outcomes among others at a personal cost to themselves, and reward those who act in a pro-social manner while punishing those who do not, even when these actions are costly."[23] The anthropologists also noted that even though human cultures vary enormously, utilizing vastly different forms of social organization and institutions, kinship systems, and environmental conditions, there nevertheless remains a set of core features of our nature that have an evolutionary basis to them, including the fact that no human group consists of purely selfish individuals, and that all humans have a sense of fairness and justice.[24]

REVENGE AND JUSTICE AS DETERRENCE

The emotions driving the need for justice evolved for a number of good evolutionary reasons, one of which is to deter others from free riding, cheating, stealing, bullying, and murder. A perpetrator who is cognizant of the possibility of retribution may hesitate or even fail to act altogether, assuming that he (or, less likely, she) has the normal response of apprehension and fear of the threat of punishment. And for those who lack such sensitivities to the feelings of others—also known as psychopaths—brute force may be the only currency they understand. So an evolved sense of retributive justice is natural, and in many cases it may have been the only course of action available to our ancestors. Indeed, had we not evolved such emotions our Paleolithic ancestral tribes might have been overrun by toughs, bullies, and killers, and that might have spelled the doom of our species.

How human groups developed systems of justice is the subject of anthropologist Christopher Boehm's enlightening—and in places shocking—book *Moral Origins: The Evolution of Virtue, Altruism, and Shame.*[25] Boehm's analysis is derived from a database that begins with 339 pure forager societies, then subtracts those that likely do not resemble our ancestors (mounted hunters, horticultural hunters, fur trade hunters, and sedentary-hierarchical hunters) to produce a working dataset of 50 Late Pleistocene Appropriate (LPA) societies from which to work. These are groups either still in existence today or groups that were studied within the past century by anthropologists, and who might reasonably be assumed to represent a close approximation to how our ancestors lived. These ethnographies, in conjunction with archaeological evidence, form the basis for theories on how our species lived before the rise of civilization.[26]

Boehm argues that *moralistic punishment* evolved, in part, to solve the problem of how altruism could evolve, and how these relatively equitable societies could be stable when free riders could game the system by taking more than they put in by, say, slacking off on a gathering expedition, hanging back during a dangerous hunt, or simply taking more than their fair share of food. Thus Boehm found that all of these societies had sanctions to deal with deviants, free riders, and bullies, which ranged from social pressure and criticism to shaming, ostracism, ejection from the group, and even—in extreme cases when nothing else worked—capital punishment. The sanctioning process begins with gossip as a private exchange of evaluative information about who is doing their fair share and who isn't, who can be trusted and who cannot, who is a good and reliable member of the group and who is a slacker, cheater, liar, or worse. Gossip permits the group to form a consensus about the deviant that can lead to a collective decision about what to do about him (and it is almost always a male). Of course, the problem with hard-core bullies such as psychopaths is that they don't care about being detected and talked about, so gossip also served as a means for subordinate band members to form coalitions based on the axiom that there is *strength in numbers*.

It's surprising to hear the term "capital punishment" used in the context of hunter-gatherer societies, but the death penalty is sometimes used to preserve group harmony in the case of an intractable thug who refuses to bend to the rules or respond to lesser sanctions. In the fifty LPA societies studied, Boehm found that twenty-four of them practiced capital punishment for crimes such as malicious sorcery, repeated murder, tyrannical behavior, psychotic behavior, theft, cheating, incest, adultery, premarital sex, a taboo violation that endangers everyone in the group, betraying the group to outsiders, "serious or shocking transgressions," and unspecified deviance. The total comes to 48 percent but is probably higher because cap-

Crimes and Sins in Traditional Societies			
TYPE OF DEVIANCE	**% OF SOCIETIES**	**% OF OFFENSES MENTIONED**	**TOTAL MENTIONS IN FIELD REPORTS**
INTIMIDATORS	100%	69%	471
MURDER	100%	37%	248
SORCERY	100%	18%	122
BEAT SOMEONE	80%	12%	79
BULLYING	70%	2%	12
DECEIVERS	100%	31%	171
STEALING	100%	15%	99
FAILURE TO SHARE	80%	6%	34
LYING	60%	7%	48
CHEATING	50%	3%	24

Figure 11-1. Crimes and Sins in Traditional Societies

The anthropologist Christopher Boehm has compiled a database of crimes and sins in traditional societies that result in punishments ranging from shunning to capital punishment.[27]

ital punishment is notoriously underreported in ethnographies, since it is generally hidden from snooping anthropologists—these modern hunter-gatherer societies know that execution is prohibited by the colonial administrators in their region, so they prudently hide it from outsiders. *Figure 11-1* shows the accounting of crimes and sins and their accompanying punishments from Boehm's database of traditional societies.

Boehm cites a gruesome example of an execution recorded by the anthropologist Richard Lee in his study of the !Kung Bushmen of Africa. The account involves a bully and murderer named /Twi, who had killed at least two people and for whom the group had decided that execution was the only solution. Lee interviewed /Twi's father, mother, sister, and brother. "All agreed he was a dangerous man. Possibly he was psychotic." The account also shows, sans modern effective weaponry, how difficult it can be to kill another human being who fights back, and the power of the collective to overcome even the strongest adversary. (Diacritical symbols represent the various phonemic "clicks" in the Bushmen dialect):

It was Xashe who attacked /Twi first. He ambushed him near the camp and shot a poisoned arrow into his hip. They grappled hand to hand, and /Twi had him down and was reaching for his knife when /Xashe's wife's mother grabbed /Twi from behind and yelled to /Xashe, "Run away! This man will kill everyone!?" And /Xashe ran away.

/Twi pulled the arrow out of his hip and went back to his hut, where he sat down. Then some people gathered and tried to help him by cutting and sucking out poison. /Twi said, "This poison is killing me. I want to piss." But instead of pissing, he deceived the people, grabbed a spear, and flailed out with it, stabbing a woman named //Kushe in the mouth, ripping open her cheek. When //Kushe's husband N!eishi came to her aid, /Twi deceived him too and shot him with a poisoned arrow in the back as he dodged. And N!eishi fell down.

Now everyone took cover, and others shot at /Twi, and no one came to his aid because all those people had decided he had to die. But he still chased after some, firing arrows, but he didn't hit any more.

Then he returned to the village and sat in the middle. The others crept back to the edge of the village and kept under cover. /Twi called out, "Hey are you all still afraid of me? Well I am finished, I have no more breath. Come here and kill me. Do you fear my weapons? Here I am putting them out of reach. I won't touch them. Come kill me."

Then they all fired on him with poisoned arrows till he looked like a porcupine. Then he lay flat. All approached him, men and women, and stabbed his body with spears even after he was dead.[28]

Anthropologists have documented many forms of disruptive behavior, which Boehm roughly separates into two categories: intimidation and deception. Intimidation includes murder, sorcery, physical violence, and bullying. Deception includes stealing, the failure to share, lying, and cheating. Of the fifty societies in the database, 100 percent reported murder, sorcery, and stealing; 90 percent said people had failed to share; 80 percent had experienced physical violence; 70 percent had bullying as a problem; 60 percent said liars were among them, and 50 percent reported cheating. All of these behaviors generated gossip in the community, which led to group decisions concerning appropriate punishments.

These traditional societies also distinguish between reversible and irreversible sanctioning of deviants. Reversible sanctions are used when the group wants to be rid of an antisocial behavior but not of the perpetrator because he is an otherwise useful member of the group. Irreversible sanctions occur in the form of either permanent expulsion (which often means death by starvation or murder by another tribe) or execution, the latter only after reversible sanctions have failed or when a bully has proven to be a serious threat to the group.

In an evolutionary context, free riders and cheaters who respond to sanctions maintain their genetic fitness and pass on their genes for modest levels of free riding and cheating, which is what we see in all societies today. Of course, as in all human traits, bullying, free riding, and cheating are results of both genes and environment in interaction, so we are talking about propensities and probabilities here. In a world in which modest free riding happens—often deceptively, making detection difficult—we evolved cheater detection tools and the propensity to gossip about those whom we think might be trying to deceive us and cheat the system. Adding it all up, Boehm concludes that "what we have is a system of social control that can drastically reduce the genetic fitness of more driven free riders whose consciences can't keep these dangerous traits under control, but that allows the more 'moderate,' would-be free riders to control themselves in matters that would otherwise bring punishment and still express their competitive tendencies in ways that are socially acceptable. It's for this reason that free riders haven't just gone away."[29]

So an evolutionary arms race has left us with cheaters and cheating detection, free riding and free riding deterrence, bullies and bullying punishment. Out of this arms race evolved another feature of the human mind—a *moral conscience*—that acts as the "inner voice" of self-control. Because social sanctions allow for individuals to reform their bad behavior (before the ultimate form of punishment—banishment or execution—is employed), conscious self-awareness of one's actions allowed for adjustments to be made of one's behavior. "It was earlier types of social control that caused a conscience to evolve, and it's an evolved conscience that makes individuals so adept at this important type of self-inhibition," Boehm writes. Why, then, are there still modern hunter-gatherers being executed, banished, ostracized, and shamed by their groups for free riding? Because, says Boehm, "they hope they can get away with it."[30] In the long run, however, most don't, but it's a long enough run that they manage to reproduce in the meantime so their genes for such cheating and free riding have made their way to modern humans.

Fortunately our individual consciences are malleable and respond to social cues, approval, opprobrium, and punishments. It is the force of this ever-expanding conscience, working in concert with reason, that has been behind the historical development of criminal justice in the West.

LADY JUSTICE: FROM THE WILD WEST TO
THE MODERN WEST

In the long history of civilization, self-help justice conducted by individuals has gradually been replaced with criminal justice conducted by the state. The former leads to higher rates of violence than the latter, due to the lack of an objective third party to oversee the process. States, for all their faults, have more checks and balances than individuals. This is why Justitia—the Roman goddess of justice—is often depicted wearing a blindfold, symbolizing blind justice and impartiality; in her left hand she carries a scale on which to weigh the evidence, a symbol for a balanced outcome, and in her right hand she wields the double-edged sword of reason and justice, symbolizing her power to enforce the law. Of course, it is not an either-or situation—nonstate vs. state justice systems; it is, rather, a sliding scale from small communities with no central authority or independent judicial system, to chiefdoms in which a single authority (the chief, or "big man") resolves conflicts, to small and weak states where individuals employ self-help justice when they feel the state's justice system has failed them, to large and strong states with relatively effective judicial systems, to totalitarian states where justice is whatever the authorities (or autocrat) say it is.

The modern West's system of criminal justice was a big improvement over the medieval system of torturing first and asking questions later. In the eighteenth century, scholars such as Jeremy Bentham and Cesare Beccaria—the criminal justice reformer we met in chapter 1—made the case that "the punishment should fit the crime," with the overall goal of "the greatest happiness of the greatest number" as the calculus for justice.[31] As Beccaria argued in his 1764 work *On Crimes and Punishments*: "It is not only the common interest of mankind that crimes should not be committed, but that crimes of every kind should be less frequent, in proportion to the evil they produce to society. Therefore, the means made use of by the legislature to prevent crimes, should be more powerful, in proportion as they are destructive of the public safety and happiness, and as the inducements to commit them are stronger. Therefore there ought to be a fixed proportion between crimes and punishments."[32]

The goal of modern Western judicial systems is to prevent citizens of a nation from using force and committing violence against one another when disputes arise. When citizens undertake self-help justice it is a net loss to the state, since it often escalates into endless cycles of violence.

Today this is conducted through two justice systems: criminal and civil. Criminal justice deals with crimes against the laws of the land that are punishable only by the state. Civil justice deals with disputes between individuals or groups, such as contract violations, property damage, or bodily injury, and the court's say is final in determining right and wrong and appropriate damages. Criminal justice involves mostly retribution. Civil justice involves both retribution and restoration (through assessed damages). For both forms of justice, states claim a monopoly on the legitimate use of force, with the goal of deterring future crimes against citizens of the society. This is why criminal cases are labeled *The State v. John Doe* or *The People v. Jane Roe*. The state becomes the injured party. My home state of California, for example, continues to pursue charges against the filmmaker Roman Polanski for raping an underage girl in 1977, even though she—now a woman in her forties—has forgiven him and has requested that the state drop the charges, and even though Polanski lives in Switzerland and has no intention of ever returning to the United States.

This is why in areas of Western nations where citizens do not feel that the law is fair to them—as in parts of the United States where the police and courts are perceived to be racist—people often take the law into their own hands. This is why it is called "self-help justice," or sometimes "frontier justice," or plain old "vigilantism." Take the case of violence in inner cities, where crime rates are much higher than elsewhere. The primary reason for this violence is gang-related illegal drug trafficking. When a product that people want is made illegal, it does not necessarily eliminate the desire for the product; instead it shifts the economic transaction from a lawful free market to an unlawful black market—think alcohol during Prohibition, or drugs today. Because drug dealers cannot turn to the state to settle disputes with other drug dealers, self-help justice is their only option. As such, criminal gangs emerge (most famously the Mafia) that enforce a different sort of criminal justice.[33]

Occasionally conditions arise in which ordinary citizens feel pressed to take the law into their own hands, as Bernhard Goetz did on December 22, 1984, when four young men approached him on a New York City subway car in what he perceived to be a threatening manner. At the time of this incident, New York City was in the throes of one of the biggest crime waves in American history, having seen its rate of violent crime skyrocket nearly fourfold from 325 to 1,100 per 100,000 people in only a decade. In fact, three years prior to this incident, three young men had robbed Goetz of some electronic equipment and then tossed him through a plate-glass door. One of his attackers was caught, but was charged only with criminal mischief for ripping Goetz's jacket and was released from the police station even sooner than Goetz was. The criminal went on to mug again, leaving Goetz skeptical of the criminal justice system and the police's capacity to protect him from harm. So to protect himself, Goetz purchased a Smith & Wesson .38-caliber handgun.

On that fateful night in 1984, the four young men boarded the subway car carrying screwdrivers, intent (they later said) on knocking off video arcade machines in Manhattan. When Goetz attempted to exit the subway train, they surrounded him and demanded money. (In their trial they claimed that they had only been "panhandling" and had merely "asked" for the money instead of demanding it.) Given his prior experience, his awareness of the crime wave, and the gun in his pocket, Goetz was conditioned for confrontation. He shot the men and exited the train.

Not long afterward, Goetz became known as the "Subway Vigilante" and was the topic of a national debate on crime and vigilantism. In response, and to show that it would not tolerate a return to the Wild West form of vigilante justice, the state criminal justice system came down hard on Goetz, charging him with four counts of attempted murder, four counts of reckless endangerment, and one count of criminal possession of a weapon. Living in what was essentially a lawless subsection of a civil society—New York City subways—Goetz reacted, in his own words, like an animal. "People are looking for a hero or they are looking for a villain. And neither is the truth. What you have here is nothing more than a vicious rat. That's all it is. It's not Clint Eastwood. It's not taking the law in your own hands. You can label that. It's not being judge, jury and executioner."[34]

In a civilized society, in fact, it is the state's duty to provide judge, jury, and executioner, but the public disagreed, most siding with Goetz, and several groups established legal defense funds on his behalf. In his criminal trial he was acquitted of the attempted murder charges but served eight months for carrying a loaded but unlicensed weapon in a public place.[35] As Goetz reflected: "What happens to me at this point is unimportant. I'm just one person. This has at least raised issues in New York. One thing I can do to show the legal system what I think of it."[36]

Why do people in civilized societies with justice systems and police forces nevertheless choose not to work within the law? The sociologist Donald Black attempted to answer this question in an article titled "Crime as Social Control," in which he takes the well-known statistic that only 10 percent of homicides belong in the category of predatory or instrumental violence, arguing that most homicides are moralistic in nature. Most murders, for example, are a form of capital punishment in which murderers are judge, jury, and executioner over a victim they perceive to have wronged them in some manner deserving of the death penalty. Black provides examples that are as common as they are disturbing: "a young man killed his brother during a heated discussion about the latter's sexual advances toward his younger sisters," another man who "killed his wife after she 'dared' him to do so during an argument about which of several bills they should pay," a woman who "killed her husband during a quarrel in which the man struck her daughter (his stepdaughter)," another woman who "killed her 21-year-old son because he had been 'fooling around with homosexuals and drugs,'" and several involving disputes over automobile parking spots.[37] Most violence, in fact, is a form of moralistic punishment.

RETRIBUTIVE JUSTICE AND RESTORATIVE JUSTICE

The theory of justice that considers proportionate punishment to be the most effective means of deterring crime is called *retributive justice*. In the context of the evolutionary origins of human emotions, retributive justice is based on the completely understandable desire for fair play. We feel, instinctively, that if people commit crimes it's only right that they get their just deserts. No one should be allowed to get away with murder—or with rape, burglary, embezzlement, kidnapping, or driving with fewer than two vehicular occupants in an HOV lane. We feel that if *we* can't take hostages or park on the steps of the Lincoln Memorial, no one else should be allowed to; and if anyone *does* get away with these things, our moral emotions go into overdrive and we naturally want to see justice served—unless, of course, we identify with the antihero of a story, say, and then we actively *want* to see Ferris Bueller get away with skipping school.

The criminal justice system as practiced by most modern societies evolved over centuries primarily under the rubric of retributive justice. And for good reason—to keep the peace and maintain a relatively smoothly functioning society, a state must maintain a monopoly on the legitimate use of force and do so through the enforcement of the law by punishing the rule breakers. However, complementary to retributive justice is *restorative justice* (also called *reparative justice*), in which the perpetrator (which might be an individual or even a nation) apologizes for the crime; attempts to set to rights the situation; and, ideally, initiates or restores good relations with the victim. Retributive justice is more emotionally driven and comes from a desire for revenge (though retribution should be distinguished from vengeance), while restorative justice is more reason-driven and grows out of the necessity to get along with our fellow group members after a crime occurs.

In the past couple of decades a restorative justice movement has been building on the foundation developed first in New Zealand and based on the indigenous Maori society's method of justice, which focuses on repair instead of punishment (captured in the Maori proverb "let shame be the punishment"). In New Zealand in the 1980s there was a crime wave sweeping through the society—as there was in most Western nations—and thousands of young people, including and especially Maori children and teenagers, were caught in the nets of law enforcement and put into foster homes or institutions. Despite New Zealand's having one of the highest juvenile incarceration rates in the world, its crime rate remained high; clearly the criminal justice system just wasn't working. Maori leaders responded by explaining how in their tradition instead of condemnation and incarceration they focused on problem solving and damage repair.

In 1989 landmark legislation was passed in the form of the Children, Young Persons, and Their Families Act, which revamped the focus and process of juvenile justice and which led to the development of Family Group Conferences (FGC), used either in addition to or instead of court, primarily focusing on the rehabilitation of troubled youths. "Organized and led by a Youth Justice Coordinator, a facilitator who is a social services professional, this approach is designed to support offenders as they take responsibility and change their behavior, to empower the offenders' families to play an important role in this process, and to address the victims' needs," explain Allan MacRae and Howard Zehr, pioneers of this system.[38] In the past two decades there have been more than one hundred thousand FGCs, with high rates of victim satisfaction reported. New Zealand's Ministry of Justice recorded a 17 percent reduction in imprisonment, a 9 percent decrease in recidivism measured after two years, and a 50 percent reduction in the seriousness of offenses where participants did reoffend.[39]

New Zealand politicians are also recognizing both the moral and the cost benefits to restorative justice programs. New Zealand's minister of finance, for example, called the country's prison system a "moral and fiscal failure," adding that prisons are "the fastest rising cost in government in the last decade and my view is that we shouldn't build any more of them." New Zealand District Court judge Fred McElrea, after applying the principles of restorative justice for twenty years, concluded, "For those seeking a more satisfying, less damaging, and cheaper form of justice, the way forward, in my view, is clear. It is not suitable in all cases, but with some principled support and seed funding, restorative justice could easily change the landscape of the criminal justice system in most common law jurisdictions."[40]

A study comparing two teenagers—one from the United States and the other from New Zealand, who each killed their abusive fathers—is highly instructive. The teen from the United States ended up in prison with a sentence of twenty-two years to life, while the teen from New Zealand (now age twenty-two), after going through New Zealand's special court for minors and the Family Group Conference, earned an education and ended up working for New Zealand's forest service as a free, contributing member of society.[41] Allan MacRae recounts another case that he mediated involving a young refugee who had come to New Zealand with his grandmother and an aunt. They had no money and survived on a tiny benefit from the New Zealand government that barely covered food and housing. In desperation, the young man assaulted his grandmother and stole their rent money for his own use. His aunt turned him in to the police, but instead of incarcerating him they referred the case to MacRae for an FGC, who set up a meeting with all affected parties, which unfolded as follows:

> The Conference started with a prayer in their native language, and all parties used interpreters to ensure full understanding. The grandmother told her story in much detail, as did the young person. As the young person began to understand the impact he was having on his grandmother, tears came to his eyes. The young man eventually told of his life in a refugee camp before the three arrived in New Zealand, what he had to do to survive, and how in his new community he felt he could not mix with others if he did not have money. Clearly, loneliness, anger, and hurt were shared by both the young man and his grandmother.

The young person agreed to pay back in full all the money he had stolen, and he was given help in finding part-time employment. He could not live with his grandmother until she felt safe being around him, and he was assigned a mentor from his own culture to help him complete his community work and attend school. "The plan was successful," MacRae writes. "The young man did no further offending, and he completed all his outcomes. Most valuable of all, both he and his grandmother found new friends and support that stayed with them, well beyond the Family Group Conference plan, and assisted them in starting their new lives in New Zealand."[42]

According to Howard Zehr—one of the innovators of the movement—restorative justice is not just about forgiveness or reconciliation (although that is a positive by-product enjoyed by many who have tried it); instead, it begins with an acknowledgment by the wrongdoer, who must take some level of responsibility for the offense, and builds from there to include the victims' losses and a plan for restoration. The stakeholders in restorative justice include the victims, their families, and the community affected by the crime. Restorative justice is meant to be a complement to retributive justice, not a replacement of it.

The problem with the criminal justice system is that crime is defined as being against the state, which often leaves the actual victims of a crime out of the loop. Recall the highly publicized criminal trial of O. J. Simpson for the murder of his wife, Nicole Brown Simpson, and her friend Ronald Goldman. O. J. was found not guilty (as in Johnny Cochran's memorable line "if the glove doesn't fit, you must acquit"), but even if the glove had fit and he'd been found guilty and sent to jail, the families of the victims would have received no compensation from either the Simpson estate or the state that tried him (California). For a modicum of restorative justice, the families had to file a lawsuit against Simpson, and in a less-publicized civil trial, O. J. was found guilty of wrongful death and battery and was ordered to pay the families $33.5 million in compensation. Naturally, in such an adversarial system, Simpson did what he could to hide assets and avoid payment. The Goldman family managed to collect a measly $500,000 from the sale of Simpson's Heisman Trophy and other personal belongings,[43] and they have pursued other avenues of collection of monies he earned from signing autographs and the sale of memorabilia.[44]

Since it is based primarily on retribution, the criminal justice system neglects the needs of victims in at least four areas that Zehr says must be addressed for a restorative system to work: (1) *Information*. Victims want to know the deeper reason for a crime—the intent by the perpetrator—and this can only come from looking into his eyes and listening to his voice with facial expressions and body language. (2) *Truth telling*. Victims feel the need to tell perpetrators how the criminal acts affected them. This is occasionally done at the end of a criminal trial when the victims or their families can face him and have their say just before the perpetrator is sent out of the room in handcuffs. (3) *Empowerment.* "Victims often feel like control has been taken away from them by the offense—control over their property, their body, their emotions, their dreams." In the criminal justice system the victims have next to no control or power, since by definition the state has taken on that role. (4) *Restitution or vindication.* "Restitution by offenders is often important to victims, sometimes because of the actual losses but just as importantly, because of the symbolic statement implied. When an offender makes an effort to correct the harm, even if only partially, it is a way of saying, 'I am taking responsibility, and you are not to blame.'"[45]

In the criminal justice system, not only are these four areas neglected, but also the adversarial nature of the legal process encourages offenders to keep their mouths shut, "lawyer-up," "take the Fifth," never acknowledge any wrongdoing, and plead guilty only if their attorney can plea-bargain them to a better outcome or spare them a death sentence. And punishment is not accountability. What pleasure a victim may get from the punishment meted out to the offender fades when the losses from the crime remain unrestored. For restorative justice to work, offenders need to acknowledge their wrongdoing and be held accountable for the losses sustained by their victims. In short, the retributive justice system is focused on what offenders *deserve*, whereas restorative justice is concerned about what victims *need*; retributive justice is about what was done wrong, whereas restorative justice is about making it right; retributive justice is *offender*-oriented, whereas restorative justice is *victim*-oriented.

Let's look more closely at how restorative justice has been practiced around the world, starting with a traditional society in Papua New Guinea.

TOK-SORI—RESTORATIVE JUSTICE IN TRADITIONAL SOCIETIES

In his epic work *The World Until Yesterday*, the evolutionary biologist Jared Diamond recounts his experiences living among traditional societies in Papua New Guinea, drawing from those experiences lessons for how we can improve our own society. In his discussion of how justice is handled among these communities, Diamond tells the story of a Papua New Guinean man named Malo, who accidentally struck and killed a young boy named Billy on a small-town road. It was an accident. Billy had darted behind the school bus he'd just exited, which blinded Malo to the boy's presence until it was too late. Billy suddenly bolted across the road in front of the bus to meet his uncle, and Malo could not stop his bus in time. Instead of waiting for the police to arrive, as we would do in the West (unless we want a hit-and-run conviction), Malo beat a hasty retreat because, as Diamond explains, "angry bystanders are likely to drag the offending driver from his car and beat him to death on the spot, even if the accident was the fault not of the driver but of the pedestrian."

Heightening tensions was the fact that Malo and Billy were from different ethnic groups (Malo was local but Billy was from the lowlands), which, according to Diamond, elevated emotions: "If Malo had stopped and gotten out to help the boy, he might well have been killed by lowlander bystanders, and possibly his passengers would have been dragged out and killed as well. But Malo had the presence of mind to drive to the local police station and surrender himself. The police locked up the passengers temporarily at the station for their own safety, and escorted Malo for his safety back to his village, where he proceeded to remain for the next several months." What happened next, Diamond says, "illustrates how New Guineans, like many other traditional peoples living largely outside the effective control of systems of justice established by state governments, nevertheless achieve justice and peacefully resolve disputes by traditional mechanisms of their own. Such mechanisms of dispute resolution probably operated throughout human prehistory, until the rise of states with their codified laws, courts, judges, and police beginning 5500 years ago."[46]

The key to restorative justice, says Diamond, is compensation. Not all wrongs can be made right through compensation, of course—death being the ultimate example—so in this case what Papua New Guineans mean by compensation is "sorry money," or payment made to the victim's family out of a deep sense of sorrow by the perpetrator. "The goal of traditional New Guinea mechanisms of justice is fundamentally different from that of state justice systems," Diamond explains. "While I agree that state justice offers big advantages and is absolutely essential for resolving many disputes between citizens of states, especially disputes between strangers, I now feel that traditional justice mechanisms have much to teach us when the disputants are not strangers but will remain locked in on-going relationships after the dispute's settlement: e.g., neighbors, people connected by a business relationship, divorcing parents of children, and siblings disputing an inheritance."[47] For several days Malo laid low, full of dread, as he anticipated the grim fallout from the accident. What happened next was remarkable. Three large men appeared at Malo's window, including the father of the deceased boy, Peti. Malo didn't know whether to face them or run. Running could result in the death of his family, so he let the men in. Diamond picks up the story there, as it was related to him by a man named Gideon, who was the office manager of the company that employed Malo and witnessed what happened next:

> For a man whose son had just been killed, and who was now confronting the killer's employer, Peti's behavior was impressive: clearly still in a state of shock, but calm respectful, and direct. Peti sat quietly for some time, and finally said to Gideon, "We understand that this was an accident, and that you didn't do it intentionally. We don't want to make any problems. We just want your help with the funeral. We ask of you a little money and food, in order to feed our relatives at the ceremony." Gideon responded by offering his sympathies on behalf of his company and its staff, and by making some vague commitment. Immediately that afternoon, he went to the local supermarket to start buying the standard food items of rice, tinned meat, sugar, and coffee.

So far so good, but there was still the matter of Billy's extended family, who would certainly be feeling the sting of Billy's death and might well seek retribution. Gideon thought Malo should go to them straightaway and offer an apology, but a senior member of the company, a man named Yaghean, who had experience in negotiating compensations, advised otherwise. "If you yourself, Gideon, go there too soon, I'm concerned that the extended family and the whole lowlander community may still have hot tempers. We should instead go through the proper compensation process. We'll send an emissary, and that will be me. I'll talk to the councilor for the ward that includes the lowlander settlement, and he in turn will talk to the lowlander community. Both he and I know how the compensation process should proceed. Only after the process has been completed can you and your staff have a say-sorry [tok-sori in Tok Pisin] ceremony with the family."

A meeting was arranged for the next day, and even though emotions were still running high, Yaghean was assured there would be no violence. Yaghean then negotiated a compensation payment of 1,000 kina (about $300) from the company to the family. Malo also arranged to give to the victim's family a pig as another form of compensation called bel kol, or "cooling the belly," intended to attenuate feelings of revenge. The next day the compensatory process began when all parties concerned met together in a tent on the grounds of the victim's family home. Diamond narrates the rest of the ceremony:

The ceremony began with an uncle speaking, to thank the visitors for coming, and to say how sad it was that Billy had died. Then Gideon, Yaghean, and other office staff talked. In describing the event to me, Gideon explained, "It felt awful, just awful, to have to give that talk. I was crying. At that time, I, too, had young children. I told the family that I was trying to imagine their level of grief. I said that I was trying to grasp it by supposing the accident to have happened instead to my own son. Their grief must have been unimaginable. I told them that the food and the money that I was giving them were nothing, mere rubbish, compared to the life of their child." . . . Billy's mother sat quietly behind the father as he spoke. A few others of Billy's uncles stood up and reiterated, "You people won't have any problems with us, we are satisfied with your response and with the compensation." Everybody—my colleagues and I, and Billy's whole family—was crying.[48]

After this the families exchanged food and ate . . . in peace. It worked, not because the perpetrators paid compensatory money and food to the victim's family (although that helped), but because the perpetrators deeply and genuinely felt and acknowledged the pain of the victim's family.

What if Billy's death was not an accident and Malo had intentionally killed him? As Diamond's interlocutors explained, in that case the compensation would have been much higher (10,000 kina instead of 1,000) and a lot more food exchanged, and if this—in conjunction with the appropriate tears of genuine remorse—did not satisfy, it could very well have degenerated into a payback killing. Likely Malo would have been targeted, but, if not he, then a close relative. This in turn could have led to a revenge killing for the payback killing, which may then have escalated into a generations-long feud, with the two sides raiding and murdering each other, possibly resulting, ultimately, in an all-out war.

TAMING THE WOOLF WITHIN—RESTORATIVE JUSTICE IN MODERN SOCIETIES

Such meetings between perpetrators and victims are exercises in the application of the *principle of interchangeable perspectives*, as each can see the crime through the eyes of the other. Consider two men named Peter Woolf and Will Riley—burglar and burgled, respectively. Woolf was a career criminal from Norfolk, England, with a drug habit that led him to theft to feed his addiction. Riley walked in on Woolf raiding his home one day, resulting in a physical struggle in which the two ended up in the street, leaving Riley traumatized and Woolf apprehended and sent to prison. When given the opportunity to confront Woolf, Riley sat across a table and unleashed upon him not hate but an impassioned explanation of how Woolf's break-in had utterly traumatized him. As he told Woolf, "You broke into my house. You destroyed my one belief that I had in my ability to protect my family, my house, from people like you, and you did it in one fell swoop." Riley went on to explain that every time he came home and turned the key in the door, his mind was flooded with fears that a criminal was on the other side.

In response, Woolf recounted, "A flood of emotions started coming out. When you hear the harm you have caused, you've got to be a very *very* bitter and twisted human being if this doesn't affect you—you've got to be a sicko. I was fully expecting these people I did a lot of harm to, to say 'lock him up . . . and throw away the key, we don't care.'" That was not how Riley responded. Instead he said of Woolf: "He was genuinely—*genuinely*—affected by what we'd said. And we start conversing and Peter starts talking from his heart, from his real core of his existence." Thus, Riley concluded about Woolf's incarceration: "You can't just leave him there. You've got to help him help himself." Woolf did get the help he needed, and he has been out of prison since 2003 and has not reoffended. He went on to pen an autobiography in 2008—appropriately titled *The Damage Done*—and toured prisons giving victim-awareness training to inmates. For his part, Riley founded Why Me?, an organization set up by victims who have bene-fited from such programs.[49] The reason this meeting worked is that Woolf was genuine in his emotional contrition. "Here was a man showing remorse," Riley explained, "not because his lawyer told him 'you've got to show remorse in order to get time off.'" Later he reflected:

> Now, six years later, it's clear the meeting wasn't simply about Peter, but had a huge impact on me too. Talking is the only way forward. People who don't talk (which is the majority of victims) are delaying and even maintaining the pain. Luckily, Peter and I are still talking. He's a great man, very clever, has a lovely sense of humour, a genuine raw presence and I'm hugely fortunate to be able to count him as a friend.[50]

Victim restoration is just the beginning. In this case, not only did both offender and offended profit from the exchange, but the community did as well as it led to other such meetings. In all, the program that brought

Woolf and Riley together has an 85 percent victim satisfaction rate, and 78 percent of them said that they would recommend it to other victims and offenders.[51] Another of the benefits can be clearly seen in this pair of statistics: two-thirds of convicts leaving prison are reconvicted within two years, but with the restorative justice program in place, rates of reoffending plummeted by half.[52] Is this moral progress? Is reducing recidivism rates by 50 percent in a community good for the people who reside there? In light of the data, the question answers itself.

Restorative justice can even work with homicide, as evidenced by the case of a twenty-one-year-old from Wyoming named Clint Haskins, who, in September 2001, while driving so drunk he couldn't even remember what had happened, plowed into a vehicle carrying eight members of a student cross-country running team, killing them all. One of those who died was Morgan McLeland. At the sentencing of Haskins—who pleaded guilty to all eight counts of homicide so he could serve his thirteen-to-twenty-year sentence for each concurrently instead of consecutively—the other families of the victims read impact statements in protest, preferring that Haskins be locked away for 104 to 160 years. But Morgan's mom, Debbie, offered a challenge to Clint instead:

> Across the court I asked him if he would be willing to come with me and address young people about the dangers of drunk-driving. When he had an opportunity to speak, he said he would like to. Finally, three years later, after a lot of hard work, I got to see Clint. I found him to be very subdued and remorseful. We both cried and I hugged him, and then we talked about what we could do together to help people make better decisions about drinking and driving. I believed in his sincerity. We first spoke to a room of 900 young people at the National Rodeo High School Finals in Gillette, where Clint had been a rodeo cowboy. It was enormously effective. Later we spoke at the University of Wyoming where all eight of the dead, and Clint, had been students. There was some opposition to this event, as some of the families didn't agree with what we were doing. I still feel bad about that. We've all had a lot of pain and I don't want to add to it. But I also truly believe that our presentations can save lives.[53]

Those other parents' desire for retribution is totally understandable, but as Debbie McLeland reflected: "I wanted to go to court for justice—not revenge," adding, "Hate is a large burden to carry." Instead, she mustered up the courage and character to forgive Haskins and turn the tragedy into something constructive. "Some people think that forgiveness is being disloyal to your loved one; that the only way to honour and remember them is to keep anger and bitterness in your heart, because negative emotion is so much more intense. But that doesn't work for me. . . . Forgiving Clint seems a logical step to me, as this tragic experience is something we both share."[54]

Debbie McLeland thinks that forgiveness is *logical,* and she's right. Forgiveness, in the sense of letting go of retribution fantasies, moving on, and trying to make something positive come from the tragedy, is the rational approach *given the right set of circumstances,* which includes genuine and heartfelt contrition by the offender and an honest attempt at righting the wrong.

THE POWER OF FORGIVENESS

On the night of July 29, 1984, in Burlington, North Carolina, a twenty-two-year-old college student named Jennifer Thompson was held at knifepoint and raped. Despite the emotional trauma of the event, as it was unfolding Thompson focused intently on the details of her rapist's face so that one day justice would be served on him. "I'm going to get this guy who did this to me," she told police investigators, who presented her with a photo lineup of suspects. She pointed to a black man named Ronald Cotton as her rapist. He was brought into the station and lined up with others for Thompson to scan through a one-way mirror. Once again she fingered Cotton. In court, Thompson was asked if she was certain that the man on trial was her rapist. She replied that she was 100 percent certain that Cotton was the perpetrator. After only forty minutes of deliberation, the jury agreed and convicted Cotton. He was handcuffed, shackled, and sent to prison for life.

You might well be thinking that this is a heartwarming, true-life tale in which a rape victim learns to forgive the man who raped her—but no. Jennifer Thompson never forgave Ronald Cotton because, as it turned out, there was nothing to forgive. If there was to be any forgiveness, it would be coming from Ronald Cotton because, in her absolute and unwavering certainty, Jennifer Thompson had accused the wrong man.

Three years later, a new prisoner, named Bobby Poole, who looked an awful lot like Cotton, was incarcerated for rape. In time, and in conversation with Poole in the prison yard, Cotton figured out what had happened and won a new trial in which Jennifer Thompson faced her real rapist for the first time. Instead of identifying Poole and setting free the wrongly convicted Cotton, when she was confronted by the original investigator on the case, Thompson said she thought, "how dare you question me, how dare you paint me as someone who could have forgotten what my rapist looked like, the one person you would never forget?" With the real memory of her rapist now erased and the new memory of Cotton as the perpetrator, Thompson made clear to the jury in no uncertain terms that the right man was in prison. Once again, the jury agreed and put Cotton away, this time for *two* life sentences.

And then something remarkable happened—DNA testing was invented. Eleven years after Ronald Cotton was sent to prison, his lawyer convinced investigators to apply the new science to a fragment of evidence left over from the crime scene. He was promptly exonerated and Poole, already in jail for raping another woman, was correctly identified as the culprit. And yet, so certain was Thompson in her memory, that when police investigators told her that the wrong man was behind bars, she was incredulous to the point of denial. She told the police investigator and the district attorney on the case, "That's not possible. I know it was Ronald Cotton who raped me." To get past the false memory, and the overwhelming, crippling shame of having put an innocent man behind bars, Thompson asked Cotton to meet with her for reconciliation and forgiveness. She was terrified and immediately began to cry when Cotton entered the room. Nevertheless: "I looked at him and said, 'Ron, if I spent every second of every minute of every hour for the rest of my life telling you how sorry I am, it wouldn't come close to how my heart feels. I'm so sorry.'" Did Ronald Cotton walk away in disgust, thanking Jennifer Thompson very much for stealing eleven years of his life? A lesser man might have, but as Thompson tells us, "Ronald just leaned down and took my hands and he looked at me and said 'I forgive you.' The minute he forgave me it was like my heart started to heal. And I thought this is what grace and mercy are all about. Here was this man that I had hated; I used to pray every day of those eleven years that he would die, that he would be raped in prison—that was my prayer to God. And here was this man with grace and mercy who forgave me."[55]

Restitution from the state for Ronald Cotton came in the form of $10,000 per year of his incarceration, or $110,000 total, and he and Thompson wrote a powerful account of their saga together called *Picking Cotton*, which has led to some success in initiating reforms in the criminal justice system.[56] In North Carolina, where the crime occurred, for example, legislation was passed requiring investigators to show pictures of possible perpetrators one by one instead of grouped together, and to add a clause emphasizing that *none* of them may be the guilty party, and to have the process be conducted by someone who doesn't know who the suspect is, or even by a computer that says "the suspect may or may not be included"— all to avoid the numerous cognitive biases inherent in the process that contaminates the memories of a victim or an eyewitness.

I met Jennifer Thompson and Ronald Cotton at a conference in Spain (see *figure 11-2*), where they spoke before a packed audience about the need for reform and, most affecting, on the power of restitution, forgiveness, and friendship to heal wounds and restore justice. When Jennifer turned to Ronald—tears in her eyes and a crackle in her voice—you could hear a pin drop in the hall when she pronounced, "Ronald Cotton is my friend. He taught me that love and hate cannot coexist in the same human heart. You cannot be an angry person and be a joyful person. You cannot be someone who lives in peace and be someone who is out looking for revenge. And it was Ronald who taught me that." This is a different form

Figure 11-2. Ronald Cotton and Jennifer Thompson

A false memory led Jennifer Thompson to pick Ronald Cotton out of a police lineup as the man who raped her, leading to his incarceration for eleven years until exonerated by DNA evidence. The two now give lectures on the need for judicial reform and the power of restitution and forgiveness.[57]

of self-help justice, and when the two of them hugged onstage after recounting their harrowing story, it was one of the most moving things I had ever witnessed, a testimony to the human spirit to rise above our baser instincts.

RETRIBUTION AND THE DEATH PENALTY

We have a long way to go if we think that restorative justice can ever be applied to psychopaths and serial killers. But if the offender is *not* a serious criminal—maybe just a kid who made a tragic mistake or who has a troubled home life—then restorative justice may be just the thing. In general, let's not abandon the goal of reducing crime and violence through better science and technology. For retributive justice to be effective, punishments must be applied judiciously and only to those who can react in a socially desirable manner. Both restorative and retributive justice can benefit from a more rigorous science of understanding how people respond to rewards and punishment (experimental psychology), internal psychological states (cognitive neuroscience), external social conditions (social psychology and sociology), positive and negative incentives (behavioral economics), the way in which incentives could be employed to deter would-be criminals from their future selves acting badly (neurocriminology), and how to prevent current criminals from becoming repeat offenders (criminology). Here again we see a smooth transition from is to ought.[58]

We are, in fact, in the middle of a profound shift in the criminal justice system, which is most noticeable in the decline of the death penalty. I've made the transition myself, from being in favor of the death penalty—primarily out of sympathy for the victims' families—to being against it. Why? In a word, power. Power corrupts in many ways, but the power of the state over its citizens can be absolutely corrosive. In the bad old days when judicial torture and cruel-and-unusual punishments were routinely meted out to suspects and convicted prisoners for dozens of different offenses—most of which today would be considered misdemeanors or less (such as robbing a rabbit warren, poaching, sodomy, gossiping, stealing cabbages, and disrespecting parents)—retributive emotions went unchecked by restorative reason.

Replacing one form of retributive justice (of individuals against each other) by another (of states against individuals) reduced the violence of self-help justice, but it also gave states an alarming amount of power over their citizens. This is yet another effect of the civilizing process brought about by the Enlightenment's emphasis on reason. Emotionally, punishment feels like it helps balance the scales of justice, but rationally it fails as a solution to a problem because it does not help restore justice to the victim. The state has been one of the major abusers of human freedom and dignity over the centuries, but fortunately the same forces that have propelled moral progress in other areas have tugged the state justice system up with it. Judicial torture was common among European states, and its abolition did not come about until the end of the eighteenth century, when the rights of the individual began to take precedence over the rights of the monarchy and, in the case of democracies, what John Adams called "the tyranny of the majority."[59] *Figure 11-3* tracks this form of moral progress over two centuries as the Rights Revolution took off and as checks on state power began.

Initially, the death penalty seemed like a good idea as a replacement for self-help justice and as a crime deterrent. It may in part have been responsible for the former, but probably not the latter, since most criminals do not think long-term and most crimes that result in the death penalty are not the type that are planned well in advance in anything like a moral calculus. As evidence that the death penalty does not deter crime,

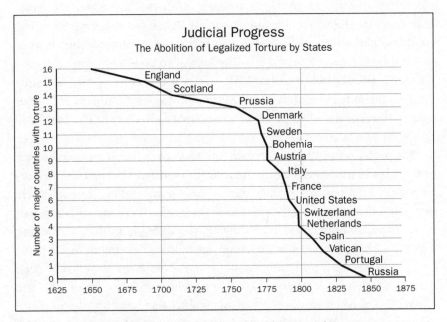

Figure 11-3. Judicial Progress on the Abolition of State-Sponsored Torture
The abolition of legalized torture by states from 1650 through 1850.[60]

consider the well-known fact that nineteenth-century public hangings in England and the United States were frequented by pickpockets working the crowd who were too preoccupied with the spectacle on the gallows to notice the crime unfolding right in front of them.[61]

An additional problem with giving the state the power over life and death was succinctly summed up by the renowned US District Court judge and legal scholar Learned Hand in 1923 when he wrote, "Our procedure has been always haunted by the ghost of the innocent man convicted. It is an unreal dream."[62] The problem is that the justice system is flawed and makes mistakes. Innocent men are convicted, and more often than Judge Hand could have known. The assault on liberty is too high a price to pay for security. Cognitive psychologists, for example, have conducted experiments in which subjects who are tasked with being jury members and who are exposed to an audio recording of an actual murder trial do not wait to form an opinion and pass judgment only after all the evidence has been presented to them. Instead, most people concoct a story in their minds about what happened, make a snap decision of guilt or innocence, and then sort through the evidence presented to them, selecting the bits that most closely fit the verdict they'd already rendered in their minds.[63] This is a quirk of human psychology known as the *confirmation bias*— where we look for and find confirming evidence for what we already believe and we ignore or rationalize away disconfirming evidence. The cognitive pathway from legitimate suspicion to confirmed guilt is very much shorter than we think. It is why kangaroo courts of autocratic governments and the old drumhead trials of battlefields—in which a drumhead was upended to serve as a makeshift bench where summary judgment against the accused could be hastily dispensed right there in the field—are no longer practiced.

The problem has spawned organizations such as the Innocence Project and the Innocent Network, which use DNA evidence to exonerate wrongfully convicted people. To date, since these organizations opened their doors and DNA labs, they have freed 311 wrongfully convicted people (all male, 70 percent minority), including 18 who were on death row awaiting execution.[64] The Justice Brandeis Innocence Project estimates that 10 percent of America's 2 million prisoners may have been wrongfully convicted, which translates to 200,000 innocent people in prison. As the University of Michigan law professor Samuel R. Gross noted about the larger problem, "If we reviewed prison sentences with the same level of care that we devote to death sentences, there would have been *over 28,500 non-death-row exonerations in the past 15 years*, rather than the 255 that have in fact occurred."[65] Such estimates were reinforced by a 2014 study published in the *Proceedings of the National Academy of Sciences* that estimated that if all prisoners who were sentenced to die remained on death row in the United States indefinitely, at least 4.1 percent of them would be exonerated. Based on this statistical method of estimation, the authors conclude that since 1973 as many as 340 American prisoners were wrongly sentenced to death. Tragically, the authors conclude, "The net result is that the great majority of innocent defendants who are convicted of capital murder in the United States are neither executed nor exonerated. They are sentenced, or resentenced to prison for life, and then forgotten."[66]

The case of Michael Morton is illustrative of the problem. Convicted in 1987 for the murder of his wife in Austin, Texas, he was sentenced to life in prison. In an interview by investigators the day after the murder, Morton's three-year-old son said he witnessed the crime, saw and described the murderer as "a monster," and most significantly said that his father *was not home* at the time of the attack. The county prosecutor, Ken Anderson, failed to provide Morton's defense attorney or the jury with the transcript of that interview, along with other exculpatory evidence that might have planted a reasonable doubt in the minds of the jurors. Tragically, while Morton was rotting away in prison for a quarter century, the real killer—a man named Mark Alan Norwood, who was later caught and convicted—struck again, murdering another woman in the same manner and in the same area of Austin, where he lived.

In 2011, Morton was exonerated based on DNA evidence found on a bandanna at the crime scene that was smeared with Norwood's blood. Morton's attorneys then asked the Texas Supreme Court to convene an inquiry into misconduct by Anderson, and on April 19, 2013, the court of inquiry ordered Anderson to be arrested, saying, "This court cannot think of a more intentionally harmful act than a prosecutor's conscious choice to hide mitigating evidence so as to create an uneven playing field for a defendant facing a murder charge and a life sentence." Nevertheless, despite the fact that Anderson was found to be in contempt of court, he was fined a mere $500 and received a sentence of ten days in county jail, of which he served only five. Compared to Morton's twenty-five years in jail and the murder of another woman, plus the devastation to all families involved, where is the justice in the criminal justice system? When asked if he harbored any ill will toward those who took almost half his life away from him, Morton replied in what has to be one of the most rational arguments ever made against vengeance, "Revenge is like drinking poison and hoping the other guy dies from it."[67]

. . .

For these reasons, and more, the death penalty has been abolished in most countries of the world, and in those where it is still legal (as in some states in these United States), it is on death row—rarely practiced, and when it is administered it is only done after lengthy, often multidecade, appeals and delays and stays of execution. In my home state of California, for example, the death penalty is automatically appealed, and prisoners on death row are more likely to die of natural causes. According to the Death Penalty Information Center, the eighty death sentences handed down in the United States in 2013 came from only 2 percent of counties, more than half of all thirty-nine executions that were carried out in 2013 took place in Texas and Florida, executions are down 60 percent since 1999, and 85 percent of all counties have not executed anyone in half a century.[68] On the rare occasion when it is dispensed, lethal injection is the primary means because it is considered humane, at least compared to hanging and the electric chair. Still, many American and European pharmaceutical companies are now refusing to sell to individual states the lethal injection drug pentobarbital, leading to a scramble to fulfill needed stockpiles, and the 2014 mishandled execution of Clayton Lockett by Oklahoma prison medical staff shows that even when the drugs are available they are not always properly used.[69]

The trend downward in the death penalty since the mid-1990s is unmistakable, and US Supreme Court justice Harry Blackmun's dissent in a 1994 case is emblematic of the tenor of the times. The case from which he was dissenting was the appeal by Bruce Edwin Callins for a stay of execution by the state of Texas, where he was on death row. He lost his appeal and was executed on February 23, 1994. Blackmun's description of what was about to take place is a haunting reminder of what is at stake when the state has the power of life or death over its citizens:

> Within days, or perhaps hours, the memory of Callins will begin to fade. The wheels of justice will churn again, and somewhere, another jury or another judge will have the unenviable task of determining whether some human being is to live or die. We hope, of course, that the defendant whose life is at risk will be represented by competent counsel—someone who is inspired by the awareness that a less than vigorous defense truly

could have fatal consequences for the defendant. We hope that the attorney will investigate all aspects of the case, follow all evidentiary and procedural rules, and appear before a judge who is still committed to the protection of defendants' rights—even now, as the prospect of meaningful judicial oversight has diminished. In the same vein, we hope that the prosecution, in urging the penalty of death, will have exercised its discretion wisely, free from bias, prejudice, or political motive, and will be humbled, rather than emboldened, by the awesome authority conferred by the State. . . .

Rather than continue to coddle the Court's delusion that the desired level of fairness has been achieved and the need for regulation eviscerated, I feel morally and intellectually obligated simply to concede that the death penalty experiment has failed. It is virtually self-evident to me now that no combination of procedural rules or substantive regulations ever can save the death penalty from its inherent constitutional deficiencies.[70]

Figures 11-4, 11-5, and *11-6* track this form of judicial progress in the decline and abolition of the death penalty around the world and within the United States.

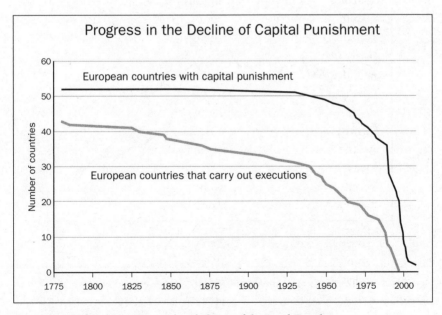

Figure 11-4. Judicial Progress on the Abolition of the Death Penalty
The decline of the death penalty by states from 1775 through 2000.[71]

Figure 11-5. The Decline in the Number of US Executions per Year
The decline of annual executions in the United States from 1977 through 2013.[72]

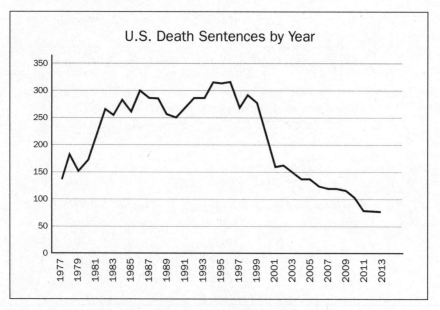

Figure 11-6. The Decline in the Number of US Death Sentences per Year
The decline of annual death sentences imposed by US states from 1977 through 2013.[73]

RESTORATIVE JUSTICE BY NATIONS

In late 2013 the United Kingdom's Queen Elizabeth II issued an official royal pardon and apology to the British scientist Alan Turing, whose work in cracking the Nazi Enigma Code arguably did more to secure the Allies' victory than any other single contribution by an individual during World War II. What was the queen apologizing for and why did Turing need a pardon? He was gay, which in England was a crime punishable by imprisonment, chemical castration, or both. After his code-breaking work and his pioneering efforts in the late 1940s and early 1950s in the nascent field of computer science, Turing was caught, tried, and convicted of "gross indecency" because he admitted to a sexual relationship with another man. In addition to being injected with a synthetic estrogen that caused the enlargement of breast tissue, the British government stripped him of his security clearance, and his career subsequently nose-dived, along with his will to live. In 1954 he took his own life at age forty-one, an incalculable loss to humanity of one of its greatest minds. "Dr. Turing deserves to be remembered and recognized for his fantastic contribution to the war effort and his legacy to science," said the British justice secretary Chris Graying. "A pardon from the queen is a fitting tribute to an exceptional man."[74]

The case is interesting for another reason related to legal justice and moral progress. The plan to pardon Turing was met with resistance from legal scholars based on the principle that no one is above the law, and at the time the law clearly stated that Turing's actions were criminal. Former justice minister Tom McNally explained the reasoning this way: "It is tragic that Alan Turing was convicted of an offense which now seems both cruel and absurd. However, the law at the time required a prosecution, and, as such, longstanding policy has been to accept that such convictions took place and, rather than trying to alter the historical context and to put right what cannot be put right, ensure instead that we never again return to those times." Agreed, but the only way to do that is to change the law, rebuke the reasoning behind it, and make right the wrongs done to those who were victims of this "cruel and absurd" law. In fact, Turing is just the most famous of those so hurt by this unjust law: scholars estimate that as many as fifty thousand men were convicted of gross indecency and that as many as fifteen thousand are still alive and carrying with them a criminal record for being homosexual.[75] True restorative justice calls for a blanket pardon and apology to everyone who was harmed.

Such apologies and admissions have a checkered history, and so it is understandable that current governments are reluctant to go down the path of restoring harms done by someone else to someone else, particularly if restoration involves reparations. After World War I, Germany was forced to sign a war-guilt clause, despite the fact that most historians agree that all of the major powers were equally to blame for the war.[76] Documenting the primordial desire for vengeance, in addition to massive payments that Germany could not afford to make, the Allies foolishly demanded 120,000 sheep, 10,000 goats, 15,000 sows, and 43.5 million tons of coal. This sowed the seeds of resentment that sprouted two decades later into the weeds of National Socialism.[77] In his many speeches as he came to power, Adolf Hitler pounded his audiences with the injustice of the Treaty of Versailles, and when he subsequently conquered France he forced his vanquished enemies to haul out the same train car in which the World War I–ending German surrender had been signed, and in moral righteousness made the French sign their surrender to him then and there.

After World War II, when the world came to grasp the full reality of the extent of the Nazi genocide, an International Law Commission was born out of the Nuremberg trials of German war criminals for their "crimes against humanity," which it defined as follows: "Murder, extermination, enslavement, deportation and other inhumane acts done against any civilian population, or persecutions on political, racial, or religious grounds, when such acts are done or such persecutions are carried on in execution of or in connection with any crime against peace or any war crime."[78] From the start, US Supreme Court justice Robert Jackson—who oversaw the initial trials of top Nazi leaders such as Hermann Goering and Rudolf Hess—insisted on a fair trial for all, noting in his opening statement, "To pass these defendants a poisoned chalice is to put it to our own lips as well."[79]

The Nuremberg trials were one of the greatest contributions to expanding the moral sphere of justice on a global scale, signaling to dictators and demagogues everywhere that the world was watching and would hold them accountable for their actions. That accountability today falls on the shoulders of the International Criminal Court (ICC) in The Hague, but as David Bosco shows in his book *Rough Justice*, the ICC has been limited in its scope and reach by the great powers of the world, who would rather not have outsiders poking around too closely into their internal affairs. (In the murky world of distant proxy fights and support of Third World dictators, Franklin Roosevelt purportedly said about Nicaragua's

Anastasio Somoza, "He may be a son of a bitch, but he's our son of a bitch.") As Bosco notes: "The court has, for the most part, become an instrument in the toolkit of major powers, responding to instability and violence in weaker states," while leaving untouched the leaders of the great powers.[80] Still, the principle, if not the practice, of holding people accountable for their actions across state and national borders is moral progress in the globalization of justice based on universal principles of human nature and morality.

Attempts at reconciliation between nations are the subject of a comprehensive study by the political scientists William Long and Peter Brecke and presented in their book *War and Reconciliation: Reason and Emotion in Conflict Resolution*. Their database consists of 114 pairs of countries that had been engaged in a conflict between 1888 and 1991, plus another 430 civil wars. Reconciliation is a mixed bag of results. Some nations who have fought one another on and off for centuries, such as Germany and France, England and France, Poland and Russia, Germany and Poland, and England and the United States, have reconciled to the point where war seems risibly ridiculous. Others, such as the eternally in conflict Israel and Palestine, seem capable of erupting into all-out war at any time. Still others, such as India and Pakistan, have managed to keep the peace but also seem quite capable of breaking it given the right (or wrong) circumstances.

Of the many factors that go into the making of a successful reconciliation after a conflict between two nations, the two most effective are similar to those that work for restorative justice between two individuals: (1) public acknowledgment of harm done and (2) acceptance of imperfect justice. "In all cases of successful reconciliation," Long and Brecke note, "retributive justice could neither be ignored nor fully achieved .... Disturbing as it may be, people appear to be able to tolerate a substantial amount of injustice wrought by amnesty in the name of social peace."[81] And as with reconciliation between individuals, forgiveness in the name of moving forward is key, even if neither side is fully satisfied with the outcome of the reconciliation efforts. The American Civil War is the prototype. At the war's end, with a death toll in excess of 650,000, when so many Northerners wanted President Lincoln to mete out a harsh peace to the Southerners, he chose instead the more conciliatory approach of welcoming the South back into the Union with these memorable words: "With malice toward none, with charity for all, with firmness in the right as God gives us to see the right, let us strive on to finish the work we are in, to bind up the nation's wounds, to care for him who shall have borne the battle and for his widow and his orphan, to do all which may achieve and cherish a just and lasting peace among ourselves and with all nations."[82]

Reconciliation between oppressor and oppressed has been progressing since World War II. In 1970, West German chancellor Willy Brandt fell to his knees at the site of the Warsaw ghetto to express the guilt, sorrow, and responsibility of Germany for the Holocaust. This was on top of compensation West Germany had provided to Holocaust survivors and to the State of Israel to help build a Jewish nation.[83] Germany goes far beyond political statements and payments. In many towns and cities one literally stumbles across what are called "Stumbling Stones" (designed by the artist Gunter Demnig) in front of homes and buildings where Jews, Roma, Sinti, and others once lived but who were relocated from their homes and perished in concentration camps.[84] They're a striking feature of the cityscape that speak volumes to remembrance and reconciliation. *Figure 11-7* shows a pair of such Stumbling Stones I happened across in the bustling city center of Köln. The translation reads:

> Here Lived
> Julius Levi
> Born 1885
> Deported 1941
> Lodz
> Murdered
>
> Here Lived
> Henriette Levi
> Maiden Schönthal
> Born 1885
> Deported 1941
> Lodz
> Murdered

Other countries express their own forms of conciliatory remembrances. In 1998 Australia launched National Sorry Day to remember and acknowledge the maltreatment of the aboriginal population by the Australian government, whose policies, among many abuses, resulted in a "Stolen Generation" of aboriginal children who were taken from their families in an attempt to convert them into "white" Australians. The movement arose when a 1997 report titled "Bringing Them Home" was tabled in

Figure 11-7. Stumbling Stones of Holocaust Remembrance in Germany
Memorials to those from Köln, Germany, who perished in the Holocaust. Thus they shall
be remembered. Source: Author collection.

Parliament and Prime Minister John Howard said he refused to apologize
to the "stolen generations" because he "did not subscribe to the black arm-
band view of history." It took a decade of activism, but finally, on February
13, 2008, Prime Minister Kevin Rudd issued a formal apology to Australian aborigines for the Australian government's laws, policies, and practices of the past regarding them.[85]

Great Britain apologized for its mistreatment of New Zealand's
Maori people and of subcontinental Indians. Canada apologized to the
First Nation (Native Americans in Canada) survivors of the residential
school fiasco, a program of deliberate cultural genocide—an attempt,
boasted the minister of Indian affairs at the time, Duncan Campbell Scott,
"to kill the Indian in the child."[86] Canadians have been especially recep-
tive to restorative justice strategies and national apologies (apologizing
also for Japanese internment camps during World War II and the Chinese
head tax), to the point where one wag wrote that Australia's National
Sorry Day "differs from Canada's National Sorry Day, in which Canada
apologizes for existing."[87]

Saying sorry is one thing; paying reparations for the sins of prior generations is quite another. Yahweh may visit "the iniquity of the fathers

upon the children unto the third and fourth generation of them that hate me" (as instructed in the Second of the Ten Commandments), but Western jurisprudence holds one responsible for one's own sins and no one else's, much less the present generation paying (literally) for the sins of generations in centuries past. As bad as most of us now feel about the decimation of Native American populations by European guns, germs, and steel, and the enslavement of Africans by American whips, chains, and shackles, there's a reason why attempts to pass an act that gives every living Native and African American a cash payoff have failed. There's a moral bright line that gets crossed in even the most enlightened of us who care deeply about rights and justice. And given the abuses and usurpations of one people against another, or one tribe, chiefdom, state, or nation against another over the past ten millennia, the apologies and reparations would never end. The best we can do is what we've been doing to fight for equal rights and opportunities for all peoples in all walks of life, and to fight against discrimination, prejudice, and injustice wherever we find it.

12

Protopia: The Future of Moral Progress

As man advances in civilization, and small tribes are united into larger communities, the simplest reason would tell each individual that he ought to extend his social instincts and sympathies to all the members of the same nation, though personally unknown to him. This point being once reached, there is only an artificial barrier to prevent his sympathies extending to the men of all nations and races.

—Charles Darwin, *The Descent of Man*, 1871[1]

I'm an optimist. I agree with Charles Darwin that reason tells us that we ought to extend our moral sympathies to people of all nations and races. One of the aims of this book, in fact, is to argue that it is reason and science more than any other force that have helped break through the artificial barriers to extending our sympathies to all peoples.

In this final chapter I want to project the moral arc out as far as I can imagine. Predictions about the future, as Yogi Berra quipped, are the hardest type to make, and I take seriously the admonition of one of the greatest scientific prophets of our age, Arthur C. Clarke (who, among other things, predicted communications satellites), "If we have learned one thing from the history of invention and discovery, it is that, in the long run—and often in the short one—the most daring prophecies seem laughingly conservative."[2] Nevertheless, if we are to take seriously the goal of furthering the moral future of humanity, we must look beyond what we can see on the immediate horizon. After all, as Robert Browning suggested, a man's reach should exceed his grasp, or what's a heaven for?

PROTOPIAN POLITICS

Prophets and prognosticators often envision what life will be like when we get "there," but this is not the right way to think about the future because there is no there there[3]—in the utopian sense of the word's Greek origin as "no place."[4] The history of attempts at putting utopian ideas into practice is strewn with the wreckage of failed societies, from Robert Owen's New Harmony in Indiana and John Humphrey Noyes's Oneida Community in New York—both relatively harmless communal experiments—to Lenin/ Stalin's Soviet Union and Mao's Communist China, which were catastrophic. Lenin said, "If you want to make an omelet, you must be willing to break a few eggs." But twenty million dead Russians and forty-five million dead Chinese are not eggs, and all those five-year plans and the Great Leap Forward failed to produce an omelet.[5]

Here we see the results of elevating the interests of the group over the interests of the individual in a moral calculus, and what happens when people are treated as means to an end rather than as ends in themselves. Recipes for disaster don't come any more foolproof than that. As one of the original citizens of Owen's New Harmony explained it, "We had tried every conceivable form of organization and government. We had a world in miniature. We had enacted the French revolution over again with despairing hearts instead of corpses as a result. It appeared that it was nature's own inherent law of diversity that had conquered us . . . our 'united interests' were directly at war with the individualities of persons and circumstances and the instinct of self-preservation."[6]

Utopias are *no place*, save for in the imagination, because they are grounded in an idealistic theory of human nature—one that assumes, quite wrongly, that perfection in the individual and social realm is a possibility. Instead of aiming for that unattainable *place* where everyone lives in perfect harmony forever, we should instead aspire to a *process* of gradual, stepwise advancement of the kind one might imagine occurring on a mountaineering expedition. It's not a straight climb up, like on a ladder; instead decisions have to be constantly made about the best route and method to get everyone farther up the mountain, one step at a time. Some people will need to be dragged up the mountain, of course, and one might reasonably ask what protective measures we ought to have in place for those who stumble and fall. And, as always, other people will race ahead; they are the visionaries and forward thinkers who have been to the mountaintop, and upon whom humanity has always relied to give direction.

A better descriptor than *utopia* for what we ought to strive for is *protopia*—a place where progress is steadfast and measured. The visionary futurist Kevin Kelly described it this way: "I call myself a protopian, not a utopian. I believe in progress in an incremental way where every year it's better than the year before but not by very much—just a micro amount."[7] Instead of the 1950s imagined jump from the jalopy to the flying car, think of the decades-long cumulative improvements that led to today's smart cars with their onboard computers and navigation systems, air bags and composite metal frames and bodies, satellite radios and hands-free phones, and electric and hybrid engines. Instead of Great Leap Forward, think Small Step Upward.[8]

The Moral Arc is a protopian work. Its prescriptions are modest and the general principle is relatively simple: try to make the world a slightly better place tomorrow than yesterday. I outlined a few specific principles of how to do that in the rational Decalogue at the end of chapter 4, and readers can readily come up with their own. All of the examples in this book of what people have done to attenuate war, abolish slavery, end torture and the death penalty, expand suffrage, build democracy, defend civil rights and liberties, legalize same-sex marriage, and protect animals are protopian measures of advancement. It is surprising how much progress can be realized by making just a few small steps upward.

Balaji Srinivasan, a Stanford University lecturer, thinks that given enough time there is no telling what technological and social solutions could be designed to solve seemingly impossible problems. He cites Bill Gates's 1998 comment that what he feared most for Microsoft was not Netscape, Sun Microsystems, Oracle, or Apple, but "someone in a garage who is devising something completely new." It was a prescient comment because on September 4 of that year Sergey Brin and Larry Page incorporated Google, then operating out of a friend's garage.[9] Srinivasan calls on Albert Hirschman's strategies of change we reviewed in chapter 2—*exit* versus *voice*—and how we can move beyond the way society has been structured for centuries. And do so peacefully. "They have aircraft carriers; we don't," he joked about the power of the nation-state. But Srinivasan is serious about experimenting with new political systems. As he pointed out in a popular talk at the Y Combinator conference in Silicon Valley, where start-up entrepreneurs gather to discuss how they're going to change the world (and make a billion dollars in the process), the peaceful revolution is already under way. iTunes revolutionized the music industry. Netflix changed how Hollywood sells movies. New media such as Twitter and blogs are challenging old media. The Khan Academy, Coursera, Udacity, and the Teaching Company are delivering world-class college courses for free or at a fraction of what brick-and-mortar universities charge. Printing in 3-D makes it nearly impossible for governments to ban physical objects because individuals can be their own manufacturers in the privacy of their homes. The Quantified Self movement, begun in the 1970s with personal use of EEG for biofeedback, and taking off in the 2000s with advanced wearable computer technology for personal health, enables people to self-monitor their food quantity, air quality, mood levels, blood oxygen levels, and physical and mental states.[10]

All of these tools—and many more—will enable those who so desire to opt out of society, or to peacefully exit an oppressive system, or just to create something new. Historically, American politics, economics, and culture have been dominated by four cities: Boston (higher education), New York (publishing, media), Los Angeles (film, television, music), and Washington, DC (laws and legislation). Today we have an opportunity to set aside parts of the planet for unregulated experimentation in which location is irrelevant because you can go anywhere you want digitally, instantly, in real time. What these Silicon Valley reformers are doing is reconfiguring this geography—not by shifting these power centers to Palo Alto, California, but to anywhere anyone wishes. In other words, there will be no power centers because power will be distributed all over the globe and placed in the hands of ordinary citizens. Distributed politics will be powered by distributed intelligence. If this happens, the very idea of political power as it has been practiced for millennia will dissolve before our very eyes.

There are good reasons to be skeptical of all such protopian schemes, but new ideas have to come from somewhere, and as long as no one is harmed in the process of such experimentation I see no reason why we should not encourage such social entrepreneurs, and even invest and participate in them where appropriate. As Srinivasan said, "The best part is this—the people who think this is weird, the people who sneer at the frontier, who hate technology, won't follow you there."[11] Those who choose to opt out of the old systems can do so in any number of ways. Watch YouTube, Hulu, or DirecTV instead of network or cable television. Log on each morning to any online news aggregators instead of reading the *New York Times* or the *Wall Street Journal* (or *USA Today* if you're in a hotel). "Hundreds of millions of people have now migrated to the cloud, spending hours per day working, playing, chatting, and laughing in real-time

HD resolution with people thousands of miles away . . . without knowing their next-door neighbors," Srinivasan noted in a 2013 article for *Wired*, which tracks such cyber-protopian technologies. "Millions of people are finding their true peers in the cloud, a remedy for the isolation imposed by the anonymous apartment complex or the remote rural location."[12]

It doesn't matter where you live geographically, because in cyberspace everyone is as close as their *geodesic distance*, or the number of degrees of separation between two nodes in a social network (similar to hubs in airline flight networks). Think of it like the popular game of "six degrees of separation" between yourself and anyone else on the planet.[13] The concept of six degrees of separation was developed as part of a "small worlds model" by the psychologist Stanley Milgram (he of the famous shock experiments) to measure how many connections between people it would take to, in their study, deliver a piece of mail randomly dropped anywhere in the country.[14] Much to everyone's surprise, the number connecting any two strangers anywhere in the country is much smaller than our intuitions would allow—to wit, six degrees.[15]

The direction of connectedness is not just from physical to digital, but can reverse from digital to physical, as people online come together in geographical space, from the personal (friends meeting at a theater or restaurant) to the political (the Occupy Wall Street movement, the Arab Spring revolution). No matter what the mother tongue of any individual might be, with a language translation program such as Google Translate the communication barrier between individuals anywhere in the world is markedly lowered. With communication services such as Skype, talking to someone else in real time anywhere else on Earth, or even in space (astronaut Chris Hadfield Skyped from the International Space Station[16])— for free—means there are no geographical barriers to being connected. How many people can be linked? There is no upper boundary. It may be 100 or 1,000, and it may last for a single meeting or for a year or more. Srinivasan projects that as the number "increases to 10,000 and 100,000 and beyond, for longer and longer durations, we may begin to see cloud towns, then cloud cities, and ultimately cloud countries materialize out of thin air."[17]

Cloud countries? Why not? Throughout history people have been coalescing into ever larger collectivities: from bands and tribes, to chiefdoms and states, to nations and empires. The historian Quincy Wright has documented that in fifteenth-century Europe there were more than five thousand independent political units. By the early seventeenth century these coalesced into five hundred political units. By 1800 there were about two hundred. Today there are fifty.[18] The political scientist Francis Fukuyama notes that in 2000 BC there were no fewer than three thousand polities in China alone, but by 221 BC there was only one.[19] The trend in unification has led ideologues on both extreme ends of the political spectrum—from fascist dictators on the far right to one-world-government dreamers on the far left—to imagine the day when there would be a single overarching Leviathan in charge. There is a certain logic to the idea—if a government with a monopoly on the legitimate use of force has been, over all, a force for good in organizing ever more people into social collectives, wouldn't extending that principle to the entire globe make sense?

No. Why? Call it the *Fixing-the-Potholes Problem*. Mostly what people want from their government is to be left alone to do their own thing while the background operations (police and courts) and general infrastructure (roads and bridges) are working properly, and to have their immediate needs met (like fixing the potholes down the street), and they don't care what other people who live thousands of miles away want, unless those needs and interests happen to coincide. The best governments are invisible, in the sense that we only notice them when something goes wrong. Otherwise, when public systems are running smoothly, we don't think about them. The problem is that bloated bureaucracies are rarely invisible because they are not optimally designed to solve many of today's problems.

Instead of bureaucracy think *adhocracy*, the futurist Alvin Toffler's term for "an organizational design whose structure is highly flexible, loosely coupled, and amenable to frequent change."[20] Bureaucracies evolved in response to rigid, hierarchical, and slow-to-change nation-states premised on the presumption that there is one right way to run an organization and that responses to problems can be standardized, copied, and implemented elsewhere. Adhocracy is premised on innovation and real-time problem solving in response to dynamic and ever-changing environments that require unique solutions to new problems. Adhocracies are decentralized and highly organic, horizontal instead of hierarchical in nature, and as the management expert Henry Mintzberg described it, the adhocracy "engages in creative effort to find a novel solution; the professional bureaucracy pigeonholes it into a known contingency to which it can apply a standard program. One engages in divergent thinking aimed at innovation; the other in convergent thinking aimed at perfection."[21]

In the public sector, NASA in the 1960s functioned as an adhocracy, and necessarily so, given that there was no ready-made instruction manual on the subject of how to put a few people on the moon, but by the space shuttle era of the 1990s, NASA had gone the way of bureaucracy. Today, DARPA (the government's Defense Advanced Research Projects Agency) is a "black box" organization tasked by Congress to identify and develop emerging science and technologies; as such, it acts like an adhocracy. Arpanet, the precursor to the Internet, was one of its inventions. In the private sector, Google X is a semi-secret adhocracy directed by Google cofounder Sergey Brin and the scientist and entrepreneur Astro Teller, whose purpose is to develop "science fiction–sounding solutions," such as Google Glass and self-driving cars, both of which are nearing fruition after only a few years in development.[22] These Google projects are often referred to as "moon shots" but not, tellingly, space shuttle shots.

· · ·

Given these trends, one day, maybe centuries hence, there will be no more nation-states, their former borders so porous economically and politically that the very concept will fall into disuse and in their stead we will see a return of smaller political units such as the city-state. Instead of power-obsessed kings and queens, vainglorious dictators and demagogues, megalomaniacal Führers and Dear Leaders, and egocentric presidents and prime ministers, perhaps the most powerful political person will be . . . the mayor. That's right, the person who cuts the ribbon at a groundbreaking ceremony for a new building, who works with the police and fire chiefs to keep crime at a minimum and disasters under control, who engages with technocrats and engineers to make sure the public buses run on time, who meets with educators to create the best environment for learning in schools, and who fixes the potholes.

Sound crazy? Not to the political scientist Benjamin Barber, whose 2013 book *If Mayors Ruled the World*—appropriately subtitled *Dysfunctional Nations, Rising Cities*—argues as much: "In the face of the most perilous challenges of our time—climate change, terrorism, poverty, and trafficking of drugs, guns, and people—the nations of the world seem paralyzed. The problems are too big, entrenched, and divisive for the nation-state. Is the nation-state, once democracy's best hope, today dysfunctional and obsolete?" In answering in the affirmative Barber includes this trenchant observation by former New York mayor Fiorello LaGuardia: "There is no Democratic or Republican way of fixing a sewer."[23]

Cities, says Barber, "are unburdened with the issues of borders and sovereignty which hobble the capacity of nation-states to work with one another." Nations and their leaders care about national issues, whereas most of us care about neighborhood issues. Mayors, not presidents (or premiers, chief executives, or federal chancellors), are best equipped to handle immediate and local problems. Thus, Barber suggests, if we need a parliament of some sort (or a senate or congress or some other gathering of people who don't know you and couldn't care less about your immediate problems), it should be a parliament of mayors: "A planet ruled by cities represents a new paradigm of global governance—of democratic glocalism rather than top-down imposition, of horizontalism rather than hierarchy, of pragmatic interdependence rather than outworn ideologies of national independence." The reason is obvious once you think about it. Cities, Barber notes, "collect garbage and collect art rather than collecting votes or collecting allies. They put up buildings and run buses rather than putting up flags and running political parties. They secure the flow of water rather than the flow of arms. They foster education and culture in place of national defense and patriotism. They promote collaboration, not exceptionalism."[24]

Former New York City mayor Michael Bloomberg explained the problem of dealing with the federal government this way: "I don't listen to Washington very much. The difference between my level of government and other levels of government is that action takes place at the city level. While national government at this time is just unable to do anything, the mayors of this country have to deal with the real world." What about terrorism? Isn't that a national problem? Not really. Terrorists don't attack a nation. They attack a specific target, such as a building or a subway. After eighteen months of training his staff at the Department of Homeland Security, Bloomberg concluded, "We're learning nothing in Washington." Ditto climate change. After next to no progress by national delegations at a climate change meeting in Mexico City in 2012, for example, representatives from 207 cities signed the Global Cities Climate Pact and pledged to pursue "strategies and actions aimed at reducing greenhouse gas emissions" at the local level.[25]

In the book *Smart Cities* the New York University urban research professor Anthony Townsend reviews the history of cities in which "buildings and infrastructure shunted the flow of people and goods in rigid, predetermined ways," but the flow's been altered by computers and the Internet, linking up people within and between them. "Smart cities are places where information technology is wielded to address problems old and new" because they can "adapt on the fly, by pulling readings from vast arrays of sensors, feeding that data into software that can see the big picture." Townsend also sees mayors as the linchpins of the future, and he shows how mayors all over the world are working with companies such as IBM, Cisco, Siemens, Google, and Apple to work on problems such as crime, pollution, garbage pickup, retail business foot traffic, high-rise corporate office space, energy consumption, housing, public venue use, parking, public transportation, and the scourge of all cities—I would know, as I live in Los Angeles—traffic.[26] And the title of his DIY chapter for city planners is perfectly protopian: "Tinkering Toward Utopia."

One solution to cities' traffic problems is what the city planner Jeff Speck calls "walkability."[27] Getting people out of their cars and on their feet are good for traffic and pollution reduction, and good for health. And if you need to cover more ground than walking might permit, the good news is that there's a city cycling revolution under way. Many cities are providing more numerous and wider bike lanes to make cycling safer, in addition to implementing bike-sharing programs in which you can pick up a bike in one location and drop it off in another. By mid-2013 there were already 535 bike-sharing programs worldwide, with more than half a million bicycles at the ready. And they work. I've tried them in a couple of cities, and while the bikes are by necessity heavy because they need to be sturdy, they often get you from point A to point B just as fast as public transportation.

Stewart Brand, the creator of the *Whole Earth Catalogue* and the Long Now Foundation, notes, "The cities did what the nations could not." They solve local problems. Brand lists more than two hundred organizations dedicated to effecting local change, including the International Union of Local Authorities, the World Association of Major Metropolises, the American League of Cities, Local Governments for Sustainability, the C40 Cities Climate Leadership Group, United Cities and Local Governments at the United Nations, the New Hanseatic League, and the Megacities Foundation.[28] Brand also points out that more than half of the world's population now lives in cities, and the percentage is growing rapidly.[29] "Cities are the human organizations with the greatest longevity but also the fastest rate of change. Just now the world is going massively and unstoppably urban . . . . In a globalized world, city-states are re-emerging as a dominant economic player."[30] He points out that in 1800 only 3 percent of the world's population lived in cities. In 1900 it had grown to 14 percent. In 2007 it reached 50 percent, and by 2030 it will exceed

60 percent. "We're becoming a city planet," he says, in which "communications and economic activities bypass national boundaries."

Something else happens when people move to cities: they have fewer babies. "Massive urbanization is stopping the population explosion cold," Brand notes. "When people move to town, their birthrate drops immediately to the replacement level of 2.1 children/women, and keeps right on dropping." Speaking specifically about women in developing nations, the head of the Global Fund for Women, Kavita Ramdas, told Brand, "In the village, all there is for a woman is to obey her husband and family elder, pound grain, and sing. If she moves to town, she can get a job, start a business, and get education for her children. Her independence goes up, and her religious fundamentalism goes down."[31]

Cities, not nations, may be the future of humanity. We are so accustomed to the nation-state as the norm we forget that its existence as a concept—depending on how it is defined (by its politics or its people)—is barely two centuries old, whereas cities date back ten millennia.[32] The Harvard economist Edward Glaeser calls the city "our greatest invention" that allows people to be richer, smarter, greener, healthier, and even happier.[33] The long-term historical trend, then, may be a U-shaped curve of lots of political units as civilization takes off, reducing in number over the millennia as smaller states coalesce into larger states, but instead of hitting the bottom of the curve at a one-world government, the curve bounces up off the bottom of the graph and rises again into much more numerous and smaller political units, each governed locally and directly by those most interested in fixing local problems.

The long-term trend toward the decline of centralized power is across the board and well documented by Moisés Naím in his book *The End of Power*. "Power is spreading, and long-established, big players are increasingly being challenged by newer and smaller ones," he writes. "And those who have power are more constrained in the way they can use it." Naím defines power as "the ability to direct or prevent the current or future actions of other groups and individuals," and in that sense power is not only "shifting from brawn to brains, from north to south and west to east, from old corporate behemoths to agile start-ups, from entrenched dictators to people in town squares and cyberspace," it is decaying as well, and is "harder to use—and easier to lose." Naím's title is a little misleading, inasmuch as it implies that power has ended, but his point is that even though "the president of the United States or China, the CEO of J. P. Morgan or Shell Oil, the executive editor of the *New York Times*, the head of the

International Monetary Fund, and the pope continue to wield immense power," they wield less power than their predecessors.[34]

Geopolitically, for example, having a massive army doesn't give you as much power as it once did. A 2001 study by Ivan Arreguín-Toft found that in militarily asymmetrical conflicts between 1800 and 1849, the smaller country realized its strategic goals only 12 percent of the time, but between 1950 and 1998 the weaker side triumphed 55 percent of the time. Think about the Vietnam War. Dictators and demagogues are also on the way out. "In 1977, a total of 89 countries were ruled by autocrats," Naím reports. "By 2011, the number had dwindled to 22." CEOs are also losing power. Among *Fortune* 500 corporations, CEOs had a 36 percent chance of keeping their jobs for five years in 1992, a 25 percent chance in 1998, and by 2005 the average CEO for all 500 companies held their position of power for a mere six years. The companies at the top of the heap are also falling by the wayside, seeing an increase from a 10 percent chance to a 25 percent chance of dropping from the top quintile within five years.[35]

To my point on the future of moral progress Naím writes, "We are on the verge of a revolutionary wave of positive political and institutional innovations" in which "power is changing in so many arenas that it will be impossible to avoid important transformations in the way humanity organizes itself to make the decisions it needs to survive and progress." He notes that nearly every political institution and principle we have today—representative democracy, political parties, independent judiciaries, judicial review, civil rights—were all invented in the eighteenth century. The next set of political innovations, Naím predicts, "will not be top-down, orderly, or quick, the product of summits or meetings, but messy, sprawling, and in fits and starts."[36]

Who knows? At this point we're all speculating, but perhaps the sixties' environmentalists got it right with their bumper-sticker slogan "Think Globally, Act Locally."

Then again, here's another slogan: "Think Historically, Act Rationally." The first informs the second. The long-term historical trend in moral progress points to the fact that we may, in fact, need a Leviathan—an overarching government of some form—to quell our inner demons and to inspire our better angels. But let's not restrict ourselves by the categorical thinking such labels as nation-state or city-state impose, as these are just ways of describing a linear process of governing groups of people of varying sizes. Such systems have taken many forms over the centuries, but in time they have narrowed their scope to include most of the following characteristics that the majority of Western peoples enjoy today (call them the Justice and Freedom Dozen):

1. A liberal democracy in which the franchise is granted to all adult citizens.
2. The rule of law defined by a constitution that is subject to change only under extraordinary circumstances and by judicial proceedings.
3. A viable legislative system for establishing fair and just laws that apply equally and fairly to all citizens regardless of race, religion, gender, or sexual orientation.
4. An effective judicial system for the equitable enforcement of those fair and just laws that employs both retributive and restorative justice.
5. Protection of civil rights and civil liberties for all citizens regardless of race, religion, gender, or sexual orientation.
6. A potent police for protection from attacks by other people within the state.
7. A robust military for protection of our liberties from attacks by other states.
8. Property rights and the freedom to trade with other citizens and companies both domestic and foreign.
9. Economic stability through a secure and trustworthy banking and monetary system.
10. A reliable infrastructure and the freedom to travel and move.
11. Freedom of speech, the press, and association.
12. Mass education, critical thinking, scientific reasoning, and knowledge available and accessible for all.

These are the elements that make up a just and free society.[37] In her book *Prosperity Unbound*, for example, the World Bank economist Elena Panaritis shows how the transition from informal to formal property rights (from what she calls "unreal estate" to real estate) in Peru in the 1990s enhanced property values of small landowners, increased trust in property transactions, and jump-started the economic development of that struggling country. With half the world's property owners living and working outside formal structures that protect property rights and their exchange, this one institution alone could be a game changer.[38]

In a similar vein, in *The Art of Community* the social anthropologist

Spencer Heath suggests alternative models for the way in which people might come together in nonpolitical voluntary communities consisting of both private and common areas. There already exist many such communities all over the world operating smoothly and efficiently. For instance, shopping centers are proprietary communities, as are condominium complexes, mobile home parks, retirement communities, industrial parks, private colleges and universities, and corporate campuses such as those of Microsoft, Apple, and Google, which are, in essence, miniature cities operating through proprietary instead of political means. The hotel is another fine example. "The hotel has its public and private areas, corridors for streets, and a lobby for its town square. In the lobby is the municipal park with its sculpture, fountains, and plantings. It has its shopping area, where restaurants and retail stores bid for patronage. Its public transit system, as it happens, operates vertically instead of horizontally."[39] When you rent a hotel room, included in the price are utilities such as water, electricity, heat and air-conditioning, and sewerage, and for an extra fee you get room service, current movies, and high-speed Internet access. Also provided are police and fire protection by security guards and sprinkler systems. Many hotels include a chapel for religious services, babysitting and play areas for children, pools for recreation and bars for imbibing, concerts and plays, and even theater shows (especially in Las Vegas). The main difference between this community and that of cities is that hotels are entirely private and organized by voluntary contract.

Could the proprietary community concept be expanded globally? In his 2014 book *Anarchy Unbound: Why Self-Governance Works Better Than You Think*, the economist Peter Leeson provides numerous examples of social self-organization in which private individuals secure social cooperation without government, and even though there is no world government, as we have seen in the decline of war and the new peace, somehow nations have found pathways toward nonviolent solutions to conflicts and disputes.[40] True enough, but critics of anarchy point out that all such proprietary communities are situated within sovereign nations that provide military protection from foreign enemies, police protection from vandals and other criminals, public roads to access their private roads, courts to adjudicate disputes over contract violations, and a monopoly on the legitimate use of force to ensure that the overarching rule of law is enforced fairly and justly. Whether the Justice and Freedom Dozen can be maintained in, say, city-states instead of nation-states, or will ultimately be replaced by other social technologies such as proprietary tools that produce the same results (e.g., mediators instead of lawyers and judges), remains to be seen. There are some social theorists who think that they can (libertarian anarchists, anarchocapitalists, market anarchists[41]), although most political scientists and economists maintain that at the very least a minimum state is necessary to avoid the inevitable conflicts of interest that arise between private individuals, corporations, and proprietary communities, as well articulated by Robert Nozick in his classic work *Anarchy, State, and Utopia*.[42] The problem seems to be that once a minimum state is established for these essentials it inexorably grows into a bloviated bureaucracy that gobbles up more and more of a nation's GDP to run, and the roughly one hundred million words of federal law and regulations make complete compliance impossible.[43]

Whatever changes are made going forward, history shows us that to succeed they should be implemented in a protopian manner, incrementally, as Thomas Jefferson wrote in his reflection on the American Revolution:

> I am not an advocate for frequent changes in laws and constitutions, but laws and institutions must go hand in hand with the progress of the human mind. As that becomes more developed, more enlightened, as new discoveries are made, new truths discovered and manners and opinions change, with the change of circumstances, institutions must advance also to keep pace with the times. We might as well require a man to wear still the coat which fitted him when a boy as civilized society to remain ever under the regimen of their barbarous ancestors.[44]

PROTOPIAN ECONOMICS

In the twenty-third-century world of Gene Roddenberry's *Star Trek*, replicator machines give you anything you want for consumption, from a four-course meal to "tea, Earl Grey, hot." In this fantasy world money isn't important because, essentially, everyone has everything they need. It is a world of abundance. How realistic is this world? A decade ago most people—myself included—would have said not very, because of the fundamental conflict between unlimited wants and limited resources. After all, economics is defined as the allocation of scarce resources that have alternative uses.[45]

However, maybe Econ 2.0 will be something entirely different from Econ 1.0. Just think for a moment the power that an individual has with a pocket-size device connected to the Web: an encyclopedia with millions of entries; detailed street maps of nearly every city in the world; instant stock quotes and weather reports; audio books, eBooks, and digital magazines and newspapers; speech recognition, dictation, and language translation; videos, films, and TV shows; games for fun, games for education, games for social interaction, and games that encourage analytical thinking; crowd funding, peer-to-peer lending, social banking, and microfinance; and millions of apps for learning languages, shopping, business, reference, news, music, travel and taxis, task managing, communication, health, restaurants, and almost anything you can think of and lots that you can't think of. And, shockingly, most of it is *free*.

Futurist Kevin Kelly imagines going back in time to the early 1990s and describing to the experts then what is available now. "You would simply be declared insane," he says. "They would say there is no economic model to make this. What is the economics of this? It doesn't make any sense, and it seems far-fetched and nearly impossible. But the next 20 years are going to make this last 20 years just pale. We're just at the beginning of the beginning of all these kind of changes. There's a sense that all the big things have happened, but relatively speaking, nothing big has happened yet."[46]

For example, "postscarcity economists" (they call their skeptical colleagues "scarcity economists") have outlined systems of resource recycling and technologically advanced automated systems (such as 3-D printers and nanotechnological molecular assemblers and nanofactories) that are able to convert raw materials into the finished products that people need.[47] It sounds like pure science fiction, but consider how far we've come in just the past half century. According to the X-Prize founder Peter Diamandis, in his optimistically titled book *Abundance,* "humanity is now entering a period of radical transformation where technology has the potential to significantly raise the basic standard of living for every man, woman and child on the planet." Diamandis projects that "Within a generation, we will be able to provide goods and services that were once reserved for the wealthy few to any and all who need them."[48]

Consider the staggering growth of information in recent decades. A Masai warrior with a smart phone and access to Google, for example, has more information than President Clinton did in the 1990s.[49] If you read a newspaper cover to cover every day for a week, you will have consumed more information than a citizen in seventeenth-century Europe would have encountered in a lifetime. That's a lot of data, but it's nothing compared to what is on the immediate horizon. By comparison, from the earliest stirrings of civilization 10,000 years ago to the year 2003, all of humankind created a grand total of about 5 exabytes of digital information. An exabyte is 1 quintillion bytes, or 1 billion gigabytes. (Your smart phone likely has 8, 16, or 32 gigabytes of storage capacity, enough to store thousands of songs, pictures, videos, and other digital information.) From 2003 through 2010 humans created 5 exabytes of digital information *every 2 days*. By 2013 we were producing 5 exabytes *every 10 minutes*. How much information is this? The 2010 total of 912 exabytes is the equivalent of 18 times the amount of information contained in all the books ever written. Make all that digital knowledge available to every person on the planet instantly through the Internet, and ideally all citizens of the world can become citizen-scientists capable of reasoning their way to solving personal, social, and moral problems.

The effect of all this information on all areas of life around the world is stunning. Education: the Khan Academy's YouTube tutorial videos on more than 2,200 topics from algebra to zoology draw more than 2 million viewings a month. Medicine: the field of personalized medicine—an industry that didn't exist before 2003—is now growing at 15 percent a year and will reach $452 billion by 2015. Poverty: the number of people living in absolute poverty has fallen since the 1950s and has dropped by more than half. At the current rate of decline it will reach zero by about 2035. Expenses: groceries today cost 13 times less than 150 years ago in inflation-adjusted dollars. Standard of living: 95 percent of Americans now living below the poverty line have electricity, Internet, water, flush toilets, a refrigerator, and a television. John D. Rockefeller and Andrew Carnegie, among the richest people on the planet, enjoyed few of these luxuries.[50]

Protopian thinkers are also doers. The CEO of SpaceX and Tesla Elon Musk, for example, envisions not just a return to space and a world of electric cars, but how we might colonize Mars in the next ten to twenty years and establish self-sustaining societies in which new forms of governance could be tried.[51] "There's no rush in the sense that humanity's doom is imminent; I don't think the end is nigh," he said. "But I do think we face some small risk of calamitous events. It's sort of like why you buy car or life insurance. It's not because you think you'll die tomorrow, but because you might." Civilizations and political systems rise and fall, Musk reflected. "There could be some series of events that cause that technology level to decline. Given that this is the first time in 4.5 billion years where it's been possible for humanity to extend life beyond Earth, it seems like we'd be wise to act while the window was open and not count on the fact it will be open a long time."[52]

Musk's Paypal cofounder Peter Thiel is another protopian who has helped bankroll the Seasteading Institute, whose goal is to establish permanent and autonomous ocean communities anchored in international waters in which people can experiment and innovate "with diverse social, political, and legal systems."[53] Futurist Ray Kurzweil thinks we can achieve digital (if not biological) immortality by 2045,[54] and he is backed by the Russian-based 2045 Initiative, whose goal is "to create technologies enabling the transfer of an individual's personality to a more advanced non-biological carrier, and extending life, including to the point of immortality."[55] Google's cofounder and CEO Larry Page launched a new project in 2013 called Calico, devoted to research on aging toward an end captured on the September 30 cover of *Time* magazine: "Can Google Solve Death?"[56]

These are just a few of the ways that a protopian economy could change our lives by the end of this century in ways that were inconceivable at the beginning of it. In his projections for what our daily lives will be like by the year 2100, the physicist and futurist Michio Kaku employed a "Delphi poll" method of surveying and visiting more than three hundred expert scientists and technologists in a wide range of fields to see what is on the horizon. By the end of this century Kaku predicts that computers will develop emotions and be self-aware, and that computer-brain implants will enable us to move objects and operate machines just by thinking. The entire Internet will be accessible through a pair of contact lenses (think Google Contacts instead of Google Glass). Nanotechnology will enable us to build or reconfigure objects at the molecular level into nearly anything we want or need. Driverless cars and robot surgeons, a moon base and a Mars base, space tourism, and the resurrection of extinct species are just a few of Kaku's projections based on current technologies. "Today, your cell phone has more computer power than all of NASA back in 1969, when it placed two astronauts on the moon," he reminds us. "The Sony PlayStation of today, which costs $300, has the power of a military supercomputer of 1997, which cost millions of dollars."[57]

Kaku envisions that all this technology will eventually unite humanity into a single planetary civilization, and as the borders of nations dissolve naturally "their power and influence will be vastly decreased as the engines of economic growth become regional, then global."[58] Nations were necessary to consolidate variegated feudal laws and to create a common currency and regulatory system that greased the wheels of commerce as capitalism arose in the eighteenth century, but by the start of the twenty-second century economic power, like technological power, will be diffused across the globe.

. . .

As a longtime skeptic of both doomsayers who project the end of the world in our generation, as well as futurists who proclaim the next big thing to revolutionize humanity and save the planet in our lifetime, I'm aware that both have been spectacularly and often embarrassingly wrong. The end of the world has not materialized (or dematerialized), and futurists who craft a utopian narrative about how one day we will live forever, colonize the galaxy in starships, and materialize food and drink in replicators are often indistinguishable from producers of science fiction and fantasy. Skeptics need hard evidence: prove it.

In response to these reasonable criticisms, Diamandis describes in specific detail how the revolution has already begun and that such abundance can be realized in our lifetime through three current forces: (1) *Do-it-yourself (DIY)* backyard tinkerers such as the aviation pioneer Burt Rutan, who won the first X-Prize for achieving privatized space flight; and the geneticist J. Craig Venter, who sequenced the human genome. Thousands of such DIYers are working away in garages and warehouses, innovating solutions in neuroscience, biology, genetics, medicine, agriculture, robotics, and numerous other areas. (2) *Techno-philanthropists* such as Bill and Melinda Gates (conquering malaria), Mark Zuckerberg (promoting education), Pierre and Pam Omidyar (generating electricity in the developing world), and many others are dedicating significant portions of their vast fortunes to solving specific problems. (3) *The bottom billion*, the poorest of the poor, as they become plugged into the global economy through microfinancing and the Internet, will lift all boats with them as we all work together toward having clean water, nutritious food, affordable housing, personalized education, top-tier medical care, and ubiquitous energy.

These trend lines are real enough, and if these principles were applied worldwide such abundance is, in principle, realizable, in the long run if not in the short. The graphs in *figure 12-1* track the exponential growth rates of prosperity progress over the centuries and into the future as measured by GDP, a standard economic metric.[59] As the economic historian Gregory Clark writes, "The average person in the world of 1800 was no

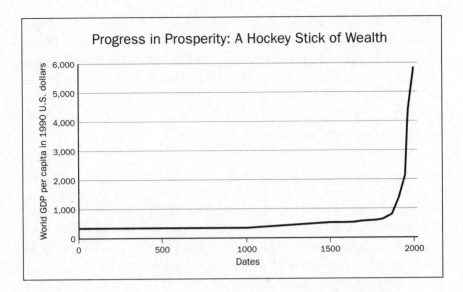

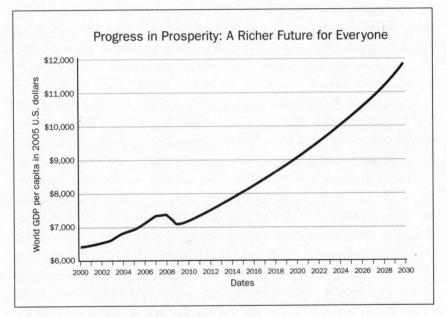

Figure 12-1. Economic Growth Graphs

Figure 12-1a. A Hockey Stick of Wealth. World GDP per capita in 1990 US dollars from the year 0 to 2000 shows that more people are more prosperous than at any time in the past.[61]
Figure 12-1b. Projected Prosperity Progress. World GDP per capita in 2005 US dollars projected from 2000 to 2030, showing the dip during the great recession of 2008. Average annual income is projected to nearly double for everyone in the world.[62] Although far too many people are still living in poverty, the trend lines are in the right direction.

better off than the average person of 100,000 BC." Thanks to science and technology applied by entrepreneurs in the Industrial Revolution, "modern economies are now ten to twenty times wealthier than the 1800 average. Moreover the biggest beneficiary of the Industrial Revolution has so far been the unskilled. There have been benefits aplenty for the typically wealthy owners of land or capital, and for the educated. But industrialized economies saved their best gifts for the poorest."[60]

What About Income Inequality?

At this point in our journey into the future, there are some who might accuse me of simply turning off the lights in the room that illuminates the elephant of income inequality. It's one thing for dot-com billionaires to wax inspirational at conferences attended by well-to-do dreamers who listen to talks about mosquito-zapping lasers and balloon-based wireless Internet access in Third World villages, but they can afford the luxury of such speculative enterprises, unlike those bottom billion who still toil to find even one square meal a day and for whom fear and violence are normal parts of daily life. While Washington, DC, invests $850 per person per year on police protection, for example, Bangladesh spends less than $1.50 per capita annually, and as a result crime, violence, and disorder rule the day there.[63] Maybe if some of those billions had trickled down to the masses in the first place some of these problems would already have been solved by the people at the bottom. That's how many people see it, anyway.

Income inequality has emerged as one of the most contentious controversies of our time,[64] particularly in the West, where so much capital has accumulated toward the top of the wealth distribution scale, most egregiously in the recent US Supreme Court decisions that have enabled the fatal cocktail of money and politics to mix even more, shifting American democracy in the direction of a plutocracy. The political scientist Martin Gilens, for example, in his aptly titled book *Affluence and Influence*, found that "The American government does respond to the public's preferences, but that responsiveness is strongly tilted toward the most affluent citizens. Indeed, under most circumstances, the preferences of the vast majority of Americans appear to have essentially no impact on which policies the government does or doesn't adopt."[65]

In a December 4, 2013, speech sponsored by the Center for American Progress, President Barack Obama called income inequality "the defining challenge of our time," noting that the trends are "bad for our economy . . .

bad for our families and social cohesion . . . bad for our democracy." Income inequality and immobility, he concluded, pose "a fundamental threat to the American dream."[66]

Do they? Maybe not. The rich *are* getting richer. That much is true and has been documented in many places, most recently in the surprise 2014 best-selling tome *Capital in the Twenty-First Century,* by Thomas Piketty.[67] As well, analyzing tax data from the Congressional Budget Office (CBO) for after-tax income trends from 1979 through 2010, the economist Gary Burtless found that the rich got richer faster than the poor and middle class got richer, as seen in *figure 12-2.*[68]

Overall, the top fifth of income earners in the United States increased their share of the national income to 50 percent in 2010 from 43 percent in 1979, and the top 1 percent increased their share of the pie to 15 percent in 2010 from 9 percent in 1979. But note what has not happened: the poor and middle classes have not gotten poorer. They've gotten richer: the income of the first three quintiles from poor to middle class increased by 49 percent, 37 percent, and 36 percent, respectively. But their prosperity growth has been slower in comparison to the very wealthiest, so it is the

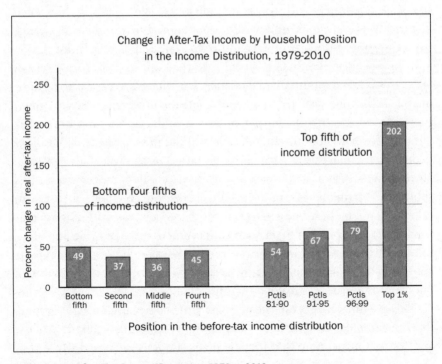

Figure 12-2. After-Tax Income Increases, 1979 to 2010

relative difference, in part, that is driving the conversation. As H. L. Mencken wisecracked, "A wealthy man is one who earns $100 a year more than his wife's sister's husband." To that extent, the spread between wives' sisters' husbands' earnings is growing.

However, the pie metaphor may not be the best way to think about this issue because a pie is of a fixed size such that if your slice is larger, then someone else's is smaller, as if this were a zero-sum game—if my slice is twice the size of yours, yours is half the size of mine, so the difference sums to zero. This is not how economies work. They grow. The pie gets larger such that you and I can both get a larger slice compared to the slices we got from last year's pie, even if your slice increase is relatively larger than mine. A report released by the Federal Reserve in early 2014, for example, noted that the overall wealth of Americans hit the highest level ever, with the net worth of US households and nonprofit organizations rising 14 percent in 2013, or an increase of almost $10 trillion to an almost unimaginable $80.7 trillion, the most ever recorded by the Fed and confirming Burtless's finding that most people in the United States are better off than they were before.[69]

Of course, on a planet with finite resources such an expansion cannot continue indefinitely with the same industries,[70] but what has happened historically is that capital and wealth production shift as industries change from, say, horticulture and agriculture to coal and steel to information and finance. Eric Beinhocker's epic historical analysis in *The Origin of Wealth* provides a mountain of evidence that "there are reasons to be optimistic . . . that the growth curve will continue for some time to come," especially when "the 2.3 billion people in India and China have fully joined the world economy this decade and the next, and the 650 million people in sub-Saharan Africa come on board in the next half century." Since the economy is a complex adaptive system, Beinhocker warns, it is subject to "tipping points, radical change, and even collapse," but it is also capable of adaptation and change in response to crises, so there are reasons for optimism.[71] Whether such shifts can continue into the far future to maintain this wealth expansion trend remains to be seen (the experts appearing in the documentary film *Surviving Progress* have their doubts[72]), but the growth curves are encouraging.

Relative difference is not the only concern people have. In a 2013 report for the Federal Reserve Bank of Richmond, the economists Kartik Athreya and Jessie Romero note that income inequality has increased while economic mobility has decreased in recent years. They observe that in 1979 the top 1 percent of households took home 7.4 percent of total after-tax income in the United States, while in 2007 that figure more than doubled, to 16.7 percent. Over that same time span "the share of income earned by households at all levels of the remaining distribution stayed flat or declined."[73] The impediments on moving up from the lower quintiles begin early in the lives of the poor, including nutritional and cognitive deficits. A Brookings Institution study, for example, found that 78 percent of children from families in the top quintile were cognitively and developmentally ready for school by age five, but that less than half (48 percent) of children from families in the bottom income quintile were so prepared.[74]

Writing in the *National Tax Journal* the economists Gerald Auten and Geoffrey Gee analyzed individual income tax returns between 1987 and 2005 and found that for individuals age twenty-five and up, "over half of taxpayers moved to a different income quintile and that roughly half of taxpayers who began in the bottom income quintile moved up to a higher income group by the end of each period." What about the rich? If you're in the top bracket there is only one direction you can move, and that is what Auten and Gee found: "those with the very highest incomes in the base year were more likely to drop to a lower income group and the median real income of these taxpayers declined in each period." In fact, they found that "60 percent of those in the top 1 percent in the beginning year of each period had dropped to a lower centile by the 10th year. Less than one-fourth of the individuals in the top 1/100th percent in 1996 remained in that group in 2005."[75]

In a follow-up study that included income data through 2010 based on the Statistics of Income available through the IRS's Compliance Data Warehouse and the Social Security Administration's records on W-2 and 1099 tax forms, the authors found that "approximately half of taxpayers in the first and fifth quintile remained in the same quintile 20 years later. About one-fourth of those in the bottom moved up one quintile, while 4.6 percent moved to the top quintile."[76] Of course, it's not like the über rich got poor; moving downward from the top 1/100th percent to, say, the top 1/10th or even the top 1 percent doesn't mean they're now eating Top Ramen noodles for dinner (not that there's anything wrong with that). Flying first class on commercial instead of private jets is not exactly a hardship that causes much hand wringing. But their point is that social mobility, while not as ideal as we would like it to be in a society that purports to champion equal opportunity, is not as fixed and immobile as it is often portrayed to be.

In any case, if history is to be our guide in this matter social mobility does not appear to be something that governments can do much about. According to the economic historian Gregory Clark—who used surnames as a proxy for social mobility in his study going back into history when last names were affixed to first—your social standing at birth accounts for more than 50 percent of your adult income and social status. Your last name—that is, the family into which you were born—more than any other variable predicts where you will end up in life in terms of your social station. Clark's book in which his massive datasets are presented is tellingly titled *The Son Also Rises*. "If all the factors that determine people's life chances are summarized by their parents' status, then these persistence rates imply that all initial advantages and disadvantages for families should be wiped out within three to five generations," he writes. But that is not what happens. Clark estimates that family legacy lasts about ten to fifteen generations, with a generation lasting thirty years. The effects of eighteenth-century wealth, he documents, can still be measured today, and those born into lower socioeconomic classes today will take centuries to climb into the higher classes.[77]

The effect is a demographic and statistical one that predicts outcomes of groups, not individuals. You or I may pull ourselves up by our bootstraps in one generation, but overall the long-run trends will prevail despite the efforts of governments to level the playing field through various measures such as mass education, the outlawing of discrimination, progressive taxes, and wealth transfers, and the immobility can be found even in the most liberal of northern European democracies, such as Sweden. (Clark is, however, in favor of most of these government actions, not to boost social mobility, but because the most talented and creative and hardworking people will rise regardless, and these measures serve other purposes.) Even Communist China—the ne plus ultra of social equality, where everyone is equal by fiat—has been unable to budge social mobility rates. "Despite Mao's best efforts," Clark concludes from his analysis of elite Chinese surnames, "'class enemies' are strongly entrenched within the current Communist government of China."[78] Encouragingly, in America Clark has identified trends among black families after the civil rights movement, but collectively he estimates that it will be 2240 before, for example, "the black population will be represented among physicians at only half the rate of the general population."[79]

Clark attributes some of this effect to what he calls "social competence," which additional studies from behavioral genetics and twin studies suggest is at least 50 percent heritable. Not just raw intelligence, but also inherited personality traits and motivational desires for success, such as the need of achievement, come to mind as factors. Technically this is called "cumulative advantage." Colloquially it is known as "the rich get richer."[80]

Whatever you call it, if the economy is expanding and most people in the West—on average—are becoming wealthier, why do so many people care that others are richer than they are? Part of the answer may be found in our evolved intuitions about folk economics,[81] derived from the study of evolutionary economics (what I call *evonomics*[82]). Our ancestors evolved in small bands and tribes of a couple dozen to a couple hundred individuals in which everyone was either genetically related or knew one another intimately, most resources were shared, wealth accumulation was almost unheard of, and excessive greed and avarice were punished. There were no capital markets, there was no economic growth, very little concentration of labor, no "invisible hand" of the marketplace, and no excessive disparity between rich and poor because everyone was, by today's standards, dirt poor; there was no accumulation of wealth because there was no wealth to accumulate.

Living on the margin means that reciprocity and food sharing were essential for survival, and this is why even modern hunter-gatherer groups employ customs and enforce mores that ensure relative equality. Without such cooperation in such a zero-sum world, the gain of one could mean the loss of another. Because of this humans are what the game theorist Martin Nowak calls "super cooperators," and it is cooperation more than competition, he argues, that led to the evolution of altruism and why we need each other to survive and succeed.[83] I think this overplays our better angels and downplays our inner demons—our super cooperator side is often counterbalanced by a super competitive side—but my point here is that today we live in a *nonzero* world in which the gain of one often means the gain of others, and thanks to science, technology, and trade we have an abundance of food and resources. But our brains operate as if we are still living in that zero-sum land of evonomics.[84]

Because we have these dual impulses of cooperation and competition, we should be careful not to portray our ancestors as living in a Stone Age hippie-dippy 1960s commune-like communities in which each gives according to his abilities and takes according to his needs in some harmonious Marxian state of nature. We saw what the application of that utopian fantasy led to in Soviet Russia and Mao's China. As the anthropologist

Patricia Drapers writes in her ethnography of the !Kung Bushmen (discussed in chapter 11) in the Kalahari Desert of Africa—a traditional society
that approximates how our evolutionary ancestors might have lived—"the
reason that goods are shared equitably and more or less continuously is
that the have-nots are so vociferous in pressing their demands. Are these
a people who live in communal harmony, happily sharing among all? Not
exactly. . . . At one level of analysis, one can show that goods circulate, that
there are no inequalities of wealth and that peaceable relations characterize
dealings within and between bands. At another level, however . . . one sees
that social action is an ongoing scrimmage—often amicable but sometimes
carried on in bitter earnest."[85]

When those small bands and tribes coalesced into chiefdoms and
states starting around seven thousand to eight thousand years ago, there
was an economic transition from the equal distribution of economic
wealth among bands, to the emergence of hierarchical wealth as a token of
status and power among tribes, to the accumulation of massive amounts
of wealth—and wealth disparity among citizens—in states. In a world of
relative poverty but in which no one has much more than anyone else, it
feels fair. But in today's Western world, in which almost everyone has
some wealth but in which some have vastly more than others, it feels
unfair. Our brains are not equipped to intuitively understand how modern economies work, so to most people such systems seem unjust. And,
frankly, throughout most of the history of civilization, economic inequalities were not results of natural differences in drive and talent between
members of a society equally free to pursue their right to prosperity;
instead, a handful of chiefs, kings, nobles, and priests exploited an unfair
and rigged social system to their personal benefit and at the cost of impoverishing the masses. Thus our natural (and understandable) response is
envy and sometimes anger, as witnessed in the Occupy Wall Street movement of 2012, seen in *figure 12-3*.

Folk economics also helps to explain the results of a 2013 study that
found people's perception of income inequality is largely overinflated
when compared to data on actual income inequality. Psychologists asked
more than five hundred people in an online survey what their perceptions
were about overall income inequality in the United States, and their own
political preferences. Results: Participants tended to *overestimate* the
number of American households that are just scraping by, believing that,
on average, about 48 percent of households make less than $35,000 a year,
when in fact census data show that 37 percent actually fall under that

Figure 12-3. Occupy Wall Street

Resentment against income inequality and the Wall Street bailout erupted in the Occupy Wall Street movement in 2011 and 2012. Photos from Zuccotti Park, New York, and a public park in Portland, Oregon. Photos by the author.

number. By contrast, participants *underestimated* the number of American households that are faring well (although not rich), believing that, on average, only about 23 percent of households make $75,000 or more a year, compared to the actual census data that 32 percent actually make more than that number. Participants also believed the income inequality gap was much larger than it actually is, estimating that the richest 20 percent make about 31 times more than the poorest 20 percent, when in fact the top 20 percent makes about 15.5 times more than the bottom 20 percent, or half as much as people's perceptions. The actual dollar difference in perception was staggering: people believe that the average annual income of the richest 20 percent of Americans is $2 million, whereas in fact it is $169,000, a perceptual difference of nearly 12 times.

That is some serious distortion of reality that shows how potent the folk economic bias is against income inequality at any noticeable level. And, predictably, the bias was strongest among self-identified liberals, who tended to overestimate the size and the growth rate of the income inequality gap significantly more than self-identified conservatives in the study. But the general effect was felt across the political spectrum, as the lead author John Chambers explained: "Almost all of our study participants—regardless of their socio-economic status, political orientation, racial ethnicity, education level, age, and gender—grossly underestimated Americans' average household incomes and overestimated the level of income inequality."[86]

The effects of income inequality have been documented by numerous scholars and scientists, such as the British social epidemiologist Richard Wilkinson in his 1996 book *Unhealthy Societies* and more recently in a coauthored book with Kate Pickett called *The Spirit Level: Why Greater Equality Makes Societies Stronger.*[87] In his TED talk Wilkinson does a massive data dump of charts strikingly similar to those I present in chapter 4 on the relationship between religiosity and societal health by the social scientist Gregory Paul (as religiosity goes up, societal health goes down); the difference is that on the horizontal axis, instead of a secular-religious spectrum, Wilkinson presents an equality-inequality scale that he believes is causing countries (and states within the United States) to have higher rates of homicides, imprisonments, teen pregnancies, and infant mortality, but lower levels of life expectancy, math and literacy, trust, obesity, mental illness (including drugs and alcohol addiction), stress, and social mobility.[88] In all of these measures, he says, more unequal societies have poorer outcomes. To their credit, Wilkinson and Pickett remain politically neutral in their suggestions for how to fix the problem, showing how both market forces that create wealth and government redistribution programs can be equally effective. The source of the equalizing forces is less important, they say, than the lowering of inequality itself.

Some of this research has been corroborated by others, such as the Stanford University biologist Robert Sapolsky, who pioneered research on the physical effects of stress on baboons caused by social inequality (and the opposite—to update his memorable book, *Why* [more socially egalitarian] *Zebras Don't Get Ulcers*). He writes, "So we've got income inequality, low social cohesion and social capital, class tensions, and lots of crime all forming an unhealthy cluster," citing both America and post-Soviet Russia as examples.[89] The anthropologists Martin Daly and Margo Wilson ran correlations between income inequality and homicide rates in a comparison between Canada and the United States, noting that "When Canadian provinces and U.S. States were considered together, local levels of income inequality appeared to be sufficient to account for the two countries' radically different national homicide rates." The authors conclude that "the degree to which resources were unequally distributed was a stronger determinant of levels of lethal violence in modern nation states than was the average level of material welfare." They attribute the effect to greater male-on-male competition in lower socioeconomic classes, which are more likely to result in violence.[90]

I'm not so sure. In the United States, anyway, income inequality was much lower in the 1970s and 1980s right in the middle of the national crime wave that spiked homicide rates almost tenfold, and then when the crime wave collapsed in the 1990s and homicide rates returned to lower, pre-1960s levels, income inequality began its steady climb upward and has continued increasing throughout the 2000s, while crime and homicide rates have remained at their historically low levels. As I said in chapter 4 in assessing Gregory Paul's studies, all of these social ills—homicides, suicides, teen pregnancies, incarceration rates, and the like—have their own unique set of causes. The chances that all of these effects have a single cause in, say, income inequality is unlikely. More likely is that income inequality is a proxy for something else that may have some effects that are highly sensitive to other causes.

For example, income inequality appears to affect the psychology of those in advantageous positions of power, thereby altering the way in which they interact with others. A poignant example was discovered by the social psychologist Paul Piff and his UC Berkeley colleagues. Subjects played a rigged game of Monopoly in which some subjects were given an unfair advantage by being allowed to roll two dice instead of one, to get $2,000 to start with instead of $1,000, and to collect double the normal amount of money each time they passed "Go." And just for good measure, the favored subjects were given the "Rolls-Royce" playing piece instead of the "Old Shoe." Even though the subjects were randomly assigned by the flip of a coin to either the normal or the advantaged position, those who pulled ahead in money and properties rationalized their good fortune as due not to luck, but to their own skills and talents. As a result, they started acting more privileged and entitled: they made more noise moving their piece around the board, they took up more physical space, they became more controlling and more demonstrative of their success in the game by the liveliness of their speech, their language became more demanding ("Give me Park Place"), and most notably they reported feeling more proud of themselves, as if they deserved to win.[91]

I asked Piff about the implications of this research. "While results are preliminary," he answered, "the main theme of the Monopoly experiment is that even in an openly rigged game, behavior changes such that 'richer players' start to behave differently—they become ruder, louder, more demanding, and less sensitive to others. And their attitudes seem to shift as well, such that people on the receiving end of an unfair advantage start to feel more deserving and entitled to that advantage."[92]

The effect is a manifestation of what is called the *fundamental attribution bias*, or the tendency to attribute different causes for our own beliefs and actions than those of others. There are several types of attribution bias. There is a *situational attribution bias*, in which we identify the cause of someone's belief or behavior in the environment ("her success is a result of luck, circumstance, and having connections") and a *dispositional attribution bias*, in which we identify the cause of someone's belief or behavior in the person as an enduring personal trait ("her success is due to her intelligence, creativity, and hard work").[93] And, thanks to the *self-serving bias*, we naturally attribute our own success to a positive disposition ("I am hardworking, intelligent, and creative") and we attribute others' success to a lucky situation ("he is successful because of circumstance and family connections").[94] In Piff's research, even people who haven't yet succeeded (such as students), but who are children of wealthier parents, feel more entitled and deserving.[95] In a 2009 paper relevantly titled "Social Class, Sense of Control, and Social Explanation," Piff and his colleagues Michael Kraus and Dacher Keltner found that social class was "closely

associated with a reduced sense of personal control and that this association would explain why lower class individuals favor contextual over dispositional explanations of social events."[96]

The effects of income inequality can be pernicious, as much as its
opposite can be salubrious.[97] For example, in another experiment Piff and
his team surreptitiously set themselves up at intersections to record cars
stopping for a pedestrian about to step into a crosswalk on a California
street (where it is the law that you must stop). Overall, 65 percent of drivers stopped, but of those who did not stop, the overwhelming majority—by
a factor of three or four times—were driving luxury automobiles (for
example, BMWs, Mercedes, Porches), a proxy for a higher socioeconomic
class.[98] In another experimental setting, as subjects were filling out a form
they were told that they could help themselves to "a few pieces" of candy
from a jar on the table, although they were also told that the candy "is
actually for children for another study." Subjects who felt wealthy took
twice as much candy as participants who felt poor.

In this study, Piff's team actually made people feel rich or poor through
a manipulation of conditions, so the results are even more revealing in
that they show that the effect is temporary and reversible. In a dice-rolling
setup in which subjects were to report the outcome of a roll that the experimenter could not see, wealthier people (those making $150,000 to
$200,000 a year) cheated four times as much as poorer people (those mak-
ing less than $15,000 a year) just to win credits for a $50 cash prize. These
effects, by the way, happened across the political spectrum from Tea Party
conservatives to Occupy Wall Street liberals. As Piff colorfully interpreted the results to me, "While having money doesn't necessarily make
anybody anything, the rich are way more likely to prioritize their own
self-interests above the interests of other people. It makes them more likely
to exhibit characteristics that we would stereotypically associate with, say,
assholes."[99]

The most common response to this state of unequal affairs is to tax the
rich and redistribute the money to the poor (often favored by liberals and,
of course, the poor).[100] The less common response is private charity, whereby
the rich give to causes that help the poor (often favored by conservatives
and, of course, the rich). Surprisingly (since he's at the infamously überliberal Berkeley campus of the University of California system), Piff ends
his TED talk about his research by first noting that there is nothing inherently fixed about the attitudes that seem to come with having a lot of
money. Merely reminding the well-off about the less well-off (through a

forty-six-second video clip about childhood poverty) caused people with money to be more willing to offer up their time to a stranger in distress presented to them in the lab. "After watching this video, later on rich people became just as generous of their time as poor people, suggesting that these differences are not innate or categorical, but are so malleable to slight changes in people's values and little nudges of compassion and bumps of empathy."

Instead of calling for a progressive tax on the rich to "do their fair share," Piff focuses on the grassroots movement of the superrich in America to *voluntarily* give away large portions of their vast fortunes to worthy causes through such organizations as the Giving Pledge, in which more than one hundred of the United States' wealthiest individuals have pledged to give away more than half their wealth to charity. Signatories include Bill Gates, Warren Buffett, Paul Allen, Michael Bloomberg, Ted Turner, Mark Zuckerberg, Elon Musk, and others of equally varied political persuasion. Then there's Resource Generation, which "organizes young people with financial wealth to leverage resources and privilege for social change."[101]

Another way of dealing with the psychological effects of income inequality is to reform capitalism itself from the inside. This may sound oxymoronic, but it is what John Mackey, the co-CEO of Whole Foods Market, wants to do. In *Conscious Capitalism* Mackey sets out to draft a new narrative of capitalism, starting with the myth that the profit motive is the sole driver of business. We all know the old narrative: *Capitalists are a bunch of cigar-chomping, money-grubbing, profit-seeking, quarterly-report-watching, "you're fired"-growling, Gordon Gekko greed-is-good gloating, coldhearted, ruthless, Machiavellian psychopaths.* Despite the fact that some capitalists *do* fit this narrative (the Gordon Gekko character in Oliver Stone's film *Wall Street* was partly modeled after junk-bond king Michael Milken, who was indicted on ninety-eight counts of racketeering and fraud), Mackey claims, "With few exceptions, entrepreneurs who start successful businesses don't do so to maximize profits. Of course they want to make money, but that is not what drives most of them. They are inspired to do something that they believe needs doing."[102]

In spite of his hippie image and vegan diet (he lives what he preaches), Mackey is no naïf when he rails against what he calls the "cancer of crony capitalism," in which crony capitalists cannot compete in the market so they turn to government to implore bureaucrats to impose regulations and duties on competitors. The treatment for the cancer of crony capitalism is *conscious capitalism*, grounded "in an ethical system based on value creation for all stakeholders," which includes not just owners, but also employees, customers, the community, the environment, and even competitors, activists, critics, unions, and the media. Mackey cites Google and Southwest Airlines as role models, and pharmaceutical companies and financial corporations as anti-role models.[103]

In a surprise pivot, Mackey lays the blame for the myth of the profit motive as the only measure of value at the feet of capitalists themselves because "they accepted as fact a narrow conceptualization of business and then proceeded to practice it in that way, creating a self-fulfilling prophecy." Mackey's goal is to write a new narrative for capitalism that asks us to care about customers and human beings instead of data points on a spreadsheet. Sounding more like John Lennon than John Galt, Mackey asks us to "Imagine a business that views its competitors not as enemies to be crushed but as teachers to learn from and fellow travelers on a journey toward excellence," one that "genuinely cares about the planet and all the sentient beings that live on it, that celebrates the glories of nature, that thinks beyond carbon and neutrality to become a healing force that nurses the ecosphere back to sustained vitality."[104]

Although Mackey provides numerous examples of businesses that practice conscious capitalism—including and especially his own Whole Foods Market, in which everyone knows what everyone else makes and top compensation is capped at nineteen times the average (compared to the average of one hundred times in other firms)—he seems to be telling his fellow capitalists that, if they don't voluntarily initiate programs that benefit all stakeholders, the government will force them to and thereby remove the moral element of conscious choice.

Government has a role in establishing and enforcing laws in which capitalism operates, but to prevent excessive top-down interventions into markets, companies should initiate voluntary bottom-up programs that find a way to turn a profit while also lifting their firm to a loftier level. When full-time employees of a company cannot earn enough to support themselves and their families at even a modest level of living, then the problem most likely resides in the business itself. The adjustment is going to be made one way or the other, voluntarily or nonvoluntarily. The voluntary solution is to lower the salaries of upper management, raise the salaries of workers, and if necessary raise the prices of products or services. If this is not done, then the nonvoluntary solution of tax transfers through higher corporate and income taxes will result in the same adjustments, but with a middle-man government bureaucracy that must also be paid, leading to an overall decrease in prosperity. According to a 2014 study published by the International Monetary Fund, "inequality may impede growth at least in part because it calls forth efforts to redistribute that themselves undercut growth. In such a situation, even if inequality is bad for growth, taxes and transfers may be precisely the wrong remedy."[105] Thus, for both moral and practical reasons, the voluntary reformation of capitalism from within is the mark at which to aim.

The economic problems of today are real but tractable. More than that, the economic trends are in the right direction, even in the most impoverished places on earth such as Africa where, if history is our guide to long-term trends, its citizens will be enjoying Western levels of wealth and prosperity that we have today well before the end of the century.[106] Globally, the postscarcity world of abundance that will ensure the survival and

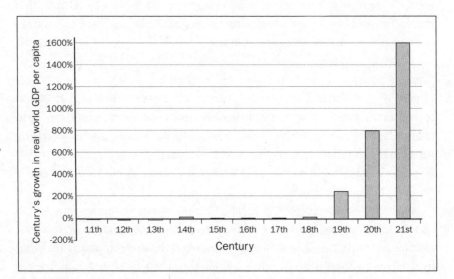

Figure 12-4. Global Economic Growth Rates by Century
Based on data and projections from the UC Berkeley economist J. Bradford DeLong, the bars denote "the relative pace of economic growth in productivity levels and material wealth for the world as a whole over the past ten centuries," estimates DeLong says are "rough and approximate" but do "not do violence to the *qualitative* picture of relative rates of economic growth."[107] The post–Industrial Revolution takeoff of world per capita GDP is staggering, at more than 200 percent in the nineteenth century, 800 percent in the twentieth century, and a projected 1,600 percent for the twenty-first century. If this happens it will mean that this century will generate more wealth and prosperity for humanity than in all previous centuries combined. That is moral progress worthy of the name.

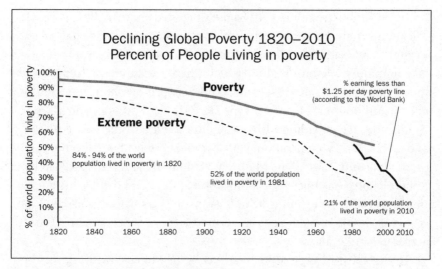

Figure 12-5. Declining Global Poverty 1820-2010

The economist Max Roser compiled data to show that in 1820 between 84% and 94% of the world's people lived in poverty. By 1981 the figure had dropped to 52%, and by 2010 the number had dropped to 21%. That's still far too many, but at this rate it will reach 0 by 2100, and possibly even by 2050. Graph source: www.ourworldindata.org/data/growth-and-distribution-of-prosperity/world-poverty. Data source: World Bank and François Bourguignon, and Christian Morrisson. 2002. "Inequality among World Citizens: 1820–1992." *The American Economic Review*, Vol. 92, No. 4, 727–744.

flourishing of nearly all sentient beings everywhere is very likely achievable by 2100 if the growth rates in *figure 12-4* hold up and the decline in poverty rates continue in *figure 12-5*.

CIVILIZATION 2.0: LOOKING INTO THE FAR FUTURE

In a 1964 article on searching for extraterrestrial civilizations, the Soviet astronomer Nikolai Kardashev suggested using radio telescopes to try to detect energy signals from other solar systems in which he suggested we might find three different types of civilizations: a Type I civilization can harvest all of the energy resources of its planet; a Type II civilization can master the power from its sun; and a Type III civilization can harness the energy from its entire galaxy.[108] Each of these types would presumably give off a different energy signature indicating intelligent life (much like astronomers elsewhere in our galaxy might detect the signatures of oxygen, methane, and other gases in our atmosphere commensurate with living organisms[109]).

In 1973 the astronomer Carl Sagan suggested another civilization classification system based on information storage, concluding that we are a
Type 0.7 civilization.[110] The physicist Michio Kaku estimates that it will
take humanity one hundred to two hundred years to reach Type I, a few
thousand years to achieve Type II, and between one hundred thousand
and one million years to attain Type III.[111] He also speculates on a Type IV
civilization, which would utilize the dark energy of the entire universe,
and even a Type V civilization, which would be able to tap into the energy
of multiple universes.[112] The Mars Society founder and aerospace engineer
Robert Zubrin has his own classification system and defines his civilizational types by the extent to which its inhabitants have dispersed: Type I
across its home planet, Type II across its home solar system, Type III
across its home galaxy.[113]

In the spirit of typing civilizations as an exercise in thinking about
progress over long periods of time, I would like to propose a typology of
civilizational progress along political, economic, and social dimensions, all
of which have implications for moral evolution. This is not even remotely
close to a detailed list of the many configurations of living systems humans
have invented. My aim is to consider in rough outline the far future based
on what has come before. Using the computer convention of 1.0 as the first
launch design of a system, I begin at Type 1.0, when our hominin ancestors
first became social primates millions of years ago. Here we can see how far
we've come and what we need to do to become a Type 2.0 civilization.

Type 1.0: *Fusion-fission* groups of hominins living in Africa, where
group membership is fluid, technology consists of primitive stone tools,
within-group conflicts are resolved through dominance hierarchy, and
between-group violence is common.

Type 1.1: *Bands* of roaming hunter-gatherers that form kinship groups,
with a mostly horizontal political system and egalitarian economy.

Type 1.2: *Tribes* of individuals linked through kinship but with a more
settled and agrarian lifestyle. The beginnings of a political hierarchy and a
primitive economic division of labor.

Type 1.3: *Chiefdoms* consisting of a coalition of tribes into a single
hierarchical political unit with a dominant leader at the top, with the
beginnings of significant economic inequalities, and a division of labor in
which lower-class members produce food and other products consumed
by nonproducing upper-class members.

Type 1.4: *City-states and feudal kingdoms* consisting of small political

bands united by land and driven by economic ties and military obligations in a hierarchical system, with confiscatory taxes (or the equivalent
thereof) to finance the bureaucracy.

Type 1.5: *Nation-states* as a political coalition with jurisdiction over a
well-defined geographical territory and its corresponding inhabitants,
with a mercantile economy that seeks a favorable balance of trade in a
win-lose game against other states.

Type 1.6: *Empires* as states that extend their control over peoples who
are not culturally or ethnically within its confined geographic jurisdiction, with a goal of economic dominance over rival empires.

Type 1.7: *Electoral democracies and republics* that divide the sources of
power over several institutions that are run by elected officials voted for by
the people. The beginnings of a market economy.

Type 1.8: *Liberal democracies and republics* that give the vote to all
citizens. Markets that begin to embrace a nonzero, win-win economic
game through free trade with other states.

Type 1.9: *Democratic capitalism,* the blending of liberal democracy and
free markets, now spreading across the globe through democratic movements in developing nations and broad trading blocs such as the European
Union, and open trade agreements such as NAFTA (North American Free
Trade Agreement).

Type 2.0: *Globalism and a planetary civilization* that includes a planetary communication system (the Internet); planetary knowledge (all
information digitized and available to anyone, anywhere, at any time); a
planetary economy practicing conscious capitalism and with markets in
which anyone can trade with anyone else without interference from states
or governments; a planetary polity consisting either of all democratic
states or small city-states with social contracts that govern anyone who
chooses to live there; a planetary energy system from renewable and sustainable resources; and a planetary culture in which tribal and ethnic differences fade into history and everyone feels they are part of one global
species.

. . .

The forces at work that are slowing us from reaching Civilization 2.0 are
primarily political and economic.[114] In his book *World 3.0,* for example,
economist Pankaj Ghemawat shows that in 2010 only 10 to 25 percent of
economic activity is international (and most of that is regional rather than
global). Distance factors (geographic and cultural) are still substantial for

a lot of people. He crunches these factors into a distance coefficient akin to Newton's law of gravitation. For example, he computes, "a 1 percent increase in the geographic distance between two locations leads to about a 1 percent decrease in trade between them," a distance sensitivity of −1. Or, he calculates, "U.S. trade with Chile is only 6 percent of what it would be if Chile were as close to the United States as Canada." Likewise, "two countries with a common language trade 42 percent more on average than a similar pair of countries that lack that link. Countries sharing membership in a trade bloc (for example, NAFTA) trade 47 percent more than otherwise similar countries that lack such shared membership. A common currency (such as the euro) increases trade by 114 percent." That actually sounds encouraging to me, but Ghemawat reminds us of our deeply ingrained tendencies to want to interact with our kin and kind, and retain our local customs and culture, which tend to balkanize people and slow globalization. A 2009 survey, for example, found that among residents of the EU-16 countries 48 percent trust their fellow nationals, 20 percent trust citizens of other EU countries, and only 13 percent trust people in the non-EU countries.[115]

The resistance by nondemocratic states to turning power over to the people is also monumental, especially in theocracies, whose leaders would prefer Type 1.3 chiefdoms (with themselves in charge, of course). The opposition toward a global economy is substantial, even in the industrialized West, where economic tribalism still dominates the thinking of most politicians, intellectuals, and citizens. (A prime example occurred in 2013 and 2014 when New Jersey, Texas, and Arizona blocked Tesla from selling electric cars directly to consumers to protect auto dealers from competition, which I equated to the absurdity of, say, typewriter manufacturers lobbying the government to block Apple Inc. from selling its products directly to consumers.[116])

For millennia we have existed in a zero-sum world in which a gain for one state or nation meant a loss for another state or nation—and our political and economic systems have been designed for use in that win-lose world. But we have the opportunity to live in a nonzero world and become a Civilization 2.0 by spreading liberal democracy, free trade, and the many other features of an advanced industrial society in which all sentient beings may flourish.

What about morality? I am assuming that we are not going to genetically engineer out of our nature greed, avarice, competitiveness, aggression, and violence, because these characteristics are part and parcel of who we are as a species and all have an evolutionary logic to them. Instead, what I foresee in the far future of civilization here on Earth (and, one day perhaps, on Mars, the moons of Jupiter and Saturn, and maybe even on exoplanets in other solar systems) are civilizations that have learned to design their political, economic, and social systems to bring out the best of our nature while holding back the worst. I envision not a monocultural civilization on Earth, but a multicultural one. And, presuming we will develop the technology to live on other planets, there will not be one civilization, but many. Given the distances and time scales involved, I foresee many species of spacefaring hominins in which each colonized planet will act like a new "founder" population from which a new species evolves reproductively isolated from other populations (the very definition of a species[117]). These civilizations will vary even more than nations on Earth varied before globalization. Instead of Civilization 3.0 or 4.0 there will be dozens, hundreds, possibly even thousands of different civilizations in which sentient beings may flourish—a vast array of peaks on the moral landscape.[118]

If this were to happen, sentient species would become immortal, inasmuch as there is no known mechanism—short of the end of the universe itself trillions of years from now in our accelerating expanding cosmos[119]— to cause the extinction of all planetary and solar systems at once.[120] In the far future, civilizations may become sufficiently advanced to colonize entire galaxies, genetically engineer new life-forms, terraform planets, and even trigger the birth of stars and new planetary solar systems through massive engineering projects.[121] Civilizations this advanced would have so much knowledge and power as to be essentially omniscient and omnipotent. (I call this extrapolation *Shermer's Last Law: Any sufficiently advanced extraterrestrial intelligence is indistinguishable from God.*[122]) Sound impossible? As long ago as 1960 the physicist Freeman Dyson showed how planets, moons, and asteroids could be torn apart and reconstructed into a giant sphere or ring surrounding the sun to capture enough solar radiation to provide limitless free energy.[123] Such a device—now called a Dyson Sphere—would enable a Type I civilization to transition into a Type II civilization. Finally, to my point in this thought experiment, any such civilization would not achieve this level of development without also being advanced morally as well. Again, by this I do not mean that we would evolve a new biological basis for morality; instead we would develop the social tools and technologies necessary to spawn civilizations more highly advanced in the moral realm.

To this end, the Italian SETI astronomer and mathematician Claudio Maccone has developed a mathematical equation to measure the "Civilization Amount" of information and entropy representing different civilizations throughout history, including Aztecs, Greece, Rome, Renaissance Italy, Portugal, Spain, France, Britain, and the United States. Maccone computes, for example, that the difference between the Aztecs and the Spaniards in their first encounter in 1519–1521 was 3.85 bits (of information) per individual. Maccone compares this figure to that of the information differences between the first living organisms 3.5 billion years ago and today's life-forms at 27.57 bits per living being. And he translates the 3.85 bits to about 50 centuries (5,000 years) of technological difference, resulting in the heavily outnumbered Spaniards easily conquering the technologically inferior Aztecs.[124] This, says Maccone, has implications for contact with an extraterrestrial civilization, which he estimates will be on the order of 10,000 bits per individual difference, or a technological difference of about 1 million years. Given how easily the Spanish conquered the Aztecs with only a 5,000-year advantage, contact with aliens 1 million years more advanced could prove catastrophic.[125] Would it?

Many prominent scientists think it would.[126] Stephen Hawking, for example, says, "We only have to look at ourselves to see how intelligent life might develop into something we wouldn't want to meet. I imagine they might exist in massive ships, having used up all the resources from their home planet. Such advanced aliens would perhaps become nomads, looking to conquer and colonize whatever planets they can reach." Given the history of encounters between earthly civilizations where the more advanced enslave or destroy the less developed, Hawking concluded, "If aliens ever visit us, I think the outcome would be much as when Christopher Columbus first landed in America, which didn't turn out very well for the Native Americans."[127] The evolutionary biologist Jared Diamond echoed these sentiments in denouncing as suicidal folly a 1974 message sent from the Arecibo radio telescope into space: "If there really are any radio civilizations within listening distance of us then for heaven's sake, let's turn off our transmitters and try to escape detection, or we're doomed."[128]

The moral character of aliens has long been a staple of science fiction, which often reflects our own moral concerns.[129] Many stories portray aliens as evil marauding conquerors intent on enslaving or exterminating humans. Examples include H. G. Wells's *The War of the Worlds* (in which invading Martians are unwittingly defeated, not by humans, but by a for-

tuitous and deadly virus), H. P. Lovecraft's *The Call of Cthulhu* (Cthulhu is
a monstrous entity who lies "dead but dreaming" in the city of R'lyeh, a
place of non-Euclidean madness sunken below the depths of the Pacific
Ocean), *Independence Day* (in which steely-eyed air force pilots gun down
the enemy alien spaceships), and Orson Scott Card's *Ender's Game* (a hive-
like enemy extraterrestrial species called the Buggers that the boy Ender
battles). Even with all his sanguine hopes for the future, Gene Roddenberry still peppered his twenty-third-century *Star Trek* galaxy with martial species such as Klingons and Romulans, and assimilating entities such
as the Borg. Perhaps the most amusing example in this genre was a *Twilight Zone* episode titled "To Serve Man," in which a nine-foot-tall alien
race called the Kanamits brings Earthlings advanced technology that ends
hunger, energy needs, and war, and offers to take people to their paradisiacal home planet, until a cryptographer discovers too late that the mysterious title message was actually the name of the aliens' cookbook.[130]

Less common are themes in which aliens are not only technologically
more advanced than we, but morally superior as well. Examples include
Doctor Who (an alien from an extraterrestrial race called the Time Lords
who acts as a custodian of humanity), and the alien race in Carl Sagan's
Contact (who provide humans with the plans for building a wormhole-
traveling spaceship). The classic film in this genre is the 1951 *The Day
the Earth Stood Still*, a Christ allegory in which an alien named Klaatu
(who takes the name "Mr. Carpenter" while visiting Earth) admonishes
humans for threatening nuclear annihilation and insists that they will not
be allowed to join the planetary community as long as they retain nuclear
weapons. A giant robot with a deadly ray beam named Gort, the enforcer
of peace, stands at the ready next to the spaceship. Klaatu wants to deliver
his message to all the leaders of Earth's nations but is refused, so (like
Jesus) he mingles among the common people, is tracked down and killed
by government authorities, and put into a tomb-like morgue. Gort
retrieves Klaatu and resurrects him,[131] after which he addresses a group of
scientists, warning them that if humans' morality does not advance to
keep pace with their destructive technology they will not be allowed to survive. "There must be security for all—or no one is secure," Klaatu lectures
his now enthralled audience. "This does not mean giving up any freedom
except the freedom to act irresponsibly. Your ancestors knew this when
they made laws to govern themselves—and hired policemen to enforce
them. For our policemen, we created a race of robots. Their function is to
patrol the planets . . . and preserve the peace. The result is that we live in

peace, without arms or armies, secure in the knowledge that we are free from aggression and war—free to pursue more profitable enterprises."[132] His message of redemption complete, Klaatu's heavenly ascension completes the allegory.

From science fiction to science conjecture, a reasonable hypothesis is that any extraterrestrial species with self-replicating molecules like DNA, and that reproduces sexually to produce genetic variation, will likely have evolved by means of natural selection. If so, then they will probably have evolved something like the moral emotions I outlined in chapter 1, in which case they will cooperate or compete based on the same game theoretic models (like Prisoner's Dilemma) that generate an array of responses to other sentient beings. Thus the odds are that ET will have a suite of moral emotions not unlike ours. The difference is that they would be morally advanced by virtue of being chronologically farther along in their social evolution.

Although we have a sample size of only one, and our species does have an unenviable track record of first contact between civilizations, the data trends for the past half millennium are encouraging: colonialism and slavery are dead, the percentage of populations that perish in wars has dramatically declined, crime and violence are way down, civil liberties are way up, and across the globe the desire for representative electoral democracies is skyrocketing, along with literacy, education, science, and technology. These trends have made our civilization more inclusive and less exploitative. Extrapolate that five-hundred-year trend out for five thousand years or five hundred thousand years and we get a sense of what an extraterrestrial intelligence might be like. Any civilization capable of extensive space travel will have moved far beyond exploitative colonialism and unsustainable energy sources such as fossil fuels.[133] (Already NASA has plans to move beyond chemical rockets to alternative means of powering space exploration. To think that a superadvanced alien spaceship would arrive on Earth to exploit our oil and gas seems ludicrous.) Enslaving the natives and harvesting their resources may be profitable in the short term for terrestrial civilizations, but such a strategy appears to be unsustainable in the long run, which is the time frame of the thousands of years needed for interstellar space travel.[134]

So any advanced civilization that survives long enough for us to make contact with it will also be a morally advanced civilization.[135] In this sense, thinking about extraterrestrial civilizations forces us to consider the nature and progress of our terrestrial civilization,[136] and offers hope that when we do make contact it will mean that at least one other intelligent species managed to reach a level where exploring space is more important than conquering land, and for which other sentient beings are more valuable alive and flourishing than conquered and dead. And any such encounter would mean that they solved the moral problem of self-extermination.[137]

. . .

Back to Civilization 1.0 and the pressing needs of the here and now: the protopian goal of incremental cumulative progress is what we can most practically work toward. It is, in fact, the strategy of the most successful social and political activists in our history, including the Reverend Dr. Martin Luther King Jr., with whose 1965 civil rights march on Montgomery, Alabama, I began this book.

Three years later, on April 3, 1968, Dr. King delivered his final speech, "I've Been to the Mountaintop," in Memphis, Tennessee, in which he exhorted his followers to work together to make America the nation its founding documents decreed it would be, foreseeing that he might not live to see the dream realized. "I've seen the Promised Land. I may not get there with you," he hinted ominously. "But I want you to know tonight, that we, as a people, will get to the promised land!"[138] The next day Dr. King was assassinated.

It is to his legacy, and the legacies of all champions of truth, justice, and freedom throughout history, that we owe our allegiance and our efforts at making the world a better place. "Each of us is two selves," Dr. King wrote. "The great burden of life is to always try to keep that higher self in command. And every time that old lower self acts up and tells us to do wrong, let us allow that higher self to tell us that we were made for the stars, created for the everlasting, born for eternity."[139]

We are, in fact, made *from* the stars. Our atoms were forged in the interior of ancient stars that ended their lives in spectacular paroxysms of supernova explosions that dispersed those atoms into space, where they coalesced into new solar systems with planets, life, and sentient beings capable of such sublime knowledge and moral wisdom. "We are stardust, we are golden, we are billion-year-old carbon. . . ."[140]

Morality is something that carbon atoms can embody given a billion years of evolution—*the* moral arc.

PROLOGUE: BENDING THE MORAL ARC

1. King, Coretta Scott. 1969. *My Life with Martin Luther King Jr.* New York: Holt, Rinehart, & Winston, 267.
2. Many accounts describe King as being either at the top of the capitol steps, on the steps, or at the bottom of the steps. There are eyewitnesses accounts in which it is claimed that King delivered his famous speech *from* the steps. For example, John N. Pawelek recalls: "When we arrived at the state capitol, the area was filled with throngs of marchers. Martin Luther King was on the steps. He gave a fiery speech which only a Baptist minister can give." (goo.gl/eNyaGX). The Alabama Byways site tells its patrons reliving the Selma to Montgomery march to "walk on the steps of the capitol, where King delivered his 'How Long, Not Long' speech to a crowd of nearly 30,000 people." (goo.gl/gnAfSX). In his book *Getting Better: Television and Moral Progress* (New Brunswick, NJ: Transaction Books, 1991, p. 48), Henry J. Perkinson writes: "By Thursday, the marchers, who now had swelled to twenty-five thousand, reached Montgomery, where the national networks provided live coverage as Martin Luther King strode up the capital [*sic*] steps with many of the movement's heroes alongside. From the top of the steps, King delivered a stunning address to the nation." Even the *Martin Luther King Encyclopedia* puts him "*on* the steps." (goo.gl/Rxw8pY).

   This is incorrect. The BBC reports of the day, for example, say that King "has taken a crowd of nearly 25,000 people to the steps of the state capitol" but was stopped from climbing the steps and so "addressed the protesters from a podium in the square." (goo.gl/7ybfKa). The *New York Times* reports that "The Alabama Freedom March from Selma to Montgomery ended shortly after noon at the foot of the Capitol steps" and that "the rally never got on to state property. It was confined to the street in front of the steps." (goo.gl/5vuJ8D). In this video, archival footage from 3:40 to 3:50 shows "The marchers make their way to the steps of the capitol building—but not beyond." (http://goo.gl/KdLEhM). The original caption to the aerial photograph included in the text, from an educational online source, reads: "King was not allowed to speak from the steps of the Capitol. Can you find the line of state troopers that blocked the way?" Finally, in this video, from 40:53 to 41:15, you can see various entertainers who preceded Dr. King and the pulpit he used, all on a flatbed truck. (http://goo.gl/zq5XG6). This is confirmed by these firsthand accounts: "A few state employees stood on the steps. They watched a construction crew building a speaker's platform on a truck bed in the street." (goo.gl/K6a8U7). And: "The speakers platform is a flatbed truck equipped with microphones and loudspeakers. The rally begins with songs by Odetta, Oscar Brand, Joan Baez, Len Chandler, Peter, Paul & Mary, and Leon Bibb. From his truck-bed podium, King can clearly see Dexter Avenue Baptist Church." (goo.gl/5HWznV).

3. The speech is commonly known as the "How Long, Not Long" speech (or sometimes "Our God Is Marching On") and is considered one of King's three most important and impactful speeches, along with "I Have a Dream" and the tragically prescient "I've Been to the Mountaintop." It can be read in its entirety at goo.gl/KcjabU. The climactic end of the speech can be seen on YouTube (goo.gl/VOKMGP).

4. Parker, Theodore. 1852/2005. *Ten Sermons of Religion.* Sermon III: Of Justice and Conscience. Ann Arbor: University of Michigan Library.

5. Pinker, Steven. 2011. *The Better Angels of Our Nature: Why Violence Has Declined.* New York: Viking, xxvi.

6. Voltaire, 1765/2005. "Question of Miracles." *Miracles and Idolotry.* New York: Penguin.

1: TOWARD A SCIENCE OF MORALITY

1. Bronowski, Jacob. 1956. *Science and Human Values.* New York: Julian Messner.

2. http://www.oed.com/

3. Damasio, Antonio R. 1994. *Descartes' Error: Emotion, Reason, and the Human Brain.* New York: Putnam.

4. Low, Philip, Jaak Panksepp, Diana Reiss, David Edelman, Bruno Van Swinderen, Philip Low, and Christof Koch. 2012. "The Cambridge Declaration on Consciousness," Francis Crick Memorial Conference on Consciousness in Human and non-Human Animals. Churchill College, University of Cambridge.

5. This is similar to the philosophical starting point of Sam Harris in his book *The Moral Landscape* on "the well-being of conscious creatures," although he does not justify the starting point with evolutionary theory.

6. Invoking group selection to explain the evolution of the moral emotions has become popular in the early twenty-first century, but most evolutionary biologists do not accept it, and those who do say it would be a minor force at most. See appendix II in my book *The Science of Good and Evil* for a history and analysis of group selection.

7. As George Williams famously observed, a fleet herd of deer is really just a herd of fleet deer. For a thoughtful analysis of recent group selection arguments see Pinker, Steven. 2012. "The False Allure of Group Selection," *Edge.org*, June 18. (http://edge.org/conversation/the-false-allure-of-group-selection).

8. Filmer, Robert. 1680. *Patriarcha, or the Natural Power of Kings.* http://www.constitution.org/eng/patriarcha.htm

9. Locke, John. 1690. *Second Treatise of Government*, chapter II. Of the State of Nature, Sec. 4. goo.gl/RJdaQB

10. "Men being, as he has been said, by nature all free, equal and independent, no one can be put of this estate and subjected to the political power of another without his own consent, which is done by agreeing with other men, to join and unite into a community for their comfortable, safe and peaceable living." Locke, John. *Second Treatise of Government*, chapter VIII. Of the Beginning of Political Societies, Sec. 95. goo.gl/MHxkoH

11. Popper, K. R. 1959. *The Logic of Scientific Discovery*. London: Hutchinson; Popper, K. R. 1963. *Conjectures and Refutations*. London: Routledge & Kegan Paul.

12. Eddington, Arthur Stanley. 1938. *The Philosophy of Physical Science*. Ann Arbor: University of Michigan Press, 9.

13. See the entries on Rationalism and Reason in: Edwards, Paul, ed. 1967. *The Encyclopedia of Philosophy*. New York: Macmillan, 7, 69–74, 83–85.

14. It is the central thesis of my book *The Believing Brain* that beliefs come first and reasons for beliefs follow. The psychologist Jonathan Haidt equates the process with a lawyer arguing a brief—the point is not to get at the truth but to win the case. (Haidt, Jonathan. 2012. *The Righteous Mind: Why Good People Are Divided by Politics and Religion*. New York: Pantheon.) Hugo Mercier and Dan Sperber call this process the "Argumentative Theory of Reasoning," in which "Reasoning can lead to poor outcomes, not because humans are bad at it, but because they systematically strive for arguments that justify their beliefs or their actions." (Mercier, Hugo, and Dan Sperber. 2011. "Why Do Humans Reason? Arguments for an Argumentative Theory." *Behavioral and Brain Sciences*, 34, no. 2, 57–74.) Other psychologists characterize it as "bounded rationality," in which our capacity to reason is bounded not only by our emotions, but also by other cognitive limitations such as memory and information processing speed, plus the finite amount of time we have to make decisions. Because of this we are often *satisficers* instead of *optimizers*—we find satisfactory (good enough) solutions rather than optimal (perfect) solutions. (Simon, Herbert. 1991. "Bounded Rationality and Organizational Learning." *Organization Science*, 2, no. 1, 125–134; Gigerenzer, Gerd, and Reinhard Selten. 2002. *Bounded Rationality*. Cambridge, MA: MIT Press; Kahneman, Daniel. 2003. "Maps of Bounded Rationality: Psychology for Behavioral Economics." *American Economic Review*, 93, no. 5, 1449–1475.)

15. Pinker, 2011, 648.

16. Liebenberg, Louis. 2013. "Tracking Science: The Origin of Scientific Thinking in Our Paleolithic Ancestors." *Skeptic*, 18, no. 3, 18–24.

17. Ibid. See also Liebenberg, L. 1990. *The Art of Tracking: The Origin of Science*. Cape Town: David Philip.

18. Pinker, Steven. 1997. *How the Mind Works*. New York: W. W. Norton.

19. Lecky, William Edward Hartpole. 1869. *History of European Morals: From Augustus to Charlemagne*. 2 vols. New York: D. Appleton, Vol. 1, Available in full at the Online Library of Liberty: http://oll.libertyfund.org

20. Lecky wasn't alone. The German philosopher Arthur Schopenhauer said, "The assumption that animals are without rights and the illusion that our treatment of them has no moral significance is a positively outrageous example of Western crudity and barbarity. Universal compassion is the only guarantee of morality." And the English social reformer Henry Salt in 1886 published *A Plea for Vegetarianism*, which apparently Gandhi picked up in pamphlet form in a restaurant in London and it had a big influence on him (the two later became friends). In 1894 Salt published

Animals' Rights: Considered in Relation to Social Progress, the book considered to be the first explicit treatment of the concept of animal rights.

21. Singer, Peter. 1981. *The Expanding Circle: Ethics, Evolution, and Moral Progress.* Princeton, NJ: Princeton University Press, 109–110.

22. Pinker, 2011, 182.

23. This is the catchphrase of Michael Medved, one of the most thoughtful and balanced of all the talk radio hosts whose shows I have been interviewed on, but his American boosterism wears thin after repetition.

24. Singer, 1981, 119.

25. I elevated Hitchens's quote to a dictum in: Shermer, Michael. 2010. "The Skeptic's Skeptic." *Scientific American,* November, 86. The quote is from: Hitchens, Christopher. 2003. "Mommie Dearest." *Slate,* October 20. goo.gl/efSMV

26. Flynn, James. 1984. "The Mean IQ of Americans: Massive Gains 1932–1978." *Psychological Bulletin,* 95, 101, 171–191.

27. Flynn, James. 2007. *What Is Intelligence?* Cambridge, UK: Cambridge University Press.

28. From the Raven's Progressive Matrices test: goo.gl/7h1akK

29. Flynn, James. 1987. "Massive IQ Gains in 14 Nations: What IQ Tests Really Measure." *Psychological Bulletin,* 101, 171–191.

30. Flynn, 2007.

31. Flynn, James. 2012. *Are We Getting Smarter?: Rising IQ in the Twenty-first Century.* Cambridge, UK: Cambridge University Press.

32. Traynor, Lee. 2014. "The Future of Intelligence: An Interview with James R. Flynn." *Skeptic,* 19, no. 1, 36–41.

33. Flynn, 2012, 135.

34. Johnson, Steven. 2006. *Everything Bad Is Good for You: How Today's Popular Culture Is Actually Making Us Smarter.* New York: Riverhead.

35. Traynor, 2014.

36. Ibid.

37. Pinker, 2011, 656.

38. Many of these studies are summarized in greater depth in Pinker, 2011, 656–670.

39. Farrington, D. P. 2007. "Origins of Violent Behavior Over the Life Span." In D. J. Flannery, A. T. Vazsonyi, and I. D. Waldman, eds., *The Cambridge Handbook of Violent Behavior and Aggression.* New York: Cambridge University Press.

40. Wilson, James Q., and Richard Herrnstein. 1985. *Crime and Human Nature.* New York: Simon & Schuster.

41. Sargent, Michael J. "Less Thought, More Punishment: Need for Cognition Predicts Support for Punitive Responses to Crime." *Personality & Social Psychology Bulletin,* 30, 1485–1493.

42. Burks, S. V., J. P. Carpenter, L. Goette, and A. Rustichini. 2009. "Cognitive Skills Affect Economic Preferences, Strategic Behavior, and Job Attainment." *Proceedings of the National Academy of Sciences,* 106, 7745–7750.

43. Jones, Garret. 2008. "Are Smarter Groups More Cooperative? Evidence from Prisoner's Dilemma Experiments, 1959–2003." *Journal of Economic Behavior & Organization,* 68, 489–497.

44. Kanazawa, S. 2010. "Why Liberals and Atheists Are More Intelligent." *Social Psychology Quarterly,* 73, 33–57.

45. Deary, Ian J., G. D. Batty, and C. R. Gale. 2008. "Bright Children Become Enlightened Adults." *Psychological Science,* 19, 1–6.

46. Caplan, Brian, and Stephen C. Miller. 2010. "Intelligence Makes People Think Like Economists: Evidence from the General Social Survey." *Intelligence*, 38, 636–647.

47. Rindermann, Heiner. 2008. "Relevance of Education and Intelligence for the Political Development of Nations: Democracy, Rule of Law, and Political Liberty." *Intelligence*, 36, 306–322.

48. Mar, Raymond, and Keith Oatley. 2008. "The Function of Fiction Is the Abstraction and Simulation of Social Experience." *Perspectives on Psychological Science*, 3, 173–192.

49. Stephens, G. J., L. J. Silbert, and U. Hasson. 2010. "Speaker-Listener Neural Coupling Underlies Successful Communication." *Proceedings of the National Academy of Sciences*, August 10, 107, no. 32, 14425–14430.

50. Hasson, Uri, Ohad Landesman, Barbara Knappmeyer, Ignacio Vallines, Nava Rubin, and David J. Heeger. 2008. "Neurocinematics: The Neuroscience of Film." *Projections*, Summer, 2, no. 1, 1–26.

51. Kidd, David Comer, and Emanuele Castano. 2013. "Reading Literary Fiction Improves Theory of Mind." *Science*, 342, no. 6156, 377–380.

52. Skousen, Tim. 2014. "The University of Sing Sing." HBO, March 31.

53. Kosko, Bart. 1992. *Fuzzy Engineering*. Englewood Cliffs, NJ: Prentice Hall.

54. Shermer, Michael. 2003. *The Science of Good and Evil*. New York: Times Books, 82–84.

55. Dawkins, Richard. 2014. "What Scientific Idea Is Ready for Retirement? Essentialism." http://www.edge.org/response-detail/25366

56. Annual Letter: 3 Myths That Block Progress for the Poor. 2014. http://annualletter.gatesfoundation.org/?cid=mg_tw_tgm0_012104#section=home

57. Hume, David. 1739. *A Treatise of Human Nature*. London: John Noon, 335.

58. Searle, John R. 1964. "How to Derive 'Ought' from 'Is.'" *Philosophical Review*, 73, no. 1, 43–58.

59. Personal correspondence, January 22, 2013.

60. Morales, Lymari. 2009. "Knowing Someone Gay/Lesbian Affects Views of Gay Issues." Gallup.com, May 29.

61. Winslow, Charles-Edward Amory. 1920. "The Untilled Fields of Public Health." *Science*, 51, no. 1306, 23–33.

62. Global Health Observatory Data Repository. 2011. World Health Organization. goo.gl/ykZKlh

63. Shermer, Michael. 2013. "The Sandy Hook Effect." *Skeptic*, 18, no. 1. goo.gl/sjTQuJ

64. See, for example, Gregg, Becca Y. 2013. "Speakers Differ on Fighting Gun Violence." *Reading Eagle*, September 19.

65. Lott, John. 2010. *More Guns, Less Crime: Understanding Crime and Gun Control Laws*, 3rd ed. Chicago: University of Chicago Press.

66. Webster, Daniel W., and Jon S. Vernick, eds. 2013. *Reducing Gun Violence in America: Informing Policy with Evidence and Analysis*. Baltimore: Johns Hopkins University Press.

67. Kellermann, Arthur L. 1998. "Injuries and Deaths Due to Firearms in the Home." *Journal of Trauma*, 45, no. 2, 263–267.

68. Branas, Charles C., Therese S. Richmond, Dennis P. Culhane, Thomas R. Ten Have, and Douglas J. Wiebe. 2009. "Investigating the Link Between Gun Possession and Gun Assault." *American Journal of Public Health*, November, 99, no. 11, 2034–2040.

69. www.fbi.gov/news/stories/2012/october/annual-crime-in-the-u.s.-report -released/annual-crime-in-the-u.s.-report-released

70. www.bradycampaign.org/facts/gunviolence/GVSuicide?s=1

71. www.bradycampaign.org/facts/gunviolence/crime?s=1

72. Eisner, Manuel. 2003. "Long-Term Historical Trends in Violent Crime." *Crime & Justice*, 30, 83–142, table 1.

73. Data from Eisner, graph rendered by Pinker, 2011, 63.

74. Dawkins, Richard. 1976. *The Selfish Gene*. New York: Oxford University Press.

75. Ibid., 66.

76. Damasio, 1994.

77. Grinde, Björn. 2002. *Darwinian Happiness: Evolution as a Guide for Living and Understanding Human Behavior*. Princeton, NJ: Darwin Press, 49. See also Grinde, Björn. 2002. "Happiness in the Perspective of Evolutionary Psychology." *Journal of Happiness Studies*, 3, 331–354.

78. Pinker, 2011, xxv, 508–509. Pinker notes that there are many taxonomies of violence, citing, for example, the four-part scheme in Baumeister, Roy. 1997. *Evil: Inside Human Violence and Cruelty*. New York: Henry Holt.

79. Boehm, Christopher. 2012. *Moral Origins: The Evolution of Virtue, Altruism, and Shame*. New York: Basic Books.

80. Daly, Martin, and Margo Wilson. 1999. *The Truth About Cinderella: A Darwinian View of Parental Love*. New Haven, CT: Yale University Press.

81. You can find the video by typing into a search engine the text string "One of the Most Powerful Videos You Will Ever See." The incident may be found at the 1:52 mark.

82. Bloom, Paul. 2013. *Just Babies: The Origins of Good and Evil*. New York: Crown.

83. Ibid., 5.

84. Ibid., 7.

85. For a general review see Tomasello, Michael. 2009. *Why We Cooperate*. Cambridge, MA: MIT Press.

86. Warneken, F., and M. Tomasello. 2006. "Altruistic Helping in Human Infants and Young Chimpanzees." *Science*, 311, 1301–1303; Warneken, F., and M. Tomasello. 2007. "Helping and Cooperation at 14 Months of Age." *Infancy*, 11, 271–294.

87. Martin, A., and K. R. Olson. 2013. "When Kids Know Better: Paternalistic Helping in 3-Year-Old Children." *Developmental Psychology*, November, 49, no. 11, 2071–2081.

88. LoBue, V., T. Nishida, C. Chiong, J. S. DeLoache, and J. Haidt. 2011. "When Getting Something Good Is Bad: Even Three-Year-Olds React to Inequality." *Social Development*, 20, 154–170.

89. Rochat, P., M. D. G. Dias, G. Liping, T. Broesch, C. Passos-Ferreira, A. Winning, and B. Berg. 2009. "Fairness in Distribution Justice in 3- and 5-Year-Olds Across Seven Cultures." *Journal of Cross-Cultural Psychology*, 40, 416–442; Fehr, E. H. Bernhard, and B. Rockenbach. 2008. "Egalitarianism in Young Children." *Nature*, 454, 1079–1083.

90. See "A Class Divided," a PBS *Frontline* documentary on Mrs. Elliott's experiment: www.pbs.org/wgbh/pages/frontline/shows/divided/

91. Bloom, 2013, 31.

92. Voltaire. 1824. *A Philosophical Dictionary*, vol. 2. London: John & H. L. Hunt, 258. ebooks.adelaide.edu.au/v/voltaire/dictionary/chapter130.html

93. Shermer, Michael. 2008. "The Doping Dilemma." *Scientific American*, April, 32–39.

94. Hamilton, Tyler, and Daniel Coyle. 2012. *The Secret Race: Inside the Hidden World of the Tour de France*. New York: Bantam Books.

95. goo.gl/iUSguA

96. Hobbes, 1651, 185.

97. Fehr, Ernst, and Simon Gachter. 2002. "Altruistic Punishment in Humans," *Nature*, 415, 137–140. See also Boyd, R., and P. J. Richerson. 1992. "Punishment Allows the Evolution of Cooperation (or Anything Else) in Sizable Groups." *Ethology and Sociobiology*, 13, 171–195.

2: THE MORALITY OF WAR, TERROR, AND DETERRENCE

1. "Arena," *Star Trek. The Original Series*. Season 1, Episode 19. January 19, 1967. Story by Fredric Brown. Teleplay by Gene L. Coon. Executive producer, Gene Rodden- berry. The episode was filmed at Vasquez Rocks on the outskirts of Los Angeles, where so many sci-fi films and shows are filmed. A transcript of this episode is available at www.chakoteya.net/startrek/19.htm For a thoughtful discussion on the morality and ethics in *Star Trek* episodes see Barad, Judity, and Ed Robertson. 2001. *The Ethics of Star Trek*. New York: HarperCollins Perennial.

2. Alexander, David. 1994. *Star Trek Creator: The Authorized Biography of Gene Roddenberry*. New York: Roc/Penguin.

3. The quote appears at the end of an episode titled "Scorched Earth" of Gene Rod- denberry's television series *Earth: Final Conflict*.

4. Zahavi, Amotz, and Avishag Zahavi. 1997. *The Handicap Principle: A Missing Piece of Darwin's Puzzle*. Oxford, UK: Oxford University Press.

5. Northcutt, Wendy. 2010. *The Darwin Awards: Countdown to Extinction*. New York: E. P. Dutton.

6. Leeson, Peter T. 2009. *The Invisible Hook*. Princeton, NJ: Princeton University Press.

7. Smith, Adam. 1776/1976. *An Inquiry into the Nature and Causes of the Wealth of Nations*, in 2 volumes, R. H. Campbell and A. S. Skinner, general editors, W. B. Todd, text editor. Oxford, UK: Clarendon Press.

8. Leeson, 2009.

9. Ibid.

10. Pinsker, Joe. 2014. "The Pirate Economy." *Atlantic*, April 16. goo.gl/sUQGNS

11. Quoted in *Cold War: MAD 1960–1972*. 1998. BBC Two Documentary. Transcript: goo.gl/etDCFg Film: goo.gl/fTLHdN

12. Brodie, Bernard. 1946. *The Absolute Weapon: Atomic Power and World Order*. New York: Harcourt, Brace, 79.

13. Kubrick, Stanley. 1964. *Dr. Strangelove or: How I Learned to Stop Worrying and Love the Bomb*. Columbia Pictures. http://youtu.be/2yfXgu37iyI

14. Ibid.

15. McNamara, Robert S. 1969. "Report Before the Senate Armed Services Committee on the Fiscal Year 1969–73 Defense Program, and 1969 Defense Budget, January 22, 1969." Washington, DC: US Government Printing Office, 11.

16. Glover, J. 1999. *Humanity: A Moral History of the Twentieth Century*. London: Jon- athan Cape, 297.

17. Quoted in obituary, *Los Angeles Times*, October 2, 2013, AA6.

18. "A Soviet Attack Scenario." 1979. *The Effects of Nuclear War*. Washington, DC: Office of Technology Assessment. Reprinted in Swedin, Eric G., ed. 2011. *Survive*

the Bomb: The Radioactive Citizen's Guide to Nuclear Survival. Minneapolis: Zenith Press, 163–177.

19. Brown, Anthony Cave, ed. 1978. *DROPSHOT: The American Plan for World War III Against Russia in 1957.* New York: Dial Press; Richelson, Jeffrey. 1986. "Population Targeting and US Strategic Doctrine." In Desmond Ball and Jeffrey Richelson, eds., *Strategic Nuclear Targeting.* Ithaca, NY: Cornell University Press, 234–249.

20. Source: Utah State Historical Society. Reprinted in Swedin, Eric G., ed. 2011. *Survive the Bomb: The Radioactive Citizen's Guide to Nuclear Survival.* Minneapolis: Zenith Press, 11.

21. For a scholarly analysis of and an alternative view to deterrence see: Kugler, Jacek. 1984. "Terror Without Deterrence: Reassessing the Role of Nuclear Weapons." *Journal of Conflict Resolution,* 28, no. 3, September, 470–506.

22. Kant, Immanuel. 1795. "Perpetual Peace: A Philosophical Sketch." In *Perpetual Peace and Other Essays.* Indianapolis: Hackett, I, 6.

23. Sagan, Carl, and Richard Turco. 1990. *A Path Where No Man Thought: Nuclear Winter and the End of the Arms Race.* New York: Random House.

24. Turco, R. P., O. B. Toon, T. P. Ackerman, J. B. Pollack, and C. Sagan. 1983. "Nuclear Winter: Global Consequences of Multiple Nuclear Explosions." *Science,* 222, 1283–1297. The acronym TTAPS from the authors' last names was used to dramatic effect in the media but most scientists rejected the hypothesis.

25. Thompson, Starley L., and Stephen H. Schneider. 1986. "Nuclear Winter Reappraised." *Foreign Affairs,* 64, no. 5, Summer, 981–1005.

26. Kearny, Cresson. 1987. *Nuclear War Survival Skills.* Cave Junction, OR: Oregon Institute of Science and Medicine, 17–19.

27. The total stockpile varies by database. The lower estimate of 16,400 comes from "Status of World Nuclear Forces." Federation of American Scientists. From most to least: Russia, 8,500; United States, 7,700; France, 300; China, 250; United Kingdom, 225; Israel, 80; Pakistan, 100–120; India, 90–110; North Korea, <10. goo.gl/CYjP1 The higher estimate of 17,200 comes from the "Nuclear Notebook" of the *Bulletin of the Atomic Scientists:* Kristensen, Hans M., and Robert S. Norris. 2013. "Global Nuclear Weapons Inventories, 1945–2013. *Bulletin of the Atomic Scientists,* 69, no. 5, 75–81.

28. Sagan and Turco, 1990, 232–233.

29. Ibid. For a striking visual demonstration by the Japanese artist Isao Hashimoto of every one of the 2,053 nuclear weapon explosions between 1945 and 1998, starting with the Trinity test in New Mexico, where in the world they happened, and who sponsored them, see bit.ly/lc9xB2M

30. Source: Federation of American Scientists. "Status of World Nuclear Forces."

31. Rhodes, Richard. 2010. *The Twilight of the Bombs: Recent Challenges, New Dangers, and the Prospects of a World Without Nuclear Weapons.* New York: Alfred A. Knopf.

32. Schlosser, Eric. 2013. *Command and Control: Nuclear Weapons, the Damascus Accident, and the Illusion of Safety.* New York: Penguin.

33. Quoted in Harris, Amy Julia. 2010. "No More Nukes Books." *Half Moon Bay Review,* August 4, goo.gl/4PsK2n

34. Shanker, Thom. 2012. "Former Commander of US Nuclear Forces Calls for Large Cut in Warheads." *New York Times,* May 16, A4.

35. Cartwright, James, et al., 2012. "Global Zero US Nuclear Policy Commission Report:

Modernizing US Nuclear Strategy, Force Structure and Posture." *Global Zero*, May. goo.gl/8fPMpa

36. Walker, Lucy, and Lawrence Bender. 2010. *Countdown to Zero*. Participant Media & Magnolia Pictures.

37. Sagan, Scott D. 2009. "The Global Nuclear Future." *Bulletin of the American Academy of Arts & Sciences*, 62, 21–23.

38. Nelson, Craig. 2014. *The Age of Radiance: The Epic Rise and Dramatic Fall of the Atomic Era*. New York: Scribner/Simon & Schuster, 370.

39. Sobek, David, Dennis M. Foster, and Samuel B. Robinson. 2012. "Conventional Wisdom? The Effects of Nuclear Proliferation on Armed Conflict, 1945–2001." *International Studies Quarterly*, 56, no. 1, 149–162.

40. Lettow, Paul. 2005. *Ronald Reagan and His Quest to Abolish Nuclear Weapons*. New York: Random House, 132–133.

41. Shultz, George. 2013. "Margaret Thatcher and Ronald Reagan: The Ultimate '80s Power Couple." *Daily Beast*, April 8. goo.gl/itZHHR

42. Shultz, George P., William J. Perry, Henry A. Kissinger, and Sam Nunn. 2007. "A World Free of Nuclear Weapons." *Wall Street Journal*, January 4, http://ow.ly /ttvGN

43. ———. 2008. "Toward a Nuclear-Free World." *Wall Street Journal*, January 15. goo.gl/iAGDQX

44. Waltz, Kenneth N. 2012. "Why Iran Should Get the Bomb: Nuclear Balancing Would Mean Stability." *Foreign Affairs*, July/August. goo.gl/2x9dF

45. Kugler, Jacek. 2012. "A World Beyond Waltz: Neither Iran nor Israel Should Have the Bomb." PBS, September 12. goo.gl/0drNe5

46. Quoted in Fathi, Nazila. 2005. "Wipe Israel 'off the Map,' Iranian Says." *New York Times*, October 27. goo.gl/9EFacw

47. Fettweis, Christopher. 2010. *Dangerous Times?: The International Politics of Great Power Peace*. Washington, DC: Georgetown University Press.

48. Quoted in Marty, Martin E. 1996. *Modern American Religion*, Vol. 3, *Under God, Indivisible, 1941–1960*. Chicago: University of Chicago Press, 117.

49. Chomsky, Noam. 1967. "The Responsibility of Intellectuals." *New York Review of Books*, 8, no. 3, goo.gl/wRPVH

50. Goldhagen, Daniel Jonah. 2009. *Worse Than War: Genocide, Eliminationism, and the Ongoing Assault on Humanity*. New York: PublicAffairs, 1, 6.

51. Lemkin, Raphael. 1946. "Genocide." *American Scholar*, 15, no. 2, 227–230.

52. United Nations General Assembly Resolution 96, no. 1. "The Crime of Genocide."

53. Katz, Steven T. 1994. *The Holocaust in Historical Perspective*. Vol. 1. New York: Oxford University Press.

54. Kugler, Tadeusz, Kyung Kook Kang, Jacek Kugler, Marina Arbetman-Rabinowitz, and John Thomas. 2013. "Demographic and Economic Consequences of Conflict." *International Studies Quarterly*, March, 57, no. 2, 1–12.

55. Toland, John. 1970. *The Rising Sun: The Decline and Fall of the Japanese Empire, 1936–1945*. New York: Random House, 731.

56. "The Cornerstone of Peace—Number of Names Inscribed." Kyushu-Okinawa Summit 2000: Okinawa G8 Summit Host Preparation Council, 2000. See also Pike, John. 2010. "Battle of Okinawa." *Globalsecurity.org*; Manchester, William. 1987. "The Bloodiest Battle of All." *New York Times*, June 14, goo.gl/d4DeVe

57. In 2002 I attended the reunion of the *Wren* crew in my father's stead and confirmed his memory.

58. Giangreco, D. M. 2009. *Hell to Pay: Operation Downfall and the Invasion of Japan, 1945–1947.* Annapolis, MD: Naval Institute Press, 121–124.

59. Giangreco, Dennis M. 1998. "Transcript of 'Operation Downfall' [US invasion of Japan]: US Plans and Japanese Counter-Measures." *Beyond Bushido: Recent Work in Japanese Military History.* https://www.mtholyoke.edu/acad/intrel/giangrec.htm See also Maddox, Robert James. 1995. "The Biggest Decision: Why We Had to Drop the Atomic Bomb." *American Heritage*, 46, no. 3, 70–77.

60. Skates, John Ray. 2000. *The Invasion of Japan: Alternative to the Bomb.* Columbia: University of South Carolina Press, 79.

61. Putnam, Frank W. 1998. "The Atomic Bomb Casualty Commission in Retrospect." *Proceedings of the National Academy of Sciences*, May 12, 95, no. 10, 5426–5431.

62. K'Olier, Franklin, ed. 1946. *United States Strategic Bombing Survey, Summary Report (Pacific War).* Washington, DC: US Government Printing Office.

63. Rhodes, Richard. 1984. *The Making of the Atomic Bomb.* New York: Simon & Schuster, 599.

64. There are at least twenty-one international organizations against nuclear weapons, and another seventy-nine antinuclear (including nuclear energy) organizations. See Rudig, Wolfgang. 1990. *Antinuclear Movements: A World Survey of Opposition to Nuclear Energy.* New York: Longman, 381–403.

65. Sagan and Turco, 1990.

66. goo.gl/o8NO55

67. Fedorov, Yuri. 2002. "Russia's Doctrine on the Use of Nuclear Weapons." Pugwash Meeting, London, November 15–17. goo.gl/yWGOpk

68. Narang, Vipin. 2010. "Pakistan's Nuclear Posture: Implications for South Asian Stability." Harvard Kennedy School, Belfer Center for Science and International Affairs Policy Brief, January 4. goo.gl/gH1Hlv

69. BBC News. 2003. "UK Restates Nuclear Threat." February 2. The Defence Secretary Geoff Hoon, said, "Saddam can be absolutely confident that in the right conditions we would be willing to use nuclear weapons." goo.gl/dqEQuZ

70. Department of Defense. 2010. "Nuclear Posture Review." April 6, http://www.web citation.org/6FY0Ol07H

71. Kang, Kyungkook, and Jacek Kugler. 2012. "Nuclear Weapons: Stability of Terror." In *Debating a Post-American World*, ed. Sean Clark and Sabrina Hoque. New York: Routledge.

72. goo.gl/AoQC3

73. Karouny, Mariam, and Ibon Villelabeitia. 2006. "Iraq Court Upholds Saddam Death Sentence." *Washington Post*, December 26. goo.gl/b9n6oW

74. goo.gl/InDEdg

75. Tannenwald, Nina. 2005. "Stigmatizing the Bomb: Origins of the Nuclear Taboo." *International Security*, Spring, 29, no. 4, 5–49.

76. Schelling, Thomas C. 1994. "The Role of Nuclear Weapons." In L. Benjamin Ederington and Michael J. Mazarr, eds., *Turning Point: The Gulf War and US Military Strategy.* Boulder, CO: Westview, 105–115.

77. Rozin, Paul, Jonathan Haidt, and C. R. McCauley. 2000. "Disgust." In *Handbook of Emotions*, ed. M. Lewis and J. M. Haviland-Jones. New York: Guilford Press, 637–653.

78. Evans, Gareth. 2014. "Nuclear Deterrence in Asia and the Pacific." *Asia & the Pacific Policy Studies*, January, 91–111.

79. Ibid.

80. Blair, B., and M. Brown. 2011. "World Spending on Nuclear Weapons Surpasses $1 Trillion per Decade." *Global Zero.* goo.gl/xDcjg5

81. Shultz et al., 2007.

82. Evans, 2014.

83. Herszenhorn, David M., and Michael R. Gordon. 2013. "US Cancels Part of Missile Defense that Russia Opposed." *New York Times*, March 16, A12.

84. Evans, 2014.

85. Shakespeare, William. 1594. *The Rape of Lucrece.* Available in full at goo.gl/MdWlsN

86. Bull, Hedley. 1995. *The Anarchical Society.* London: Macmillan, 234; Bull, Hedley. 1961. *The Control of the Arms Race.* London: Institute of Strategic Studies.

87. Quoted in Maclin, Beth. 2008. "A Nuclear Weapon-Free World Is Possible, Nunn Says." Belfer Center for Science and International Affairs, Harvard University, October 20. goo.gl/60XTeM

88. Quoted in Perez-Rivas, Manuel. 2001. "Bush Vows to Rid the World of 'Evil-Doers.'" CNN Washington Bureau. September 16, goo.gl/zrZMCV

89. Mueller, John, and Mark G. Stewart. 2013. "Hapless, Disorganized, and Irrational." *Slate*, April 22. goo.gl/j0cqUl

90. Atran, Scott. 2010. "Black and White and Red All Over." *Foreign Policy*, April 22. goo.gl/SyhiEu

91. Scahill, Jeremy. 2013. *Dirty Wars: The World Is a Battlefield.* Sundance Selects.

92. Ibid.

93. The 9/11 Commission Report, 2004. xvi. http://www.9-11commission.gov/report /911Report.pdf

94. Abrahms, Max. 2013. "Bottom of the Barrel." *Foreign Policy*, April 24. goo.gl /hj4J1h

95. Krueger, Alan B. 2007. *What Makes a Terrorist: Economics and the Roots of Terrorism.* Princeton, NJ: Princeton University Press, 3.

96. Bailey, Ronald. 2011. "How Scared of Terrorism Should You Be?" *Reason*, September 6. goo.gl/3ZvkR

97. Quoted in Levi, Michael S. 2003. "Panic More Dangerous than WMD." *Chicago Tribune*, May 26. goo.gl/tlVogl

98. Levi, Michael S. 2011. "Fear and the Nuclear Terror Threat." *USA Today*, March 24, 9A.

99. Harper, George W. 1979. "Build Your Own A-Bomb and Wake Up the Neighborhood." *Analog*, April, 36–52.

100. Levi, Michael S. 2009. *On Nuclear Terrorism.* Cambridge, MA: Harvard University Press, 5.

101. 2012. "Fact Sheet on Dirty Bombs." US Nuclear Regulatory Commission, December. goo.gl/hXbFlx See also goo.gl/Rf1mhO

102. Abrahms, Max. 2006. "Why Terrorism Does Not Work." *International Security*, 31, 42–78. http://ow.ly/ttvYv

103. Abrahms, Max, and Matthew S. Gottfried. 2014. "Does Terrorism Pay? An Empirical Analysis." *Terrorism and Political Violence.* goo.gl/ZWdAP1

104. Cronin, Audrey. 2011. *How Terrorism Ends: Understanding the Decline and Demise of Terrorist Campaigns.* Princeton, NJ: Princeton University Press.

105. Global Research News. 2014. "US Wars in Afghanistan, Iraq to Cost $6 Trillion." Global Research, February 12. goo.gl/7ERWOl

106. goo.gl/douNPn

107. See the response to Snowden's TED appearance by NSA deputy director Richard Ledgett: goo.gl/ljBCUo

108. Hirschman, Albert O. 1970. *Exit, Voice, and Loyalty: Responses to Decline in Firms, Organizations, and States.* Cambridge, MA: Harvard University Press.

109. Eisner, Manuel. 2011. "Killing Kings: Patterns of Regicide in Europe, 600–1800." *British Journal of Criminology*, 51, 556–577.

110. Zedong, Mao. 1938. "Problems of War and Strategy." *Selected Works*, Vol. II, 224.

111. Stephan, Maria J., and Erica Chenoweth. 2008. "Why Civil Resistance Works: The Strategic Logic of Nonviolent Conflict." *International Security*, 33, no. 1, 7–44. See also Chenoweth, Erica, and Maria J. Stephan. 2011. *Why Civil Resistance Works: The Strategic Logic of Nonviolent Conflict.* New York: Columbia University Press.

112. Stephan and Chenoweth, 2008.

113. Chenoweth, Erica. 2013. "Nonviolent Resistance." TEDx Boulder. goo.gl /xqTy5P

114. Ibid.

115. Graph rendered from data in Stephan and Chenoweth, 2008, op cit., Chenoweth and Stephan, 2011, op cit., and Chenoweth, 2013, op cit.

116. Ibid.

117. Low, Bobbi. 1996. "Behavioral Ecology of Conservation in Traditional Societies." *Human Nature*, 7, no. 4, 353–379.

118. Edgerton, Robert. 1992. *Sick Societies: Challenging the Myth of Primitive Harmony.* New York: Free Press.

119. Keeley, Lawrence. 1996. *War Before Civilization: The Myth of the Peaceful Savage.* New York: Oxford University Press.

120. Leblanc, Steven, and Katherine E. Register. 2003. *Constant Battles: The Myth of the Peaceful, Noble Savage.* New York: St. Martin's Press, 125, 224–228.

121. Ibid., 202.

122. Photographs by the author.

123. Personal correspondence, July 28, 2011.

124. Ibid.

125. To which Colbert sardonically responded: "Don't patronize me!" goo.gl/RlzMz7

126. Pinker, 2011, 49, rendered from data published in Bowles, S. 2009. "Did Warfare Among Ancestral Hunter-Gatherers Affect the Evolution of Human Social Behaviors?" *Science*, 324, 1293–1298; Keeley, L. H. 1996. *War Before Civilization: The Myth of the Peaceful Savage.* New York: Oxford University Press; Gat, A. 2006. *War in Human Civilization.* New York: Oxford University Press; White, M. 2011. *The Great Big Book of Horrible Things: The Definitive Chronicle of History's 100 Worst Atrocities.* New York: W. W. Norton; Harris, M. 1975. *Culture, People, Nature*, 2nd ed. New York: Crowell, Lacina, B., and N. P. Gleditsch. 2005. "Monitoring Trends in Global Combat: A New Dataset in Battle Deaths." *European Journal of Population*, 21, 145–166; Sarkees, M. R. 2000. "The Correlates of War Data on War." *Conflict Management and Peace Science*, 18, 123–144.

127. Van der Dennen, J. M. G. 1995. *The Origin of War: The Evolution of a Male-Coalitional Reproductive Strategy.* Groningen, Neth.: Origin Press; Van der Dennen, J. M. G. 2005. *Querela Pacis*: Confession of an Irreparably Benighted Researcher on War and Peace. An Open Letter to Frans de Waal and the "Peace and Harmony Mafia." Groningen: University of Groningen.

128. Horgan, John. 2014. "Jared Diamond, Please Stop Propagating the Myth of the Savage Savage!" *Scientific American Blogs*, January 20.

129. For a brief history of the evolution wars see appendix I in Shermer, Michael. 2004. *The Science of Good and Evil.* New York: Times Books; for a book-length treat-

ment see Segerstråle, U. 2000. *Defenders of the Truth: The Battle for Science in the Sociobiology Debate and Beyond*. New York: Oxford University Press.

130. Ferguson, R. Brian. 2013. "Pinker's List: Exaggerating Prehistoric War Mortality." In Douglas P. Fry, ed., *War, Peace, and Human Nature*. New York: Oxford University Press, 112–129.

131. Azar Gat notes that many of the peace and harmony scholars have "silently forfeited" their position that "the aboriginal human condition before agriculture and the state was fundamentally characterized by little or no violent killing among people," and that there are now three strands of "Rousseauism" fighting a rearguard action against the onslaught of data in recent decades. Gat, Azar. In press. "Rousseauism I, Rousseauism II, and Rousseauism of Sorts: Shifting Perspectives on Aboriginal Human Peacefulness and Their Implications for the Future of War." *Journal of Peace Research*.

132. Fry, Douglas, and Patrik Söderberg. 2013. "Lethal Aggression in Mobile Forager Bands and Implications for the Origins of War." *Science*, July 19, 270–273.

133. Bowles, Samuel. 2013. "Comment on Fry and Söderberg 'Lethal Aggression in Mobile Forager Bands and Implications for the Origins of War.'" July 19, tuvalu .santafe.edu/~bowles/

134. Ibid.

135. Miroff, Nick, and William Booth. 2012. "Mexico's Drug War Is at a Stalemate as Calderon's Presidency Ends." *Washington Post*, November 27. goo.gle/pu2uJ; Reuters. 2012. "Desplazdos, ragedia Silenciosa en Mexico." *El Economista*, January 7.

136. Bowles, 2013.

137. Levy, Jack S., and William R. Thompson. 2011. *The Arc of War: Origins, Escalation, and Transformation*. Chicago: University of Chicago Press, 1.

138. Ibid., 3.

139. Ibid., 51–53.

140. Arkush, Elizabeth, and Charles Stanish. 2005. "Interpreting Conflict in the Ancient Andes." *Current Anthropology*, 46, no. 1, February, 3–28. goo.gl/TrELvz

141. Milner, George R., Jane E. Buikstra, and Michael D. Wiant. 2009. "Archaic Burial Sites in the American Midcontinent." In *Archaic Societies*, ed. Thomas E. Emerson, Dale L. McElrath, and Andres C. Fortier. Albany: State University of New York Press, 128.

142. Wrangham, Richard, and Dale Peterson. 1996. *Demonic Males: Apes and the Origins of Human Violence*. Boston: Houghton Mifflin.

143. Wrangham, Richard W., and Luke Glowacki. 2012. "Intergroup Aggression in Chimpanzees and War in Nomadic Hunter-Gatherers." *Human Nature*, 23, 5–29.

144. Glowacki, Luke, and Richard W. Wrangham. 2013. "The Role of Rewards in Motivating Participation in Simple Warfare." *Human Nature* 24, 444–460. A modern example of the cultural rewards war-risk hypothesis may be found in the research of the anthropologist Scott Atran, who has documented precisely how the process works in al Qaeda cells with new recruits. The families of successful suicide bombers (successful meaning they blew themselves to smithereens) are taken care of financially and their lost son (or, more rarely, a daughter, in the case of the "black widow") is posterized like a sports star. Religion and ideology may not always be enough motivation for would-be martyrs and holy warriors to overcome their aversion to injury and death (Atran thinks that the highly publicized "seventy-two virgins" belief is greatly exaggerated), so the "band of brothers" mentality is reinforced through social groups, soccer clubs, and other social activities that

reinforce core cultural values. Atran, Scott. 2011. *Talking to the Enemy: Religion, Brotherhood, and the (Un)Making of Terrorists.* New York: Ecco.

145. Wrangham and Glowacki, 2012.

146. Keeley, Lawrence. 2000. *Warless Societies and the Origins of War.* Ann Arbor: University of Michigan Press, 4.

147. Coe, Michael D. 2005. *The Maya,* 7th ed. London: Thames & Hudson, 161.

148. The Harvard Hawks have also been accused of data-mining examples of violence— while allegedly ignoring the cases of peaceful hunter-gatherers—so I queried the archaeologist Lawrence Keeley on this charge, to which he responded, "I included every case I could find. That's why there are several cases that counted *no* war deaths over two hundred years. Regarding accuracy, Paul Roscoe, an ethnographer of New Guinea, has checked the war death figures I found for that region, going to the original reports, and found my figures low/conservative," adding that "LeBlanc and I are archaeologists; we've excavated unequivocal war victims and fortifications—lots of them. Prehistoric warfare is a whole intertwined set of physical facts." Keeley's recommendation to the peace-and-harmony mafia is to "Admit that war was common over the past 10,000s of years but not constant; try to explain why this is so; ask how different cultures conducted war or maintained peace." Personal correspondence, February 5, 2014.

149. Keegan, John. 1994. *A History of Warfare.* New York: Random House, 59.

150. Goldstein, Joshua. 2011. *Winning the War on War: The Decline of Armed Conflict Worldwide.* New York: Dutton, 328. See also Payne, James. 2004. *A History of Force: Exploring the Worldwide Movement Against Habits of Coercion, Bloodshed, and Mayhem.* Sandpoint, ID: Lytton Publishing.

151. Lebow, Richard Ned. 2010. *Why Nations Fight: Past and Future Motives for War.* Cambridge, UK: Cambridge University Press.

152. Ibid., 206–207.

153. Human Security Report 2013. 2014. *The Decline in Global Violence: Evidence, Explanation, and Contestation.* Human Security Report Project, Simon Fraser University, Canada, 11, 48.

154. www.visionofhumanity.org

155. Personal correspondence, February 1, 2014.

3: WHY SCIENCE AND REASON ARE THE DRIVERS OF MORAL PROGRESS

1. In "Answer to the Abbe Raynal." In Paine, Thomas. 1796. *The Works of Thomas Paine.* Google eBook.

2. Taylor, John M. 1908. *The Witchcraft Delusion in Colonial Connecticut (1647–1697).* Read at Project Gutenberg: goo.gl/UtPyLk

3. Transcripts of Steve Martin as Theodoric of York are available online: goo.gl/fbHsUp

4. Voltaire, 1765/2005. "Question of Miracles," in *Miracles and Idolotry.* New York: Penguin.

5. The trolley car thought experiment was first proposed by the philosopher Phillipa Foot in: Foot, Phillipa. 1967. "The Problem of Abortion and the Doctrine of Double Effect." *Oxford Review,* 5, 5–15. The extensive research utilizing the trolley car scenario has been summarized in many works, most recently in Edmonds, David. 2013. *Would You Kill the Fat Man?* Princeton, NJ: Princeton University Press. See also Petrinovich, L., P. O'Neill, and M. J. Jorgensen. 1993. "An Empirical Study

of Moral Intuitions: Towards an Evolutionary Ethics." *Ethology and Sociobiology*, 64, 467–478.

6. I review these theories of the European witch craze in Shermer, Michael. 1997. *Why People Believe Weird Things*. New York: W. H. Freeman, chap. 7.

7. Diamond, Jared. 2013. *The World Until Yesterday: What Can We Learn from Traditional Societies?* New York: Viking, 345.

8. Senft, Gunter. 1997. "Magical Conversation on the Trobriand Islands." *Anthropos*, 92, 369–371.

9. Malinowski, Bronislaw. 1954. *Magic, Science, and Religion*. Garden City, NY: Doubleday, 139–140.

10. Evans-Pritchard, E. E. 1976. *Witchcraft, Oracles and Magic Among the Azande*. New York: Oxford University Press, 18.

11. Ibid., 23.

12. Kieckhefer, Richard. 1994. "The Specific Rationality of Medieval Magic." *American Historical Review*, 99, no. 3, 813–836.

13. Hutchinson, Roger. 1842. *The Works of Roger Hutchinson*, ed. J. Bruce. New York: Cambridge University Press, 140–141.

14. Quoted in Walker, D. P. 1981. *Unclean Spirits: Possession and Exorcism in France and England in the Late Sixteenth and Early Seventeenth Centuries*. Philadelphia: University of Pennsylvania Press, 71.

15. Hedesan, Jo. 2011. "Witch Hunts in Papua New Guinea and Nigeria." *The International*, October 1. goo.gl/Lrkq6N

16. Tortora, Bob. 2010. "Witchcraft Believers in Sub-Saharan Africa Rate Lives Worse." Gallup, August 25. goo.gl/ZlrvEJ

17. Pollak, Sorcha. 2013. "Woman Burned Alive for Witchcraft in Papua New Guinea." *Time*, February 7. goo.gl/t4yR19

18. Oxfam New Zealand. 2012. "Protecting the Accused: Sorcery in PNG." goo.gl/mEevih

19. Napier, William. 1851. *History of General Sir Charles Napier's Administration of Scinde*. London: Chapman & Hall, 35.

20. Mackay, Charles. 1841/1852/1980. *Extraordinary Popular Delusions and the Madness of Crowds*. New York: Crown, 559.

21. Ibid., 560.

22. Levak, Brian. 2006. *The Witch-hunt in Early Modern Europe*. New York: Routledge.

23. Llewellyn Barstow, Anne. 1994. *Witchcraze: A New History of the European Witch-hunts*. New York: HarperCollins.

24. Thomas, Keith. 1971. *Religion and the Decline of Magic*. New York: Charles Scribner's Sons, 643.

25. Ibid., 643–644.

26. Cited in Thomas, 1971, 1–21 passim.

27. La Roncière, Charles de. 1988. "Tuscan Notables on the Eve of the Renaissance." In *A History of Private Life: Revelations of the Medieval World*. Cambridge, MA: Harvard University Press, 171.

28. Snell, Melissa. "The Medieval Child." *Medieval History*. goo.gl/6uhDiR

29. Blackmore, S., and R. Moore. 1994. "Seeing Things: Visual Recognition and Belief in the Paranormal." *European Journal of Parapsychology*, 10, 91–103; Musch, J., and K. Ehrenberg. 2002. "Probability Misjudgment, Cognitive Ability, and Belief in the Paranormal." *British Journal of Psychology*, 93, 169–177; Brugger, P., T. Landis, and M. Regard. 1990. "A 'Sheep-Goat Effect' in Repetition Avoidance: Extra-Sensory

Perception as an Effect of Subjective Probability?" *British Journal of Psychology*, 81, 455–468.

Whitson, Jennifer A., and Adam D. Galinsky. 2008. "Lacking Control Increases Illusory Pattern Perception." *Science*, 322, 115–117.

30. Cited in Thomas, 1971, 16.
31. Ibid., 177.
32. Ibid.
33. Ibid., 668.
34. Ibid., 91.
35. Shermer, Michael. 2011. *The Believing Brain*. New York: Times Books, chap. 13.
36. Quoted in Cohen, I. Bernard. 1985. *Revolution in Science*. Cambridge, MA: Harvard University Press.
37. Bacon, F. 1620/1939. *Novum Organum*. In Burtt, E. A., ed., *The English Philosophers from Bacon to Mill*. New York: Random House.
38. goo.gl/vSBMO5
39. Olson, Richard. 1990. *Science Deified and Science Defied: The Historical Significance of Science in Western Culture*. Berkeley: University of California Press, 15–40.
40. These and many other assessments of Newton and his work come from Westfall, Richard. 1980. *Never at Rest: A Biography of Isaac Newton*. Cambridge, UK: Cambridge University Press.
41. Cohen, I. Bernard. 1985. *Revolution in Science*. Cambridge, MA: Harvard University Press, 174, 175.
42. All quotes are from Olson, 1990, 191–202 passim. See also Hankins, Thomas L. 1985. *Science and the Enlightenment*. Cambridge, UK: Cambridge University Press, 161–163.
43. Olson, Richard. 1990, 183–189.
44. Smith, Adam. 1795/1982. "The History of Astronomy." In *Essays on Philosophical Subjects*, ed. W. P. D. Wightman and J. C. Bryce. Vol. III of *Glasgow Edition of the Works and Correspondence of Adam Smith*. Indianapolis: Liberty Fund, 2.
45. ———. 1759. *The Theory of Moral Sentiments*. London: A. Millar, I., I., 1.
46. Hobbes, Thomas. 1839. *The English Works of Thomas Hobbes*, ed. William Molesworth, Vol. 1, ix–1.
47. ———. 1642. *De Cive, or the Citizen*. New York: Appleton-Century-Crofts, 15.
48. Olson, 1990, 45, 47.
49. A succinct summary of Hobbes's theory and how Hobbes and his Enlightenment colleagues perceived what they were doing is what we would call science may be found in Olson, 1990, 51–58.
50. Hobbes, Thomas. 1651/1968. *Leviathan, or The Matter, Forme and Power of a Common Wealth Ecclesiasticall and Civil*, ed. C. B. Macpherson. New York: Penguin, 76.
51. As Olson so well taught his students, including me, science does not happen in a social vacuum, and he shows that Hobbes was motivated to make the case for the necessity of a strong state because of the looming English civil war. *Leviathan* was to be the third in a trilogy, "a grand scientific system" that would cover physics, humanity, and society. But as Hobbes later explained (*De Cive*, 15): "my country some few years before the civil war did rage, was boiling hot with questions concerning the rights of dominion and the obedience due from subjects, the true forerunners of an approaching war; and this was the cause which ripened and plucked from me this third part. Therefore, what was last in order, is yet come forth first in time."
52. Hume, David. 1748/1902. *An Enquiry Concerning Human Understanding*. Cambridge, UK: Cambridge University Press, 165.

53. Walzer, Michael. 1967. "On the Role of Symbolism in Political Thought." *Political Science Quarterly*, 82, 201.

54. Wright, Quincy. 1942. *A Study of War*, 2nd ed. Chicago: University of Chicago Press; Gat, A. 2006. *War in Human Civilization*. New York: Oxford University Press; Fukuyama, Francis. 2011. *The Origins of Political Order: From Prehuman Times to the French Revolution*. New York: Farrar, Straus & Giroux.

55. Shermer, Michael. 2007. *The Mind of the Market: How Biology and Psychology Shape Our Economic Lives*. New York: Times Books, 252.

56. Ibid., 256.

57. Henrich, Joseph, et al. 2010. "Markets, Religion, Community Size, and the Evolution of Fairness and Punishment." *Science*, March 19, 327, 1480–1484.

58. Russett, Bruce, and John Oneal. 2001. *Triangulating Peace: Democracy, Interdependence, and International Organizations*. New York: W. W. Norton.

59. Ibid., 108–111.

60. Ibid., 145–148. See also McDonald, P. J. 2010. "Capitalism, Commitment, and Peace." *International Interactions*, 36, 146–168.

61. McDonald, 148. See also Gartzke, E., and J. J. Hewitt. 2010. "International Crises and the Capitalist Peace." *International Interactions*, 36, 115–145.

62. Marshall, Monty. 2009. "Major Episodes of Political Violence, 1946–2009." Vienna, VA: Center for Systematic Peace, November 9, www.systemicpeace.org/warlist.htm

63. ———. 2009. *Polity IV Project: Political Regime Characteristics and Transitions, 1800–2008*. Fairfax, VA: Center for Systematic Peace, George Mason University. http://www.systemicpeace.org/polityproject.html

64. Graph from Russett, Bruce. 2008. *Peace in the Twenty-first Century? The Limited but Important Rise of Influences on Peace*. New Haven, CT: Yale University Press.

65. Levy, Jack S. 1989. "The Causes of War." In: Philip E Tetlock, et al. (eds.) *Behavior, Society, and Nuclear War*. New York: Oxford University Press, 209–313. For a dissenting view see: Barbieri, Katherine. 2002. *The Liberal Illusion: Does Trade Promote Peace?* Ann Arbor: University of Michigan Press.

66. Dorussen, Han, and Hugh Ward. 2010. "Trade Networks and the Kantian Peace." *Journal of Peace Research*, 47, no. 1, 229–242; Hegre, Håvard. 2014. "Democracy and Armed Conflict." *Journal of Peace Research* 51, no. 2, 159–172.

67. Cited in: Wyne, Ali. 2014. "Disillusioned by the Great Illusion: The Outbreak of Great War." *War on the Rocks*, January 29. All quotes come from this article, available at goo.gl/wRio4K

68. Ibid.

69. Ibid.

70. Graph from Russett, Bruce. 2008. Based on data in Lacina, B., N. P. Gleditsch, and B. Russett. 2006. "The Declining Risk of Death in Battle." *International Studies Quarterly*, 50, 673–680.

71. "Estimated Annual Deaths from Political Violence, 1939–2011." Source: fig. 7 at goo.gl/8V8L5L

72. Ibid., fig. 9.

73. Hobbes, 1651, 110–140.

74. Madison, James. 1788. "*The Federalist* No. 51: The Structure of the Government Must Furnish the Proper Checks and Balances Between the Different Departments." *Independent Journal*, February 6.

75. Burke, Edmund. 1790. *Reflections on the Revolution in France*. In *The Works of Edmund Burke*, 3 vols. New York: Harper & Brothers, 1860, 481–483. Available online at goo.gl/e8aTD

76. In *The Science of Liberty* the science writer Timothy Ferris notes of the architects of the United States: "The founders often spoke of the new nation as an 'experiment,'" Ferris writes. "Procedurally, it involved deliberations about how to facilitate both liberty and order, matters about which the individual states experimented considerably during the eleven years between the Declaration of Independence and the Constitution." Ferris, Timothy. 2010. *The Science of Liberty: Democracy, Reason, and the Laws of Nature.* New York: HarperCollins.

77. Jefferson, Thomas. 1804. Letter to Judge John Tyler Washington, June 28. *The Letters of Thomas Jefferson, 1743–1826.* goo.gl/hn6qNP

78. Isaacson, Walter. 2004. *Benjamin Franklin: An American Life.* New York: Simon & Schuster, 311–312.

79. Koonz, Claudia. 2005. *The Nazi Conscience.* Cambridge, MA: Harvard University Press.

80. Kiernan, Ben. 2009. *Blood and Soil: A World History of Genocide and Extermination from Sparta to Darfur.* New Haven, CT: Yale University Press.

81. Quote in Koonz, 2005, 2.

82. Eaves, L. J., H. J. Eysenck, and N. G. Martin. 1989. *Genes, Culture, and Personality: An Empirical Approach.* San Diego: Academic Press. Since this book there have been numerous studies confirming that about half the variation in political attitudes are accounted for by genes. See, for example, Hatemi, Peter K., and Rose McDermott. 2012. "The Genetics of Politics: Discovery, Challenges, and Progress." *Trends in Genetics,* October, 28, no. 10, 525–533; Hatemi, Peter K., and Rose McDermott, eds., 2011. *Man Is by Nature a Political Animal.* Chicago: University of Chicago Press.

83. Gilbert, W. S. 1894. "The Contemplative Sentry." Available online at goo.gl/Z6lXK6

84. Hibbing, John R., Kevin B. Smith, and John R. Alford. 2013. *Predisposed: Liberals, Conservatives, and the Biology of Political Differences.* New York: Routledge.

85. Tuschman, Avi. 2013. *Our Political Nature: The Evolutionary Origins of What Divides Us.* Amherst, NY: Prometheus Books, 402–403.

86. Pedro, S. A., and A. J. Figueredo. 2011. "Fecundity, Offspring Longevity, and Assortative Mating: Parametric Tradeoffs in Sexual and Life History Strategy." *Biodemography and Social Biology,* 57, no. 2, 172.

87. Sowell, Thomas. 1987. *A Conflict of Visions: Ideological Origins of Political Struggles.* New York: Basic Books, 24–25.

88. Pinker, Steven. 2002. *The Blank Slate: The Modern Denial of Human Nature.* New York: Viking, 290–291.

89. Shermer, Michael. 2011. *The Believing Brain.* New York: Times Books, chap. 11.

90. In Jost, J. T., C. M. Federico, and J. L. Napier. 2009. "Political Ideology: Its Structures, Functions, and Elective Affinities." *Annual Review of Psychology,* 60, 307–337.

91. See also Richard Olson's broader definition of ideology as "any sets of assumptions, values, and goals which direct the actions of members of a community. These assumptions, values, and goals need not be explicitly expressed, and they are rarely subject to critical analysis within the community that shares them." Olson, Richard. 1993. *The Emergence of the Social Sciences 1642–1792.* New York: Twayne, 4–5.

92. Smith, Christian. 2003. *Moral, Believing Animals: Human Personhood and Culture.* Oxford, UK: Oxford University Press.

93. Russell, Bertrand. 1946. *History of Western Philosophy.* London: George Allen & Unwin, 8.

94. Levin, Yuval. 2013. *The Great Debate: Edmund Burke, Thomas Paine, and the Birth of Left and Right.* New York: Basic Books.

95. Burke, Edmund. 1790/1967. *Reflections on the Revolution in France*. London: J. M. Dent & Sons.
96. Ibid.
97. Burke, Edmund. 2009. *The Works of the Right Honorable Edmund Burke*, ed. Charles Pedley. Ann Arbor: University of Michigan Library, 377.
98. Paine, Thomas. 1776. *Common Sense*. Available online at www.constitution.org/civ/comsense.htm
99. ———. 1794. *The Age of Reason*. Available online at http://www.gutenberg.org/ebooks/31270
100. ———. 1795. *Dissertation on First Principles of Government*. Available online at www.gutenberg.org/ebooks/31270 Add to *reason* and *interchangeable perspectives* the principle of *probable deterrence* and *costly signaling theory* and you've got a recipe for relative peace, as Paine noted in his "Thoughts on Defensive War" (Paine, Thomas. 1795. *Thoughts on Defensive War*. Available online at www.gutenberg.org/ebooks/31270):

    The supposed quietude of a good man allures the ruffian; while on the other hand, arms like laws discourage and keep the invader and the plunderer in awe, and preserve order in the world as well as property. The balance of power is the scale of peace. The same balance would be preserved were all the world destitute of arms, for all would be alike; but since some will not, others dare not lay them aside. And while a single nation refuses to lay them down, it is proper that all should keep them up. Horrid mischief would ensue were one half the world deprived of the use of them; for while avarice and ambition have a place in the heart of man, the weak will become a prey to the strong.

101. Mill, John Stuart. 1859. *On Liberty*. Chap. 2. goo.gl/AWjszJ
102. Mencken, H. L. 1927. "Why Liberty?" *Chicago Tribune*, January 30. goo.gl/1Csn6h
103. Clemens, Walter C., Jr. 2013. *Complexity Science and World Affairs*. Albany: State University of New York Press.
104. Reported in: Knight, Richard. 2012. "Are North Koreans Really Three Inches Shorter Than South Koreans?" *BBC News Magazine*, April 22. goo.gl/dWWGxp
105. Electric power consumption (kWh per capita). World Bank. goo.gl/kVlq9
106. Graph based on data generated by Agnus Maddison: goo.gl/lK9xmF. See also Maddison, Agnus. 2006. *The World Economy*. Washington, DC: OECD Publishing; Maddison, Agnus. 2007. *Contours of the World Economy 1–2030 AD: Essays in Macro-Economic History*. New York: Oxford University Press.
107. earthobservatory.nasa.gov/
108. Harris, Sam. 2010. *The Moral Landscape: How Science Can Determine Human Values*. New York: Free Press.

4: WHY RELIGION IS NOT THE SOURCE OF MORAL PROGRESS

1. I was a born-again Christian for seven years and I still have close friends who are deeply religious. And I wrote an entire book on the psychology and power of religious belief, *How We Believe*. So I am sensitive to the fact that for many religious believers an analysis like this is irrelevant because what they get out of religion is social community and personal comfort. *"Yea, though I walk through the valley of the shadow of death, I will fear no evil: for thou art with me; thy rod and thy staff they comfort me."* (Psalms 23, 4). In my book I reported the findings of a study I conducted that found that one of the main reasons people give for why they believe in God is the "personal experience of God" in their lives, and that they think the primary reason that other people believe in God is "comfort

and purpose to life." So I understand on both an emotional and an intellectual level that there are many profoundly personal and deeply meaningful reasons for religion and belief in God that seem wholly removed from the type of scientific analysis that I am effecting in this book. But this is a work of science, so as I continue to dismantle the myth that religion was the driving force behind moral progress, just know that it is not my intent to offend, only to understand.

2. Sagan, Carl. 1990. "Preserving and Cherishing the Earth: An Appeal for Joint Commitment in Science and Religion." Statement signed by thirty-two Nobel laureates and presented to the Global Forum of Spiritual and Parliamentary Leaders Conferences in Moscow, Russia. goo.gl/fO9PQY

3. Hartung, J. 1995. "Love Thy Neighbor: The Evolution of In-Group Morality." *Skeptic*, 3, no. 4, 86–99.

4. Krakauer, Jon. 2004. *Under the Banner of Heaven: A Story of Violent Faith*. New York: Anchor.

5. Darwin, Charles. 1871. *The Descent of Man and Selection in Relation to Sex*. London: John Murray, 571.

6. Betzig, Laura. 2005. "Politics as Sex: The Old Testament Case." *Evolutionary Psychology*, 3: 326.

7. Ibid., 327.

8. 1 Kings 4, 11–14; 11, 3; 1 Chron 3, 10–24.

9. Sweeney, Julia. 2006. *Letting Go of God*. Book transcript of monologue. Indefatigable Inc., 24.

10. Ibid., 26.

11. Dawkins, Richard. 2006. *The God Delusion*. New York: Houghton Mifflin, 31.

12. Sheer, Robert. 1976. "Jimmy Carter Interview." *Playboy*. goo.gl/CiSTUp

13. Reagan used the metaphor many times in his political career. You can watch his farewell address here: goo.gl/lhnSh4 and read several of his quotes here: www.pbs .org/wgbh/americanexperience/features/general-article/reagan-quotes/

14. Kennedy, John F. 1961. "Address of President-elect John F. Kenney Delivered to a Joint Convention of the General Court of the Commonwealth of Massachusetts." January 9. goo.gl/W6f2LZ

15. D'Souza, Dinesh. 2008. *What's So Great About Christianity*. Carol Stream, IL: Tyndale House.

16. Ibid., 34–35.

17. Roberts, J. M. 2001. *The Triumph of the West*. New York: Sterling.

18. D'Souza, 2008, 36.

19. Stark, Rodney. 2005. *The Victory of Reason*. New York: Random House, xii–xiii.

20. For lengthy treatments of the many factors that went into the development of democracy and capitalism in Western Europe see Fukuyama, Francis. 2011. *The Origins of Political Order: From Prehuman Times to the French Revolution*. New York: Farrar, Straus & Giroux; Morris, Ian. 2013. *The Measure of Civilization: How Social Development Decides the Fate of Nations*. Princeton, NJ: Princeton University Press; Morris, Ian. 2011. *Why the West Rules—for Now: The Patterns of History and What They Reveal About the Future*. New York: Farrar, Straus & Giroux; Beinhocker, Eric. 2006. *The Origin of Wealth*. Cambridge, MA: Harvard Business School Press.

21. The religion editor for *Skeptic* magazine, Tim Callahan, assessed this hypothesis and concluded: "Geography then, far more than a Christian mind-set, enabled

democracies to come into being and survive as Western Civilization evolved. Geography, plus chance also made it possible for capitalism to develop where it did in Europe. This was aided by the Western European, rather than exclusively Christian outlook. Separation of church and state in Western Christendom was entirely serendipitous. The failure of Eastern Christianity to produce separation of church and state, capitalism or democracy demonstrates that, whatever impact the Christian outlook might have had, it only worked as part of a specific synthesis that, aided by geography and chance, produced Western Civilization." Callahan, Tim. 2012. "Is Ours a Christian Nation?" *Skeptic*, 17, no. 3, 31–55. See also Tim Callahan's review of *What's so Great About Christianity* in *Skeptic*, 14, no. 1, 68–71.

22. Gebauer, Jochen, Andreas Nehrlich, Constantine Sedikides, and Wiebke Neberich. 2013. "Contingent on Individual-Level and Culture-Level Religiosity." *Social Psychological and Personality Science*, 4, no. 5, 569–578.

23. Bowler, Kate. 2013. *Blessed: A History of the American Prosperity Gospel*. New York: Oxford University Press.

24. Hitchens, Christopher. 1995. *The Missionary Position: Mother Teresa in Theory and Practice*. Brooklyn, NY: Verso.

25. D'Souza, 2008, 54.

26. 1992. *The Interpreter's Bible: The Holy Scriptures in the King James and Revised Standard Versions with General Articles and Introduction, Exegesis, Exposition for Each Book of the Bible*. Nashville: Abingdon Press.

27. Ibid.

28. Ibid.

29. Letter to Henry Lee of May 8, 1825. goo.gl/TLOk4U

30. Brooks, Arthur C. 2006. *Who Really Cares: The Surprising Truth About Compassionate Conservatism*. New York: Basic Books.

31. Ibid., 5–10 passim.

32. Ibid., 142–144 passim.

33. Ibid., 8.

34. Ibid., 55.

35. Ibid., 182–183.

36. Lindgren, James. 2006. "Concerns About Arthur Brooks's 'Who Really Cares.'" November 20. goo.gl/iGK9M0

37. Paul, Gregory S. 2009. "The Chronic Dependence of Popular Religiosity upon Dysfunctional Psychosociological Conditions." *Evolutionary Psychology*, 7, no. 3, 398–441.

38. Rerendered by Pat Linse from graphs in: Paul, 2009, 320–441 passim.

39. Personal correspondence, September 25, 2013.

40. Norris, Pippa, and Ronald Inglehart. 2004. *Sacred and Secular*. New York: Cambridge University Press.

41. Putnam, Robert. 2000. *Bowling Alone: The Collapse and Revival of American Community*. New York: Simon & Schuster, 19.

42. Ibid., 20–21.

43. Norris and Inglehart, op cit.

44. Referencing Paul's international study, Sulloway added, "At the between-country level, this is a big, or macro effect," adding, "Now it is also true that, *within* each particular country, people who pray and believe in God obtain solace and seem, in some studies, to derive modest health benefits. So, at the purely individual level, you could score one for conservatives, because the causal relationship here

is presumptively that religion causes small health benefits (which, however, are generally swamped at the macro level, or group level, by other between-country effects). Hence arises the paradox that religion is good for the individual, in terms of health and social benefits, but appears to be bad for the individual when the data are analyzed on a country-by-country basis." Nevertheless, Sulloway does not see a conflict in the data cited here: "As for the matter of charitable behavior, religion may indeed be causing more charitable giving. But this is not inconsistent with Paul's data on health and religion, because this is primarily a within-country effect, with data being presented here at the individual level. In addition, the presumptive causal relationship here is that religion does motivate giving. In short, at the individual level, the causal direction is consistent with the result that religion leads to social and health benefits." Personal correspondence, September, 2006.

45. Hitchens, Christopher. 2007. *God Is Not Great: How Religion Poisons Everything*. New York: Twelve.

46. Hall, Harriet. 2013. "Does Religion Make People Healthier?" *Skeptic*, 19, no. 1.

47. McCullough, M. E., W. T. Hoyt, D. B. Larson, H. G. Koenig, and C. E. Thoresen. 2000. "Religious Involvement and Mortality: A Meta-Analytic Review." *Health Psychology*, 19, 211–222.

48. McCullough, M. E., and B. L. B. Willoughby. 2009. "Religion, Self-Regulation, and Self-Control: Associations, Explanations, and Implications." *Psychological Bulletin*, 125, 69–93.

49. Baumeister, Roy, and John Tierney. 2011. *Willpower: Rediscovering the Greatest Human Strength*. New York: Penguin.

50. Mischel, Walter, Ebbe B. Ebbesen, and Antonette Raskoff Zeiss. 1972. "Cognitive and Attentional Mechanisms in Delay of Gratification." *Journal of Personality and Social Psychology*, 21, no. 2, 204–218.

51. Ibid., 180.

52. Ibid., 181.

53. Quoted in ibid., 187–188.

54. The Ten Commandments are stated in two books of the Old Testament: Exodus, 20, 1–17, and Deuteronomy, 5, 4–21. I quote from Exodus, King James Version.

55. Hitchens, Christopher. 2010. "The New Commandments." *Vanity Fair*, April. goo .gl/lcXo After demolishing the Decalogue in his inimitable style, Hitchens proffered his own list of commandments: "*Do not* condemn people on the basis of their ethnicity or color. *Do not* ever use people as private property. Despise those who use violence or the threat of it in sexual relations. Hide your face and weep if you dare to harm a child. *Do not* condemn people for their inborn *nature*—why would God create so many homosexuals only in order to torture and destroy them? Be aware that you too are an animal and dependent on the web of nature, and think and act accordingly. *Do not* imagine that you can escape judgment if you rob people with a false prospectus rather than with a knife. Turn off that fucking cell phone— you have no idea how *un*important your call is to us. Denounce all jihadists and crusaders for what they are: psychopathic criminals with ugly delusions. Be willing to renounce any god or any religion if any holy commandments should contradict any of the above." Hitchens caps his list with this summary judgment: "In short: Do not swallow your moral code in tablet form." That's a rational prescription.

56. Source: Freedom in the World report from Freedom House. goo.gl/GFeA9 See also goo.gl/8ocNM

5: SLAVERY AND A MORAL SCIENCE OF FREEDOM

1. Bannerman, Helen. 1899. *The Story of Little Black Sambo*. London: Grant Richards.
2. goo.gl/LyWnOZ
3. Overbea, Luix. 1981. "Sambo's Fast-Food Chain, Protested by Blacks Because of Name, Is Now Sam's in 3 States." *Christian Science Monitor*, April 22. goo.gl/6y227h
4. The name was not the company's only problem. It also fumbled on some business practices as it expanded rapidly, recounted in Bernstein, Charles. 1984. *Sambo's: Only a Fraction of the Action: The Inside Story of a Restaurant Empire's Rise and Fall*. Burbank, CA: National Literary Guild.
5. A Google search turns up hundreds of images of book covers over the decades, along with a 1935 cartoon film depicting the story in what by today's standards would be maximally offensive and racist, but at that time was commonly accepted as a way to portray blacks: goo.gl/ZQ8lzQ
6. The Code of Hammurabi. fordham.edu/halsall/ancient/hamcode.asp
7. goo.gl/wUyei
8. goo.gl/f5l2Pb
9. White, Matthew. 2011. *The Great Big Book of Horrible Things: The Definitive Chronicle of History's 100 Worst Atrocities*. New York: W. W. Norton, 161. Rubinstein, W. D. 2004. *Genocide: A History*. New York: Pearson Education, 78.
10. Full text of an English translation available here: goo.gl/7JmVNp
11. Hochschild, Adam. 2005. *Bury the Chains: The British Struggle to Abolish Slavery*. London: Macmillan.
12. *The Parliamentary History of England from the Earliest Period to the Year 1803*. Vol. XXIX. 1817. London: T. C. Hansard, 278.
13. Festinger, Leon, Henry W. Riecken, and Stanley Schachter. 1964. *When Prophecy Fails: A Social and Psychological Study*. New York: Harper & Row, 3.
14. Trivers, Robert. 2011. *The Folly of Fools: The Logic of Deceit and Self-Deception in Human Life*. New York: Basic Books; Tavris, Carol, and Elliot Aronson. 2007. *Mistakes Were Made (but Not By Me)*. New York: Mariner Books.
15. Genovese, Eugene D., and Elizabeth Fox-Genovese. 2011. *Fatal Self-Deception: Slaveholding Paternalism in the Old South*. New York: Cambridge University Press, 1. For a related but different take on the subject see Jones, Jacqueline. 2013. *A Dreadful Deceit: The Myth of Race from the Colonial Era to Obama's America*. New York: Basic Books.
16. Clarke, Lewis Garrard. 1845. *Narrative of the Sufferings of Lewis Clarke, During a Captivity of More Than Twenty-Five Years, Among the Algerines of Kentucky, One of the So Called Christian States of North America*. Boston: David H. Ela, Printer. See also John W. Blassingame, ed. 1977. *Slave Testimony: Two Centuries of Letters, Speeches, Interviews, and Autobiographies*. Baton Rouge: Courier Dover Publications, 6–8.
17. Olmsted, Frederick Law. 1856. *A Journey in the Seaboard Slave States*. New York; London: Dix and Edwards; Sampson Low, Son & Co., 58–59. goo.gl/7rixHj
18. Troup, George M. 1824. "First Annual Message to the State Legislature of Georgia." In Hardin, Edward J. 1859. *The Life of George M. Troup*. Savannah, GA. goo.gl/j9sgPw.
19. Fearn, Frances. 1910. *Diary of a Refugee*, ed. Rosalie Urquart. University of North Carolina at Chapel Hill, 7–8.
20. Pollard, E. A. 1866. *Southern History of the War*, 2 vols. New York: Random House Value Publishing, I, 202.
21. In Genovese and Fox-Genovese, op cit., 93.

22. Cobb, T. R. R. 1858 (1999). *An Inquiry into the Law of Negro Slavery in the United States*. Reprint University of Georgia, ccxvii.

23. Harper, William. 1853. "Harper on Slavery." In *The Pro-Slavery Argument, as Maintained by the Most Distinguished Writers of the Southern States*. Philadelphia: Lippincott, Grambo, and Co., 94.

24. Jackman, Mary R. 1994. *The Velvet Glove: Paternalism and Conflict in Gender, Class, and Race Relations*. Berkeley: University of California Press, 13.

25. McDuffie, George. 1835. *Governor McDuffie's Message on the Slavery Question*. 1893. New York: A. Lovell, 8.

26. Thomas, Hugh. 1997. *The Slave Trade: The Story of the Atlantic Slave Trade, 1440–1870*. New York: Simon & Schuster, 451.

27. Ibid., 454–455.

28. Ibid., 459.

29. Ibid., 457.

30. Ibid., 462–463.

31. Ibid., 464.

32. Lloyd, Christopher. 1968. *The Navy and the Slave Trade: The Suppression of the African Slave Trade in the Nineteenth Century*. London: Cass, 118.

33. Voltaire. *Complete Works of Voltaire*, ed. Theodore Besterman. Banbury: Voltaire Foundation, 1974, 117, 374.

34. Montesquieu. *Oeuvres Complètes*, ed. Édouard Laboulaye. 1877. Paris, Vol. iv, I, 330.

35. *Encyclopédie*, 1765. Vol. xvi, 532.

36. Rousseau, J. J. *Du Contrat Social*. In *Oeuvres Complètes*, ed. Pléide. Vol. I, iv.

37. Hutcherson, Francis. 1755. *A System of Moral Philosophy*. London: A. Millar, II, 213. goo.gl/410LUK

38. Smith, Adam. 1759. *The Theory of Moral Sentiments*. London: A. Millar, 402. goo.gl/DhWCB

39. Blackstone, William. 1765. *Commentaries on the Laws of England*, I, 411–412.

40. Lincoln, Abraham. 1858. In *The Collected Works of Abraham Lincoln*, 1953, ed. Roy P. Basler. Vol. II, August 1, 532.

41. Rawls, J. 1971. *A Theory of Justice*. Cambridge, MA: Belknap Press.

42. Lincoln, Abraham. 1854. Fragment on Slavery. July 1. goo.gl/ux6b9Z

43. Quoted in Hecht, Jennifer Michael. 2013. "The Last Taboo." *Politico.com*, goo.gl/loOSDz

44. For a book-length defense of this connection between Lincoln and Euclid see Hirsch, David, and Dan Van Haften. 2010. *Abraham Lincoln and the Structure of Reason*. New York: Savas Beatie.

45. Douglas, Stephen. 1858. In Harold Holzer, ed., 1994. *The Lincoln-Douglas Debates: The First Complete, Unexpurgated Text*. New York: Fordham University Press, 55.

46. Ibid.

47. Lincoln, Abraham. 1864. Letter to Albert G. Hodges. Library of Congress. goo.gl/HpHoJJ The line appears in the opening of a letter to the editor of the Frankfort, Kentucky, *Commonwealth*, Albert G. Hodges, who had journeyed from Kentucky to meet with Lincoln to discuss the recruitment of slaves as soldiers in Kentucky, which was a border state and thus the Emancipation Proclamation did not apply. Nevertheless, slaves who entered the military could gain their freedom. Lincoln wrote: "I am naturally anti-slavery. If slavery is not wrong, nothing is wrong. I can not remember when I did not so think, and feel. And yet I have never understood that the Presidency conferred upon me an unrestricted right to act officially upon this judgment and feeling."

48. http://en.wikipedia.org/wiki/Abolition_of_slavery_timeline

49. www.endslaverynow.com

50. See, for example, Agustin, Laura Maria. 2007. *Sex at the Margins: Migration, Labour Markets and the Rescue Industry.* London: Zed Books; Bernstein, Elizabeth. 2010. "Militarized Humanitarianism Meets Carceral Feminism: The Politics of Sex, Rights, and Freedom in Contemporary Antitrafficking Campaigns." *Signs,* Autumn, 45–71; Weitzer, Ronald. 2012. *Legalizing Prostitution: From Illicit Vice to Lawful Business.* New York: New York University Press.

51. www.globalslaveryindex.org/findings/ " 'Slavery' refers to the condition of treating another person as if they were property—something to be bought, sold, traded or even destroyed. 'Forced labour' is a related but not identical concept, referring to work taken without consent, by threats or coercion. 'Human trafficking' is another related concept, referring to the process through which people are brought, through deception, threats or coercion, into slavery, forced labour or other forms of severe exploitation." Whatever term is used, the significant characteristic of all forms of modern slavery is that it involves one person depriving another people of their freedom: their freedom to leave one job for another, their freedom to leave one workplace for another, their freedom to control their own body.

52. goo.gl/Rtunu

53. goo.gl/lstp

54. www.freetheslaves.net/

55. Sutter, John D., and Edythe McNamee. 2012. "Slavery's Last Stronghold." CNN, March. goo.gl/BTv6N

6: A MORAL SCIENCE OF WOMEN'S RIGHTS

1. Lecky, William Edward Hartpole. 1869. *History of European Morals: From Augustus to Charlemagne,* 2 vols. New York: D. Appleton and Co., Vol. 1, 274.

2. Wollstonecraft, Mary. 1792. *A Vindication of the Rights of Woman: With Strictures on Political and Moral Subjects.* Boston: Peter Edes. Available online at www.bartleby.com/144/

3. Mill, John Stuart (and possibly coauthored with Harriet Taylor Mill). 1869. *The Subjection of Women.* London: Longmans, Green, Reader, & Dyer. Available online at www.constitution.org/jsm/women.htm

4. Lecky, 1869, op cit.

5. goo.gl/zO11V2

6. For a comprehensive account see Flexner, Eleanor. 1959/1996. *Century of Struggle.* Cambridge, MA: Belknap Press.

7. Purvis, June. 2002. *Emmeline Pankhurst: A Biography.* London: Routledge, 354.

8. Ibid., 354.

9. Stevens, Doris. 1920/1995. *Jailed for Freedom: American Women Win the Vote,* ed. Carol O'Hare. Troutdale, OR: New Sage Press, 18–19.

10. Source: Library of Congress. George Grantham Bain Collection. Original caption reads: Inez Milholland Boissevain, wearing white cape, seated on white horse at the National American Woman Suffrage Association parade, March 3, 1913, Washington, D.C. LC-DIG-ppmsc-00031 (digital file from original photograph) LC-USZ62-77359 goo.gl/6oKi9p

11. Ibid., 19.

12. Adams, Katherine H., and Michael L. Keene. 2007. *Alice Paul and the American Suffrage Campaign.* Champaign: University of Illinois Press, 206–208.

13. Ibid., 211.
14. goo.gl/6dKTtG
15. Ibid.
16. The Wikipedia entry for "Women's Suffrage" has a complete list of every country and when they legalized the franchise for women: goo.gl/CJEj2Q
17. goo.gl/YLXph
18. goo.gl/Otxf1v
19. Murray, Sara. 2013. "BM's Barra a Breakthrough." *Wall Street Journal*, December 11, B7.
20. goo.gl/EqBHZs
21. Wang, Wendy, Kim Parker, and Paul Taylor. 2013. "Breadwinner Moms." Pew Research, Social and Demographic Trends, May 29. goo.gl/jtCac See also Rampell, Catherine. 2013. "US Women on the Rise as Family Breadwinner." *New York Times*, May 29. goo.gl/o9igft
22. Kumar, Radha. 1993. *The History of Doing: An Account of Women's Rights and Feminism in India*. Bhayana Neha, 2011. "Indian Men Lead in Sexual Violence, Worst on Gender Equality." *Times of India*, March 7.
23. Daniel, Lisa. 2012. "Panetta, Dempsey Announce Initiatives to Stop Sexual Assault." *American Forces Press Service*, April 16.
24. Botelho, Greg, and Marlena Baldacci. 2014. "Brigadier General Accused of Sex Assault Must Pay over $20,000; No Jail Time." *CNN*. goo.gl/EqBHZs
25. Planty, Michael, Lynn Langton, Christopher Krebs, Marcus Berzofsky, and Hope Smiley-McDonald. 2013. "Female Victims of Sexual Violence, 1994–2010." US Department of Justice. Office of Justice Programs. Bureau of Justice Statistics, March. goo.gl/7GUWXp
26. A 2014 report issued by the White House Council on Women and Girls confirmed that most rape victims know their assailants, and that poor, homeless, and minority women are especially at risk. goo.gl/J3ABNW
27. Yung, Corey Rayburn. 2014. "How to Lie with Rape Statistics: America's Hidden Rape Crisis." *Iowa Law Review*, 99, 1197–1255.
28. Ibid., 1240.
29. According to the Rape, Abuse, & Incest National Network (RAINN), the largest and most influential activist organization in America fighting all forms of sexual violence, "In the last few years, there has been an unfortunate trend towards blaming 'rape culture' for the extensive problem of sexual violence on campus. While it is helpful to point out the systemic barriers to addressing the problem, it is important not to lose sight of a simple fact: Rape is caused not by cultural factors but by the conscious decisions, of a small percentage of the community, to commit a violent crime." RAINN recommends that rape should be treated as a serious crime and prosecuted as such, rather than bypass the law and allow internal campus judicial boards to whip up moral panics on college campuses that "has the paradoxical effect of making it harder to stop sexual violence, since it removes the focus from the individual at fault, and seemingly mitigates personal responsibility for his or her own actions." See Kitchens, Caroline. 2014. "It's Time to End 'Rape Culture' Hysteria," *Time*, March 20. goo.gl/lWffXq See also Hamblin, James. 2014. "How Not to Talk About a Culture of Sexual Assault." *Atlantic*, March 29. goo.gl/mwqxiJ MacDonald, Heather. 2008. "The Campus Rape Myth." *City Journal*, Winter, 18, no. 1. goo.gl/dDuaR
30. Buss, David. 2003. *The Evolution of Desire: Strategies of Human Mating*. New York:

Basic Books, 266. For a general overview of evolutionary psychology see Buss, David. 2011. *Evolutionary Psychology: The New Science of the Mind*. New York: Pearson.

31. Hrdy, Sara. 2000. "The Optimal Number of Fathers: Evolution, Demography, and History in the Shaping of Female Mate Preference." *Annals of the New York Academy of Science*, 907, 75–96.

32. Goetz, Aaron T., Todd K. Shackelford, Steven M. Platek, Valerie G. Starratt, and William F. McKibbin. 2007. "Sperm Competition in Humans: Implications for Male Sexual Psychology, Physiology, Anatomy, and Behavior." *Annual Review of Sex Research*, 18, no. 1, 1–22.

33. Scelza, Brooke A. 2011. "Female Choice and Extra-Pair Paternity in a Traditional Human Population." *Biology Letters*, December 23, 7, no. 6, 889–891.

34. Larmuseau, M. H. D., J. Vanoverbeke, A. Van Geystelen, G. Defraene, N. Vander-heyden, K. Matthys, T. Wenseleers, and R. Decorte. 2013. "Low Historical Rates of Cuckoldry in a Western European Human Population Traced by Y-Chromosome and Genealogical Data." *Proceedings of the Royal Society B*, December, 280, no. 1772.

35. Anderson, Kermyt G. 2006. "How Well Does Paternity Confidence Match Actual Paternity?" *Current Anthropology*, 47, no. 3, June, 513–520.

36. Baker, R. Robin, and Mark A. Bellis. 1995. *Human Sperm Competition: Copulation, Masturbation, and Infidelity*. London: Chapman & Hall.

37. Anderson, 2006, 516.

38. Personal correspondence, December 17, 2013.

39. Pillsworth, Elizabeth, and Martie Haselton. 2006. "Male Sexual Attractiveness Predicts Differential Ovulatory Shifts in Female Extra-Pair Attraction and Male Mate Retention." *Evolution and Human Behavior*, 27, 247–258.

40. Buss, David. 2001. *The Dangerous Passion: Why Jealousy Is as Necessary as Love and Sex*. New York: Free Press.

41. Laumann, E. O., J. H. Gagnon, R. T. Michael, and S. Michaels. 1994. *The Social Organization of Sexuality: Sexual Practices in the United States*. Chicago: University of Chicago Press.

42. Tafoya, M. A., and B. H. Spitzberg. 2007. "The Dark Side of Infidelity: Its Nature, Prevalence, and Communicative Functions." In B. H. Spitzberg and W. R. Cupach, eds., *The Dark Side of Interpersonal Communication*, 2nd ed., 201–242. Mahwah, NJ: Lawrence Erlbaum Associates.

43. Gangestad, S. W., and Randy Thornhill. 1997. "The Evolutionary Psychology of Extra-Pair Sex: The Role of Fluctuating Asymmetry." *Evolution and Human Behavior*, 18, no. 28, 69–88.

44. Buss, David. 2002. "Human Mate Guarding." *NeuroEndocrinology Letters*, December, 23, no. 4, 23–29.

45. Schmitt, D. P., and David M. Buss. 2001. "Human Mate Poaching: Tactics and Temptations for Infiltrating Existing Mateships." *Journal of Personality and Social Psychology*, 80, 894–917.

46. Schmitt, D. P., L. Alcalay, J. Allik, A. Angleitner, L. Ault, et al. 2004. "Patterns and Universals of Mate Poaching Across 53 Nations: The Effects of Sex, Culture, and Personality on Romantically Attracting Another Person's Partner." *Journal of Personality and Social Psychology*, 86, 560–584.

47. Kellerman, A. L., and J. A. Mercy. 1992. "Men, Women, and Murder: Gender-Specific Differences in Rates of Fatal Violence and Victimization." *Journal of Trauma*, July, 33, no. 1, 1–5. www.ncbi.nlm.nih.gov/pubmed/1635092

48. UN Office on Drugs and Crime. 2011. *Global Study on Homicide: Trends, Contexts, Data*. goo.gl/Hz2ie

49. Williamson, Laura. 1978. "Infanticide: An Anthropological Analysis." In M. Kohl, ed., *Infanticide and the Value of Life*. Buffalo, NY: Prometheus Books.

50. See, for example, any of my debates with the theologian and religious philosopher Doug Geivett.

51. Quoted in Milner, Larry. 2000. *Hardness of Heart, Hardness of Life: The Stain of Human Infanticide*. Lanham, MD: University Press of America.

52. Daly, Martin, and Margo Wilson. 1988. *Homicide*. New York: Aldine De Gruyter.

53. Ranke-Heinemann, Uta. 1991. *Eunuchs for the Kingdom of Heaven: Women, Sexuality and the Catholic Church*. New York: Penguin.

54. Milner, 2000.

55. Deschner, Amy, and Susan A. Cohen. 2003. "Contraceptive Use Is Key to Reducing Abortion Worldwide." *The Guttmacher Report on Public Policy*, October, 6, no. 4. goo.gl/Ovgge See also Marston, Cicely, and John Cleland. 2003. "Relationships Between Contraception and Abortion: A Review of the Evidence." *International Family Planning Perspectives*, March, 29, no. 1, 6–13.

56. Ibid.

57. Senlet, Pinar, Levent Cagatay, Julide Ergin, and Jill Mathis. 2001. "Bridging the Gap: Integrating Family Planning with Abortion Services in Turkey." *International Family Planning Perspectives*, June, 27, no. 2.

58. Kohler, Pamela K., Lisa E. Manhart, and William E. Lafferty. 2008. "Abstinence-Only and Comprehensive Sex Education and the Initiation of Sexual Activity and Teen Pregnancy." *Journal of Adolescent Health*, April, 42, no. 4, 344–351.

59. Herring, Amy H., Samantha M. Attard, Penny Gordon-Larsen, William H. Joyner, and Carolyn T. Halpern. 2013. "Like a Virgin (mother): Analysis of Data from a Longitudinal, US Population Representative Sample Survey." *British Medical Journal*, December 17. 347, goo.gl/bkMVnJ

60. Nelson, Charles A., Nathan A. Fox, and Charles H. Zeanah. 2014. *Romania's Abandoned Children: Deprivation, Brain Development, and the Struggle for Recovery*. Cambridge, MA: Harvard University Press.

61. Deschner and Cohen. 2003. op cit.

62. goo.gl/i1Hc

63. See, for example, the *Amici Curiae Brief in Support of Appellees*. 1988. *William L. Webster et al., Appellants, v. Reproductive Health Services et al., Appellees*.

64. See, for example, Pleasure, J. R., M. Dhand, and M. Kaur. 1984. "What Is the Lower Limit of Viability?" *American Journal of Diseases of Children*, 138, 783; R. D. G. Milner and R. V. Beard. 1984. "Limit of Fetal Viability." *Lancet*, 1, 1079; B. L. Koops, L. J. Morgan and F. C. Battaglia. 1982. "Neonatal Mortality Risk in Relation to Birth Weight and Gestational Age: Update." *Journal of Pediatrics*, 101, 969–977.

65. Beddis, I. R., P. Collins, S. Godfrey, N. M. Levy, and M. Silverman. 1979. "New Technique for Servo-Control of Arterial Oxygen Tension in Preterm Infants." *Archives of Disease in Childhood*, 54, 278–280.

66. Flower, M. 1989. "Neuromaturation and the Moral Status of Human Fetal Life." In E. Doerr and J. Prescott, eds., *Abortion Rights and Fetal Personhood*. Centerline Press, 65–75.

67. "4-Sided Battle in Court for Child." 1914. *Los Angeles Times*, October 31.

68. Most of this story has been carefully documented by Ann Marie Batesole, a private detective and my cousin—our grandmother was Christine, Aunt Fanc's mother.

69. Source: Author's collection.

7: A MORAL SCIENCE OF GAY RIGHTS

1. Friends of the gay rights movement can be found in the Episcopal Church (which elected Gene Robinson as its first gay bishop in 2003), the Unitarian Universalists, the United Church of Canada, Conservative and Reform Judaism, Native American religions, some liberal Hindus and Buddhists, Wiccans, and others. Though most religions have come late to the table, photographic evidence shows that at least some ministers have been marching in gay pride parades since the early 1970s, with signs that say things like, "Ministers for human rights" and "Jesus does not condemn gay people." goo.gl/FYXGOS-francisco-pride-parade-the-first-two-decades-photos/

2. getequal.org/press/ The First Annual "Anita Bryant Award for Unbridled and Unparalleled Bigotry," given by GetEQUAL—a national LGBT civil rights organization—was awarded to Maggie Gallagher, chairman of the antigay National Organization for Marriage, accompanied by these words: "At a time when Americans overwhelmingly support marriage equality, it takes a very special person like Ms. Gallagher to stand up and fight for discrimination and bigotry."

3. goo.gl/Tz3xWO

4. goo.gl/dX66lF

5. For an excellent documentary that shows how the Mormons turned words from the pulpit into policy, see *8—The Mormon Proposition*. goo.gl/iZLsl

6. goo.gl/muiQ9a

7. goo.gl/NkxuBd

8. goo.gl/T3BKc

9. Hamer, Dean, and P. Copeland. 1994. *The Science of Desire: The Search for the Gay Gene and the Biology of Behavior*. New York: Simon & Schuster; Baily, Michael. 2003. *The Man Who Would Be Queen: The Science of Gender-Bending and Transsexualism*. Washington, DC: National Academies Press; LeVay, Simon. 2010. *Gay, Straight, and the Reason Why: The Science of Sexual Orientation*. New York: Oxford University Press.

10. goo.gl/5CLnXm

11. Trudeau was minister of justice at the time and he spoke these words in defense of bill C-150, which, among other things (including the legalization of contraception and abortion), aimed to decriminalize homosexuality. The bill passed, 149 votes to 55. goo.gl/iFYe5n

12. See David K. Johnson. 2004. *The Lavender Scare: The Cold War Persecution of Gays and Lesbians in the Federal Government*. Chicago: University of Chicago Press.

13. Lingeman, Richard R. 1973. "There Was Another Fifties." *New York Times Magazine*, June 17, 27.

14. goo.gl/WGzw5D

15. Davis, Kate, and David Heilbroner, directors. 2010. *Stonewall Uprising*. Documentary. Based on the book: Carter, David. 2004. *Stonewall: The Riots That Sparked the Gay Revolution*. New York: St. Martin's Press.

16. From *Stonewall Uprising*.

17. Ibid. Transcript: goo.gl/1jnTdM

18. Wolf, Sherry. "Stonewall: The Birth of Gay Power." *International Socialist Review*. goo.gl/F1otuO

19. goo.gl/wjVLB See also McCormack, Mark. 2013. *The Declining Significance of Homophobia*. Oxford, UK: Oxford University Press.

20. goo.gl/xKWn4X The National School Climate Survey says that homophobic remarks are on the decline and that there is a "decrease in victimization based on

sexual orientation." glsen.org/sites/default/files/2011%20National%20School%20 Climate%20Survey%20Full%20Report.pdf

21. Nicholson, Alexander. 2012. *Fighting to Serve: Behind the Scenes in the War to Repeal "Don't Ask, Don't Tell."* Chicago: Chicago Review Press, 12.

22. www.ncbi.nlm.nih.gov/pubmed/22670652 Canadian Olympian and gold-medal winner Mark Tewksbury explains why there are few openly gay athletes: www.you tube.com/watch?v=XS3jevs3l5o

23. Associated Press. 2013. "Obama Names Billie Jean King as one of Two Gay Sochi Olympic Delegates." December 17. www.theguardian.com/sport/2013/dec/18/ obama-names-gay-delegates-sochi-olympics

24. Goessling, Ben. 2014. "86 Percent OK with Gay Teammate." ESPN.com, February 17. goo.gl/W27892

25. Beech, Richard. 2014. "Thomas Hitzlsperger Comes Out as Being Gay." *Mirror,* January 8. goo.gl/j4pq9L

26. From *Stonewall Uprising.*

27. New Mexico: Santos, Fernanda. 2013. "New Mexico Becomes 17th State to Allow Gay Marriage." *New York Times,* December 19. goo.gl/hNv2WY

28. 2014. "Court Ruling: Germany Strengthens Gay Adoption Rights." *Spiegel Online International,* February 19. goo.gl/xShdV

29. Source: Pinker, 2011, 452, combines surveys from Gallup, 2001, 2008, and 2010, and the General Social Survey: goo.gl/5yHDK6 Gallup 2001: "American Attitudes Toward Homosexuality Continue to Become More Tolerant." Gallup 2002: "Acceptance of Homosexuality: A Youth Movement." Gallup 2008: "America Evenly Divided on Morality of Homosexuality."

30. Source: Pew Research Center's Forum on Religion and Public Life. June 2013. "Changing Attitudes on Gay Marriage." goo.gl/vkpbm2

31. Ibid.

32. Magnier, Mark, and Tanvi Sharma. 2013. "India Court Makes Homosexuality a Crime Again." *Los Angeles Times,* December 11, A7.

33. Cowell, Alan. 2013. "Ugandan Lawmakers Pass Measure Imposing Harsh Penalties on Gays." *New York Times,* December 20, A8. goo.gl/IN9cLL

34. The law has been expanded so that people from countries that allow same-sex marriage may no longer adopt Russian children. goo.gl/HClnyL

35. goo.gl/2Xh5Pi

36. goo.gl/aT8Z9T

37. goo.gl/SfvA37

38. goo.gl/ZQQiZ

39. goo.gl/lh60Na

40. "Married Gays to Tour Drought-hit Countries." 2014. *The Daily Mash.* January 20. goo.gl/gicu0I

41. goo.gl/LRB6v

42. Jorge Valencia, president and executive director of the Trevor Project from 2001 to 2006, as quoted in the award-winning documentary *For the Bible Tells Me So*: goo.gl/9GYGaN

43. *For the Bible Tells Me So.* Documentary. 2007. Written by Daniel G. Karslake and Nancy Kennedy. Directed by Daniel G. Karslake. goo.gl/9GYGaN

44. Spitzer's study: goo.gl/QkdK4H

45. goo.gl/my7v8h goo.gl/MNHID

46. goo.gl/B2vs5

47. goo.gl/QSPvO

48. goo.gl/mxWaO
49. goo.gl/nQObLY
50. goo.gl/XgY9JG
51. Statement made by Pope Francis on September 29, 2013, reported throughout the media, for example: goo.gl/tMfXhz
52. Statement made by Pope Francis on July 29, 2013, on a plane from Brazil to the Vatican during a wide-ranging interview. goo.gl/EVDvlW
53. September 29, 2013, statement.
54. Klemesrud, Judy. 1977. "Equal Rights Plan and Abortion Are Opposed by 15,000 at Rally." *New York Times*, November 20, A32.
55. King, Neil. 2013. "Evangelical Leader Preaches a Pullback from Politics, Culture Wars." *Wall Street Journal*, October 22, A1, 14.
56. Rauch, Jonathan. 2013. "The Case for Hate Speech." *Atlantic*. October 23. goo.gl /vNJYRF
57. Rauch, 2013. For a fuller defense of this position see Rauch, Jonathan. 2004. *Gay Marriage: Why It Is Good for Gays, Good for Straights, and Good for America*. New York: Henry Holt.
58. Fry tells the story here in his excellent two-part series on homosexuality and homophobia: goo.gl/UpCQCO
59. goo.gl/jUoUNT
60. As Savage claims here: goo.gl/AU7jWg
61. *It Gets Better*. itgetsbetter.org/ Note Dan Savage's remarks from 19:45 on the subject of the *It Gets Better* project. goo.gl/8DzP9Q
62. Wilcox, Ella Wheeler. 1914/2012. "Protest," *Poems of Problems*. Chicago: W. B. Conkey, 154.

8: A MORAL SCIENCE OF ANIMAL RIGHTS

1. Darwin, Charles. 1859. *On the Origin of Species by Means of Natural Selection*. London: John Murray, 488–489.
2. It is a myth that Darwin discovered evolution or its mechanism of natural selection while in the Galápagos. After Darwin returned home to England he began his notebooks outlining his ideas that would eventually develop into the full-blown theory, but that was a year after he visited the islands. The myth was debunked by Frank Sulloway, whose historical reconstruction of the development of Darwin's evolutionary thinking can be found in a number of papers: Sulloway, Frank. 1982. "Darwin and His Finches: The Evolution of a Legend." *Journal of the History of Biology*, 15, 1–53; Sulloway, Frank. 1982. "Darwin's Conversion: The *Beagle* Voyage and Its Aftermath." *Journal of the History of Biology*, 15, 325–396; Sulloway, Frank. 1984. "Darwin and the Galápagos." *Biological Journal of the Linnean Society*, 21, 29–59.
3. Cruz, F., V. Carrion, K. J. Campbell, C. Lavoie, and C. J. Donlan. 2009. "Bio-Economics of Large-Scale Eradication of Feral Goats from Santiago Island, Galápagos." *Wildlife Management*, 73, 191–200.
4. Weisman, Alan. 2007. *The World Without Us*. New York: St. Martin's Press. See also his sequel: Weisman, Alan. 2013. *Countdown: Our Last Best Hope for a Future on Earth?* Boston: Little, Brown.
5. After Darwin and his shipmates ate their tortoises they threw the carapaces overboard, which might have delayed the development of Darwin's theory in that he had a harder time piecing together an evolutionary tree for the tortoises since he

could not recall on which islands they were found. See Shermer, Michael. 2006. *Why Darwin Matters*. New York: Henry Holt/Times Books.

6. The animal rights scholar and attorney Steven M. Wise makes a similar argument in his book *Drawing the Line*, in which he argues for four categories of animal rights: Category 1 includes species "who clearly possess sufficient autonomy for basic liberty rights," including the great apes; Category 2 includes species that might qualify for basic legal rights, depending on what other criteria we might consider; Category 3 includes species for which we do not have enough knowledge to determine what rights they should have; and Category 4 includes those species that lack sufficient autonomy for basic liberty rights. Wise, S. M. 2002. *Drawing the Line: Science and the Case for Animal Rights*. Boston: Perseus Books, 241. See also Wise, S. M. 2000. *Rattling the Cage: Toward Legal Rights for Animals*. Boston: Perseus.

7. Marino, L. 1988. "A Comparison of Encephalization Between Odontocete Cetaceans and Anthropoid Primates." *Brain, Behavior, and Evolution*, 51, 230.
Ridgway, S. H. 1986. "Physiological Observations on Dolphin Brains." In R. J. Schusterman et al., eds., *Dolphin Cognition and Behavior: A Comparative Approach*. Hillsdale, NJ: Lawrence Erlbaum Associates, 32–33.
Herman, L. M., and P. Morrel-Samuels. 1990. "Knowledge Acquisition and Asymmetry Between Language Comprehension and Production: Dolphins and Apes as General Models for Animals." In M. Bekoff and D. Jamieson, eds., *Interpretation and Explanation in the Study of Animal Behavior*. Boulder, CO: Westview Press.

8. Reiss, Diana. 2012. *The Dolphin in the Mirror: Exploring Dolphin Minds and Saving Dolphin Lives*. Boston: Mariner Books.

9. goo.gl/dG2xK

10. Gregg, Justin. 2013. "No, Flipper Doesn't Speak Dolphinese." *Wall Street Journal*, December 21, C3; Gregg, Justin. 2013. *Are Dolphins Really Smart? The Mammal Behind the Myth*. New York: Oxford University Press.

11. Ibid. My friend Jack Horner, the famed dinosaur paleontologist, tells me that we have an additional bias in assuming brains will be in heads. He is imaging dino skeletons and constructing dino anatomy models that show sizable neural control centers in the pelvic region of some of the largest dinosaurs, which makes sense given how far it is from head to tail, and how adaptive it could be to distribute your intelligence throughout the body rather than focus it all in the head. He sites experiments with chickens who have their heads cut off but can nonetheless survive and even right themselves after falling over, indicating that vestibular control over balance must be located somewhere other than the head.

12. Amsterdam, Beulah. 1972. "Mirror Self-Image Reactions Before Age Two." *Developmental Psychobiology* 5, no. 4, 297–305; Lewis, M., and J. Brooks-Gunn. 1979. *Social Cognition and the Acquisition of Self*. New York: Plenum Press, 296; Gopnik, Alison. 2009. *The Philosophical Baby: What Children's Minds Tell Us About Truth, Love, and the Meaning of Life*. New York: Farrar, Straus & Giroux.

13. Animal language is a contentious subject because of the questionable quality of much of the research. Koko, for example, has learned hundreds of symbolic language signs with which she can answer questions and attempt to deceive her handlers, and she has even apparently attempted to teach language signs to other gorillas. But this and other ape language research have been challenged by skeptics of animal language and cognition—we have published several articles in *Skeptic* magazine, for example, with one animal psychologist, Clive Wynne, suggesting this little test any of us can conduct (referencing a bonobo named Kanzi): "Next time you see Kanzi or one of his kind on a television documentary, turn down the

sound so you can just watch what he is doing without interpretation from the ape's trainers. See if that really appears to be language. Somewhere in the history of our kind there must have been the first beings who could rearrange tokens to create new meanings, to distinguish *Me Banana* from *Banana Me*. But the evidence from many years of training apes to press buttons or sign in ASL is that this must have happened sometime after we split off from chimps, bonobos, and gorillas. Since then we have been talking to ourselves." Wynne, Clive. 2007. "Aping Language: A Skeptical Analysis of the Evidence for Nonhuman Primate Language." *Skeptic*, 13, no. 4, 10–14.

14. Plotnik, Joshua M., and Frans de Waal. 2014. "Asian Elephants (*Elephas maximus*) Reassure Others in Distress." *PeerJ*, 2: e278. peerj.com/articles/278/

15. Smet, Anna F., and Richard W. Byrne. 2013. "African Elephants Can Use Human Pointing Cues to Find Hidden Food." *Current Biology*, 23, 1–5.

16. Quoted in Collins, Katie. 2013. "Study: Elephants Found to Understand Human Pointing Without Training." www.wired.co.uk/news/archive/2013-10/10/elephant-pointing

17. Smet and Byrne, op cit.

18. The dates are based on DNA extracted from eighteen prehistoric dog bones found in caves in Eurasia and the New World and compared to the DNA from forty-nine modern wolves and from seventy-seven different modern dog breeds.

19. Thalmann, O., et al. 2013. "Complete Mitochrondrial Genomes of Ancient Canids Suggest a European Origin of Domestic Dogs." *Science*, November 15, 342 no. 6160, 871–874. Older studies put the date of dog domestication at about thirteen thousand years ago: Savolainen, P., Y. Zhang, J. Luo, J. Lundeberg, and T. Leitner. 2002. "Genetic Evidence for an East Asian Origin of Domestic Dogs." *Science*, November 22, 298, 1610–1612. Leonard, J. A., R. K. Wayne, J. Wheeler, R. Valadez, S. Guillén, and C. Vilá. 2002. "Ancient DNA Evidence for Old World Origin of New World Dogs." *Science*, November 22, 298, 1613–1615.

20. Teglas, E., A. Gergely, K. Kupan, A. Miklosi, and J. Topal. 2012. "Dogs' Gaze Following Is Tuned to Human Communicative Signals." *Current Biology* 22, 209–212.
    B. Hare and M. Tomasello. 2005. "Human-like Social Skills in Dogs?" *Trends in Cognitive Science* 9, 439–444.
        Hare, B., M. Brown, C. Williamson, and M. Tomasello. 2002. "The Domestication of Social Cognition in Dogs." *Science*, November 22, 298, 1634–1636.

21. Berns, Gregory, Andrew Brooks, and Mark Spivak. 2012. "Functional MRI in Awake Unrestrained Dogs." *PLoS ONE* 7, no. 5.

22. Ibid.

23. Berns, Gregory. 2013. *How Dogs Love Us: A Neuroscientist and His Adopted Dog Decode the Canine Brain*. New York: New Harvest, 226–227.

24. Berns, Gregory. 2013. "Dogs Are People Too." *New York Times*, October 6, SR5.

25. Ibid.

26. Marsh, James, director. 2011. *Project Nim*. A documentary film.

27. It was, Terrace determined, a "Clever Hans" effect in primates. Clever Hans was a horse who became world famous by performing arithmetic problems given to him by his trainer, one Mr. von Osten, the answers of which Hans could tap out with his hooves. But in 1907 the German psychologist Oskar Pfungst determined that the horse was picking up on the unconscious body language signals of his human trainer—when Hans reached the correct number of hoof taps the trainer would shift or move in some detectable manner that signaled to Hans to stop tapping his foot. Now known as the Clever Hans effect, it is, in fact, another experiment in cross-species

social signaling for which we should include horses, along with dogs and elephants. Pfungst, O. 1911. *Clever Hans (The Horse of Mr. von Osten): A Contribution to Experimental Animal and Human Psychology*, trans. C. L. Rahn. New York: Henry Holt.

28. Marsh's film shuttles between talking-head interviews with all the major players in the project (including Terrace himself) and original footage shot throughout the experiment.

29. To be fair, the trainers and handlers in the film seem to be caring, loving people who did the best they could under the circumstances, but they had little say in the long-term course of Nim's existence. Terrace, by contrast, who ran the show and called the shots, comes across as an almost psychopathic manipulator, an alpha male egotist who, in his own words on camera, spoke of Nim's suffering in cold, clinical language.

30. Here is a short sampling of the rich literature on animal cognition and emotion:

Bekoff, M., ed. 2000. *The Smile of a Dolphin: Remarkable Accounts of Animal Emotions*. New York: Crown Books.

Bonvillian, J. D., and F. G. P. Patterson. 1997. "Sign Language Acquisition and the Development of Meaning in a Lowland Gorilla." In C. Mandell and A. McCabe, eds., *The Problem of Meaning: Behavioral and Cognitive Perspectives*. Amsterdam: Elsevier.

Byrne, R. 1995. *The Thinking Ape: Evolutionary Origins of Intelligence*. Oxford, UK: Oxford University Press.

Dawkins, M. S. 1993. *Through Our Eyes Only: The Search for Animal Consciousness*. New York: W. H. Freeman.

Galdikas, B. M. F. 1995. *Reflections of Eden: My Years with the Orangutans of Borneo*. Boston: Little, Brown.

Griffin, D. R. 2001. *Animal Minds: Beyond Cognition to Consciousness*. Chicago: University of Chicago Press.

Miles, H. L. 1994. "ME CHANTEK: The Development of Self-Awareness in a Signing Orangutan." In S. T. Parker et al., eds., *Self-Awareness in Animals and Humans: Developmental Perspectives*. Cambridge, UK: Cambridge University Press.

Miles, H. L. 1996. "Simon Says: The Development of Imitation in an Encultured Orangutan." In *Reaching into Thought: The Minds of the Great Apes*, ed. A. E. Russon et al. Cambridge, UK: Cambridge University Press.

Moussaieff, M., and S. McCarthy. 1995. *When Elephants Weep: The Emotional Lives of Animals*. New York: Delacorte Press.

Parker, S. T., and M. L. McKinney, eds. 1994. *Self-Awareness in Animals and Humans: Developmental Perspectives*. Cambridge, UK: Cambridge University Press.

———. 1999. *The Mentalities of Gorillas and Orangutans*. Cambridge, UK: Cambridge University Press.

Patterson, F. G. P. 1993. "The Case for the Personhood of Gorillas." In P. Cavalieri and P. Singer, eds., *The Great Ape Project: Equality Beyond Humanity*. New York: St. Martin's Press.

Patterson, F. G. P., and E. Linden. 1981. *The Education of Koko*. New York: Holt, Rinehart, & Winston.

Pepperberg, I. 1999. *The Alex Studies: Cognitive and Communicative Abilities of Parrots*. Cambridge, MA: Harvard University Press.

Pryor, K., and K. S. Norris, eds. 2000. *Dolphin Societies: Discoveries and Puzzles*. Chicago: University of Chicago Press.

Reiss, D., and L. Marino. 2001. "Mirror Self-Recognition in the Bottlenose Dolphin: A Case of Cognitive Convergence." *Proceedings of the National Academy of Sciences*, 8, 5937–5942.

    Rogers, L. J. 1998. *Minds of Their Own: Thinking and Awareness in Animals.* Boulder, CO: Westview Press.

    Ryder, R. D. 1989. *Animal Revolution: Changing Attitudes Toward Speciesism.* London: Basil Blackwell.

    Sorabji, R. 1993. *Animal Minds and Human Morals: The Origin of the Western Debate*. Ithaca, NY: Cornell University Press.

31. Bentham, Jeremy. 1823. *Introduction to the Principles of Morals and Legislation*, Chap. XVII, for 122. See full text copy: www.econlib.org/library/Bentham/bnth PML18.html

32. Singer, Peter. 1989. "All Animals Are Equal." In Tom Regan and Peter Singer, eds., *Animal Rights and Human Obligations*. Englewood Cliffs, NJ: Prentice Hall, 148–162. See Singer's book-length treatment of these arguments: Singer, Peter. 1975. *Animal Liberation: Towards an End to Man's Inhumanity to Animals*. New York: Harper & Row.

33. Personal correspondence, October 3, 2013. The term "speciesism" has been in use for some time. Singer used it in his 1989 paper, and there he credits Richard Ryder. See Ryder, Richard. 1971. "Experiments on Animals" in Stanley and Roslind Godlovitch and John Harris, eds., *Animals, Men and Morals*. London: Victor Gollancz.

34. Cohen, Carl, and Tom Regan. 2001. *The Animal Rights Debate*. Lanham, MD: Rowman & Littlefield.

35. Personal correspondence, October 3, 2013.

36. Morell, Virginia. 2013. *Animal Wise: The Thoughts and Emotions of Our Fellow Creatures*. New York: Crown, 261.

37. Grandin, Temple, and Catherine Johnson. 2006. *Animals in Translation: Using the Mysteries of Autism to Decode Animal Behavior*. New York: Harcourt. Grandin's TED talk is available here: goo.gl/KBan

38. Personal correspondence, October 3, 2013.

39. Pollan, Michael. 2007. *The Omnivore's Dilemma*. New York: Penguin; Pollan, Michael. 2009. *In Defense of Food: An Eater's Manifesto*. New York: Penguin.

40. Scully, Matthew. 2003. *Dominion: The Power of Man, the Suffering of Animals, and the Call to Mercy*. New York: St. Martin's Press.

41. Pollan, Michael. 2002. "An Animal's Place." *New York Times Magazine*, November 10. goo.gl/OlsKp After wrestling with Peter Singer's arguments in *Animal Liberation* while devouring a medium-rare rib eye steak at a fine-dining steakhouse, Pollan reflected, "This is where I put down my fork. If I believe in equality, and equality is based on interests rather than characteristics, then either I have to take the interests of the steer I'm eating into account or concede that I am a speciesist. For the time being, I decided to plead guilty as charged. I finished my steak."

42. Scully, 2003, 303–304.

43. Pinker, 2011, op. cit., 509.

44. On July 16, 1964, in his speech accepting the Republican presidential nomination, Barry Goldwater gave voice to one of the most memorable one-liners in the history of politicking: "Extremism in the defense of liberty is no vice. Moderation in the pursuit of justice is no virtue." I hold that in most cases this is precisely backward.

45. Carlton, Jim. 2013. "A Winter Without Walruses." *Wall Street Journal*, October 4, A4.

46. Patterson, Charles. 2002. *Eternal Treblinka: Our Treatment of Animals and the Holocaust*. Herndon, VA: Lantern Books.

47. Singer, Isaac Bashevis. 1980. *The Seance and Other Stories*. New York: Farrar, Straus & Giroux, 270.
48. Shermer, Michael. 2000. *Denying History*. Berkeley: University of California Press.
49. Personal correspondence, October 7, 2013.
50. *Guide for the Care and Use of Laboratory Animals*, 8th ed. 2011. Institute for Laboratory Animal Research, Division on Earth and Life Studies, National Research Council of the National Academies. Washington, DC: National Academies Press, 123–124.
51. *My Cousin Vinny* script. 1992. goo.gl/S9ucO9
52. goo.gl/rhsP2U
53. Conover, Ted. 2013. "The Stream: Ted Conover Goes Undercover as a USDA Meat Inspector." *Harper's*. April 15. goo.gl/6sP9M
54. *Earthlings*. 2005. Directed by Shaun Monson. Narrated by Joaquin Phoenix. Available for viewing for free on many sites online, such as goo.gl/PhSys7
55. Much of the narration was taken from animal rights works by Peter Singer, Tom Regan, and others. A transcript of *Earthlings* is available here: goo.gl/o8ls25
56. Emerson, Ralph Waldo. 1860. "Fate." In his *The Conduct of Life*. goo.gl/lxcf25
57. Moral progress: On March 31, 2014, the United Nations' International Court of Justice ruled that Japan must discontinue its annual whale hunt, rejecting the country's rationalization for this barbarous act that it was done in the name of scientific research. goo.gl/bgkdD9
58. Dennett, Daniel. 1997. *Kinds of Minds*. New York: Basic Books.
59. Personal correspondence, October 3, 2013.
60. Wise, Steven M. 2002. *Drawing the Line: Science and the Case for Animal Rights*. Boston: Perseus Books.
61. Davis, D. B. 1984. *Slavery and Human Progress*. New York: Oxford University Press.
62. Francione, Gary. 2006. "The Great Ape Project: Not So Great." In *Animal Rights: The Abolition Approach*. goo.gl/ojbptB
63. "India Bans Captive Dolphin Shows as 'Morally Unacceptable.'" 2013. *Environment News Service*. May 20.
64. Genesis 1, 28, King James Version.
65. Aquinas, Thomas. Book 3-2, chap. CVII. "That Rational Creatures Are Governed for Their Own Sake, and Other Creatures, as Directed to Them." *The Summa Contra Gentiles*. goo.gl/JilOo9
66. Payne, J. L. 2004. *A History of Force: Exploring the Worldwide Movement Against Habits of Coercion, Bloodshed, and Mayhem*. Sandpoint, ID: Lytton Publishing.
67. Spencer, Colin. 1995. *The Heretic's Feast: A History of Vegetarianism*. Lebanon, NH: University Press of New England, 215.
68. Moore, David. 2003. "Public Lukewarm on Animal Rights." Gallup News Service, May 21. goo.gl/9hj5lP
69. Ibid.
70. goo.gl/8aYvp
71. goo.gl/vfY4f
72. Updated from Pinker, 2011, 471.
73. Bittman, Mark. 2012. "We're Eating Less Meat. Why?" *New York Times*, January 10. goo.gl/Tp2Si4
74. "Vegetarianism in the United States: A Summary of Quantitative Research." 2007. Humane Research Council. humaneresearch.org/content/vegetarianism-us -summary-quantitative-research

75. Newport, Frank. 2008. "Post-Derby Tragedy, 38% Support Banning Animal Racing." goo.gl/iEYxdN

76. US Fish and Wildlife Service. 2006. National Survey of Fishing, Hunting, and Wildlife-Associated Recreation.

77. goo.gl/mlvOUU

78. Miller, Gerri. 2011. "Animal Safety Was Spielberg's Top Concern on 'War Horse.'" goo.gl/mYYWtK Of course, Hollywood does have a reputation for championing liberal causes, and this was on display in 2013 when the city itself banned circuses and other entertainment acts involving wild and exotic animals, to protect them "from cruel and inhumane treatment." The ordinance followed others in the same vein in Los Angeles, including the banning of steel-jaw traps and animal testing of cosmetics in 1989, declawing animals in 2003, the retail sale of dogs and cats in 2010, and the sale of fur in 2011. Mullins, Alisa. 2013. "Why We, but Especially Elephants, Love West Hollywood." goo.gl/ml NH6h

79. Pollan, 2002, op cit.

80. For counterarguments to even this middle-ground position, and in general the ethics of eating, see Foer, Jonathan Safran. 2010. *Eating Animals*. Boston: Back Bay Books, and Singer, Peter, and Jim Mason. 2007. *The Ethics of What We Eat*. Emmaus, PA: Rodale Press.

81. *Free Range: A Short Documentary*. goo.gl/ANCvon The farm featured is Sunny Day Farms in Texas: www.sunnydayfarms.com/

82. Cooper, Rob. 2013. "Free Range Eggs Outsell Those from Caged Hens for First Time." *Daily Mail Online*. goo.gl/tpTXio

83. goo.gl/Ll64Bl

84. I have dined several times with Mackey and can assure readers that he practices what he preaches, even to the point of bringing his own salad dressing to one of the finest restaurants in Las Vegas, where we dine together every year at Freedomfest, the world's largest annual gathering of free market–loving libertarians.

85. For a sobering look at the business of food production see Robert Kenner's documentary *Food, Inc.*, which shows the extent to which the factory food industry goes to maintain the pastoral fantasy of an agrarian America, even as that America disappears off the landscape: Kenner, Robert, and Melissa Robledo. 2008. *Food, Inc.* Participant Media.

86. Lutz, Wolfgang, and Sergei Scherbov. 2008. "Exploratory Extension of IIASA's World Population Projections: Scenarios to 2300. goo.gl/4lQtEO

87. See, for example, Andras Forgacs's TED talk "Leather and Meat Without Killing Animals." goo.gl/AigbB8

88. Conover, op cit.

89. Augustine of Hippo, *Confessions*, 8, 17. The entire passage in which he is addressing God reads: "I in my great worthlessness had begged You for chastity, saying 'Grant me chastity and continence, but not yet.' For I was afraid that You would hear my prayer too soon, and too soon would heal me from disease of which I wanted satisfied rather than extinguished."

90. The original French sentence, from the final chapter of Hugo's book *Histoire d'un Crime (The History of a Crime)*, is: *"On resiste à l'invasion des armées; on ne resiste pas à l'invasion des idées."* goo.gl/Gfd2Y

91. Darwin, 1859, op cit., 489.

9: MORAL REGRESS AND PATHWAYS TO EVIL

1. Milgram, Stanley. 1969. *Obedience to Authority: An Experimental View*. New York: Harper & Row.
2. Interview with Phil Zimbardo conducted by the author on March 26, 2007.
3. Milgram, 1969.
4. The replication was conducted on October 8–9, 2009. Setup: See all our experiments at www.msnbc.msn.com, keywords "What Were You Thinking?" Or "Did You See That?" goo.gl/dEDmSl "What Were You Thinking?" goo.gl/KDTp06
5. A transcript of the complete NBC Dateline Special is available here: goo.gl/slM8FU
6. In screening this segment of the NBC special in public talks I am occasionally asked how we got this replication passed by an Institutional Review Board (IRB), which is required for scientific research. We didn't. This was for network television, not an academic laboratory, so the equivalent of an IRB was review by NBC's legal department, which approved it. This seems to surprise—even shock—many academics, until I remind them of what people do to one another on reality television programs in which subjects are stranded on a remote island and left to fend for themselves in various contrivances that resemble a Hobbesian world of a war of all against all.
7. Pinker, Steven. 2002. *The Blank Slate: The Modern Denial of Human Nature*. New York: Viking.
8. Milgram, 1969.
9. Ibid.
10. Burger, Jerry. 2009. "Replicating Milgram: Would People Still Obey Today?" *American Psychologist*, 64, 1–11.
11. The Trial of Adolf Eichmann, Session 95, July 13, 1961. goo.gl/YghtTS
12. Cesarani, David. 2006. *Becoming Eichmann: Rethinking the Life, Crimes, and Trial of a "Desk Murderer."* New York: Da Capo Press. Von Trotta, Margarethe, director. 2012. *Hannah Arendt*. Zeitgeist Films. See also Lipstadt, Deborah E. 2011. *The Eichmann Trial*. New York: Schocken.
13. Young, Robert. 2010. *Eichmann*. Regent Releasing, Here! Films. October.
14. Quoted in Goldhagen, Daniel Jonah. 2009. *Worse Than War: Genocide, Eliminationism, and the Ongoing Assault on Humanity*. New York: PublicAffairs, 158.
15. Lozowick, Yaacov. 2003. *Hitler's Bureaucrats: The Nazi Security Police and the Banality of Evil*. New York: Continuum, 279.
16. Quoted in Lifton, Robert Jay. 1989. *The Nazi Doctors: Medical Killing and the Psychology of Genocide*. New York: Basic Books, 337.
17. Ibid., *Genocide*. New York: Basic Books, 382–383.
18. Levi, Primo. 1989. *The Drowned and the Saved*. New York: Vintage Books, 56. For a thoughtful discussion on personality and morality see Doris, John Michael. 2002. *Lack of Character: Personality and Moral Behavior*. New York: Cambridge University Press.
19. Baumeister, R. F. 1997. *Evil: Inside Human Violence and Cruelty*. New York: Henry Holt.
20. Ibid., 379.
21. ———. 1990. "Victim and Perpetrator Accounts of Interpersonal Conflict: Autobiographical Narratives About Anger." *Journal of Personality and Social Psychology*, 59, no. 5, 994–1005.
22. Richardson, Lewis Fry. 1960. *Statistics of Deadly Quarrels*. Pittsburgh: Boxwood Press, xxxv. Translation of the French portion of Richardson's quote by Steven Pinker.

23. McMillan, Dan. 2014. *How Could This Happen?: Explaining the Holocaust*. New York: Basic Books, 213.

24. Quoted in Broszat, Martin. 1967. "Nationalsozialistiche Konzentrationslager 1933–1945." In H. Bucheim, ed., *Anatomie des SS-Staates*. 2 vols. Munich: Deutscher Taschenbuchverlag, 143.

25. goo.gl/2uNiMX

26. goo.gl/iOiOgA

27. Quoted in Snyder, L. 1981. *Hitler's Third Reich*. Chicago: Nelson-Hall, 29.

28. Quoted in Jäckel, Eberhard. 1993. *Hitler in History*. Lebanon, NH: Brandeis University Press/University Press of New England, 33.

29. Quoted in Snyder, 1981, 521.

30. Quoted in Friedlander, 1995, 97.

31. Ibid., 284.

32. This account also helps us understand another mystery—the "missing" order from Hitler to exterminate the Jews. In the opinion of my coauthor Alex Grobman and me, one of the reasons there exists no written order from Hitler is because he once authorized in writing the euthanasia of handicapped patients and this came back to haunt him with negative press. Furthermore, we know that as a general principle Hitler always tried not to sign orders himself. There is no order by Hitler, for example, to start the Second World War. Herein lies the key to understanding the contingent evolution of the extermination camps—they evolved out of concentration camps and work camps, utilizing techniques for mass murder developed for the euthanasia program, disguised to maximize secrecy and security.

33. Navarick, Douglas J. 2013. "Moral Ambivalence: Modeling and Measuring Bivariate Evaluative Processes in Moral Judgment." *Review of General Psychology*, 17, no. 4, 443–452.

34. See my chapter on animal rights in this book for my current moral position on such research, which has changed considerably since my days working in Doug's lab running experiments with rats and pigeons.

35. Janoff-Bulman, R., S. Sheikh, and S. Hepp. 2009. "Proscriptive versus Prescriptive Morality: Two Faces of Moral Regulation." *Journal of Personality and Social Psychology*, 96, 521–537.

36. Ibid.

37. Navarick, 2013, 444.

38. See Pinker, 2011, and his discussion with many examples of post–World War II Germans who have marched against militarism, nuclear weapons, the US war in Iraq, and many others, including the Parker Brothers German version of the board game Risk, which the German government tried to censor because the game involves players trying to conquer the world.

39. Klee, Ernst, Wili Dressen, Volker Riess, and Hugh Trevor-Roper. 1996. *"The Good Old Days": The Holocaust As Seen by Its Perpetrators and Bystanders*. New York: William S. Konecky Associates, 163–171.

40. Navarick, Douglas. 1979. *Principles of Learning: From Laboratory to Field*. Reading, MA: Addison-Wesley.

41. Breiter, Hans, N. Etcoff, P. Whalen, W. Kennedy, S. Rauch, R. Buckner, M. Srauss, S. Hyman, and B. Rosen. 1996. "Response and Habituation of the Human Amygdala During Visual Processing of Facial Expression." *Neuron*, November, 17, 875–887; Blackford, Jennifer, A. Allen, R. Cowan, and S. Avery. 2012. "Amygdala and Hippocampus Fail to Habituate to Faces in Individuals with an Inhibited Temperament." *Social Cognitive and Affective Neuroscience*, January, 143–150.

42. *The Waffen-SS.* 2002. Film documentary directed by Christian Frey, written by Mark Halliley, produced by Guido Knopp. Story House Production for Channel 4 in the United Kingdom and the History Channel in the United States. youtube .com/watch?v=Fzyx6DbOhec

43. Navarick, Douglas J. 2012. "Historical Psychology and the Milgram Paradigm: Tests of an Experimentally Derived Model of Defiance Using Accounts of Massacres by Nazi Reserve Police Battalion 101." *Psychological Record*, 62, 133–154. Navarick notes the importance of social scientists and historians working together to understand the past: "Psychological science has a role to play in ensuring that humanity remembers—and learns from—the past."

44. Quoted in Navarick, 2013.

45. Navarick, 2012. "Historical Psychology," op cit.

46. Navarick, Douglas J. 2009. "Reviving the Milgram Obedience Paradigm in the Era of Informed Consent." *Psychological Record*, 59, 155–170.

47. Theweleit, Klaus. 1989. *Male Fantasies.* Vol. 2, *Male Bodies.* Minneapolis: University of Minnesota Press, 301.

48. *The Waffen-SS.* 2002. Film documentary.

49. Ibid.

50. Browning, Christopher. 1991. *The Path to Genocide: Essays on Launching the Final Solution.* Cambridge, UK: Cambridge University Press, 143.

51. Le Bon, Gustave. 1896. *The Crowd: A Study of the Popular Mind.* New York: Macmillan.

52. Sherif, Muzafer, O. J. Harvey, B. Jack White, William R. Hood, and Carolyn W. Sherif. 1961. *Intergroup Conflict and Cooperation: The Robbers Cave Experiment.* Norman: University of Oklahoma Press.

53. Hofling, Charles K., E. Brotzman, S. Dalrymple, N. Graves, and C. M. Pierce. 1966. "An Experimental Study in Nurse-Physician Relationships." *Journal of Nervous and Mental Disease*, 143, 171–180.

54. Krackow, A., and T. Blass. 1995. "When Nurses Obey or Defy Inappropriate Physician Orders: Attributional Differences." *Journal of Social Behavior and Personality*, 10, 585–594.

55. Haslam, S. Alexander, Stephen D. Reicher, and Joanne R. Smith. 2012. "Working Toward the Experimenter: Reconceptualizing Obedience Within the Milgram Paradigm as Identification-Based Followership." *Perspectives on Psychological Science.* 7, no. 4, 315–324. doi:10.1371/journal.pbio.1001426

56. Haslam, S. Alexander, and Stephen D. Reicher. 2012. "Contesting the 'Nature' of Conformity: What Milgram and Zimbardo's Studies Really Show." *PLoS Biol* 10, no. 11, November 20: e1001426. doi:10.1371/journal.pbio.1001426

57. Zimbardo, Philip. 2007. *The Lucifer Effect: Understanding How Good People Turn Evil.* New York: Random House.

58. Reicher, S. D., and S. A. Haslam. 2006. "Rethinking the Psychology of Tyranny: The BBC Prison Study." *British Journal of Social Psychology* 45, 1–40. doi: 10.1348/014466605X48998

59. Asch, Solomon E. 1951. "Studies of Independence and Conformity: A Minority of One Against a Unanimous Majority." *Psychological Monographs*, 70, no. 416. See also Asch, Solomon E. 1955. "Opinions and Social Pressure." *Scientific American*, November, 31–35.

60. Berns, Gregory, et al. 2005. "Neurobiological Correlates of Social Conformity and Independence During Mental Rotation." *Biological Psychiatry*, 58, August 1, 245–253.

61. Perdue, Charles W., John F. Dovidio, Michael B. Gurtman, and Richard B. Tyler.

1990. "Us and Them: Social Categorization and the Process of Intergroup Bias." *Journal of Personality and Social Psychology*, 59, 475–486.

62. Grossman, Dave. 2009. *On Killing*. Boston: Little, Brown.

63. Malmstrom, Frederick V., and David Mullin. "Why Whistleblowing Doesn't Work: Loyalty Is a Whole Lot Easier to Enforce Than Honesty." *Skeptic*, 19, no. 1, 30–34. goo.gl/BGdm47

64. Prentice, D. A., and D. T. Miller. 1993. "Pluralistic Ignorance and Alcohol Use on Campus: Some Consequences of Misperceiving the Social Norm." *Journal of Personality and Social Psychology*, February, 64, no. 2, 243–256. goo.gl/W2Cjek

65. Lambert, Tracy A., Arnold S. Kahn, and Kevin J. Apple. 2003. "Pluralistic Ignorance and Hooking Up." *Journal of Sex Research*, 40, no. 2, May, 129–133.

66. Russell, Jeffrey B. 1982. *A History of Witchcraft: Sorcerers, Heretics and Pagans*. London: Thames & Hudson; Briggs, Robin. 1996. *Witches and Neighbors: The Social and Cultural Context of European Witchcraft*. New York: Viking.

67. Quoted in Glover, J. 1999. *Humanity: A Moral History of the Twentieth Century*. London: Jonathan Cape. See also Solzhenitsyn, Aleksandr. 1973. *The Gulag Archipelago*. New York: Harper & Row.

68. Macy, Michael W., Robb Willer, and Ko Kuwabara. 2009. "The False Enforcement of Unpopular Norms." *American Journal of Sociology*, 115, no. 2, September, 451–490.

69. O'Gorman, Hubert J. 1975. "Pluralistic Ignorance and White Estimates of White Support for Racial Segregation." *Public Opinion Quarterly*, 39, no. 3, Autumn, 313–330.

70. Boven, Leaf Van. 2000. "Pluralistic Ignorance and Political Correctness: The Case of Affirmative Action." *Political Psychology*, 21, no. 2, 267–276.

71. Dostoevsky, Fyodor. 1864/1918. *Notes from the Underground*. New York: Vintage. Available at www.classicreader.com/book/414/12/

72. Prentice and Miller, 1993.

73. Macy et al., 2009.

74. goo.gl/FVUpmH

10: MORAL FREEDOM AND RESPONSIBILITY

1. Personal correspondences. Out of respect for the victims and families of the victims of these crimes, I will give no further identifying information so as not to give any personal satisfaction to these prisoners of seeing their names or accounts in print.

2. Harris, Sam. 2012. *Free Will*. New York: Free Press, 5.

3. Libet, Benjamin. 1985. "Unconscious Cerebral Initiative and the Role of Conscious Will in Voluntary Action." *Behavior and Brain Sciences*, 8, 529–566.

4. Haynes, J. D. 2011. "Decoding and Predicting Intentions." *Annals of the New York Academy of Sciences*, 1224, no. 1, 9–21.

5. Fried, I., R. Mukamel, and G. Kreimann. 2011. "Internally Generated Preactivation of Single Neurons in Human Medial Frontal Cortex Predicts Volition." *Neuron*, 69, 548–562. See also Haggard, P. 2011. "Decision Time for Free Will." *Neuron*, 69, 404–406.

6. Kurzban, Robert. 2012. *Why Everyone (Else) Is a Hypocrite*. Princeton, NJ: Princeton University Press.

7. Ultimatum game research and applications are reviewed here: Camerer, Colin. 2003. *Behavioral Game Theory*. Princeton, NJ: Princeton University Press.

8. I consider the fate of *Homo economicus* in a book-length analysis: Shermer, Michael. 2007. *The Mind of the Market*. New York: Times Books.

9. Dennett, Daniel. 2003. *Freedom Evolves*. New York: Viking.

10. Brass, Marcel, and Patrick Haggard. 2007. "To Do or Not to Do: The Neural Signature of Self-Control." *Journal of Neuroscience*, 27, no. 34, 9141–9145.

11. Ibid., 9143.

12. Ibid., 9144.

13. Libet, Benjamin. 1999. "Do We Have Free Will?" *Journal of Consciousness Studies*, 6, no. 809, 47–57.

14. For a discussion of how the brain operates to make economic decisions that feel "free" to the decision maker see Glimcher, P. W. 2003. *Decisions, Uncertainty, and the Brain: The Science of Neuroeconomics*. Cambridge, MA: MIT Press. See also Steven Pinker's excellent discussion on free will and determinism in Pinker, Steven. 2002. *The Blank Slate: The Modern Denial of Human Nature*. New York: Viking, 175.

15. Dennett, Daniel. 2003.

16. Scheb, John M., and John M. Scheb II. 2010. *Criminal Law and Procedure*, 7th ed. Stamford, CT: Cengage Learning.

17. Hare, Robert. 1991. *Without Conscience: The Disturbing World of the Psychopaths Among Us*. New York: Guilford Press; Baron-Cohen, Simon. 2011. *The Science of Evil: On Empathy and the Origins of Cruelty*. New York: Basic Books; Dutton, Kevin. 2012. *The Wisdom of Psychopaths: What Saints, Spies, and Serial Killers Teach Us About Success.* New York: Farrar, Straus & Giroux.

18. Personal interview, July 23, 2012.

19. Quoted in Dutton, 2012.

20. Fallon, James. 2013. *The Psychopath Inside: A Neuroscientist's Personal Journey into the Dark Side of the Brain*. New York: Current, 1.

21. Ibid., 190.

22. Ibid., 206.

23. Dutton, Kevin. 2012, 200.

24. www.heroicimagination.org

25. Dutton, 2012, 222.

26. goo.gl/VYaGA

27. UPI press release. 1966. "Sniper in Texas U. Tower Kills 12, Hits 33." *New York Times*, August 2, 1.

28. Letter dictated by Whitman on Sunday, July 31, 1966, 6:45 p.m. In the collections of the Austin History Center. goo.gl/muBEJ8

29. Report to the Governor, Medical Aspects, Charles J. Whitman Catastrophe. 1966. Whitman Archives. *Austin American-Statesman*, September 8.

30. Heatly, Maurice. 1966. "Whitman Case Notes. Whitman Archives." *Austin American-Statesman*, March 29.

31. Raine, Adrian. 2013. *The Anatomy of Violence: The Biological Roots of Crime*. New York: Pantheon.

32. Ibid., 309–310.

33. Ibid., 67.

34. Ibid., 69.

35. Ibid., 326.

36. Kiehl, Kent A., et al. 2001. "Limbic Abnormalities in Affective Processing by Criminal Psychopaths as Revealed by Functional Magnetic Resonance Imaging." *Biological Psychiatry*, 50, 677–684.

37. Aharoni, Eyal, Gina Vincent, Carla Harenski, Vince Calhoun, Walter Sinnott-Armstrong, Michael Gazzaniga, and Kent Kiehl. 2013. "Neuroprediction of Future Rearrest." *PNAS*, 110, no. 15, 6223–6228.

38. *2011 Global Study on Homicide*. UN Office on Drugs and Crime, 63–70. goo.gl /Hz2ie
39. Ibid., 73.
40. Kellerman, A. L., and J. A. Mercy. 1992. "Men, Women, and Murder: Gender-Specific Differences in Rates of Fatal Violence and Victimization." *Journal of Trauma*, July, 33, no. 1, 1–5. goo.gl/ie5isw

11: MORAL JUSTICE: RETRIBUTION AND RESTORATION

1. www.nelsonmandela.org
2. Buss, David. 2005. *The Murderer Next Door: Why the Mind Is Designed to Kill*. New York: Penguin.
3. Ibid., 70.
4. Ibid.
5. Ibid., 106.
6. Ibid.
7. Ibid.
8. Collins, Randall. 2008. *Violence: A Micro-Sociological Theory*. Princeton, NJ: Princeton University Press.
9. Buss, 2005.
10. De Waal, Frans. 1982. *Chimpanzee Politics: Sex and Power Among the Apes*. Baltimore: Johns Hopkins University Press, 203, 207.
11. ———. 1989. *Peacemaking Among Primates*. Cambridge, MA: Harvard University Press.
12. ———. 2005. *Our Inner Ape*. New York: Riverhead Books, 175.
13. De Waal, Frans, and Frans Lanting. 1998. *Bonobo: The Forgotten Ape*. Berkeley: University of California Press; Kano, Takayoshi, and Evelyn Ono Vineberg. 1992. *The Last Ape: Pygmy Chimpanzee Behavior and Econology*. Ann Arbor, MI: University Microfilms International.
14. De Waal, Frans B. M. 1997. "Food Transfers Through Mesh in Brown Capuchins." *Journal of Comparative Psychology*, 111, 370–378.
15. Brosnan, Sarah F., and Frans de Waal. 2003. "Monkeys Reject Unequal Pay." *Nature*, 425, September 18, 297–299.
16. Cords, M., and S. Thurnheer, 1993. "Reconciling with Valuable Partners by Long-Tailed Macaques." *Behaviour*, 93, 315–325.
17. De Waal, Frans. 1996. *Good Natured: The Origins of Right and Wrong in Humans and Other Animals*. Cambridge, MA: Harvard University Press.
18. Koyama, N. F., and E. Palagi. 2007. "Managing Conflict: Evidence from Wild and Captive Primates." *International Journal of Primatology*, 27, no. 5, 1235–1240. Koyama, N. F., C. Caws, and F. Aureli. 2007. "Interchange of Grooming and Agonistic Support in Chimpanzees." *International Journal of Primatology*, 27, no. 5, 1293–1309.
19. I document the evidence for this thesis in *The Science of Good and Evil*, and the logic of why actually believing you are a moral person rather than faking it was worked out by Robert Trivers in Trivers, Robert. 2011. *The Folly of Fools: The Logic of Deceit and Self-Deception in Human Life*. New York: Basic Books.
20. In support of this claim see de Waal, Frans. 2008. "How Selfish an Animal? The Case of Primate Cooperation." In Paul Zak, ed., *Moral Markets: The Critical Role of Values in the Economy*. Princeton, NJ: Princeton University Press.
21. Brosnan, Sarah F. 2008. "Fairness and Other-Regarding Preferences in Nonhuman

Primates." In Paul Zak, ed., *Moral Markets: The Critical Role of Values in the Economy*. Princeton, NJ: Princeton University Press.

22. Henrich, Joseph, Robert Boyd, Sam Bowles, Colin Camerer, Herbert Gintis, Richard McElreath, and Ernst Fehr. 2001. "In Search of *Homo economicus*: Experiments in 15 Small-Scale Societies." *American Economic Review*, 91, no. 2, 73–79.

23. Henrich, Joseph, Robert Boyd, Sam Bowles, Colin Camerer, Ernst Fehr, and Herbert Gintis. 2004. *Foundations of Human Sociality*. New York: Oxford University Press, 8.

24. Gintis, Herbert, Samuel Bowles, Robert Boyd, and Ernst Fehr. 2005. *Moral Sentiments and Material Interests*. Cambridge, MA: MIT Press.

    Boyd, Robert, and Peter J. Richerson. 2005. *The Origin and Evolution of Cultures*. New York: Oxford University Press.

25. Boehm, Christopher. 2012. *Moral Origins: The Evolution of Virtue, Altruism, and Shame*. New York: Basic Books.

26. There are critics of this assumption—usually cultural anthropologists and sociologists who place a much stronger emphasis on the role of learning, culture, and the environment than I think is perhaps warranted by the evidence—but almost no one today denies that we have an evolved nature, and that facts about this nature may be gleaned from these numerous sources.

27. Figure based on table III in Boehm, 2012, 196; see also table 1 in Boehm, Christopher. 2014. "The Moral Consequences of Social Selection." *Behaviour*, 151, 167–183.

28. Lee, Richard B. 1979. *The !Kung San: Men, Women, and Work in a Foraging Society*. New York: Cambridge University Press, 394–395.

29. Boehm, 2012, 201. Boehm also makes the argument for group selection—in addition to individual selection—in his book, which I think is unnecessary in making his case for the evolution of altruism and the problem of free riding and bullying. The group may seem coherent and cohesive, but it is still a group of individuals, per my discussion of the issue in chapter 1.

30. Ibid., 201.

31. Bentham, J. 1789/1948. *The Principles of Morals and Legislation*. New York: Macmillan.

32. goo.gl/iUSguA

33. Cooney, Mark. 1997. "The Decline of Elite Homicide." *Criminology*, 35, 381–407.

34. Transcript of Goetz's videotape confession available here: goo.gl/oRHSxJ See also the documentary film "The Confessions of Bernhard Goetz": goo.gl/iLYt5W

35. Fletcher, George P. 1999. *A Crime of Self-Defense: Bernhard Goetz and the Law on Trial*. Chicago: University of Chicago Press. Fletcher writes: "There is a way to deal with the problem as a moral question. The reason that Goetz was charged with a crime of self-defense was precisely that the law enforcement agencies and the district attorney felt strongly that he had overreacted, that he was blameworthy for having overreacted, and that he ought to be punished." Quoted in the documentary film *The Confessions of Bernhard Goetz*: goo.gl/iLYt5W

36. Quoted in the documentary film *The Confessions of Bernhard Goetz*: goo.gl/iLYt5W

37. Black, Donald. 1983. "Crime as Social Control." *American Sociological Review*, 48, 34–45.

38. MacRae, Allan, and Howard Zehr. 2011. "Right Wrongs the Maori Way." *Yes! Magazine*, July 8. goo.gl/PBd67u

39. New Zealand Ministry of Justice. "Child Offending and Youth Justice Processes." goo.gl/m7HYht

40. McElrea, Fred W. M. 2012. "Twenty Years of Restorative Justice in New Zealand." January 10. goo.gl/24mtP

41. goo.gl/JrPmw6 See as well this from the New Zealand Ministry of Justice: goo.gl /m7HYht And here a judge talks about restorative justice in New Zealand: goo. gl/24mtP

42. MacRae and Zehr. 2011. "Right Wrongs the Maori Way."

43. McCann, Michael. 2007. "No Easy Answers." SI.Com. goo.gl/lWkvwU

44. The collection process became even more problematic when O. J. was subsequently incarcerated for another crime in which he illegally entered a hotel room brandishing a gun to retrieve what he said were stolen items from his football memorabilia collection. He was charged with criminal conspiracy, kidnapping, assault, robbery, and use of a deadly weapon. This too was considered to be a crime against the state—instead of the holders of said memorabilia—and for taking the law into his own hands Simpson was sent to prison, where he remains to this day. See *State of Nevada v. O. J. Simpson et al.* goo.gl/wHkBc8

45. Zehr, Howard, and Ali Gohar. 2003. *The Little Book of Restorative Justice.* Intercourse, PA: Good Books, 10–14. Available online as a PDF: goo.gl/ssvHRl

46. Diamond, Jared. 2012. *The World Until Yesterday: What Can We Learn from Traditional Societies?* New York: Viking.

47. Ibid.

48. Ibid.

49. Woolf and Riley are featured together in a short film titled *The Woolf Within.* goo.gl/NVMRt

50. goo.gl/u1VSma

51. Richardson, Lucy. 2013. "Restorative Justice Does Work, Says Career Burglar Who Has Turned Life Around on Teesside." *Darlington and Stockton Times,* May 1. goo.gl/Vjpr6t

52. Reported in *The Woolf Within* film.

53. McLeland, Debbie. 2010, March 29. goo.gl/eW9Huv

54. Ibid.

55. Stahl, Lesley. 2011. "Eyewitness." *60 Minutes,* Shari Finkelstein, producer. CBS.

56. Cannino-Thompson, Jennifer, Ronald Cotton, and Erin Torneo. 2010. *Picking Cotton.* New York: St. Martin's Press.

57. Source: Photo by the author.

58. See, for example, McCullough, Michael. 2008. *Beyond Revenge: The Evolution of the Forgiveness Instinct.* San Francisco: Jossey-Bass. The neuroscientist Daniel Reisel draws three lessons from his fifteen years of work with serial killers and psychopaths: (1) We need to change our mind-set about incarceration. "The moment we speak about prisons, it's like we're back in Dickensian—if not medieval—times. For too long we've allowed ourselves to be persuaded of the false notion that human beings can't change, and, as a society, it's costing us dearly." (2) We need transdisciplinary research from multiple fields to work on the problem: "We need people from different disciplines, lab-based scientists, clinicians, social workers and policy makers, to work together." (3) We need to change our mind-set on prisoners. If we see psychopaths as irredeemable, how are they ever going to see themselves as any different? Wouldn't it be better for psychopaths to spend their time in jail training their amygdalas and generating new brain cells? goo.gl/VYaGA

59. Adams, John. 1788. *A Defence of the Constitutions of Government of the United States of America,* Vol. 3, 291. goo.gl/rkuuDi

60. Graph rendered by Pinker, 2011, 149, based on data in Hunt, Lynn. 2007. *Inventing*

Human Rights: A History. New York: W. W. Norton, 76, 179; Mannix, D. P. 1964. *The History of Torture.* Sparkford, UK: Sutton, 137–138.

61. "Public Executions." Boone, NC: Department of Government and Justice Studies, Appalachian State University. goo.gl/gdEpj8

62. Quoted in "An Unreal Dream: The Michael Morton Story." *CNN Films,* December 8, 2013. Morton's compensation reported in the *Houston Chronicle:* goo.gl/SpP6Kd

63. Kuhn, Deanna, M. Weinstock, and R. Flaton. 1994. "How Well Do Jurors Reason? Competence Dimensions of Individual Variation in a Juror Reasoning Task." *Psychological Science,* 5, 289–296.

64. goo.gl/PTSY

65. goo.gl/9omAH2

66. Gross, Samuel R., Barbara O'Brien, Chen Hu, and Edward H. Kennedy. 2014. "Rate of False Conviction of Criminal Defendants Who Are Sentenced to Death." *Proceedings of the National Academy of Sciences,* April 28. goo.gl/mljR2M

67. "An Unreal Dream: The Michael Morton Story." *CNN Films,* December 8, 2013. Eventually Morton was awarded compensation of $1,973,333.33 by the state, and shortly thereafter Texas governor Rick Perry passed into law the Michael Morton Act, requiring prosecutors to turn evidence over to defense lawyers in criminal cases upon the defendant's request and without the need for a court order. It went into effect on January 1, 2014. Remarkably, according to the Innocence Project, before the Morton case no prosecutor had ever been criminally punished for failing to turn over exculpatory evidence. That's progress.

68. goo.gl/Jx89yr

69. Eckholm, Erik, and John Schwartz. 2014. "Timeline Describes Frantic Scene at Oklahoma Execution." *New York Times,* May 1, goo.gl/QtvkKJ

70. Blackmun, Harry. 1994. Dissent. *Bruce Edwin Callins, Petitioner, v. James A. Collins, Director,* Texas Department of Criminal Justice, Institutional Division. Supreme Court of the United States. No. 93-7054. goo.gl/P5sKv4

71. Graph rendered by Pinker, 2011, 150, based on data in French Ministry of Foreign Affairs. 2007. *The Death Penalty in France.* goo.gl/4P7vlX *The End of Capital Punishment in Europe.* goo.gl/BkMfs2 Amnesty International. 2010. *Abolitionist and Retentionist Countries.* goo.gl/mfw5RF

72. goo.gl/Jx89yr

73. goo.gl/Jx89yr

74. Chu, Henry. 2013. "Gay British Scientist Gets Posthumous Royal Pardon." *Los Angeles Times,* December 25, A1, 7.

75. Ibid.

76. Ferguson, Niall. 2000. *The Pity of War: Explaining World War I.* New York: Basic Books; Tuchman, Barbara. 1963. *The Guns of August.* New York: Dell.

77. Cited in Grossman, Richard S. 2013. *Wrong: Nine Economic Policy Disasters and What We Can Learn from Them.* New York: Oxford University Press.

78. This was based on a new set of legal and moral principles, such as Principle I: "Any person who commits an act which constitutes a crime under international law is responsible therefore and liable to punishment." And Principle II: "The fact that internal law does not impose a penalty for an act which constitutes a crime under international law does not relieve the person who committed the act from responsibility under international law." Nuremberg Trial Proceedings, vol. 1. Charter of the International Military Tribunal. goo.gl/wkaTs

79. Conot, Robert E. 1993. *Justice at Nuremberg.* New York: Basic Books.

80. Bosco, David. 2014. *Rough Justice: The International Criminal Court in a World of Power Politics*. New York: Oxford University Press.

81. Long, William, and Peter Brecke. 2003. *War and Reconciliation: Reason and Emotion in Conflict Resolution*. Cambridge, MA: MIT Press, 70–71.

82. Lincoln, Abraham. 1865. Second Inaugural Address. March 4. goo.gl/a48frS

83. Holocaust deniers argue that the six million figure for the number of Jews killed in the genocide is exaggerated by Israel to increase reparation payments from Germany, but in fact the accounting is based on the number of survivors, not on the number exterminated, so by denier reasoning the six million figure is, if anything, an underreporting. See Shermer, Michael. 2000. *Denying History*. Berkeley: University of California Press.

84. Since 1990 when the "Here Lived-Stumbling Stones" project was launched by Gunter Demnig, there are as of the end of 2013 1,909 stumbling blocks in Cologne and 43,500 total in 915 places in the Federal Republic of Germany of people who were relocated. goo.gl/4r9cDS

85. Torpey, John C. 2006. *Making Whole What Has Been Smashed: On Reparations Politics*. Cambridge, MA: Harvard University Press.

86. goo.gl/cwozGi

87. goo.gl/Zwjn7Z

12: PROTOPIA: THE FUTURE OF MORAL PROGRESS

1. Darwin, Charles. 1871. *The Descent of Man and Selection in Relation to Sex*. Vol. 1. London: John Murray, 69.

2. Clarke, Arthur C. 1951. *The Exploration of Space*. Frederick, MD: Wonder Book.

3. The phrase was introduced by Gertrude Stein in her autobiography in describing her childhood home of Oakland where she famously declared "there is no there there." It's not clear what she meant, although it appears to reference changing identities (one's home city and one's self). Stein, Gertrude. 1937. *Gertrude Stein, Everybody's Autobiography*. New York: Random House, 289.

4. οὐ ("not") and τόπος ("place"): "no place"

5. Rayfield, Donald. 2005. *Stalin and His Hangmen: The Tyrant and Those Who Killed for Him*. New York: Random House; White, Matthew. 2011. *The Great Big Book of Horrible Things: The Definitive Chronicle of History's 100 Worst Atrocities*. New York: W. W. Norton, 382–392; Akbar, Arifa. 2010. "Mao's Great Leap Forward 'Killed 45 Million in Four Years.'" *Independent* (London), September 17; Becker, Jasper. 1998. *Hungry Ghosts: Mao's Secret Famine*. New York: Henry Holt; Pipes, Richard. 2003. *Communism: A History*. New York: Modern Library. See also goo.gl/ryHSYd

6. The description of the failure of New Harmony was made by the individualist anarchist Josiah Warren in his 1856 *Periodical Letter II*. Quoted in Brown, Susan Love, ed. 2002. *Intentional Community: Anthropological Perspective*. Albany: State University of New York Press, 156.

7. Kelly, Kevin. 2014. "The Technium. A Conversation with Kevin Kelly by John Brockman." goo.gl/LhfbMS

8. In researching his 2010 book *What Technology Wants*, for example, Kelly recalls that he went through back issues of *Time* and *Newsweek*, plus early issues of *Wired* (which he cofounded and edited), to see what everyone was predicting for the Web. "Generally, what people thought, including to some extent myself, was it was going to be better TV, like TV 2.0. But, of course, that missed the entire real

revolution of the Web, which was that most of the content would be generated by the people using it. The Web was not better TV, it was the Web. Now we think about the future of the Web, we think it's going to be the better Web; it's going to be Web 2.0, but it's not. It's going to be as different from the Web as Web was from TV." How does this type of technological improvement translate into moral progress? Kelly explains (goo.gl/LhfbMS):

> One way to think about this is if you imagine the very first tool made, say, a stone hammer. That stone hammer could be used to kill somebody, or it could be used to make a structure, but before that stone hammer became a tool, that possibility of making that choice did not exist. Technology is continually giving us ways to do harm and to do well; it's amplifying both . . . but the fact that we also have a new choice each time is a new good. That, in itself, is an unalloyed good—the fact that we have another choice and that additional choice tips that balance in one direction towards a net good. So you have the power to do evil expanded. You have the power to do good expanded. You think that's a wash. In fact, we now have a choice that we did not have before, and that tips it very, very slightly in the category of the sum of good.

 9. Srinivasan, Balaji. 2013. "Silicon Valley's Ultimate Exit Strategy." Startup School 2013 speech. goo.gl/mkvG2J
10. Hill, Kashmir. 2011. "Adventures in Self-Surveillance, aka The Quantified Self, aka Extreme Navel-Gazing." *Forbes*, April 7. goo.gl/JSRkAS
11. Ibid.
12. Srinivasan, Balaji. 2013. "Software Is Reorganizing the World." *Wired*, November. goo.gl/0Oxa3s
13. The most popular target is the actor Kevin Bacon. I am two degrees away through a bike-racing friend of mine who worked with Bacon on the film *Quicksilver* (in which the actor played a New York City bike messenger), so my Bacon number is 2.
14. Milgram, Stanley. 1967. "The Small World Problem." *Psychology Today*, 2, 60–67.
15. Travers, Jeffrey, and Stanley Milgram. 1969. "An Experimental Study of the Small World Problem." *Sociometry* 32, no. 4, December, 425–443.
16. Hadfield, Chris. 2013. *An Astronaut's Guide to Life on Earth*. New York: Little, Brown; goo.gl/m7TI2M
17. Srinivasan, 2013, *Wired*.
18. Wright, Quincy. 1942. *A Study of War*, 2nd ed. Chicago: University of Chicago Press; Gat, A. 2006. *War in Human Civilization*. New York: Oxford University Press.
19. Fukuyama, Francis. 2011. *The Origins of Political Order: From Prehuman Times to the French Revolution*. New York: Farrar, Straus & Giroux, 98.
20. goo.gl/vWBn1N
21. Mintzberg, Henry. 1989. *Mintzberg on Management: Inside Our Strange World of Organizations*. New York: Free Press. The Internet is a tool of adhocracy because it enables instant real-time communication through virtual communities that are spontaneously self-organizing online.
22. Stone, Brad. 2013. "Inside Google's Secret Lab." *Business Week*, May 22. goo.gl/BSwk7
23. Barber, Benjamin. 2013. *If Mayors Ruled the World: Dysfunctional Nations, Rising Cities*. New Haven, CT: Yale University Press.
24. Quoted in Barber, 2013.

25. Ibid.
26. Townsend, Anthony M. 2013. *Smart Cities: Big Data, Civic Hackers, and the Quest for a New Utopia*. New York: W. W. Norton, xii–xiii.
27. Speck, Jeff. 2012. *Walkable City: How Downtown Can Save America, One Step at a Time*. New York: Farrar, Straus & Giroux.
28. Brand, Stewart. 2013. "City-Based Global Governance." The Long Now Foundation. goo.gl/FbS5E
29. goo.gl/hVdX1E
30. goo.gl/G9eyKc
31. Ibid.
32. Konvitz, Josef W. 1985. *The Urban Millennium: The City-Building Process from the Early Middle Ages to the Present*. Carbondale, IL: Southern Illinois University Press; Kostof, Spiro. 1991. *The City Shaped: Urban Patterns and Meanings Through History*. Boston: Little, Brown; Jacobs, Jane. 1961. *The Death and Life of Great American Cities*. New York: Random House.
33. Glaeser, Edward. 2011. *The Triumph of the City: How Our Greatest Invention Makes Us Richer, Smarter, Greener, Healthier, and Happier*. New York: Penguin.
34. Naím, Moisés. 2013. *The End of Power: From Boardrooms to Battlefields and Churches to States: Why Being in Charge Isn't What It Used to Be*. New York: Basic Books, 16, 1–2.
35. Ibid., 7.
36. Ibid., 243–244.
37. The closest label that summarizes these characteristics is "classical liberal," and it follows John Locke's model for the protection of natural rights that people possess by virtue of their humanity. For a good discussion of the US Constitution reflecting these values see Epstein, Richard A. 2014. *The Classical Liberal Constitution: The Uncertain Quest for Limited Government*. Cambridge, MA: Harvard University Press.
38. Panaritis, Elena. 2007. *Prosperity Unbound: Building Property Markets with Trust*. New York: Palgrave Macmillan.
39. MacCallum, Spencer Heath. 1970. *The Art of Community*. Menlo Park, CA: Institute for Humane Studies, 2. See also Heath, Spencer. 1957. *Citadel, Market and Altar: Emerging Society*. Baltimore: Science of Society Foundation.
40. Leeson, Peter. 2014. *Anarchy Unbound: Why Self-Governance Works Better Than You Think*. Cambridge, UK: Cambridge University Press.
41. See, for example, Casey, Gerard. 2012. *Libertarian Anarchy: Against the State*. New York: Continuum International Publishing; Morris, Andrew. 2008. "Anarcho-Capitalism." In Hamowy, Ronald, ed., *The Encyclopedia of Libertarianism*. Thousand Oaks, CA: Sage; Rothbard, Murray. 1962. *Man, Economy, and State*. New York: D. Van Nostrand. For a good debate in one volume see Duncan, Craig, and Tibor R. Machan. 2005. *Libertarianism: For and Against*. Lanham, MD: Rowman & Littlefield.
42. Nozick, Robert. 1973. *Anarchy, State, and Utopia*. New York: Basic Books.
43. Howard, Philip K. 2014. *The Rule of Nobody: Saving America from Dead Laws and Broken Government*. New York: W. W. Norton.
44. Jefferson, Thomas. 1804. Letter to Judge John Tyler Washington, June 28. *The Letters of Thomas Jefferson, 1743–1826*. goo.gl/hn6qNp
45. Robbins, Lionel. 1945. *An Essay on the Nature and Significance of Economic Science*. London: Macmillan, 16. See also Sowell, Thomas. 2010. *Basic Economics:*

A Common Sense Guide to the Economy, 4th ed. New York: Basic Books, 5; Mankiw, Gregory. 2011. *Principles of Economics*, 6th ed. Stamford, CT: Cengage Learning, 11.

46. goo.gl/LhfbMS
47. Drexler, Eric K. 1986. *Engines of Creation*. New York: Anchor Books.
48. Diamandis, Peter, and Steven Kotler. 2012. *Abundance: The Future Is Better Than You Think*. New York: Free Press, 8.
49. Diamandis and Kotler, 9.
50. Ibid.
51. Musk, Elon. 2014. "Here's How We Can Fix Mars and Colonize It." *Business Insider*, January 2. goo.gl/iapxDx
52. Carroll, Rory. 2013. "Elon Musk's Mission to Mars." *Guardian*, July 17. goo.gl /lF1sXP
53. goo.gl/7BdRJ
54. Kurzweil, Ray. 2006. *The Singularity Is Near: When Humans Transcend Biology*. New York: Penguin.
55. 2045.com/faq/
56. McCracken, Harry, and Lev Grossman. 2013. "Google vs. Death." *Time*, September 30. time.com/574/google-vs-death/
57. Kaku, Michio. 2011. *Physics of the Future: How Science Will Shape Human Destiny and Our Daily Lives by the Year 2100*. New York: Doubleday, 21.
58. Ibid., 337.
59. Even though GDP is the most common metric of economic growth, it has its limitations and critics. In mid-2014 the Bureau of Economic Analysis released a new economic statistic called Gross Output (GO), a measure of total sales volumes at all stages of production—from the procurement of raw materials through all the intermediate stages of production and distribution, to the final stages of retail sales. This was the first major upgrade in economic measurement since GDP was introduced half a century ago. See: Skousen, Mark. 2013. "Beyond GDP: Get Ready for a New Way to Measure the Economy." *Forbes*, December 16. goo.gl/xwICMV
60. Clark, Gregory. 2007. *A Farewell to Alms: A Brief Economic History*. Princeton, NJ: Princeton University Press, 2–3.
61. Based on the graph in Clark, 2007, 2. See also Maddison, Agnus. 2006. *The World Economy*. Washington, DC: OECD Publishing.
62. Graph from data produced by the Economic Research Service of the US Department of Agriculture. "Historical and Projected Gross Domestic Product Per Capita." goo.gl/CWcmHt
63. Haugen, Gary A., and Victor Boutros. 2014. *The Locust Effect: Why the End of Poverty Requires the End of Violence*. New York: Oxford University Press, 137.
64. Stiglitz, Joseph E. 2013. *The Price of Inequality: How Today's Divided Society Endangers Our Future*. New York: W. W. Norton.
65. Gilen, Martin. 2012. *Affluence and Influence: Economic Inequality and Political Power in America*. Princeton, NJ: Princeton University Press, 1.
66. Quoted in Hiltzik, Michael. 2013. "A Huge Threat to Social Mobility." *Los Angeles Times*, December 22, B1.
67. Piketty, Thomas. 2014. *Capital in the Twenty-first Century*. Cambridge, MA: Belknap Press.
68. Burtless, Gary. 2014. "Income Growth and Income Inequality: The Facts May Surprise You." Washington, DC: Brookings Institution. goo.gl/g4vTt6

69. Shah, Neil. 2014. "US Household Net Worth Hits Record High." *Wall Street Journal*, March 6, A1.

70. See, for example, Rubin, Jeff. 2012. *The End of Growth*. New York: Random House.

71. Beinhocker, Eric. 2006. *The Origin of Wealth: Evolution, Complexity, and the Radical Remaking of Economics*. Cambridge, MA: Harvard Business School Press, 453.

72. For a contrary view of this proposition see the documentary film *Surviving Progress* by Mathieu Roy and Harold Crooks: goo.gl/uTVf2c

73. Athreya, Kartik, and Jessie Romero. 2013. "Land of Opportunity? Economic Mobility in the United States." Federal Reserve Bank of Richmond, July. goo.gl/7KFc6H

74. Sawhill, Isabel V., Scott Winship, and Kerry Searle Grannis. 2012. "Pathways to the Middle Class: Balancing Personal and Public Responsibilities." Washington, DC: Brookings Institution Center on Children and Families, September.

75. Auten, Gerald, and Geoffrey Gee. 2009. "Income Mobility in the United States: New Evidence from Income Tax Data." *National Tax Journal*, June, 301–328. ntj.tax.org/

76. Auten, Gerald, Geoffrey Gee, and Nicholas Turner. 2013. "Income Inequality, Mobility and Turnover at the Top in the United States, 1987–2010." Paper presented at the Allied Social Science Association's annual meeting, San Diego, January 4.

77. Clark, Gregory. 2014. *The Son Also Rises: Surnames and the History of Social Mobility*. Princeton, NJ: Princeton University Press, 5.

78. Ibid., 180.

79. Ibid., 58.

80. Here's a simple example of how it works. Joe the Plumber owns 10 shares of Apple Computer stock trading at $500 a share. That $5,000 is a good chunk of change for his retirement, and when the stock goes up $10 in a day (which it is wont to do with some regularity), Joe's portfolio just increased $100. By contrast, Bob the Banker has 10,000 shares of Apple stock, and on the same day Joe made $100 Bob netted a sweet $1 million. Bob could almost retire on a single day of trading on the stock market, whereas Joe's retirement is nowhere on the horizon.

81. Rubin, Paul. H. 2003. "Folk Economics." *Southern Economic Journal*, 70, no. 1, 157–171.

82. Shermer, Michael. 2007. *The Mind of the Market: How Biology and Psychology Shape Our Economic Lives*. New York: Times Books.

83. Nowak, Martin, and Roger Highfield. 2012. *SuperCooperators: Altruism, Evolution, and Why We Need Each Other to Succeed*. New York: Free Press. See also: Nowak, Martin A., and Sarah Coakley, eds. 2013. *Evolution, Games, and God: The Principle of Cooperation*. Cambridge, MA: Harvard University Press.

84. For a history of the transition from the zero-sum interactions of our ancestors to the nonzero world of today see Wright, Robert. 2000. *Nonzero: The Logic of Human Destiny*. New York: Pantheon.

85. Draper, Patricia. 1978. "The Learning Envrionment for Aggression and Anti-Social Behavior among the !Kung." In A. Montagu, ed., *Learning Non-Aggression: The Experience of Non-literate Societies*. New York: Oxford University Press, 46.

86. Chambers, John R., Lawton K. Swan, and Martin Heesacker. 2013. "Better Off Than We Know: Distorted Perceptions of Incomes and Income Inequality in America." *Psychological Science*, 1–6.

87. Wilkinson, Richard G. 1996. *Unhealthy Societies: The Afflictions of Inequality*. New York: Routledge; Pickett, Kate, and Richard Wilkinson. 2011. *The Spirit Level: Why Greater Equality Makes Societies Stronger*. New York: Bloomsbury Press.

88. ———. 2011. "How Economic Inequality Harms Societies." goo.gl/B4hsrW

89. Sapolsky, Robert. 1995. *Why Zebras Don't Get Ulcers: A Guide to Stress, Stress-Related Diseases, and Coping.* New York: W. H. Freeman, 381.

90. Daly, Martin, Margo Wilson, and Shawn Vasdev. 2001. "Income Inequality and Homicide Rates in Canada and the United States." *Canadian Journal of Criminology*, 43, no. 2, 219–236.

91. Piff, Paul K. 2013. "Does Money Make You Mean?" TED talk, posted December 20: goo.gl/sfrJY9 See also the excellent *PBS NewsHour* report "Exploring the Psychology of Wealth, 'Pernicious' Effects of Economic Inequality." PBS, June 13. goo.gl/6kiOQ

92. Personal correspondence, December 22, 2013.

93. Ross, M., and F. Sicoly. 1979. "Egocentric Biases in Availability and Attribution." *Journal of Personality and Social Psychology*, 37, 322–336.

   Arkin, R. M., H. Cooper, and T. Kolditz. 1980. "A Statistical Review of the Literature Concerning the Self-Serving Bias in Interpersonal Influence Situations." *Journal of Personality*, 48, 435–448.

   Davis, M. H., and W. G. Stephan. 1980. "Attributions for Exam Performance." *Journal of Applied Social Psychology*, 10, 235–248.

94. Nisbett, R. E., and L. Ross. 1980. *Human Inference: Strategies and Shortcomings of Social Judgment.* Englewood Cliffs, NJ: Prentice-Hall.

95. Piff, Paul K. 2013. "Wealth and the Inflated Self: Class, Entitlement, and Narcissism." *Personality and Social Psychology Bulletin*, August, 1–10.

96. Kraus, Michael W., Paul K. Piff, and Dacher Keltner. 2009. "Social Class, Sense of Control, and Social Explanation." *Journal of Personality and Social Pschology*, 97, no. 6, 992–1004.

97. Keltner, Dacher, Aleksandr Kogan, Paul K. Piff, and Sarina Saturn. 2014. "The Sociocultural Appraisals, Values, and Emotions (SAVE) Framework of Prosociality: Core Processes from Gene to Meme." *Annual Review of Psychology*, 65, no. 25, 1–25.

98. Piff, Paul K., Daniel M. Stancato, Stephane Cote, Rodolfo Mendoza-Denton, and Dacher Keltner. 2012. "Higher Social Class Predicts Increased Unethical Behavior." *Proceedings of the National Academy of Sciences*, March 13, 109, no. 11, 4086–4091.

99. Quoted in Miller, Lisa. 2012. "The Money-Empathy Gap." *New York*, July 1. goo.gl/nCOc6

100. Piketty, Thomas. 2014. *Capital in the Twenty-first Century.* Cambridge, MA: Belknap Press.

101. goo.gl/BBo6kz

102. Mackey, John, and Raj Sisodia. 2013. *Conscious Capitalism: Liberating the Heroic Spirit of Business.* Cambridge, MA: Harvard Business Review Press, 20.

103. Ibid., 21.

104. Ibid., 31.

105. Ostry, Jonathan D., Andrew Berg, and Charalambos G. Tsangarides. 2014. "Redistribution, Inequality and Growth." Washington, DC: International Monetary Fund. February, 4. goo.gl/4xTwcP

106. Gates, Bill, and Melinda Gates. 2014. Annual Letter of the Bill and Melinda Gates Foundation. annualletter.gatesfoundation.org/ See also Penn World Table. Philadelphia: Center for International Comparisons at the University of Pennsylvania. pwt.sas.upenn.edu/

107. DeLong, J. Bradford. 2000. "Cornucopia: The Pace of Economic Growth in the Twentieth Century." Working Paper 7602. Washington, DC: National Bureau of

Economic Research. goo.gl/PLNJTG See also DeLong, J. Bradford. 1998. "Estimating World GDP, One Million BC–Present." goo.gl/7ttdjw

108. Kardashev, Nikolai. 1964. "Transmission of Information by Extraterrestrial Civilizations." *Soviet Astronomy*, 8, 217. Kardashev computed the energy levels of the three types to be: Type I (~4×10^{19} ergs/second), Type II (~4×10^{33} ergs/second), and Type III (~4×10^{44} ergs/second).

109. Heidmann, Jean. 1992. *Extraterrestrial Intelligence*. New York: Cambridge University Press, 210–212.

110. Sagan, Carl. 1973. *The Cosmic Connection: An Extraterrestrial Perspective*. Garden City, NY: Anchor Books/Doubleday, 233–234. Sagan's information storage capacity metric increases one order of magnitude at each step, starting with $A = 10^6$ bits of information, $B = 10^7$ bits ... $Z = 10^{31}$ bits. Sagan estimated we were at 10^{13} bits in 1973, making us a Type 0.7H civilization. My own calculations lead me to conclude that in 2014 we were at 10^{21} bits—a zettabyte—thus we are presently a Type 0.7P civilization. I based my estimate on the 2010 figure cited earlier from Diamandis that at the end of 2010 we were producing 912 exabytes of information. An exabyte is 1 quintillion bytes. A quintillion is 10^{18}. I estimate that we have exceeded 1,000 exabytes by 2014. A thousand exabytes is a zettabyte, or 10^{21}, or a 1 followed by 21 zeros.

As the Kardashev scale is logarithmic—where each increase in power consumption on the scale requires a huge leap in production—we have a ways to go. Fossil fuels won't get us there. Renewable sources such as solar, wind, and geothermal are a good start, but to achieve Type 1.0 status in this typology we need to go nuclear—for example, fusing 1,000 kg of hydrogen into helium per second, a rate of 3×10^{10} kg/year. As a cubic kilometer of water contains about 10^{11} kg of hydrogen, and our oceans contain about 1.3×10^9 km of water, this would give us ample time to make the transition to the next level.

111. Kaku, 2011, op cit. See also Kaku, Michio. 2010. "The Physics of Interstellar Travel: To One Day Reach the Stars." goo.gl/TBExNt

112. Kaku, Michio. 2005. *Parallel Worlds: The Science of Alternative Universes and Our Future in the Cosmos*. New York: Doubleday, 317.

113. Zubrin, Robert. 2000. *Entering Space: Creating a Spacefaring Civilization*. New York: Putnam, x.

114. See, for example, Rapaille, Clotaire, and Andres Roemer. 2013. *Move Up*. Mexico City: Taurus.

115. Ghemawat, Pankaj. 2011. *World 3.0: Global Prosperity and How to Achieve It*. Cambridge, MA: Harvard Business Review Press.

116. Shermer, Michael. 2014. "The Car Dealers' Racket." *Los Angeles Times*, March 17. goo.gl/sjTQwJ

117. The evolutionary biologist Ernst Mayr defined a species as "a group of actually or potentially interbreeding natural populations reproductively isolated from other such populations." Ernst Mayr. 1957. "Species Concepts and Definitions," in *The Species Problem*. Washington, DC: American Association for the Advancement of Science Publication 50. Mayr offers this expanded definition: "A species consists of a group of populations which replace each other geographically or ecologically and of which the neighboring ones intergrade or hybridize wherever they are in contact or which are potentially capable of doing so (with one or more of the populations) in those cases where contact is prevented by geographical or ecological barriers." See also Mayr, Ernst. 1976. *Evolution and the Diversity of Life*. Cambridge, MA: Harvard University Press; Mayr, Ernst.

1988. *Toward a New Philosophy of Biology*. Cambridge, MA: Harvard University Press.

118. Harris, Sam. 2010. *The Moral Landscape: How Science Can Determine Human Values*. New York: Free Press.

119. Smolin, Lee. 1997. *The Life of the Cosmos*. New York: Oxford University Press; Liddle, Andrew, and Jon Loveday. 2009. *The Oxford Companion to Cosmology*. New York: Oxford University Press; Weinberg, Stephen. 2008. *Cosmology*. New York: Oxford University Press.

120. Dyson, Freeman. 1979. "Time Without End: Physics and Biology in an Open Universe." *Reviews of Modern Physics*, 51, no. 3, July, 447. goo.gl/FM6ezU

121. Pollack, James, and Carl Sagan. 1993. "Planetary Engineering." In J. Lewis, M. Matthews, and M. Guerreri, eds., *Resources of Near Earth Space*. Tucson: University of Arizona Press; Niven, Larry. 1990. *Ringworld*. New York: Ballantine; Stapledon, Olaf. 1968. *The Starmaker*. New York: Dover.

122. Shermer, Michael. 2002. "Shermer's Last Law." *Scientific American*, January, 33.

123. Dyson, Freeman J. 1960. "Search for Artificial Stellar Sources of Infra-Red Radiation." *Science*, 1311, no. 3414, 1667–1668.

124. Maccone, Claudio. 2013. "SETI, Evolution and Human History Merged into a Mathematical Model." *International Journal of Astrobiology*, 12, no. 3, 218–245. goo.gl/zsSkZv

125. Maccone, Claudio. 2014. "Evolution and History in a New 'Mathematical SETI' Model." *Acta Astronautica*, 93, 317–344.

126. Brin, David. 2006. "Shouting at the Cosmos . . . Or How SETI Has Taken a Worrisome Turn into Dangerous Territory." September. goo.gl/ywun4f

127. Hawking, Stephen. 2010. *Into the Universe with Stephen Hawking*. Discovery Channel; Jonathan Leake, "Don't Talk to Aliens, Warns Stephen Hawking." *Sunday Times* (London), April, 25, 2010. See also Shostak, Seth. 1998. *Sharing the Universe: Perspectives on Extraterrestrial Life*. Berkeley, CA: Berkeley Hills Books.

128. Diamond, Jared M. 1991. *The Third Chimpanzee: The Evolution and Future of the Human Animal*. New York: HarperPerennial, 214. For an overview on the risks of contact with extraterrestrial intelligences from the perspective of risk management assessment, see Neal, Mark. 2014. "Preparing for Extraterrestrial Contact." *Risk Management*, 16, no. 2, 6387.

129. For more on this possibility, and its alternative that we might be alone, see Davies, Paul. 1995. *Are We Alone?* New York: Basic Books; Davies, Paul. 2010. *The Eerie Silence: Renewing Our Search for Alien Intelligence*. Boston: Houghton Mifflin Harcourt; Morris, Simon Conway. 2003. *Life's Solution: Inevitable Humans in a Lonely Universe*. Cambridge, UK: Cambridge University Press.

130. Grams, Martin. 2008. *The Twilight Zone: Unlocking the Door to a Television Classic*. Churchville, MD: OTR Publishing.

131. Gort is instructed to get the deceased Klaatu after the Mary Magdalene–like character who has developed a friendship with Klaatu instructs Gort "*Klaatu barada nikto*," one of the most famous lines in all science fiction. In the original script, after Klaatu is resurrected he explains to the astonished onlooker that this is the future power of science and technology, but the Motion Picture Association of America's censor for the film, Joseph Breen, determined that this was unacceptable and forced the producer to add the line "that power is reserved to the Almighty Spirit." See Blaustein, Julian, Robert Wise, Patricia Neal, and Billy Gray. 1995. *Making the Earth Stand Still*. DVD Extra, 20th Century–Fox Home Entertainment.

132. North, Edmund H. 1951. Script for *The Day the Earth Stood Still*. February 21. goo.gl /E885Gi

133. Michael, George. 2011. "Extraterrestrial Aliens: Friends, Foes, or Just Curious?" *Skeptic*, 16, no. 3. goo.gl/0fbAj

134. Harrison, Albert A. 2000. "The Relative Stability of Belligerent and Peaceful Societies: Implications for SETI." *Acta Astronautica*, 46, nos. 10–12, 707–712; Brin, David. 2009. "The Dangers of First Contact: The Moral Nature of Extraterrestrial Intelligence and a Contrarian Perspective on Altruism." *Skeptic*, 15, no. 3, 1–7.

135. Zubrin, Robert. 2000. *Entering Space: Creating a Spacefaring Civilization*. New York: Penguin Putnam; Michaud, Michael. 2007. *Contact with Alien Civilizations: Our Hopes and Fears About Encountering Extraterrestrials*. New York: Copernicus Books.

136. Peters, Ted. 2011. "The Implications of the Discovery of Extra-Terrestrial Life for Religion." *Philosophical Transactions of the Royal Society A*, February, 369, no. 1936, 644–655.

137. Shklovskii, Iosif, and Carl Sagan. 1964. *Intelligent Life in the Universe*. New York: Dell.

138. Delivered at Bishop Charles Mason Temple on April 3, 1968. Full text available here: goo.gl/Zlcom

139. Warren, Mervyn A. 2001. *King Came Preaching: The Pulpit Power of Dr. Martin Luther King Jr.* Downers Grove, IL: Varsity Press, 193–194.

140. From the lyrics of "Woodstock" by Joni Mitchell, also performed by Crosby, Stills, Nash, and Young. The astronomer Carl Sagan was fond of saying that we are made of "starstuff" (one word). According to Sagan's co-author, collaborator, and wife, Ann Druyan, the originator of the idea was the astronomer Harlow Shapley, who first referenced it in a 1926 publication called *The Universe of Stars*, which was a transcript of a lecture series first broadcast on WEEI radio in which he said: "We are therefore made out of star stuff . . . we feed upon sunbeams, we are kept warm by the radiation of the sun and we are made out of the same materials that constitute the stars."

BIBLIOGRAPHY

Abrahms, Max. 2006. "Why Terrorism Does Not Work." *International Security*, 31, 42–78.

———. 2013. "Bottom of the Barrel." *Foreign Policy*, April 24. goo.gl/hj4J1h

Abrahms, Max, and Matthew S. Gottfried. 2014. "Does Terrorism Pay? An Empirical Analysis." *Terrorism and Political Violence*, forthcoming.

Adams, John. 1788. *A Defence of the Constitutions of Government of the United States of America*, Vol. 3. goo.gl/rkuuDi

Adams, Katherine H., and Michael L. Keene. 2007. *Alice Paul and the American Suffrage Campaign*. Champaign: University of Illinois Press, 206–208.

Agustin, Laura Maria. 2007. *Sex at the Margins: Migration, Labour Markets and the Rescue Industry*. London: Zed Books.

Aharoni, Eyal, Gina Vincent, Carla Harenski, Vince Calhoun, Walter Sinnott-Armstrong, Michael Gazzaniga, and Kent Kiehl. 2013. "Neuroprediction of Future Rearrest." *PNAS*, 110, no. 15, 6223–6228.

Akbar, Arifa. 2010. "Mao's Great Leap Forward 'Killed 45 Million in Four Years.'" *Independent* (London), September 17. goo.gl/uZki.

Alexander, David. 1994. *Star Trek Creator: The Authorized Biography of Gene Roddenberry*. New York: Roc/Penguin.

Amsterdam, Beulah. 1972. "Mirror Self-Image Reactions Before Age Two." *Developmental Psychobiology*, 5, no. 4, 297–305.

Anderson, Kermyt G. 2006. "How Well Does Paternity Confidence Match Actual Paternity?" *Current Anthropology*, 47, no. 3, June, 513–520.

Arkin, R. M., H. Cooper, and T. Kolditz. 1980. "A Statistical Review of the Literature Concerning the Self-Serving Bias in Interpersonal Influence Situations." *Journal of Personality*, 48, 435–448.

Arkush, Elizabeth, and Charles Stanish. 2005. "Interpreting Conflict in the Ancient Andes." *Current Anthropology*, 46, no. 1, February, 3–28.

Asch, Solomon E. 1951. "Studies of Independence and Conformity: A Minority of One Against a Unanimous Majority." *Psychological Monographs*, 70, no. 416.

———. 1955. "Opinions and Social Pressure." *Scientific American*, November, 31–35.

Athreya, Kartik, and Jessie Romero. 2013. "Land of Opportunity? Economic Mobility in the United States." Federal Reserve Bank of Richmond, July.

Atran, Scott. 2010. "Black and White and Red All Over." *Foreign Policy*, April 22. goo.gl /GQePL

———. 2011. *Talking to the Enemy: Religion, Brotherhood, and the (Un)Making of Terrorists*. New York: Ecco.

Auten, Gerald, and Geoffrey Gee. 2009. "Income Mobility in the United States: New Evidence from Income Tax Data." *National Tax Journal*, June, 301–328.

Auten, Gerald, Geoffrey Gee, and Nicholas Turner. 2013. "Income Inequality, Mobility and Turnover at the Top in the United States, 1987–2010." Paper presented at the Allied Social Science Association's annual meeting, San Diego, January 4.

Bacon, F. 1620/1939. *Novum Organum*. In E. A. Burtt, ed., *The English Philosophers from Bacon to Mill*. New York: Random House.

Baily, Michael. 2003. *The Man Who Would Be Queen: The Science of Gender-Bending and Transsexualism*. Washington, DC: National Academies Press.

Bailey, Ronald. 2011. "How Scared of Terrorism Should You Be?" *Reason*, September 6.

Baker, R. Robin, and Mark A. Bellis. 1995. *Human Sperm Competition: Copulation, Masturbation, and Infidelity*. London: Chapman & Hall.

Bannerman, Helen. 1899. *The Story of Little Black Sambo*. London: Grant Richards.

Barad, Judity, and Ed Robertson. 2001. *The Ethics of Star Trek*. New York: Perennial/ HarperCollins.

Barber, Benjamin. 2013. *If Mayors Ruled the World: Dysfunctional Nations, Rising Cities*. New Haven CT: Yale University Press.

Barbieri, Katherine. 2002. *The Liberal Illusion: Does Trade Promote Peace?* Ann Arbor: University of Michigan Press.

Baron-Cohen, Simon. 2011. *The Science of Evil: On Empathy and the Origins of Cruelty*. New York: Basic Books.

Baumeister, Roy. 1990. "Victim and Perpetrator Accounts of Interpersonal Conflict: Autobiographical Narratives About Anger." *Journal of Personality and Social Psychology*, 59, no. 5, 994–1005.

———. 1997. *Evil: Inside Human Violence and Cruelty*. New York: Henry Holt.

Baumeister, Roy, and John Tierney. 2011. *Willpower: Rediscovering the Greatest Human Strength*. New York: Penguin.

Becker, Jasper. 1998. *Hungry Ghosts: Mao's Secret Famine*. New York: Henry Holt.

Beddis, I. R., P. Collins, S. Godfrey, N. M. Levy, and M. Silverman. 1979. "New Technique for Servo-Control of Arterial Oxygen Tension in Preterm Infants." *Archives of Disease in Childhood*, 54, 278–280.

Beech, Richard. 2014. "Thomas Hitzlsperger Comes Out as Being Gay." *Mirror*, January 8.

Beinhocker, Eric. 2006. *The Origin of Wealth: Evolution, Complexity, and the Radical Remaking of Economics*. Cambridge, MA: Harvard Business School Press, 453.

Bekoff, M., ed. 2000. *The Smile of a Dolphin: Remarkable Accounts of Animal Emotions*. New York: Crown Books.

Bentham, Jeremy. 1789/1948. *The Principles of Morals and Legislation*. New York: Macmillan.

Berns, Gregory. 2013. "Dogs Are People Too." *New York Times*, October 6, SR5.

———. 2013. *How Dogs Love Us: A Neuroscientist and His Adopted Dog Decode the Canine Brain*. New York: New Harvest, 226–227.

Berns, Gregory, Andrew Brooks, and Mark Spivak. 2012. "Functional MRI in Awake Unrestrained Dogs." *PLoS ONE* 7, no. 5.

Berns, Gregory, et al. 2005. "Neurobiological Correlates of Social Conformity and Independence During Mental Rotation." *Biological Psychiatry*, 58, August 1, 245–253.

Bernstein, Charles. 1984. *Sambo's: Only a Fraction of the Action: The Inside Story of a Restaurant Empire's Rise and Fall*. Burbank, CA: National Literary Guild.

Bernstein, Elizabeth. 2010. "Militarized Humanitarianism Meets Carceral Feminism: The Politics of Sex, Rights, and Freedom in Contemporary Antitrafficking Campaigns." *Signs*, Autumn, 45–71.

Betzig, Laura. 2005. "Politics as Sex: The Old Testament Case." *Evolutionary Psychology*, 3, 326.

Bhayana, Neha. 2011. "Indian Men Lead in Sexual Violence, Worst on Gender Equality." *Times of India*, March 7.

Bittman, Mark. 2012. "We're Eating Less Meat. Why?" *New York Times*, January 10.

Black, Donald. 1983. "Crime as Social Control." *American Sociological Review*, 48, 34–45.

Blackford, Jennifer; A. Allen, R. Cowan, and S. Avery. 2012. "Amygdala and Hippocampus Fail to Habituate to Faces in Individuals with an Inhibited Temperament." *Social Cognitive and Affective Neuroscience*, January, 143–150.

Blackmore, S., and R. Moore. 1994. "Seeing Things: Visual Recognition and Belief in the Paranormal." *European Journal of Parapsychology*, 10, 91–103.

Blackmun, Harry. 1994. Dissent. *Bruce Edwin Callins, Petitioner, v. James A. Collins, Director*, Texas Department of Criminal Justice, Institutional Division. Supreme Court of the United States, no. 93–7054.

Blackstone, William. 1765. *Commentaries on the Laws of England*, I, 411–412.

Blair, B., and M. Brown. 2011. "World Spending on Nuclear Weapons Surpasses $1 Trillion per Decade." *Global Zero*. goo.gl/xDcjg5

Bloom, Paul. 2013. *Just Babies: The Origins of Good and Evil*. New York: Crown.

Boehm, Christopher. 2012. *Moral Origins: The Evolution of Virtue, Altruism, and Shame*. New York: Basic Books.

———. 2014. "The Moral Consequences of Social Selection." *Behaviour*, 151, 167–183.

Bonvillian, J. D., and F. G. P. Patterson. 1997. "Sign Language Acquisition and the Development of Meaning in a Lowland Gorilla." In C. Mandell and A. McCabe, eds., *The Problem of Meaning: Behavioral and Cognitive Perspectives*. Amsterdam: Elsevier.

Bosco, David. 2014. *Rough Justice: The International Criminal Court in a World of Power Politics*. New York: Oxford University Press.

Botelho, Greg, and Marlena Baldacci. 2014. "Brigadier General Accused of Sex Assault Must Pay over $20,000; No Jail Time." *CNN*.

Boven, Leaf Van. 2000. "Pluralistic Ignorance and Political Correctness: The Case of Affirmative Action." *Political Psychology*, 21, no. 2, 267–276.

Bowler, Kate. 2013. *Blessed: A History of the American Prosperity Gospel*. New York: Oxford University Press.

Bowles, S. 2009. "Did Warfare Among Ancestral Hunter-Gatherers Affect the Evolution of Human Social Behaviors?" *Science*, 324, 1293–1298.

Boyd, Robert, and Peter J. Richerson. 1992. "Punishment Allows the Evolution of Cooperation (or Anything Else) in Sizable Groups." *Ethology and Sociobiology*, 13, 171–195.

———. 2005. *The Origin and Evolution of Cultures*. New York: Oxford University Press.

Branas, Charles C., Therese S. Richmond, Dennis P. Culhane, Thomas R. Ten Have, and Douglas J. Wiebe. 2009. "Investigating the Link Between Gun Possession and Gun Assault." *American Journal of Public Health*, November, 99, no. 11, 2034–2040.

Brand, Stewart. 2013. *City-Based Global Governance*. The Long Now Foundation. goo.gl/FbS5E

Brass, Marcel, and Patrick Haggard. 2007. "To Do or Not to Do: The Neural Signature of Self-Control." *Journal of Neuroscience*, 27, no. 34, 9141–9145.

Breiter, Hans, N. Etcoff, P. Whalen, W. Kennedy, S. Rauch, R. Buckner, M. Srauss, S. Hyman, and B. Rosen. 1996. "Response and Habituation of the Human Amygdala During Visual Processing of Facial Expression." *Neuron*, November, 17, 875–887.

Briggs, Robin. 1996. *Witches and Neighbors: The Social and Cultural Context of European Witchcraft*. New York: Viking.

Brin, David. 2006. "Shouting at the Cosmos . . . Or How SETI Has Taken a Worrisome Turn into Dangerous Territory." September. goo.gl/ywun4f

———. 2009. "The Dangers of First Contact: The Moral Nature of Extraterrestrial Intelligence and a Contrarian Perspective on Altruism." *Skeptic*, 15, no. 3, 1–7.

Brodie, Bernard. 1946. *The Absolute Weapon: Atomic Power and World Order*. New York: Harcourt, Brace.

Bronowski, Jacob. 1956. *Science and Human Values*. New York: Julian Messner.

Brooks, Arthur C. 2006. *Who Really Cares: The Surprising Truth About Compassionate Conservatism*. New York: Basic Books.

Brosnan, Sarah F. 2008. "Fairness and Other-Regarding Preferences in Nonhuman Primates." In Paul Zak, ed., *Moral Markets: The Critical Role of Values in the Economy*. Princeton, NJ: Princeton University Press.

Brosnan, Sarah F., and Frans de Waal. 2003. "Monkeys Reject Unequal Pay." *Nature*, 425, September 18, 297–299.

Broszat, Martin. 1967. "Nationalsozialistiche Konzentrationslager 1933–1945." In H. Bucheim, ed., *Anatomie des SS-Staates*. 2 vols. Munich: Deutscher Taschenbuchverlag.

Brown, Anthony Cave, ed. 1978. *DROPSHOT: The American Plan for World War III Against Russia in 1957*. New York: Dial Press.

Brown, Susan Love, ed. 2002. *Intentional Community: Anthropological Perspective*. Albany: State University of New York Press, 156.

Browning, Christopher. 1991. *The Path to Genocide: Essays on Launching the Final Solution*. Cambridge, UK: Cambridge University Press, 143.

Brugger, P., T. Landis, and M. Regard. 1990. "A 'Sheep-Goat Effect' in Repetition Avoidance: Extra-Sensory Perception as an Effect of Subjective Probability?" *British Journal of Psychology*, 81, 455–468.

Bull, Henry. 1961. *The Control of the Arms Race*. London: Institute of Strategic Studies.

———. 1995. *The Anarchical Society*. London: Macmillan.

Burger, Jerry. 2009. "Replicating Milgram: Would People Still Obey Today?" *American Psychologist*, 64, 1–11.

Burke, Edmund. 1790/1967. *Reflections on the Revolution in France*. London: J. M. Dent & Sons.

———. 2009. *The Works of the Right Honorable Edmund Burke*, ed. Charles Pedley. Ann Arbor: University of Michigan Library.

Burks, S. V., J. P. Carpenter, L. Goette, and A. Rustichini. 2009. "Cognitive Skills Affect Economic Preferences, Strategic Behavior, and Job Attainment." *Proceedings of the National Academy of Sciences*, 106, 7745–7750.

Burtless, Gary. 2014. "Income Growth and Income Inequality: The Facts May Surprise You." Washington, DC: Brookings Institution.

Buss, David. 2001. *The Dangerous Passion: Why Jealousy Is as Necessary as Love and Sex.* New York: Free Press.

——. 2002. "Human Mate Guarding." *NeuroEndocrinology Letters*, December, 23, no. 4, 23–29.

——. 2003. *The Evolution of Desire: Strategies of Human Mating.* New York: Basic Books, 266.

——. 2005. *The Murderer Next Door: Why the Mind Is Designed to Kill.* New York: Penguin.

——. 2011. *Evolutionary Psychology: The New Science of the Mind.* New York: Pearson.

Byrne, R. 1995. *The Thinking Ape: Evolutionary Origins of Intelligence.* Oxford, UK: Oxford University Press.

Callahan, Tim. 2012. "Is Ours a Christian Nation?" *Skeptic*, 17, no. 3, 31–55.

Camerer, Colin. 2003. *Behavioral Game Theory.* Princeton, NJ: Princeton University Press.

Cannino-Thompson, Jennifer, Ronald Cotton, and Erin Torneo. 2010. *Picking Cotton.* New York: St. Martin's Press.

Caplan, Brian, and Stephen C. Miller. 2010. "Intelligence Makes People Think Like Economists: Evidence from the General Social Survey." *Intelligence*, 38, 636–647.

Carlton, Jim. 2013. "A Winter Without Walruses." *Wall Street Journal*, October 4, A4.

Carroll, Rory. 2013. "Elon Musk's Mission to Mars." *Guardian*, July 17. goo.gl/IFsXP

Carter, David. 2004. *Stonewall: The Riots That Sparked the Gay Revolution.* New York: St. Martin's Press.

Cartwright, James, et al. 2012. "Global Zero US Nuclear Policy Commission Report: Modernizing US Nuclear Strategy, Force Structure and Posture." *Global Zero*, May.

Casey, Gerard. 2012. *Libertarian Anarchy: Against the State.* New York: Continuum International Publishing.

Cesarani, David. 2006. *Becoming Eichmann: Rethinking the Life, Crimes, and Trial of a "Desk Murderer."* New York: Da Capo Press.

Chambers, John R., Lawton K. Swan, and Martin Heesacker. 2013. "Better Off Than We Know: Distorted Perceptions of Incomes and Income Inequality in America." *Psychological Science*, 1–6. goo.gl/U5cyNR

Chenoweth, Erica. 2013. "Nonviolent Resistance." TEDx Boulder. goo.gl/t00JSa

Chenoweth, Erica, and Maria J. Stephan. 2011. *Why Civil Resistance Works: The Strategic Logic of Nonviolent Conflict.* New York: Columbia University Press.

Chomsky, Noam. 1967. "The Responsibility of Intellectuals." *New York Review of Books*, 8, no. 3. goo.gl/wRPVH

Chu, Henry. 2013. "Gay British Scientist Gets Posthumous Royal Pardon." *Los Angeles Times*, December 25, A1, 7.

Clark, Gregory. 2007. *A Farewell to Alms: A Brief Economic History.* Princeton, NJ: Princeton University Press, 2–3.

——. 2014. *The Son Also Rises: Surnames and the History of Social Mobility.* Princeton, NJ: Princeton University Press, 5.

Clemens, Walter C. Jr. 2013. *Complexity Science and World Affairs.* Albany: State University of New York Press.

Cobb, T. R. R. 1858 (1999). *An Inquiry into the Law of Negro Slavery in the United States.* Reprint University of Georgia.

Coe, Michael D. 2005. *The Maya*, 7th ed. London: Thames & Hudson.

Cohen, I. Bernard. 1985. *Revolution in Science.* Cambridge, MA: Harvard University Press.

Cohen, Carl, and Tom Regan. 2001. *The Animal Rights Debate.* Lanham, MD: Rowman & Littlefield.

Collins, Katie. 2013. "Study: Elephants Found to Understand Human Pointing Without Training." *Wired.* goo.gl/4WdLVu

Collins, Randall. 2008. *Violence: A Micro-Sociological Theory.* Princeton, NJ: Princeton University Press.

Conover, Ted. 2013. "The Stream: Ted Conover Goes Undercover as a USDA Meat Inspector." *Harper's,* April 15. goo.gl/6sP9M

Conot, Robert E. 1993. *Justice at Nuremberg.* New York: Basic Books.

Cooney, Mark. 1997. "The Decline of Elite Homicide." *Criminology,* 35, 381–407.

Cords, M., and S. Thurnheer, 1993. "Reconciling with Valuable Partners by Long-Tailed Macaques." *Behaviour,* 93, 315–325.

Cowell, Alan. 2013. "Ugandan Lawmakers Pass Measure Imposing Harsh Penalties on Gays." *New York Times,* December 20, A8.

Cronin, Audrey. 2011. *How Terrorism Ends: Understanding the Decline and Demise of Terrorist Campaigns.* Princeton, NJ: Princeton University Press.

Cruz, F., V. Carrion, K. J. Campbell, C. Lavoie, and C. J. Donlan. 2009. "Bio-Economics of Large-Scale Eradication of Feral Goats from Santiago Island, Galápagos." *Wildlife Management,* 73, 191–200.

Daly, Martin, and Margo Wilson. 1988. *Homicide.* New York: Aldine De Gruyter.

———. 1999. *The Truth About Cinderella: A Darwinian View of Parental Love.* New Haven, CT: Yale University Press.

Daly, Martin, Margo Wilson, and Shawn Vasdev. 2001. "Income Inequality and Homicide Rates in Canada and the United States." *Canadian Journal of Criminology,* 43, no. 2, 219–236.

Damasio, Antonio R. 1994. *Descartes' Error: Emotion, Reason, and the Human Brain.* New York: Putnam.

Daniel, Lisa. 2012. "Panetta, Dempsey Announce Initiatives to Stop Sexual Assault." *American Forces Press Service,* April 16. goo.gl/n90vmq

Darwin, Charles. 1859. *On the Origin of Species by Means of Natural Selection.* London: John Murray, 488–489.

———. 1871. *The Descent of Man and Selection in Relation to Sex.* London: John Murray, 69, 571.

Davies, Paul. 1995. *Are We Alone?* New York: Basic Books.

———. 2010. *The Eerie Silence: Renewing Our Search for Alien Intelligence.* Boston: Houghton Mifflin Harcourt.

Davis, D. B. 1984. *Slavery and Human Progress.* New York: Oxford University Press.

Davis, Kate, and David Heilbroner, directors. 2010. *Stonewall Uprising.* Documentary.

Davis, M. H., and W. G. Stephan. 1980. "Attributions for Exam Performance." *Journal of Applied Social Psychology,* 10, 235–248.

Dawkins, Marion S. 1993. *Through Our Eyes Only: The Search for Animal Consciousness.* New York: W. H. Freeman.

Dawkins, Richard. 1976. *The Selfish Gene.* New York: Oxford University Press.

———. 2006. *The God Delusion.* New York: Houghton Mifflin, 31.

———. 2014. "What Scientific Idea Is Ready for Retirement? Essentialism." edge.org /response-detail/25366

Deary, Ian J., G. D. Batty, and C. R. Gale. 2008. "Bright Children Become Enlightened Adults." *Psychological Science,* 19, 1–6.

DeLong, J. Bradford. 1998. "Estimating World GDP, One Million B.C.–Present." goo.gl/7ttdjw

———. 2000. "Cornucopia: The Pace of Economic Growth in the Twentieth Century." Working Paper 7602. National Bureau of Economic Research. goo.gl/PLNJTG

De Waal, Frans. 1982. *Chimpanzee Politics: Sex and Power Among the Apes.* Baltimore: Johns Hopkins University Press, 203, 207.

———. 1989. *Peacemaking Among Primates.* Cambridge, MA: Harvard University Press.

———. 1996. *Good Natured: The Origins of Right and Wrong in Humans and Other Animals.* Cambridge, MA: Harvard University Press.

———. 1997. "Food Transfers Through Mesh in Brown Capuchins." *Journal of Comparative Psychology,* 111, 370–378.

———. 2005. *Our Inner Ape.* New York: Riverhead Books, 175.

———. 2008. "How Selfish an Animal? The Case of Primate Cooperation." In Paul Zak, ed., *Moral Markets: The Critical Role of Values in the Economy.* Princeton, NJ: Princeton University Press.

De Waal, Frans, and Frans Lanting. 1998. *Bonobo: The Forgotten Ape.* Berkeley: University of California Press.

Dennett, Daniel. 1997. *Kinds of Minds.* New York: Basic Books.

———. 2003. *Freedom Evolves.* New York: Viking.

Deschner, Amy, and Susan A. Cohen. 2003. "Contraceptive Use Is Key to Reducing Abortion Worldwide." *The Guttmacher Report on Public Policy,* October, 6, no. 4, goo.gl/Ovgge

Diamandis, Peter, and Steven Kotler. 2012. *Abundance: The Future Is Better Than You Think.* New York: Free Press, 8.

Diamond, Jared M. 1991. *The Third Chimpanzee: The Evolution and Future of the Human Animal.* New York: HarperPerennial, 214.

———. 2012. *The World Until Yesterday: What Can We Learn from Traditional Societies?* New York: Viking.

Doris, John Michael. 2002. *Lack of Character: Personality and Moral Behavior.* New York: Cambridge University Press.

Dorussen, Han, and Hugh Ward. 2010. "Trade Networks and the Kantian Peace." *Journal of Peace Research,* 47, no. 1, 229–242.

Douglas, Stephen. 1858. In Harold Holzer, ed. 1994. *The Lincoln-Douglas Debates: The First Complete, Unexpurgated Text.* New York: Fordham University Press, 55.

Draper, Patricia. 1978. "The Learning Environment for Aggression and Anti-Social Behavior among the !Kung." In *Learning Non-Aggression: The Experience of Nonliterate Societies,* ed. A. Montagu. New York: Oxford University Press, 46.

Drexler, Eric K. 1986. *Engines of Creation.* New York: Anchor Books.

D'Souza, Dinesh. 2008. *What's So Great About Christianity.* Carol Stream, IL: Tyndale House.

Duncan, Craig, and Tibor R. Machan. 2005. *Libertarianism: For and Against.* Lanham, MD: Rowman & Littlefield.

Dutton, Kevin. 2012. *The Wisdom of Psychopaths: What Saints, Spies, and Serial Killers Teach Us About Success.* New York: Farrar, Straus & Giroux.

Dyson, Freeman J. 1960. "Search for Artificial Stellar Sources of Infra-Red Radiation." *Science,* 1311, no. 3414, 1667–1668.

———. 1979. "Time Without End: Physics and Biology in an Open Universe." *Reviews of Modern Physics,* 51, no. 3, July, 447.

Eaves, L. J., H. J. Eysenck, and N. G. Martin. 1989. *Genes, Culture and Personality: An Empirical Approach*. San Diego: Academic Press.

Eckholm, Erik, and John Schwartz. 2014. "Timeline Describes Frantic Scene at Oklahoma Execution." *New York Times*, May 1, A1.

Eddington, Arthur Stanley. 1938. *The Philosophy of Physical Science*. Ann Arbor: University of Michigan Press.

Edgerton, Robert. 1992. *Sick Societies: Challenging the Myth of Primitive Harmony*. New York: Free Press.

Edmonds, David. 2013. *Would You Kill the Fat Man?* Princeton, NJ: Princeton University Press.

Eisner, Manuel. 2003. "Long-Term Historical Trends in Violent Crime." *Crime & Justice*, 30, 83–142.

———. 2011. "Killing Kings: Patterns of Regicide in Europe, 600–1800." *British Journal of Criminology*, 51, 556–577.

Emerson, Ralph Waldo. 1860. "Fate." In *The Conduct of Life*. Boston: Ticknor and Fields.

Epstein, Richard A. 2014. *The Classical Liberal Constitution: The Uncertain Quest for Limited Government*. Cambridge, MA: Harvard University Press.

Evans, Gareth. 2014. "Nuclear Deterrence in Asia and the Pacific." *Asia & the Pacific Policy Studies*, January, 91–111.

Evans-Pritchard, E. E. 1976. *Witchcraft, Oracles and Magic Among the Azande*. New York: Oxford University Press.

Fallon, James. 2013. *The Psychopath Inside: A Neuroscientist's Personal Journey into the Dark Side of the Brain*. New York: Current, 1.

Farrington, D. P. 2007. "Origins of Violent Behavior Over the Life Span." In D. J. Flannery, A. T. Vazsonyi, and I. D. Waldman, eds., *The Cambridge Handbook of Violent Behavior and Aggression*. New York: Cambridge University Press.

Fathi, Nazila. 2005. "Wipe Israel 'off the map' Iranian Says." *New York Times*, October 27. goo.gl/9EFacw

Fearn, Frances. 1910. *Diary of a Refugee*, ed. Rosalie Urquart. University of North Carolina at Chapel Hill.

Fedorov, Yuri. 2002. "Russia's Doctrine on the Use of Nuclear Weapons." Pugwash Meeting, London, November 15–17.

Fehr, Ernst, H. Bernhard, and B. Rockenbach. 2008. "Egalitarianism in Young Children." *Nature*, 454, 1079–1083.

Fehr, Ernst, and Simon Gachter. 2002. "Altruistic Punishment in Humans," *Nature*, 415, 137–140.

Ferguson, Niall. 2000. *The Pity of War: Explaining World War I*. New York: Basic Books.

Ferguson, R. Brian. 2013. "Pinker's List: Exaggerating Prehistoric War Mortality." In *War, Peace, and Human Nature*. ed. Douglas P. Fry. New York: Oxford University Press, 112–129.

Ferris, Timothy. 2010. *The Science of Liberty: Democracy, Reason, and the Laws of Nature*. New York: HarperCollins.

Festinger, Leon, Henry W. Riecken, and Stanley Schachter. 1964. *When Prophecy Fails: A Social and Psychological Study*. New York: Harper & Row, 3.

Fettweis, Christopher. 2010. *Dangerous Times?: The International Politics of Great Power Peace*. Washington, DC: Georgetown University Press.

Filmer, Robert. 1680. *Patriarcha, or the Natural Power of Kings*. www.constitution.org/eng/patriarcha.htm

Fletcher, George P. 1999. *A Crime of Self-Defense: Bernhard Goetz and the Law on Trial*. Chicago: University of Chicago Press.

Flexner, Eleanor. 1959/1996. *Century of Struggle*. Cambridge, MA: Belknap Press.

Flower, M. 1989. "Neuromaturation and the Moral Status of Human Fetal Life." In *Abortion Rights and Fetal Personhood*. E. Doerr and J. Prescott, eds. Centerline Press, 65–75.

Flynn, James. 1984. "The Mean IQ of Americans: Massive Gains 1932–1978." *Psychological Bulletin*, 101, 171–191.

———. 1987. "Massive IQ Gains in 14 Nations: What IQ Tests Really Measure." *Psychological Bulletin*, 101, 171–191.

———. 2007. *What Is Intelligence?* Cambridge, UK: Cambridge University Press.

———. 2012. *Are We Getting Smarter?: Rising IQ in the Twenty-first Century*. Cambridge, UK: Cambridge University Press.

Foer, Jonathan Safran. 2010. *Eating Animals*. Boston: Back Bay Books.

Foot, Phillipa. 1967. "The Problem of Abortion and the Doctrine of Double Effect." *Oxford Review*, 5, 5–15.

Francione, Gary. 2006. "The Great Ape Project: Not So Great." *Animal Rights: The Abolitionist Approach*. goo.gl./ojbptB

Fried, I., R. Mukamel, and G. Kreimann. 2011. "Internally Generated Preactivation of Single Neurons in Human Medial Frontal Cortex Predicts Volition." *Neuron*, 69, 548–562.

Fry, Douglas, and Patrik Söderberg. 2013. "Lethal Aggression in Mobile Forager Bands and Implications for the Origins of War." *Science*, July 19, 270–273.

Fukuyama, Francis. 2011. *The Origins of Political Order: From Prehuman Times to the French Revolution*. New York: Farrar, Straus & Giroux.

Galdikas, B. M. F. 1995. *Reflections of Eden: My Years with the Orangutans of Borneo*. Boston: Little, Brown.

Gangestad, S. W., and Randy Thornhill. 1997. "The Evolutionary Psychology of Extra-Pair Sex: The Role of Fluctuating Asymmetry." *Evolution and Human Behavior*, 18, no. 28, 69–88.

Gartzke, E., and J. J. Hewitt. 2010. "International Crises and the Capitalist Peace." *International Interactions*, 36, 115–145.

Gat, Azar. 2006. *War in Human Civilization*. New York: Oxford University Press.

———. In press. "Rousseauism I, Rousseauism II, and Rousseauism of Sorts: Shifting Perspectives on Aboriginal Human Peacefulness and their Implications for the Future of War." *Journal of Peace Research*.

Gates, Bill, and Melinda Gates. 2014. Annual Letter of the Bill and Melinda Gates Foundation. annual letter. gatesfoundation.org.

Gebauer, Jochen, Andreas Nehrlich, Constantine Sedikides, and Wiebke Neberich. 2013. "Contingent on Individual-Level and Culture-Level Religiosity." *Social Psychological and Personality Science*, 4, no. 5, 569–578.

Genovese, Eugene D., and Elizabeth Fox-Genovese. 2011. *Fatal Self-Deception: Slaveholding Paternalism in the Old South*. New York: Cambridge University Press, 1.

Ghemawat, Pankaj. 2011. *World 3.0: Global Prosperity and How to Achieve it*. Cambridge, MA: Harvard Business Review Press.

Giangreco, Dennis M. 1998. "Transcript of 'Operation Downfall' [US Invasion of Japan]: US Plans and Japanese Counter-Measures." *Beyond Bushido: Recent Work in Japanese Military History*. goo.gl/ORyRQT

———. 2009. *Hell to Pay: Operation Downfall and the Invasion of Japan 1945–1947*. Annapolis, MD: Naval Institute Press.

Gigerenzer, Gerd, and Reinhard Selten. 2002. *Bounded Rationality*. Cambridge, MA: MIT Press.

Gilen, Martin. 2012. *Affluence and Influence: Economic Inequality and Political Power in America*. Princeton, NJ: Princeton University Press, 1.

Gintis, Herbert, Samuel Bowles, Robert Boyd, and Ernst Fehr. 2005. *Moral Sentiments and Material Interests*. Cambridge, MA: MIT Press.

Glaeser, Edward. 2011. *The Triumph of the City: How Our Greatest Invention Makes Us Richer, Smarter, Greener, Healthier, and Happier*. New York: Penguin.

Glimcher, P. W. 2003. *Decisions, Uncertainty, and the Brain: The Science of Neuroeconomics*. Cambridge, MA: MIT Press.

Glover, J. 1999. *Humanity: A Moral History of the Twentieth Century*. London: Jonathan Cape.

Glowacki, Luke, and Richard W. Wrangham. 2013. "The Role of Rewards in Motivating Participation in Simple Warfare." *Human Nature*, 24, 444–460.

Goessling, Ben. 2014. "86 Percent OK with Gay Teammate." ESPN.com, February 17.

Goetz, Aaron T., Todd K. Shackelford, Steven M. Platek, Valerie G. Starratt, and William F. McKibbin. 2007. "Sperm Competition in Humans: Implications for Male Sexual Psychology, Physiology, Anatomy, and Behavior." *Annual Review of Sex Research*, 18, no. 1, 1–22.

Goldhagen, Daniel Jonah. 2009. *Worse Than War: Genocide, Eliminationism, and the Ongoing Assault on Humanity*. New York: PublicAffairs, 158.

Goldstein, Joshua. 2011. *Winning the War on War: The Decline of Armed Conflict Worldwide*. New York: E. P. Dutton.

Gopnik, Alison. 2009. *The Philosophical Baby: What Children's Minds Tell us About Truth, Love, and the Meaning of Life*. New York: Farrar, Straus & Giroux.

Grandin, Temple, and Catherine Johnson. 2006. *Animals in Translation: Using the Mysteries of Autism to Decode Animal Behavior*. New York: Harcourt.

Gregg, Becca Y. 2013. "Speakers Differ on Fighting Gun Violence." *Reading Eagle*, September 19.

Gregg, Justin. 2013. *Are Dolphins Really Smart? The Mammal Behind the Myth*. New York: Oxford University Press.

———. "No, Flipper Doesn't Speak Dolphinese." *Wall Street Journal*, December 21, C3.

Griffin, D. R. 2001. *Animal Minds: Beyond Cognition to Consciousness*. Chicago: University of Chicago Press.

Grinde, Björn. 2002. *Darwinian Happiness: Evolution as a Guide for Living and Understanding Human Behavior*. Princeton, NJ: Darwin Press.

———. 2002. "Happiness in the Perspective of Evolutionary Psychology." *Journal of Happiness Studies*, 3, 331–354.

Gross, Samuel R., Barbara O'Brien, Chen Hu, and Edward H. Kennedy. 2014. "Rate of False Conviction of Criminal Defendants Who Are Sentenced to Death." *Proceedings of the National Academy of Sciences*. April 28. goo.gl/mIjR2M

Grossman, Dave. 2009. *On Killing*. Boston: Little, Brown.

Grossman, Richard S. 2013. *Wrong: Nine Economic Policy Disasters and What We Can Learn from Them*. New York: Oxford University Press.

Hadfield, Chris. 2013. *An Astronaut's Guide to Life on Earth*. New York: Little Brown.

Haggard, P. 2011. "Decision Time for Free Will." *Neuron*, 69, 404–406.

Haidt, Jonathan. 2012. *The Righteous Mind: Why Good People Are Divided by Politics and Religion*. New York: Pantheon.

Hall, Harriet. 2013. "Does Religion Make People Healthier?" *Skeptic*, 19, no. 1.

Ham, Paul. 2014. *Hiroshima Nagasaki: The Real Story of the Atomic Bombing and Their Aftermath*. New York: St. Martin's Press.

Hamblin, James. 2014. "How Not to Talk About a Culture of Sexual Assault." *Atlantic*, March 29. goo.gl/mwqxiJ

Hamer, Dean, and P. Copeland. 1994. *The Science of Desire: The Search for the Gay Gene and the Biology of Behavior*. New York: Simon & Schuster.

Hamilton, Tyler, and Daniel Coyle. 2012. *The Secret Race: Inside the Hidden World of the Tour de France*. New York: Bantam Books.

Hankins, Thomas L. 1985. *Science and the Enlightenment*. Cambridge, UK: Cambridge University Press, 161–163.

Hare, B., and M. Tomasello. 2005. "Human-like Social Skills in Dogs?" *Trends in Cognitive Science*, 9, 439–444.

Hare, B., M. Brown, C. Williamson, and M. Tomasello. 2002. "The Domestication of Social Cognition in Dogs." *Science*, November 22, 298, 1634–1636.

Hare, Robert. 1991. *Without Conscious: The Disturbing World of the Psychopaths Among Us*. New York: Guilford Press.

Harper, George W. 1979. "Build Your Own A-Bomb and Wake Up the Neighborhood." *Analog*, April, 36–52.

Harper, William. 1853. "Harper on Slavery." In *The Pro-Slavery Argument, as Maintained by the Most Distinguished Writers of the Southern States*. Philadelphia: Lippincott, Grambo, & Co., 94. goo.gl/kdggN9

Harris, Amy Julia. 2010. "No More Nukes Books." *Half Moon Bay Review*, August 4. goo.gl/4PsK2n

Harris, Judith Rich. 1998. *The Nurture Assumption: Why Children Turn Out the Way They Do*. New York: Free Press.

Harris, Marvin. 1975. *Culture, People, Nature*, 2nd ed. New York: Crowell.

Harris, Sam. 2010. *The Moral Landscape: How Science Can Determine Human Values*. New York: Free Press.

———. 2012. *Free Will*. New York: Free Press, 5.

Harrison, Albert A. 2000. "The Relative Stability of Belligerent and Peaceful Societies: Implications for SETI." *Acta Astronautica*, 46, nos. 10–12, 707–712.

Hartung, J. 1995. "Love Thy Neighbor: The Evolution of In-Group Morality." *Skeptic*, 3, no. 4, 86–99.

Haslam, S. Alexander, and Stephen D. Reicher. 2012. "Contesting the 'Nature' of Conformity: What Milgram and Zimbardo's Studies Really Show." *PLoS Biol* 10, no. 11, November 20: e1001426. doi:10.1371/journal.pbio.1001426

Haslam, S. Alexander, Stephen D. Reicher, and Joanne R. Smith. 2012. "Working Toward the Experimenter: Reconceptualizing Obedience Within the Milgram Paradigm as Identification-Based Followership." *Perspectives on Psychological Science*, 7, no. 4, 315–324. doi:10.1371/journal.pbio.1001426

Hasson, Uri, Ohad Landesman, Barbara Knappmeyer, Ignacio Vallines, Nava Rubin, and David J. Heeger. 2008. "Neurocinematics: The Neuroscience of Film." *Projections*, Summer, 2, no. 1, 1–26.

Hatemi, Peter K., and Rose McDermott, eds., 2011. *Man Is by Nature a Political Animal*. Chicago: University of Chicago Press.

Hatemi, Peter K., and Rose McDermott. 2012. "The Genetics of Politics: Discovery, Challenges, and Progress." *Trends in Genetics*, October, 28, no. 10, 525–533.

Haugen, Gary A., and Victor Boutros. 2014. *The Locust Effect: Why the End of Poverty Requires the End of Violence*. New York: Oxford University Press, 137.

Hawking, Stephen. 2010. *Into the Universe with Stephen Hawking*. Discovery Channel.

Haynes, J. D. 2011. "Decoding and Predicting Intentions." *Annals of the New York Academy of Sciences*, 1224, no. 1, 9–21.

Heath, Spencer. 1957. *Citadel, Market and Altar: Emerging Society.* Baltimore: Science of Society Foundation.

Heatly, Maurice. 1966. "Whitman Case Notes. The Whitman Archives." *Austin American-Statesman*, March 29.

Hecht, Jennifer Michael. 2013. "The Last Taboo." *Politico.com*, goo.gl/loOSDz

Hedesan, Jo. 2011. "Witch Hunts in Papua New Guinea and Nigeria." *The International*, October 1. goo.gl/Lrkg6N

Heidmann, Jean. 1992. *Extraterrestrial Intelligence.* New York: Cambridge University Press, 210–212.

Henrich, Joseph, et al. 2010. "Markets, Religion, Community Size, and the Evolution of Fairness and Punishment." *Science*, March 19, 327, 1480–1484.

Henrich, Joseph, Robert Boyd, Sam Bowles, Colin Camerer, Ernst Fehr, and Herbert Gintis. 2004. *Foundations of Human Sociality.* New York: Oxford University Press, 8.

Henrich, Joseph, Robert Boyd, Sam Bowles, Colin Camerer, Herbert Gintis, Richard McElreath, and Ernst Fehr. 2001. "In Search of *Homo economicus*: Experiments in 15 Small-Scale Societies." *American Economic Review*, 91, no. 2, 73–79.

Herman, L. M., and P. Morrel-Samuels. 1990. "Knowledge Acquisition and Asymmetry Between Language Comprehension and Production: Dolphins and Apes as General Models for Animals." In M. Bekoff and D. Jamieson, eds., *Interpretation and Explanation in the Study of Animal Behavior.* Boulder, CO: Westview Press.

Herring, Amy H., Samantha M. Attard, Penny Gordon-Larsen, William H. Joyner, and Carolyn T. Halpern. 2013. "Like a Virgin (mother): Analysis of Data from a Longitudinal, US Population Representative Sample Survey." *British Medical Journal*, December 17. 347. goo.gl/bkMVnJ

Herszenhorn, David M., and Michael R. Gordon. 2013. "US Cancels Part of Missile Defense That Russia Opposed." *New York Times*, March 16, A12.

Hibbing, John R., Kevin B. Smith, and John R. Alford. 2013. *Predisposed: Liberals, Conservatives, and the Biology of Political Differences.* New York: Routledge.

Hill, Kashmir. 2011. "Adventures in Self-Surveillance, aka The Quantified Self, aka Extreme Navel-Gazing." *Forbes*, April 7. goo.gl/Euyn

Hiltzik, Michael. 2013. "A Huge Threat to Social Mobility." *Los Angeles Times*, December 22, B1.

Hirsch, David, and Dan Van Haften. 2010. *Abraham Lincoln and the Structure of Reason.* New York: Savas Beatie.

Hirschman, Albert O. 1970. *Exit, Voice, and Loyalty: Responses to Decline in Firms, Organizations, and States.* Cambridge, MA: Harvard University Press.

Hitchens, Christopher. 1995. *The Missionary Position: Mother Teresa in Theory and Practice.* Brooklyn, NY: Verso.

———. 2003. "Mommie Dearest." *Slate*, October 20. goo.gl/efSMV

———. 2007. *God Is Not Great: How Religion Poisons Everything.* New York: Twelve.

———. 2010. "The New Commandments." *Vanity Fair*, April. goo.gl/lcXo

Hobbes, Thomas. 1642/1998. *De Cive, or The Citizen.* New York: Cambridge University Press.

———. 1651/1968. *Leviathan, or The Matter, Forme and Power of a Common Wealth Ecclesiasticall and Civil*, ed. C. B. Macpherson. New York: Penguin, 76.

———. 1839. *The English Works of Thomas Hobbes*, Vol. 1, ed. William Molesworth, ix–1.

Hochschild, Adam. 2005. *Bury the Chains: The British Struggle to Abolish Slavery.* London: Macmillan.

Hofling, Charles K., E. Brotzman, S. Dalrymple, N. Graves, and C. M. Pierce. 1966. "An Experimental Study in Nurse-Physician Relationships." *Journal of Nervous and Mental Disease*, 143, 171–180.

Horgan, John. 2014. "Jared Diamond, Please Stop Propagating the Myth of the Savage Savage!" *Scientific American Blogs*, January 20.

Howard, Philip K. 2014. *The Rule of Nobody: Saving America from Dead Laws and Broken Government*. New York: W. W. Norton.

Hrdy, Sara. 2000. "The Optimal Number of Fathers: Evolution, Demography, and History in the Shaping of Female Mate Preference." *Annals of the New York Academy of Science*, 907, 75–96.

Hume, David. 1739. *A Treatise of Human Nature*. London: John Noon.

———. 1748/1902. *An Enquiry Concerning Human Understanding*. Cambridge, UK: Cambridge University Press.

Hunt, Lynn. 2007. *Inventing Human Rights: A History*. New York: W. W. Norton, 76, 179.

Hutcherson, Francis. 1755. *A System of Moral Philosophy*. London, II, 213.

Hutchinson, Roger. 1842. *The Works of Roger Hutchinson*, ed. J. Bruce. Cambridge, UK: Cambridge University Press.

Isaacson, Walter. 2004. *Benjamin Franklin: An American Life*. New York: Simon & Schuster, 311–312.

Jäckel, Eberhard. 1993. *Hitler in History*. Lebanon, NH: Brandeis University Press/University Press of New England.

Jackman, Mary R. 1994. *The Velvet Glove: Paternalism and Conflict in Gender, Class, and Race Relations*. Berkeley: University of California Press, 13.

Jacobs, Jane. 1961. *The Death and Life of Great American Cities*. New York: Random House.

Janoff-Bulman, R., S. Sheikh, and S. Hepp. 2009. "Proscriptive versus Prescriptive Morality: Two Faces of Moral Regulation." *Journal of Personality and Social Psychology*, 96, 521–537.

Jefferson, Thomas. 1804. Letter to Judge John Tyler Washington, June 28. *The Letters of Thomas Jefferson, 1743–1826*. goo.gl/hn6qNp

Johnson, David K. 2004. *The Lavender Scare: The Cold War Persecution of Gays and Lesbians in the Federal Government*. Chicago: University of Chicago Press.

Johnson, Steven. 2006. *Everything Bad Is Good for You: How Today's Popular Culture Is Actually Making Us Smarter*. New York: Riverhead.

Jones, Garret. 2008. "Are Smarter Groups More Cooperative? Evidence from Prisoner's Dilemma Experiments, 1959–2003." *Journal of Economic Behavior & Organization*, 68, 489–497.

Jones, Jacqueline. 2013. *A Dreadful Deceit: The Myth of Race from the Colonial Era to Obama's America*. New York: Basic Books.

Jost, J. T., C. M. Federico, and J. L. Napier. 2009. "Political Ideology: Its Structures, Functions, and Elective Affinities." *Annual Review of Psychology*, 60, 307–337.

Kahneman, Daniel. 2003. "Maps of Bounded Rationality: Psychology for Behavioral Economics." *American Economic Review*, 93, no. 5, 1449–1475.

Kaku, Michio. 2005. *Parallel Worlds: The Science of Alternative Universes and Our Future in the Cosmos*. New York: Doubleday, 317.

———. 2010. "The Physics of Interstellar Travel: To One Day Reach the Stars." goo.gl/TBExNt

———. 2011. *Physics of the Future: How Science Will Shape Human Destiny and Our Daily Lives by the Year 2100*. New York: Doubleday, 21.

Kanazawa, S. 2010. "Why Liberals and Atheists Are More Intelligent." *Social Psychology Quarterly*, 73, 33–57.

Kang, Kyungkook, and Jacek Kugler. 2012. "Nuclear Weapons: Stability of Terror." In Sean Clark and Sabrina Hoque, eds., *Debating a Post-American World*. New York: Routledge.

Kano, Takayoshi, and Evelyn Ono Vineberg. 1992. *The Last Ape: Pygmy Chimpanzee Behavior and Econology*. Ann Arbor, MI: University Microfilms International.

Kant, Immanuel. 1795/1983. "Perpetual Peace: A Philosophical Sketch." In Ted Humphrey (trans.), *Perpetual Peace and Other Essays*. Indianapolis: Hackett.

Kardashev, Nikolai. 1964. "Transmission of Information by Extraterrestrial Civilizations." *Soviet Astronomy*, 8, 217.

Karouny, Mariam, and Ibon Villelabeitia. 2006. "Iraq Court Upholds Saddam Death Sentence." *Washington Post*, December 26. goo.gl/b9n6oW

Katz, Steven T. 1994. *The Holocaust in Historical Perspective*, Vol. 1. New York: Oxford University Press.

Kearny, Cresson. 1987. *Nuclear War Survival Skills*. Cave Junction, OR: Oregon Institute of Science and Medicine.

Keegan, John. 1994. *A History of Warfare*. New York: Random House.

Keeley, Lawrence. 1996. *War Before Civilization: The Myth of the Peaceful Savage*. New York: Oxford University Press.

Kellerman, Arthur L. 1998. "Injuries and Deaths Due to Firearms in the Home." *Journal of Trauma*, 45, no. 2, 263–267.

Kellerman, Arthur L., and J. A. Mercy. 1992. "Men, Women, and Murder: Gender-Specific Differences in Rates of Fatal Violence and Victimization." *Journal of Trauma*, July, 33, no. 1, 1–5.

Kelly, Kevin. 2014. "The Technium. A Conversation with Kevin Kelly by John Brockman." Edge.org.

Kelly, Lawrence. 2000. *Warless Societies and the Origins of War*. Ann Arbor: University of Michigan Press.

Keltner, Dacher, Aleksandr Kogan, Paul K. Piff, and Sarina Saturn. 2014. "The Sociocultural Appraisals, Values, and Emotions (SAVE) Framework of Prosociality: Core Processes from Gene to Meme." *Annual Review of Psychology*, 65, no. 25, 1–25.

Kenner, Robert, and Melissa Robledo. 2008. *Food, Inc*. Participant Media.

Kidd, David Comer, and Emanuele Castano. 2013. "Reading Literary Fiction Improves Theory of Mind." *Science*, 342, no. 6156, 377–380.

Kieckhefer, Richard. 1994. "The Specific Rationality of Medieval Magic." *American Historical Review*, 99, no. 3, 813–836.

Kiehl, Kent A., et al. 2001. "Limbic Abnormalities in Affective Processing by Criminal Psychopaths as Revealed by Functional Magnetic Resonance Imaging." *Biological Psychiatry*, 50, 677–684.

Kiernan, Ben. 2009. *Blood and Soil: A World History of Genocide and Extermination from Sparta to Darfur*. New Haven, CT: Yale University Press.

King, Coretta Scott. 1969. *My Life with Martin Luther King Jr*. New York: Holt, Rinehart, & Winston.

King, Neil. 2013. "Evangelical Leader Preaches a Pullback from Politics, Culture Wars." *Wall Street Journal*, October 22, A1, 14.

Kitchens, Caroline. 2014. "It's Time to End 'Rape Culture' Hysteria." *Time*, March 20. goo.gl/lWffXq

Klee, Ernst, Wili Dressen, Volker Riess, and Hugh Trevor-Roper. 1996. *"The Good Old Days": The Holocaust As Seen by Its Perpetrators and Bystanders.* New York: William S. Konecky Associates, 163–171.

Klemesrud, Judy. 1977. "Equal Rights Plan and Abortion Are Opposed by 15,000 at Rally." *New York Times*, November 20, A32.

Knight, Richard. 2012. "Are North Koreans Really Three Inches Shorter than South Koreans?" *BBC News Magazine*, April 22. goo.gl/dWWGxp

Kohler, Pamela K., Lisa E. Manhart, and William E. Lafferty. 2008. "Abstinence-Only and Comprehensive Sex Education and the Initiation of Sexual Activity and Teen Pregnancy." *Journal of Adolescent Health*, April, 42, no. 4, 344–351.

K'Olier, Franklin, ed. 1946. *United States Strategic Bombing Survey, Summary Report (Pacific War).* Washington, DC: US Government Printing Office.

Konvitz, Josef W. 1985. *The Urban Millennium: The City-Building Process from the Early Middle Ages to the Present.* Carbondale, IL: Southern Illinois University Press.

Koonz, Claudia. 2005. *The Nazi Conscience.* Cambridge, MA: Harvard University Press.

Koops, B. L., L. J. Morgan and F. C. Battaglia. 1982. "Neonatal Mortality Risk in Relation to Birth Weight and Gestational Age: Update." *Journal of Pediatrics*, 101, 969–977.

Kosko, Bart. 1992. *Fuzzy Engineering.* Englewood Cliffs, NJ: Prentice Hall.

Kostof, Spiro. 1991. *The City Shaped: Urban Patterns and Meanings Through History.* Boston: Little, Brown.

Koyama, N. F., C. Caws, and F. Aureli. 2007. "Interchange of Grooming and Agonistic Support in Chimpanzees." *International Journal of Primatology* 27, no. 5, 1293–1309.

Koyama, N. F., and E. Palagi. 2007. "Managing Conflict: Evidence from Wild and Captive Primates." *International Journal of Primatology*, 27, no. 5, 1235–1240.

Krackow, A., and T. Blass. 1995. "When Nurses Obey or Defy Inappropriate Physician Orders: Attributional Differences." *Journal of Social Behavior and Personality*, 10, 585–594.

Krakauer, Jon. 2004. *Under the Banner of Heaven: A Story of Violent Faith.* New York: Anchor.

Kraus, Michael W., Paul K. Piff, and Dacher Keltner. 2009. "Social Class, Sense of Control, and Social Explanation." *Journal of Personality and Social Pschology*, 97, no. 6, 992–1004.

Kristensen, Hans M., and Robert S. Norris. 2013. "Global Nuclear Weapons Inventories, 1945–2013." *Bulletin of the Atomic Scientists*, 69, no. 5, 75–81.

Krueger, Alan B. 2007. *What Makes a Terrorist: Economics and the Roots of Terrorism.* Princeton, NJ: Princeton University Press.

Kubrick, Stanley. 1964. *Dr. Strangelove or: How I Learned to Stop Worrying and Love the Bomb.* Columbia Pictures.

Kugler, Jacek. 1984. "Terror Without Deterrence: Reassessing the Role of Nuclear Weapons." *Journal of Conflict Resolution*, 28, no. 3, September, 470–506.

———. 2012. "A World Beyond Waltz: Neither Iran nor Israel Should Have the Bomb." PBS, September 12.

Kugler, Tadeusz, Kyung Kook Kang, Jacek Kugler, Marina Arbetman-Rabinowitz, and John Thomas. 2013. "Demographic and Economic Consequences of Conflict." *International Studies Quarterly*, March, 57, no. 1, 1–12.

Kuhn, Deanna, M. Weinstock, and R. Flaton. 1994. "How Well Do Jurors Reason? Competence Dimensions of Individual Variation in a Juror Reasoning Task." *Psychological Science*, 5, 289–296.

Kurzban, Robert. 2012. *Why Everyone (Else) Is a Hypocrite*. Princeton, NJ: Princeton University Press.

Kurzweil, Ray. 2006. *The Singularity Is Near: When Humans Transcend Biology*. New York: Penguin.

Lacina, B., and N. P. Gleditsch. 2005. "Monitoring Trends in Global Combat: A New Dataset in Battle Deaths." *European Journal of Population*, 21, 145–166.

Lacina, B., N. P. Gleditsch, and B. Russett. 2006. "The Declining Risk of Death in Battle." *International Studies Quarterly*, 50, 673–680.

Lambert, Tracy A., Arnold S. Kahn, and Kevin J. Apple. 2003. "Pluralistic Ignorance and Hooking Up." *Journal of Sex Research*, 40, no. 2, May, 129–133.

Larmuseau, M. H. D., J. Vanoverbeke, A. Van Geystelen, G. Defraene, N. Vanderheyden, K. Matthys, T. Wenseleers, and R. Decorte. 2013. "Low Historical Rates of Cuckoldry in a Western European Human Population Traced by Y-Chromosome and Genealogical Data." *Proceedings of the Royal Society B*, December, 280, no. 1772.

La Roncière, Charles de. 1988. "Tuscan Notables on the Eve of the Renaissance." In Philippe Ariès and George Duby (eds.), *A History of Private Life: Revelations of the Medieval World*. Cambridge, MA: Harvard University Press.

Laumann, E. O., J. H. Gagnon, R. T. Michael, and S. Michaels. 1994. *The Social Organization of Sexuality: Sexual Practices in the United States*. Chicago: University of Chicago Press.

Leake, Jonathan. 2010. "Don't Talk to Aliens, Warns Stephen Hawking." *Sunday Times* (London), April 25. goo.gl/yxZlah

Leblanc, Steven, and Katherine E. Register. 2003. *Constant Battles: The Myth of the Peaceful, Noble Savage*. New York: St. Martin's Press.

Le Bon, Gustave. 1896. *The Crowd: A Study of the Popular Mind*. New York: Macmillan.

Lebow, Richard Ned. 2010. *Why Nations Fight: Past and Future Motives for War*. Cambridge, UK: Cambridge University Press.

Lecky, William Edward Hartpole. 1869. *History of European Morals: From Augustus to Charlemagne*, 2 vols. New York: D. Appleton, Vol. 1.

Lee, Richard B. 1979. *The !Kung San: Men, Women, and Work in a Foraging Society*. New York: Cambridge University Press.

Leeson, Peter T. 2009. *The Invisible Hook*. Princeton, NJ: Princeton University Press.

———. 2014. *Anarchy Unbound: Why Self-Governance Works Better Than You Think*. Cambridge, UK: Cambridge University Press.

Lemkin, Raphael. 1946. "Genocide." *American Scholar*, 15, no. 2, 227–230.

Leonard, J. A., R. K. Wayne, J. Wheeler, R. Valadez, S. Guillén, and C. Vilá. 2002. "Ancient DNA Evidence for Old World Origin of New World Dogs." *Science*, November 22, 298, 1613–1615.

Lettow, Paul. 2005. *Ronald Reagan and His Quest to Abolish Nuclear Weapons*. New York: Random House.

Levak, Brian. 2006. *The Witch-Hunt in Early Modern Europe*. New York: Routledge.

LeVay, Simon. 2010. *Gay, Straight, and the Reason Why: The Science of Sexual Orientation*. New York: Oxford University Press.

Levi, Michael S. 2003. "Panic More Dangerous than WMD." *Chicago Tribune*, May 26. goo.gl/tlVogl

———. 2009. *On Nuclear Terrorism*. Cambridge, MA: Harvard University Press.

——. 2011. "Fear and the Nuclear Terror Threat." *USA Today*, March 24, 9A.

Levi, Primo. 1989. *The Drowned and the Saved*. New York: Vintage Books, 56.

Levin, Yuval. 2013. *The Great Debate: Edmund Burke, Thomas Paine, and the Birth of Left and Right*. New York: Basic Books.

Levy, Jack S., and William R. Thompson. 2011. *The Arc of War: Origins, Escalation, and Transformation*. Chicago: University of Chicago Press.

Lewis, J., M. Matthews, and M. Guerreri, eds., *Resources of Near Earth Space*. Tucson: University of Arizona Press.

Lewis, M., and J. Brooks-Gunn. 1979. *Social Cognition and the Acquisition of Self*. New York: Plenum Press, 296.

Libet, Benjamin. 1985. "Unconscious Cerebral Initiative and the Role of Conscious Will in Voluntary Action." *Behavior and Brain Sciences*, 8, 529–566.

——. 1999. "Do We Have Free Will?" *Journal of Consciousness Studies*, 6, no. 809, 47–57.

Liddle, Andrew, and Jon Loveday. 2009. *The Oxford Companion to Cosmology*. New York: Oxford University Press.

Liebenberg, Louis. 1990. *The Art of Tracking: The Origin of Science*. Cape Town: David Philip.

——. 2013. "Tracking Science: The Origin of Scientific Thinking in Our Paleolithic Ancestors." *Skeptic*, 18, no. 3, 18–24.

Lifton, Robert Jay. 1989. *The Nazi Doctors: Medical Killing and the Psychology of Genocide*. New York: Basic Books.

Lincoln, Abraham. 1854. Fragment on Slavery. July 1. goo.gl/ux6b9Z

——. 1858. In Roy P. Basler, ed., *The Collected Works of Abraham Lincoln*. 1953, Vol. II, August 1, 532.

Lindgren, James. 2006. "Concerns About Arthur Brooks's 'Who Really Cares.'" November 20. goo.gl/iGk9M0

Lingeman, Richard R. June 17, 1973. "There Was Another Fifties." *New York Times Magazine*, 27.

Lipstadt, Deborah E. 2011. *The Eichmann Trial*. New York: Schocken.

Llewellyn Barstow, Anne. 1994. *Witchcraze: A New History of the European Witch Hunts*. New York: HarperCollins.

Lloyd, Christopher. 1968. *The Navy and the Slave Trade: The Suppression of the African Slave Trade in the Nineteenth Century*. London: Cass, 118.

LoBue, V., T. Nishida, C. Chiong, J. S. DeLoache, and J. Haidt. 2011. "When Getting Something Good Is Bad: Even Three-Year-Olds React to Inequality." *Social Development*, 20, 154–170.

Locke, John. 1690. *Second Treatise of Government*, chap. II. Of the State of Nature, sec. 4. goo.gl/RJdaQB

Long, William, and Peter Brecke. 2003. *War and Reconciliation: Reason and Emotion in Conflict Resolution*. Cambridge, MA: MIT Press, 70–71.

Lott, John. 2010. *More Guns, Less Crime: Understanding Crime and Gun Control Laws*, 3rd ed. Chicago: University of Chicago Press.

Low, Bobbi. 1996. "Behavioral Ecology of Conservation in Traditional Societies." *Human Nature* 7, no. 4: 353–379.

Low, Philip, Jaak Panksepp, Diana Reiss, David Edelman, Bruno Van Swinderen, and Christof Koch. 2012. "The Cambridge Declaration on Consciousness." Francis Crick Memorial Conference on Consciousness in Human and non-Human Animals. Cambridge, UK: Churchill College, University of Cambridge.

Lozowick, Yaacov. 2003. *Hitler's Bureaucrats: The Nazi Security Police and the Banality of Evil*. New York: Continuum, 279.

MacCallum, Spencer Heath. 1970. *The Art of Community*. Menlo Park, CA: Institute for Humane Studies.

Maccone, Claudio. 2013. "SETI, Evolution and Human History Merged into a Mathematical Model." *International Journal of Astrobiology*, 12, no. 3, 218–245.

———. 2014. "Evolution and History in a New 'Mathematical SETI' Model." *Acta Astronautica*, 93, 317–344.

MacDonald, Heather. 2008. "The Campus Rape Myth." *City Journal*, Winter, 18, no. 1. goo.gl/dDuaR

Mackay, Charles. 1841/1852/1980. *Extraordinary Popular Delusions and the Madness of Crowds*. New York: Crown.

Mackey, John, and Raj Sisodia. 2013. *Conscious Capitalism: Liberating the Heroic Spirit of Business*. Cambridge, MA: Harvard Business Review Press, 20.

Maclin, Beth. 2008. "A Nuclear Weapon-Free World Is Possible, Nunn Says." Belfer Center for Science and International Affairs, Harvard University, October 20.

MacRae, Allan, and Howard Zehr. 2011. "Right Wrongs the Maori Way." *Yes! Magazine*, July 8. goo.gl/PBd67u

Macy, Michael W., Robb Willer, and Ko Kuwabara. 2009. "The False Enforcement of Unpopular Norms." *American Journal of Sociology*, 115, no. 2, September, 451–490.

Maddison, Agnus. 2006. *The World Economy*. Washington, DC: OECD Publishing.

———. 2007. *Contours of the World Economy 1–2030 AD: Essays in Macro-Economic History*. New York: Oxford University Press.

Maddox, Robert James. 1995. "The Biggest Decision: Why We Had to Drop the Atomic Bomb." *American Heritage*, 46, no. 3, 70–77.

Madison, James. 1788. "The Federalist No. 51: The Structure of the Government Must Furnish the Proper Checks and Balances Between the Different Departments." *Independent Journal*, February 6.

Magnier, Mark, and Tanvi Sharma. 2013. "India Court Makes Homosexuality a Crime Again." *Los Angeles Times*, December 11, A7.

Malinowski, Bronislaw. 1954. *Magic, Science, and Religion*. Garden City, NY: Doubleday, 139–140.

Malmstrom, Frederick V., and David Mullin. "Why Whistleblowing Doesn't Work: Loyalty Is a Whole Lot Easier to Enforce Than Honesty." *Skeptic*, 19, no. 1, 30–34.

Manchester, William. 1987. "The Bloodiest Battle of All." *New York Times*, June 14. goo.gl/d4DcVe

Mankiw, Gregory. 2011. *Principles of Economics*. 6th ed. Stamford, CT: Cengage Learning, 11.

Mannix, D. P. 1964. *The History of Torture*. Sparkford, UK: Sutton, 137–138.

Mar, Raymond, and Keith Oatley. 2008. "The Function of Fiction Is the Abstraction and Simulation of Social Experience." *Perspectives on Psychological Science*, 3, 173–192.

Marino, L. 1988. "A Comparison of Encephalization Between Odontocete Cetaceans and Anthropoid Primates." *Brain, Behavior, and Evolution*, 51, 230.

Marsh, James, director. 2011. *Project Nim*. Documentary film.

Marshall, Monty. 2009. "Major Episodes of Political Violence, 1946–2009." Vienna, VA: Center for Systematic Peace, November 9, www.systemicpeace.org/warlist.htm

———. 2009. *Polity IV Project: Political Regime Characteristics and Transitions, 1800–2008*. Fairfax, VA: Center for Systematic Peace, George Mason University. http://www.systemicpeace.org/polityproject.html

Marston, Cicely, and John Cleland. 2003. "Relationships Between Contraception and Abortion: A Review of the Evidence." *International Family Planning Perspectives*, March, 29, no. 1, 6–13.

Martin, A., and K. R. Olson. 2013. "When Kids Know Better: Paternalistic Helping in 3-Year-Old Children." *Developmental Psychology*, November, 49, no. 11, 2071–2081.

Marty, Martin E. 1996. *Modern American Religion*, Vol 3: *Under God, Indivisible, 1941–1960*. Chicago: University of Chicago Press.

Mayr, Ernst. 1957. "Species Concepts and Definitions," in *The Species Problem*. Washington DC: American Association for the Advancement of Science Publication 50.

———. 1976. *Evolution and the Diversity of Life*. Cambridge, MA: Harvard University Press.

———. 1988. *Toward a New Philosophy of Biology*. Cambridge, MA: Harvard University Press.

McCormack, Mark. 2013. *The Declining Significance of Homophobia*. Oxford, UK: Oxford University Press.

McCracken, Harry, and Lev Grossman. 2013. "Google vs. Death." *Time*, September 30. time.com/574/google-vs-death/

McCullough, Michael. 2008. *Beyond Revenge: The Evolution of the Forgiveness Instinct*. San Francisco: Jossey-Bass.

McCullough, M. E., W. T. Hoyt, D. B. Larson, H. G. Koenig, and C. E. Thoresen. 2000. "Religious Involvement and Mortality: A Meta-Analytic Review." *Health Psychology*, 19, 211–222.

McCullough, M. E., and B. L. B. Willoughby. 2009. "Religion, Self-Regulation, and Self-Control: Associations, Explanations, and Implications." *Psychological Bulletin*, 125, 69–93.

McDonald, P. J. 2010. "Capitalism, Commitment, and Peace." *International Interactions*, 36, 146–168.

McDuffie. George. 1835. *Governor McDuffie's Message on the Slavery Question*. 1893. New York: A. Lovell & Co., 8.

McMillan, Dan. 2014. *How Could This Happen?: Explaining the Holocaust*. New York: Basic Books, 213.

McNamara, Robert S. 1969. "Report Before the Senate Armed Services Committee on the Fiscal Year 1969–73 Defense Program, and 1969 Defense Budget, January 22, 1969." Washington, DC: US Government Printing Office.

Mencken, H. L. 1927. "Why Liberty?" *Chicago Tribune*, January 30. goo.gl/1Csn6h

Mercier, Hugo, and Dan Sperber. 2011. "Why Do Humans Reason? Arguments for an Argumentative Theory." *Behavioral and Brain Sciences*, 34, no. 2, 57–74.

Michael, George. 2011. "Extraterrestrial Aliens: Friends, Foes, or Just Curious?" *Skeptic*, 16, no. 3. goo.gl/0fbAj

Michaud, Michael. 2007. *Contact with Alien Civilizations: Our Hopes and Fears About Encountering Extraterrestrials*. New York: Copernicus Books.

Miles, H. L. 1994. "ME CHANTEK: The Development of Self-Awareness in a Signing Orangutan." In S. T. Parker et al., eds., *Self-Awareness in Animals and Humans: Developmental Perspectives*. Cambridge, UK: Cambridge University Press.

———. 1996. "Simon Says: The Development of Imitation in an Encultured Orangutan." In A. E. Russon et al., eds., *Reaching into Thought: The Minds of the Great Apes*. Cambridge, UK: Cambridge University Press.

Milgram, Stanley. 1967. "The Small World Problem." *Psychology Today*, 2, 60–67.

———. 1969. *Obedience to Authority: An Experimental View*. New York: Harper & Row.

Mill, John Stuart. 1859. *On Liberty*. Chapter 2. www.bartleby.com/130/2.html

———. 1869. *The Subjection of Women*. London: Longmans, Green, Reader, & Dyer.

Miller, Lisa. 2012. "The Money-Empathy Gap." *New York*, July 1. goo.gl/nC0c6

Milner, George R., Jane E. Buikstra, and Michael D. Wiant. 2009. "Archaic Burial Sites in the American Midcontinent." In Thomas E. Emerson, Dale L. McElrath, and Andres C. Fortier, eds., *Archaic Societies*. Albany: State University of New York Press.

Milner, Larry. 2000. *Hardness of Heart, Hardness of Life: The Stain of Human Infanticide*. Lanham, MD: University Press of America.

Milner, R. D. G., and R. W. Beard. 1984. "Limit of Fetal Viability." *Lancet*, 1: 1097.

Mintzberg, Henry. 1989. *Mintzberg on Management: Inside Our Strange World of Organizations*. New York: Free Press.

Miroff, Nick, and William Booth. 2012. "Mexico's Drug War Is at a Stalemate as Calderon's Presidency Ends." *Washington Post*, November 27. goo.gl/pu2uJ

Mischel, Walter, Ebbe B. Ebbesen, and Antonette Raskoff Zeiss. 1972. "Cognitive and Attentional Mechanisms in Delay of Gratification." *Journal of Personality and Social Psychology*, 21, no. 2, 204–218.

Moore, David. 2003. "Public Lukewarm on Animal Rights." Gallup News Service, May 21. goo.gl/9hj5lP

Morales, Lymari. 2009. "Knowing Someone Gay/Lesbian Affects Views of Gay Issues." Gallup.com, May 29.

Morell, Virginia. 2013. *Animal Wise: The Thoughts and Emotions of Our Fellow Creatures*. New York: Crown, 261.

Morris, Andrew. 2008. "Anarcho-Capitalism." In Ronald Hamowy, ed., *The Encyclopedia of Libertarianism*. Thousand Oaks, CA: Sage.

Morris, Ian. 2011. *Why the West Rules—for Now: The Patterns of History and What They Reveal About the Future*. New York: Farrar, Straus & Giroux.

———. 2013. *The Measure of Civilization: How Social Development Decides the Fate of Nations*. Princeton, NJ: Princeton University Press.

Morris, Simon Conway. 2003. *Life's Solution: Inevitable Humans in a Lonely Universe*. Cambridge, UK: Cambridge University Press.

Moussaieff, M., and S. McCarthy. 1995. *When Elephants Weep: The Emotional Lives of Animals*. New York: Delacorte Press.

Mueller, John, and Mark G. Stewart. 2013. "Hapless, Disorganized, and Irrational." *Slate*, April 22. goo.gl/j0cqUl

Murray, Sara. 2013. "BM's Barra a Breakthrough." *Wall Street Journal*, December 11, B7.

Musch, J., and K. Ehrenberg. 2002. "Probability Misjudgment, Cognitive Ability, and Belief in the Paranormal." *British Journal of Psychology*, 93, 169–177.

Musk, Elon. 2014. "Here's How We Can Fix Mars and Colonize It." *Business Insider*, January 2. goo.gl/iapxDx

Naím, Moisés. 2013. *The End of Power: From Boardrooms to Battlefields and Churches to States: Why Being in Charge Isn't What It Used to Be*. New York: Basic Books, 16, 1–2.

Napier, William. 1851. *History of General Sir Charles Napier's Administration of Scinde*. London: Chapman & Hall.

Narang, Vipin. 2010. "Pakistan's Nuclear Posture: Implications for South Asian Stability." Cambridge, MA: Harvard Kennedy School, Belfer Center for Science and International Affairs Policy Brief, January 4. goo.gl/gH1Hlv

Navarick, Douglas J. 1979. *Principles of Learning: From Laboratory to Field*. Reading, MA: Addison-Wesley.

———. 2009. "Reviving the Milgram Obedience Paradigm in the Era of Informed Consent." *Psychological Record*, 59, 155–170.

———. 2012. "Historical Psychology and the Milgram Paradigm: Tests of an Experimentally Derived Model of Defiance Using Accounts of Massacres by Nazi Reserve Police Battalion 101." *Psychological Record*, 62, 133–154.

———. 2013. "Moral Ambivalence: Modeling and Measuring Bivariate Evaluative Processes in Moral Judgment." *Review of General Psychology*, 17, no. 4, 443–452.

Nelson, Charles A., Nathan A. Fox, and Charles H. Zeanah. 2014. *Romania's Abandoned Children: Deprivation, Brain Development, and the Struggle for Recovery.* Cambridge, MA: Harvard University Press.

Nelson, Craig. 2014. *The Age of Radiance: The Epic Rise and Dramatic Fall of the Atomic Era.* New York: Scribner/Simon & Schuster, 370.

Nicholson, Alexander. 2012. *Fighting to Serve: Behind the Scenes in the War to Repeal "Don't Ask, Don't Tell."* Chicago: Chicago Review Press, 12.

Nisbett, R. E. and L. Ross. 1980. *Human Inference: Strategies and Shortcomings of Social Judgment.* Englewood Cliffs, NJ: Prentice-Hall.

Niven, Larry. 1990. *Ringworld.* New York: Ballantine.

Norris, Pippa, and Ronald Inglehart. 2004. *Sacred and Secular.* New York: Cambridge University Press.

Northcutt, Wendy. 2010. *The Darwin Awards: Countdown to Extinction.* New York: E. P. Dutton.

Nowak, Martin, and Roger Highfield. 2012. *SuperCooperators: Altruism, Evolution, and Why We Need Each Other to Succeed.* New York: Free Press.

Nowak, Martin A., and Sarah Coakley, eds. 2013. *Evolution, Games, and God: The Principle of Cooperation.* Cambridge, MA: Harvard University Press.

Nozick, Robert. 1973. *Anarchy, State, and Utopia.* New York: Basic Books.

O'Gorman, Hubert J. 1975. "Pluralistic Ignorance and White Estimates of White Support for Racial Segregation." *Public Opinion Quarterly*, 39, no. 3, Autumn, 313–330.

Olmsted, Frederick Law. 1856. *A Journey in the Seaboard Slave States.* goo.gl/7rixHj

Olson, Richard. 1990. *Science Deified and Science Defied: The Historical Significance of Science in Western Culture.* Berkeley: University of California Press, 15–40.

———. 1993. *The Emergence of the Social Sciences 1642–1792.* New York: Twayne, 4–5.

Ostry, Jonathan D., Andrew Berg, and Charalambos G. Tsangarides. 2014. "Redistribution, Inequality and Growth." Washington, DC: International Monetary Fund. February, 4.

Overbea, Luix. 1981. "Sambo's Fast-Food Chain, Protested by Blacks Because of Name, Is Now Sam's in 3 States." *Christian Science Monitor*, April 22. goo.gl /6y227h

Paine, Thomas. 1776. *Common Sense.* Available online at http://www.constitution.org /civ/comsense.htm

———. 1794. *The Age of Reason.* Available online at http://www.gutenberg.org/ebooks /31270

———. 1795. *Dissertation on First Principles of Government.* Available online at http:// www.gutenberg.org/ebooks/31270

———. 1795. *Thoughts on Defensive War.* Available online at http://www.gutenberg.org /ebooks/31270

Palmer, Tom G. (ed.) 2014. *Peace, Love, and Liberty: War Is Not Inevitable.* Ottawa, IL: Jameson Books, Inc.

Panaritis, Elena. 2007. *Prosperity Unbound: Building Property Markets with Trust.* New York: Palgrave Macmillan.

Parker, S. T., and M. L. McKinney, eds. 1994. *Self-Awareness in Animals and Humans: Developmental Perspectives.* Cambridge, UK: Cambridge University Press.

———. 1999. *The Mentalities of Gorillas and Orangutans.* Cambridge, UK: Cambridge University Press.

Parker, Theodore. 1852/2005. *Ten Sermons of Religion*. Sermon III: Of Justice and Conscience. Ann Arbor: University of Michigan Library.

Patterson, Charles. 2002. *Eternal Treblinka: Our Treatment of Animals and the Holocaust*. Herndon, VA: Lantern Books.

Patterson, F. G. P. 1993. "The Case for the Personhood of Gorillas." In P. Cavalieri and P. Singer, eds., *The Great Ape Project: Equality Beyond Humanity*. New York: St. Martin's Press.

Patterson, F. G. P., and E. Linden. 1981. *The Education of Koko*. New York: Holt, Rinehart, & Winston.

Paul, Gregory S. 2009. "The Chronic Dependence of Popular Religiosity upon Dysfunctional Psychosociological Conditions." *Evolutionary Psychology*, 7, no. 3, 398–441.

Payne, James. 2004. *A History of Force: Exploring the Worldwide Movement Against Habits of Coercion, Bloodshed, and Mayhem*. Sandpoint, ID: Lytton Publishing.

Pedro, S. A., and A. J. Figueredo. 2011. "Fecundity, Offspring Longevity, and Assortative Mating: Parametnric Tradeoffs in Sexual and Life History Strategy." *Biodemography and Social Biology*, 57, no. 2, 172.

Pepperberg, I. 1999. *The Alex Studies: Cognitive and Communicative Abilities of Parrots*. Cambridge, MA: Harvard University Press.

Perdue, Charles W., John F. Dovidio, Michael B. Gurtman, and Richard B. Tyler. 1990. "Us and Them: Social Categorization and the Process of Intergroup Bias." *Journal of Personality and Social Psychology*, 59, 475–486.

Perez-Rivas, Manuel. 2001. "Bush Vows to Rid the World of 'Evil-Doers.'" CNN Washington Bureau, September 16.

Peters, Ted. 2011. "The Implications of the Discovery of Extra-Terrestrial Life for Religion." *Philosophical Transactions of the Royal Society A*, February, 369, no. 1936, 644–655.

Petrinovich, L., P. O'Neill, and M. J. Jorgensen. 1993. "An Empirical Study of Moral Intuitions: Towards an Evolutionary Ethics." *Ethology and Sociobiology*, 64, 467–478.

Pfungst, O. 1911. *Clever Hans (The Horse of Mr. von Osten): A Contribution to Experimental Animal and Human Psychology*, trans. C. L. Rahn. New York: Henry Holt.

Pickett, Kate, and Richard Wilkinson. 2011. *The Spirit Level: Why Greater Equality Makes Societies Stronger*. New York: Bloomsbury Press.

Piff, Paul K. 2013. "Does Money Make You Mean?" TED talk. goo.gl/sfrJY9

———. 2013. "Wealth and the Inflated Self: Class, Entitlement, and Narcissism." *Personality and Social Psychology Bulletin*, August, 1–10.

Piff, Paul K., Daniel M. Stancato, Stephane Cote, Rodolfo Mendoza-Denton, and Dacher Keltner. 2012. "Higher Social Class Predicts Increased Unethical Behavior." *Proceedings of the National Academy of Sciences*, March 13, 109, no. 11, 4086–4091.

Pike, John. 2010. "Battle of Okinawa." *Globalsecurity.org*

Piketty, Thomas. 2014. *Capital in the Twenty-first Century*. Cambridge, MA: Belknap Press.

Pillsworth, Elizabeth, and Martie Haselton. 2006. "Male Sexual Attractiveness Predicts Differential Ovulatory Shifts in Female Extra-Pair Attraction and Male Mate Retention." *Evolution and Human Behavior*, 27, 247–258.

Pinker, Steven. 1997. *How the Mind Works*. New York: W. W. Norton.

———. 2002. *The Blank Slate: The Modern Denial of Human Nature*. New York: Viking, 175, 290–291.

———. 2011. *The Better Angels of Our Nature: Why Violence Has Declined*. New York: Viking.

———. 2012. "The False Allure of Group Selection." *Edge.org*, June 18. edge.org/conversation/the-false-allure-of-group-selection

Pinsker, Joe. 2014. "The Pirate Economy." *Atlantic*, April 16.

Pipes, Richard. 2003. *Communism: A History*. New York: Modern Library.

Planty, Michael, Lynn Langton, Christopher Krebs, Marcus Berzofsky, and Hope Smiley-McDonald. 2013. "Female Victims of Sexual Violence, 1994–2010." US Department of Justice, Office of Justice Programs. Bureau of Justice Statistics. March.

Pleasure, J. R., M. Dhand, and M. Kaur. 1984. "What Is the Lower Limit of Viability?" *American Journal of Diseases of Children*, 138, 783;

Plotnik, Joshua M., and Frans de Waal. 2014. "Asian Elephants (*Elephas maximus*) Reassure Others in Distress." *PeerJ*, 2.

Pollack, James, and Carl Sagan. 1993. "Planetary Engineering." In *Resources of Near Earth Space* (J. Lewis, M. Matthews, and M. Guerreri, eds.) Tucson: University of Arizona Press.

Pollak, Sorcha. 2013. "Woman Burned Alive for Witchcraft in Papua New Guinea." *Time*, February 7.

Pollan, Michael. 2002. "An Animal's Place." *New York Times Magazine*, November 10.

———. 2007. *The Omnivore's Dilemma*. New York: Penguin.

———. 2009. *In Defense of Food: An Eater's Manifesto*. New York: Penguin.

Pollard, E. A. 1866. (1990) *Southern History of the War*, 2 vols. New York: Random House Value Publishing.

Popper, K. R. 1959. *The Logic of Scientific Discovery*. London: Hutchinson.

———. 1963. *Conjectures and Refutations*. London: Routledge & Kegan Paul.

Prentice, D. A., and D. T. Miller. 1993. "Pluralistic Ignorance and Alcohol Use on Campus: Some Consequences of Misperceiving the Social Norm." *Journal of Personality and Social Psychology*, February 64, no. 2, 243–256.

Pryor, K., and K. S. Norris, eds. 2000. *Dolphin Societies: Discoveries and Puzzles*. Chicago: University of Chicago Press.

Purvis, June. 2002. *Emmeline Pankhurst: A Biography*. London: Routledge, 354.

Putnam, Frank W. 1998. "The Atomic Bomb Casualty Commission in Retrospect." *Proceedings of the National Academy of Sciences*, 95, no. 10, 5426–5431.

Putnam, Robert. 2000. *Bowling Alone: The Collapse and Revival of American Community*. New York: Simon & Schuster, 19.

Raine, Adrian. 2013. *The Anatomy of Violence: The Biological Roots of Crime*. New York: Pantheon.

Rampell, Catherine. 2013. "US Women on the Rise as Family Breadwinner." *New York Times*, May 29. goo.gl/o9igft

Ranke-Heineman, Uta. 1991. *Eunuchs for the Kingdom of Heaven: Women, Sexuality and the Catholic Church*. New York: Penguin.

Rapaille, Clotaire, and Andres Roemer. 2013. *Move Up*. Mexico City: Taurus.

Rauch, Jonathan. 2004. *Gay Marriage: Why It Is Good for Gays, Good for Straights, and Good for America*. New York: Henry Holt.

———. 2013. "The Case for Hate Speech." *Atlantic*, October 23. goo.gl/vNJYRF

Rawls, J. 1971. *A Theory of Justice*. Cambridge, MA: Belknap Press.

Rayfield, Donald. 2005. *Stalin and His Hangmen: The Tyrant and Those Who Killed for Him*. New York: Random House.

Reicher, S. D., and S. A. Haslam. 2006. "Rethinking the Psychology of Tyranny: The BBC Prison Study." *British Journal of Social Psychology*, 45, 1–40.

Reiss, D., and L. Marino. 2001. "Mirror Self-Recognition in the Bottlenose Dolphin: A Case of Cognitive Convergence." *Proceedings of the National Academy of Sciences*, 8: 5937–5942.

Reiss, Diana. 2012. *The Dolphin in the Mirror: Exploring Dolphin Minds and Saving Dolphin Lives*. Boston: Mariner Books.

Rhodes, Richard. 1984. *The Making of the Atomic Bomb*. New York: Simon & Schuster.
———. 2010. *Twilight of the Bombs: Recent Challenges, New Dangers, and the Prospects of a World Without Nuclear Weapons*. New York: Alfred A. Knopf.
Richardson, Lewis Fry. 1960. *Statistics of Deadly Quarrels*. Pittsburgh: Boxwood Press, xxxv.
Richardson, Lucy. 2013. "Restorative Justice Does Work, Says Career Burglar Who has Turned Life Around on Teesside." *Darlington and Stockton Times*, May 1. goo.gl /Vjpr6t
Richelson, Jeffrey. 1986. "Population Targeting and US Strategic Doctrine." In Desmond Ball and Jeffrey Richelson, eds., *Strategic Nuclear Targeting*. Ithaca, NY: Cornell University Press, 234–249.
Ridgway, S. H. 1986. "Physiological Observations on Dolphin Brains." In R. J. Schusterman et al., eds., *Dolphin Cognition and Behavior: A Comparative Approach*. Hillsdale, NJ: Lawrence Erlbaum Associates, 32–33.
Rindermann, Heiner. 2008. "Relevance of Education and Intelligence for the Political Development of Nations: Democracy, Rule of Law, and Political Liberty." *Intelligence*, 36, 306–322.
Roberts, J. M. 2001. *The Triumph of the West*. New York: Sterling.
Robbins, Lionel. 1945. *An Essay on the Nature and Significance of Economic Science*. London: Macmillan, 16.
Rochat, P., M. D. G. Dias, G. Liping, T. Broesch, C. Passos-Ferreira, A. Winning, and B. Berg. 2009. "Fairness in Distribution Justice in 3- and 5-Year-Olds Across Seven Cultures." *Journal of Cross-Cultural Psychology*, 40, 416–442.
Rogers, L. J. 1998. *Minds of Their Own: Thinking and Awareness in Animals*. Boulder, CO: Westview Press.
Ross, M., and F. Sicoly. 1979. "Egocentric Biases in Availability and Attribution." *Journal of Personality and Social Psychology*, 37, 322–336.
Rothbard, Murray. 1962. *Man, Economy, and State*. New York: D. Van Nostrand.
Rousseau, J. J. *Du Contrat Social*. In *Oeuvres Complètes*. Pléide (ed.), I, iv. goo.gl/G5UCc
Rozin, Paul, Jonathan Haidt, and C. R. McCauley. 2000. "Disgust." In *Handbook of Emotions*, ed. M. Lewis and J. M. Haviland-Jones. New York: Guilford Press, 637–653.
Rubin, Jeff. 2012. *The End of Growth*. New York: Random House.
Rubin, Paul H. 2003. "Folk Economics." *Southern Economic Journal*, 70, no. 1, 157–171.
Rubinstein, W. D. 2004. *Genocide: A History*. New York: Pearson Education, 78.
Rudig, Wolfgang. 1990. *Anti-Nuclear Movements: A World Survey of Opposition to Nuclear Energy*. New York: Longman.
Russell, Bertrand. 1946. *History of Western Philosophy*. London: George Allen & Unwin, 8.
Russell, Jeffrey B. 1982. *A History of Witchcraft: Sorcerers, Heretics and Pagans*. London: Thames & Hudson.
Russett, Bruce. 2008. *Peace in the Twenty-first Century? The Limited but Important Rise of Influences on Peace*. New Haven, CT: Yale University Press.
Russett, Bruce, and John Oneal. 2001. *Triangulating Peace: Democracy, Interdependence, and International Organizations*. New York: W. W. Norton.
Ryder, Richard. 1971. "Experiments on Animals," in Stanley and Roslind Godlovitch and John Harris, eds., *Animals, Men and Morals*. London: Victor Gollancz.
———. 1989. *Animal Revolution: Changing Attitudes Toward Speciesism*. London: Basil Blackwell.
Sagan, Carl. 1973. *The Cosmic Connection: An Extraterrestrial Perspective*. Garden City, NY: Anchor Books/Doubleday, 233–234.

———. 1990. "Preserving and Cherishing the Earth: An Appeal for Joint Commitment in Science and Religion." Statement signed by thirty-two Nobel laureates and presented to the Global Forum of Spiritual and Parliamentary Leaders Conferences in Moscow, Russia.

Sagan, Carl, and Richard Turco. 1990. *A Path Where No Man Thought: Nuclear Winter and the End of the Arms Race*. New York: Random House.

Sagan, Scott D. 2009. "The Global Nuclear Future." *Bulletin of the American Academy of Arts & Sciences*, 62, 21–23.

Sapolsky, Robert. 1995. *Why Zebras Don't Get Ulcers: A Guide to Stress, Stress-Related Diseases, and Coping*. New York: W. H. Freeman, 381.

Sargent, Michael J. "Less Thought, More Punishment: Need for Cognition Predicts Support for Punitive Responses to Crime." *Personality & Social Psychology Bulletin*, 30, 1485–1493.

Sarkees, M. R. 2000. "The Correlates of War Data on War." *Conflict Management and Peace Science*, 18, 123–144.

Savolainen, P., Y. Zhang, J. Luo, J. Lundeberg, and T. Leitner. 2002. "Genetic Evidence for an East Asian Origin of Domestic Dogs." *Science*, November 22, 298: 1610–1612.

Sawhill, Isabel V., Scott Winship, and Kerry Searle Grannis. 2012. "Pathways to the Middle Class: Balancing Personal and Public Responsibilities." Washington, DC: Brookings Institution Center on Children and Families, September.

Scahill, Jeremy. 2013. *Dirty Wars: The World Is a Battlefield*. Documentary film. Sundance Selects.

Scelza, Brooke A. 2011. "Female Choice and Extra-Pair Paternity in a Traditional Human Population." *Biology Letters*, December 23, 7, no. 6, 889–891.

Scheb, John M., and John M. Scheb II. 2010. *Criminal Law and Procedure*, 7th ed. Stamford, CT: Cengage Learning.

Schelling, Thomas C. 1994. "The Role of Nuclear Weapons." In L. Benjamin Ederington and Michael J. Mazarr, eds., *Turning Point: The Gulf War and US Military Strategy*. Boulder, CO: Westview, 105–115.

Schlosser, Eric. 2013. *Command and Control: Nuclear Weapons, the Damascus Accident, and the Illusion of Safety*. New York: Penguin.

Schmitt, D. P., and David M. Buss. 2001. "Human Mate Poaching: Tactics and Temptations for Infiltrating Existing Mateships." *Journal of Personality and Social Psychology*, 80, 894–917.

Schmitt, D. P., L. Alcalay, J. Allik, A. Angleitner, L. Ault, et al. 2004. "Patterns and Universals of Mate Poaching Across 53 Nations: The Effects of Sex, Culture, and Personality on Romantically Attracting Another Person's Partner." *Journal of Personality and Social Psychology*, 86, 560–584.

Scully, Matthew. 2003. *Dominion: The Power of Man, the Suffering of Animals, and the Call to Mercy*. New York: St. Martin's Press.

Searle, John R. 1964. "How to Derive 'Ought' from 'Is.'" *Philosophical Review*, 73, no. 1, 43–58.

Segerstråle, U. 2000. *Defenders of the Truth: The Battle for Science in the Sociobiology Debate and Beyond*. New York: Oxford University Press.

Senft, Gunter. 1997. "Magical Conversation on the Trobriand Islands." *Anthropos*, 92, 369–391.

Senlet, Pinar, Levent Cagatay, Julide Ergin, and Jill Mathis. 2001. "Bridging the Gap: Integrating Family Planning with Abortion Services in Turkey." *International Family Planning Perspectives*, June, 27, no. 2. goo.gl/T4Izo

Shah, Neil. 2014. "US Household Net Worth Hits Record High." *Wall Street Journal*, March 6, A1.

Shanker, Thom. 2012. "Former Commander of US Nuclear Forces Calls for Large Cut in Warheads." *New York Times*, May 16, A4.

Sherif, Muzafer, O. J. Harvey, B. Jack White, William R. Hood, and Carolyn W. Sherif. 1961. *Intergroup Conflict and Cooperation: The Robbers Cave Experiment*. Norman: University of Oklahoma Press.

Shermer, Michael. 1997. *Why People Believe Weird Things*. New York: W. H. Freeman.

———. 2000. *Denying History*. Berkeley: University of California Press.

———. 2002. "Shermer's Last Law." *Scientific American*, January, 33.

———. 2003. *The Science of Good and Evil*. New York: Times Books.

———. 2006. *Why Darwin Matters*. New York: Henry Holt/Times Books.

———. 2007. *The Mind of the Market: How Biology and Psychology Shape Our Economic Lives*. New York: Times Books.

———. 2008. "The Doping Dilemma." *Scientific American*, April, 32–39.

———. 2010. "The Skeptic's Skeptic." *Scientific American*, November, 86.

———. 2011. *The Believing Brain*. New York: Times Books, chapter 13.

———. 2013. "The Sandy Hook Effect." *Skeptic*, 18, no. 1. goo.gl/sjTQuJ

———. 2014. "The Car Dealers' Racket." *Los Angeles Times*, March 17. goo.gl/KfiU6j

Shklovskii, Iosif, and Carl Sagan. 1964. *Intelligent Life in the Universe*. New York: Dell.

Shostak, Seth. 1998. *Sharing the Universe: Perspectives on Extraterrestrial Life*. Berkeley, CA: Berkeley Hills Books.

Shultz, George. 2013. "Margaret Thatcher and Ronald Reagan: The Ultimate '80s Power Couple." *Daily Beast*, April 8.

Shultz, George P., William J. Perry, Henry A. Kissinger, and Sam Nunn. 2007. "A World Free of Nuclear Weapons." *Wall Street Journal*, January 4. goo.gl/7x1cyG

———. 2008. "Toward a Nuclear-Free World." *Wall Street Journal*, January 15. goo.gl /iAGDQX

Simon, Herbert. 1991. "Bounded Rationality and Organizational Learning." *Organization Science*, 2, no. 1, 125–134.

Singer, Isaac Bashevis. 1980. *The Seance and Other Stories*. New York: Farrar, Straus & Giroux, 270.

Singer, Peter. 1975. *Animal Liberation: Towards an End to Man's Inhumanity to Animals*. New York: Harper & Row.

———. 1981. *The Expanding Circle: Ethics, Evolution, and Moral Progress*. Princeton, NJ: Princeton University Press.

———. 1989. "All Animals Are Equal." In Tom Regan and Peter Singer, eds., *Animal Rights and Human Obligations*. Englewood, Cliffs, NJ: Prentice Hall, 148–162.

Singer, Peter, and Jim Mason. 2007. *The Ethics of What We Eat*. Emmaus, PA: Rodale Press.

Skates, John Ray. 2000. *The Invasion of Japan: Alternative to the Bomb*. Columbia: University of South Carolina Press.

Skousen, Mark. 2013. "Beyond GDP: Get Ready for a New Way to Measure the Economy." *Forbes*, December 16.

Skousen, Tim. 2014. "The University of Sing Sing." HBO, March 31.

Smet, Anna F., and Richard W. Byrne. 2013. "African Elephants Can Use Human Pointing Cues to Find Hidden Food." *Current Biology*, 23, 1–5.

Smith, Adam. 1759. *The Theory of Moral Sentiments*. London: A. Millar, I., I., 1.

———. 1776/1976. *An Inquiry into the Nature and Causes of the Wealth of Nations*. 2

vols., R. H. Campbell and A. S. Skinner, gen. eds., W. B. Todd, text ed. Oxford, UK: Clarendon Press.

———. 1795/1982. "The History of Astronomy." In *Essays on Philosophical Subjects*, ed. W. P. D. Wightman and J. C. Bryce. Vol. III of the *Glasgow Edition of the Works and Correspondence of Adam Smith*. Indianapolis: Liberty Fund, 2.

Smith, Christian. 2003. *Moral, Believing Animals: Human Personhood and Culture*. Oxford, UK: Oxford University Press.

Smolin, Lee. 1997. *The Life of the Cosmos*. New York: Oxford University Press.

Snyder, L. 1981. *Hitler's Third Reich*. Chicago: Nelson-Hall, 29.

Sobek, David, Dennis M. Foster, and Samuel B. Robinson. 2012. "Conventional Wisdom? The Effects of Nuclear Proliferation on Armed Conflict, 1945–2001." *International Studies Quarterly*, 56, no. 1, 149–162.

Solzhenitsyn, Aleksandr. 1973. *The Gulag Archipelago*. New York: Harper & Row.

Sorabji, R. 1993. *Animal Minds and Human Morals: The Origin of the Western Debate*. Ithaca, NY: Cornell University Press.

Sowell, Thomas. 1987. *A Conflict of Visions: Ideological Origins of Political Struggles*. New York: Basic Books.

———. 2010. *Basic Economics: A Common Sense Guide to the Economy*, 4th ed. New York: Basic Books, 24–25.

Speck, Jeff. 2012. *Walkable City: How Downtown Can Save America, One Step at a Time*. New York: Farrar, Straus & Giroux.

Spencer, Colin. 1995. *The Heretic's Feast: A History of Vegetarianism*. Lebanon, NH: University Press of New England, 215.

Srinivasan, Balaji. 2013. "Silicon Valley's Ultimate Exit Strategy." Startup School 2013 speech.

———. 2013. "Software Is Reorganizing the World." *Wired*, November. goo.gl /PVjBCW.

Stahl, Lesley. 2011. "Eyewitness." *60 Minutes*. Shari Finkelstein, producer. CBS.

Stapledon, Olaf. 1968. *The Starmaker*. New York: Dover.

Stark, Rodney. 2005. *The Victory of Reason*. New York: Random House, xii–xiii.

Stephan, Maria J., and Erica Chenoweth. 2008. "Why Civil Resistance Works: The Strategic Logic of Nonviolent Conflict." *International Security*, 33, no. 1, 7–44.

Stephens, G. J., L. J. Silbert, and U. Hasson. 2010. "Speaker-Listener Neural Coupling Underlies Successful Communication." *Proceedings of the National Academy of Sciences*. August 10, 107, no. 32, 14425–14430.

Stevens, Doris. 1920/1995. *Jailed for Freedom: American Women Win the Vote*, ed. Carol O'Hare. Troutdale, OR: New Sage Press, 18–19.

Stiglitz, Joseph E. 2013. *The Price of Inequality: How Today's Divided Society Endangers Our Future*. New York: W. W. Norton.

Stone, Brad. 2013. "Inside Google's Secret Lab." *Business Week*, May 22. goo.gl /BSwk7

Sulloway, Frank. 1982. "Darwin and His Finches: The Evolution of a Legend." *Journal of the History of Biology*, 15, 1–53.

———. 1982. "Darwin's Conversion: The *Beagle* Voyage and Its Aftermath." *Journal of the History of Biology*, 15, 325–396.

———. 1984. "Darwin and the Galápagos." *Biological Journal of the Linnean Society*, 21, 29–59.

Sutter, John D., and Edythe McNamee. 2012. "Slavery's Last Stronghold." CNN, March. goo.gl/BTv6N

Swedin, Eric G., ed. 2011. *Survive the Bomb: The Radioactive Citizen's Guide to Nuclear Survival*. Minneapolis: Zenith.

Sweeney, Julia. 2006. *Letting Go of God*. Book transcript of monologue. Indefatigable Inc., 24.

Tafoya, M. A. and B. H. Spitzberg. 2007. "The Dark Side of Infidelity: Its Nature, Prevalence, and Communicative Functions." In B. H. Spitzberg and W. R. Cupach, eds., *The Dark Side of Interpersonal Communication*, 2nd ed., 201–242. Mahwah, NJ: Lawrence Erlbaum Associates.

Tannenwald, Nina. 2005. "Stigmatizing the Bomb: Origins of the Nuclear Taboo." *International Security*, Spring, 29, no. 4, 5–49.

Tavris, Carol, and Elliott Aronson. 2007. *Mistakes Were Made (but Not By Me)*. New York: Mariner Books.

Taylor, John M. 1908. *The Witchcraft Delusion in Colonial Connecticut (1647–1697)*. Project Gutenberg. goo.gl/UtPyLk

Teglas, E., A. Gergely, K. Kupan, A. Miklosi, and J. Topal. 2012. "Dogs' Gaze Following Is Tuned to Human Communicative Signals." *Current Biology*, 22, 209–212.

Thalmann, O., et al. 2013. "Complete Mitochrondrial Genomes of Ancient Canids Suggest a European Origin of Domestic Dogs." *Science*, November 15, 342, no. 6160, 871–874.

Theweleit, Klaus. 1989. *Male Fantasies*. Vol. 2, *Male Bodies*. Minneapolis: University of Minnesota Press, 301.

Thomas, Hugh. 1997. *The Slave Trade: The Story of the Atlantic Slave Trade, 1440–1870*. New York: Simon & Schuster, 451.

Thomas, Keith. 1971. *Religion and the Decline of Magic*. New York: Charles Scribner's Sons, 643.

Thompson, Starley L., and Stephen H. Schneider. 1986. "Nuclear Winter Reappraised." *Foreign Affairs*, 64, no. 5, Summer, 981–1005.

Toland, John. 1970. *The Rising Sun: The Decline and Fall of the Japanese Empire, 1936–1945*. New York: Random House.

Tomasello, Michael. 2009. *Why We Cooperate*. Cambridge, MA: MIT Press.

Torpey, John C. 2006. *Making Whole What Has Been Smashed: On Reparations Politics*. Cambridge, MA: Harvard University Press.

Tortora, Bob. 2010. "Witchcraft Believers in Sub-Saharan Africa Rate Lives Worse." Gallup, August 25.

Townsend, Anthony M. 2013. *Smart Cities: Big Data, Civic Hackers, and the Quest for a New Utopia*. New York: W. W. Norton, xii–xiii.

Travers, Jeffrey, and Stanley Milgram. 1969. "An Experimental Study of the Small World Problem." *Sociometry* 32, no. 4, December, 425–443.

Traynor, Lee. 2014. "The Future of Intelligence: An Interview with James R. Flynn." *Skeptic*, 19, no. 1, 36–41.

Trivers, Robert. 2011. *The Folly of Fools: The Logic of Deceit and Self-Deception in Human Life*. New York: Basic Books.

Troup, George M. 1824. "First Annual Message to the State Legislature of Georgia." In Hardin, Edward J. 1859. *The Life of George M. Troup*. Savannah, GA.

Tuchman, Barbara. 1963. *The Guns of August*. New York: Dell.

Turco, R. P., O. B. Toon, T. P. Ackerman, J. B. Pollack, and C. Sagan. 1983. "Nuclear Winter: Global Consequences of Multiple Nuclear Explosions." *Science*, 222, 1283–1297.

Tuschman, Avi. 2013. *Our Political Nature: The Evolutionary Origins of What Divides Us*. Amherst, NY: Prometheus Books.

Van der Dennen, J. M. G. 1995. *The Origin of War: The Evolution of a Male-Coalitional Reproductive Strategy*. Groningen, Neth.: Origin Press.

———. 2005. *Querela Pacis*: Confession of an Irreparably Benighted Researcher on War and Peace. An Open Letter to Frans de Waal and the "Peace and Harmony Mafia." Groningen: University of Groningen.

Voltaire, 1765/2005. "Question of Miracles." *Miracles and Idolotry*. New York: Penguin.

———. 1824/2013. *A Philosophical Dictionary*, vol. 2. London: John and H. L. Hunt.

———. 1877. *Complete Works of Voltaire*, ed. Theodore Besterman. Banbury: Voltaire Foundation, 1974, 117, 374.

Von Trotta, Margarethe, director. 2012. *Hannah Arendt*. Zeitgeist Films.

Walker, D. P. 1981. *Unclean Spirits: Possession and Exorcism in France and England in the Late Sixteenth and Early Seventeenth Centuries*. Philadelphia: University of Pennsylvania Press.

Walker, Lucy, and Lawrence Bender. 2010. *Countdown to Zero*. Participant Media & Magnolia Pictures.

Waltz, Kenneth N. 2012. "Why Iran Should Get the Bomb: Nuclear Balancing Would Mean Stability." *Foreign Affairs*, July/August. goo.gl/2x9dF

Walzer, Michael. 1967. "On the Role of Symbolism in Political Thought." *Political Science Quarterly*, 82, 201.

Wang, Wendy, Kim Parker, and Paul Taylor. 2013. "Breadwinner Moms." Pew Research, Social & Demographic Trends, May 29.

Warneken, F., and M. Tomasello. 2006. "Altruistic Helping in Human Infants and Young Chimpanzees." *Science*, 311, 1301–1303.

———. 2007. "Helping and Cooperation at 14 Months of Age." *Infancy*, 11, 271–294.

Warren, Mervyn A. 2001. *King Came Preaching: The Pulpit Power of Dr. Martin Luther King, Jr.* Downers Grove, IL: Varsity Press, 193–194.

Webster, Daniel W., and Jon S. Vernick, eds. 2013. *Reducing Gun Violence in America: Informing Policy with Evidence and Analysis*. Baltimore: Johns Hopkins University Press.

Weinberg, Stephen. 2008. *Cosmology*. New York: Oxford University Press.

Weisman, Alan. 2007. *The World Without Us*. New York: St. Martin's Press.

———. 2013. *Countdown: Our Last Best Hope for a Future on Earth?* Boston: Little, Brown.

Weitzer, Ronald. 2012. *Legalizing Prostitution: From Illicit Vice to Lawful Business*. New York: New York University Press.

Westfall, Richard. 1980. *Never at Rest: A Biography of Isaac Newton*. Cambridge, UK: Cambridge University Press.

White, Matthew. 2011. *The Great Big Book of Horrible Things: The Definitive Chronicle of History's 100 Worst Atrocities*. New York: W. W. Norton, 161.

Whitson, Jennifer A., and Adam D. Galinsky. 2008. "Lacking Control Increases Illusory Pattern Perception." *Science*, 322, 115–117.

Wilcox, Ella Wheeler. 1914/2012. "Protest," in *Poems of Problems*. Chicago: W. B. Conkey, 154.

Wilkinson, Richard G. 1996. *Unhealthy Societies: The Afflictions of Inequality*. New York: Routledge.

Williamson, Laura. 1978. "Infanticide: An Anthropological Analysis." In M. Kohl, ed., *Infanticide and the Value of Life*. Buffalo, NY: Prometheus Books.

Wilson, James Q., and Richard Herrnstein. 1985. *Crime and Human Nature*. New York: Simon & Schuster.

Winslow, Charles-Edward Amory. 1920. "The Untilled Fields of Public Health." *Science*, 51, no. 1306, 23–33.

Wise, Steven M. 2000. *Rattling the Cage: Toward Legal Rights for Animals.* Boston: Perseus Books.

———. 2002. *Drawing the Line: Science and the Case for Animal Rights.* Boston: Perseus Books, 24.

Wolf, Sherry. "Stonewall: The Birth of Gay Power." *International Socialist Review.* goo.gl /F1otuO

Wollstonecraft, Mary. 1792. *A Vindication of the Rights of Woman: With Strictures on Political and Moral Subjects.* Boston: Peter Edes.

Wrangham, Richard, and Dale Peterson. 1996. *Demonic Males: Apes and the Origins of Human Violence.* Boston: Houghton Mifflin.

Wrangham, Richard W., and Luke Glowacki. 2012. "Intergroup Aggression in Chimpanzees and War in Nomadic Hunter-Gatherers." *Human Nature*, 23, 5–29.

Wright, Quincy. 1942. *A Study of War*, 2nd ed. Chicago: University of Chicago Press.

Wright, Robert. 2000. *Nonzero: The Logic of Human Destiny.* New York: Pantheon.

Wyne, Ali. 2014. "Disillusioned by the Great Illusion: The Outbreak of Great War." *War on the Rocks*, January 29.

Wynne, Clive. 2007. "Aping Language: A Skeptical Analysis of the Evidence for Nonhuman Primate Language." *Skeptic*, 13, no. 4, 10–14.

Young, Robert. 2010. *Eichmann.* Regent Releasing, Here! Films. October.

Yung, Corey Rayburn. 2014. "How to Lie with Rape Statistics: America's Hidden Rape Crisis." *Iowa Law Review*, 99, 1197–1255.

Zahavi, Amotz, and Avishag Zahavi. 1997. *The Handicap Principle: A Missing Piece of Darwin's Puzzle.* Oxford, UK: Oxford University Press.

Zedong, Mao. 1938. "Problems of War and Strategy." *Selected Works*, II, 224.

Zehr, Howard, and Ali Gohar. 2003. *The Little Book of Restorative Justice.* Intercourse, PA: Good Books, 10–14.

Zimbardo, Philip. 2007. *The Lucifer Effect: Understanding How Good People Turn Evil.* New York: Random House.

Zubrin, Robert. 2000. *Entering Space: Creating a Spacefaring Civilization.* New York: Penguin Putnam, x.

ACKNOWLEDGMENTS

To the extent that morality entails choosing to do the right thing by our fellow humans when one could have done otherwise, and often the right thing involves providing support, assistance, and especially honest, constructive feedback, many moral people did just that for me in the research, writing, and production of this book. First and foremost I acknowledge the editors and readers of the manuscript for their careful review of every sentence and for providing exceptional critical commentary, feedback, references, and source materials that improved the book beyond measure. To Pat Linse, for rendering all of the data graphs and charts into visual masterpieces easy to read and understand, and for her patience in the endless refinements and corrections. To Steven Pinker, not only for writing one of the most important social science books ever published—*The Better Angels of Our Nature*—but also for the many suggestions, additions, sources, corrections, and clarifications that have helped elevate this book to a level higher than it would have otherwise been; but most importantly for being such a good friend—tailwinds always. And to my editor Serena Jones, who understood what I wanted to do in integrating science, history, and morality, and to Allison Adler, for her many excellent edits and suggestions.

As with all of my books I thank my agents Katinka Matson, John

Brockman, Max Brockman, Russell Weinberger, and the staff of Brockman, Inc., which does far more than any other literary agency I know of in crowd-sourcing a sizable community of world-class scientists, philosophers, and scholars to share their thoughts and to generate new ideas about the biggest topics one can imagine through the Edge.org web community.

I have been with the publisher of this book, Macmillan/Henry Holt/ Times Books, since it published my first book, *Why People Believe Weird Things*, in 1997, and that long-term relationship means a lot to me as an author in knowing that my words will be edited and produced with the greatest of care. To that end I thank Paul Golob for his big vision on how my books fit into the larger publishing industry; Rita Quintas, my remarkable production editor, who read through the manuscript line by line and improved my prose beyond measure; and I want to appreciate the designer of the book, Kelly Too, whose design, layout, and typography turned digital word files into an elegant book; thanks too go to Maggie Richards in sales and marketing and Carolyn O'Keefe in publicity for the final step in the birth of a book in introducing it to the world.

Much of this book was written at my office, so I also wish to recognize the many fine people working at or associated with the Skeptics Society and *Skeptic* magazine, including Nicole McCullough, Ann Edwards, Daniel Loxton, William Bull, Jerry Friedman, and most especially my partner Pat Linse. Our many volunteers that make such an organization run smoothly deserve acknowledgment: senior editor Frank Miele; senior scientists David Naiditch, Bernard Leikind, Liam McDaid, Claudio Maccone, Thomas McDonough, and Donald Prothero; contributing editors Tim Callahan, Harriet Hall, Karen Stollznow, and Carol Tavris; editors Sara Meric and Kathy Moyd; photographer David Patton and videographer Brad Davies; and our many volunteers: Jaime Botero, Bonnie Callahan, Tim Callahan, Cliff Caplan, Michael Gilmore, Diane Knutdson, and Teresa Lavelle. Thanks as well for the institutional support for the Skeptics Society at the California Institute of Technology go to Eric Wood, Hall Daily, and Laurel Auchampaugh.

As well, I am also in debt to my lecture agent (who is now also my friend) Scott Wolfman and his team at Wolfman Productions (Diane Thompson and Miriam Pachniuk) for their contributions in bringing science and skepticism to the speakers' circuit. And special thanks to my graduate student and teaching and research assistant Anondah Saide, who, along with Kevin McCaffree, helped me think through many of the moral issues this book considers.

Of all the writing I do none means more to me than my monthly column in *Scientific American*, which I began in April 2001. My editor, Mariette DiChristina, is in fact the editor in Chief and the first woman to hold that august post in this, the longest continuously published magazine in American history (175 years and counting). To her and my new editor Fred Guterl, I owe a deep debt of gratitude.

Finally, special thanks go to my daughter Devin Shermer, who is a tribute to the Flynn Effect in which IQ scores increase three points every ten years, which is about right in tracking how much smarter she is than I am. The love of a child by a parent is the ultimate in moral connectedness. And to my wife Jennifer—forever—to whom this book is dedicated, for the cover idea, for her many important suggestions in the text, and especially for enriching my life beyond measure.

INDEX

Page numbers in italics refer to illustrations.

ABOUT THE AUTHOR

Dr. Shermer received his BA in psychology from Pepperdine University (1976), MA in experimental psychology from California State University, Fullerton (1978), and his PhD in the history of science from Claremont Graduate University (1991). He is a Presidential Fellow at Chapman University, where he teaches a general course on critical thinking entitled Skepticism 101 and honors courses on history, science, and moral progress. He has been a college professor since 1979, also teaching psychology, evolution, and the history of science at Occidental College and Glendale College. As a public intellectual he regularly contributes opinion editorials, book reviews, and essays to the *Wall Street Journal*, the *Los Angeles Times*, *Science, Nature*, and other publications. He has appeared on such shows as the *Colbert Report, 20/20, Dateline, Charlie Rose, Larry King Live, Oprah, Unsolved Mysteries*, and other shows, as well as interviews in countless science and history documentaries aired on PBS, A&E, Discovery, the History Channel, the Science Channel, and the Learning Channel. Dr. Shermer was the cohost and coproducer of the thirteen-hour Family Channel television series *Exploring the Unknown*. He is the founding publisher of *Skeptic* magazine, the executive director of the Skeptics Society (www.skeptic.com), and a monthly columnist for *Scientific American*.